Nickel-Hydrogen Batteries
Principles and Practice

Nickel-Hydrogen Batteries
Principles and Practice

Albert H. Zimmerman

The Aerospace Press • El Segundo, California
American Institute of Aeronautics and Astronautics, Inc. • Reston, Virginia

The Aerospace Press
2350 E. El Segundo Boulevard
El Segundo, California 90245-4691

American Institute of Aeronautics and Astronautics, Inc.
1801 Alexander Bell Drive
Reston, Virginia 20191-4344

Library of Congress Cataloging-in-Publication Data

Zimmerman, Albert H.
  Nickel-hydrogen batteries : principles and practice / Albert H. Zimmerman.
     p. cm.
  Includes bibliographical references and index.
  ISBN 1-884989-20-9 (alk. paper)
  1. Nickel-hydrogen batteries. 2. Space vehicles--Batteries. I. Title.

TK2945.N53Z56 2008
621.31'2423--dc22
                              2008042210

Cover design by John Griffith; photo by Eric Hamburg

Copyright © 2009 by The Aerospace Corporation
All rights reserved

Printed in the United States of America. No part of this publication may be reproduced, distributed, or transmitted in any form or by any means, or stored in a database or retrieval system, without the prior written permission of the publishers.

Data and information appearing in this book are for informational purposes only. The publishers and the authors are not responsible for any injury or damage resulting from use or reliance, nor do the publishers or the author warrant that use or reliance will be free from privately owned rights.

The material in this book was reviewed by the Air Force Space and Missile Systems Center, and approved for public release.

To my loving wife, Susie, for her patience and support.

Contents

Preface .... **xi**
Organization of the Book xii
Acknowledgments .... **xv**
Sources xvi

Part I: Introduction to Nickel-Hydrogen Technology........ 1

1 Overview of Nickel-Hydrogen Cell Technology 3
1.1 Position of Nickel-Hydrogen Technology in Space Power 5

2 The Historical Evolution of Nickel-Hydrogen Cell Designs . . 11
2.1 Background of Nickel-Hydrogen Development.......................11
2.2 Overview of Nickel-Hydrogen Development History 13
2.3 First-Generation Nickel-Hydrogen Cells 15
2.4 The ManTech Program..21
2.5 Second-Generation Cell Designs 22
2.6 NASA Advanced Design Matrix 28
2.7 Single Pressure Vessel Cells and Batteries..........................30
2.8 Nickel-Hydrogen Cell Design Variants 31
2.9 Experimental Nickel-Hydrogen Cell Designs 32
2.10 References..35

3 Electrical and Thermal Performance................... 37
3.1 Steady-State Voltage Behavior....................................37
3.2 Cell Power Response 44
3.3 Cell Capacity Behavior..45
3.4 Impedance Behavior...48
3.5 Self-Discharge Behavior...52
3.6 Secondary-Plateau Discharge Behavior 53
3.7 Charge Efficiency Behavior 54
3.8 Pressure Behavior...55
3.9 Heat Generation and Thermal Behavior of Cells 59
3.10 Precharge..63

3.11 Cycle-Life Behavior 65
3.12 Acceptance and Qualification Test Procedures 65
3.13 References 67

Part II: Fundamental Principles 69

4 The Nickel Electrode 71
4.1 Construction and Design of Nickel Electrodes 71
4.2 Chemistry of the Nickel Electrode 87
4.3 Role of Additives and Contaminants in the Nickel Electrode 120
4.4. Interactions of Hydrogen Gas with the Nickel Electrode 137
4.5. Storage of Nickel Electrodes 142
4.6 Advanced Nickel Electrode Concepts 144
4.7 References 147

5 The Hydrogen Electrode 149
5.1 Design and Function of the Hydrogen Electrode 149
5.2. Hydrogen Electrode Fabrication 150
5.3. Gas and Electrolyte Management 151
5.4. Chemistry of the Hydrogen Electrode 153
5.5. Negative Electrode Thermodynamics 157
5.6. Negative Electrode Contaminants 158
5.7. Pore Size Distributions 159
5.8 References 161

6 Separators and Electrolyte 163
6.1 Roles of the Separator in a Nickel-Hydrogen Cell 163
6.2 Zircar Separator 164
6.3 Asbestos Separator 166
6.4 Cell Electrolyte Properties 170
6.5 References 176

7 Cell Dynamics 177
7.1 Charge Efficiency 177
7.2 Hydrogen Gas Management 180
7.3 Oxygen Gas Management 185
7.4 Thermal Management 192
7.5 Electrolyte Management 201

7.6 The Effect of Pressure Fluctuations on Cell Performance 202
7.7 References. 203

8 Nickel-Hydrogen Cell Modeling 205

8.1 First-Principles Models of Cell Performance 205
8.2 Nickel-Hydrogen Cell Storage Model 251
8.3 Wear-Out Models for Predicting Cell Cycle Life. 263
8.4 Summary and Status of Nickel-Hydrogen Cell Modeling 284
8.5 References. 285

Part III: Application and Practice. 287

9 Charge Management for Nickel-Hydrogen Cells and Batteries 289

9.1 Theory of Charge Management 289
9.2 Temperature-Compensated Voltage Charge Control 291
9.3 Recharge Ratio Charge Control 293
9.4 Pressure-Based Charge Control 294
9.5 Adaptive or Self-Optimizing Charge Control. 297
9.6 Ratchet Charging 300
9.7 Reconditioning 302
9.8 Management in Storage 304
9.9 Other Charge-Management Tools and Methods. 305
9.10 References. 307

10 Thermal Management and Reliability 309

10.1 Thermal Management. 309
10.2 Other Battery Components 317
10.3 Cell Reliability 321
10.4 Cell Design Guidelines 324
10.5 Cell Safety. 327
10.6 References. 329

11 Cell and Battery Test Experience 331

11.1 Cell Discharge Voltage Signatures. 332
11.2 Cell Recharge Behavior 340
11.3 Cell Pressure Behavior 347
11.4 Charge-Control Methods and Cell Lifetime. 351

11.5 Temperature and Cell Lifetime................................353
11.6 Depth of Discharge and Cell Lifetime 356
11.7 Electrolyte Concentration and Cell Lifetime 356
11.8 Accelerated *vs*. Real-Time Testing............................359
11.9 Effects of Cell Storage on Cycle Life...........................371
11.10 Effects of Specialized Cell Design Features on Lifetime...........375
11.11 Specialized Cell Life Tests...................................379
11.12 References...381

12 Degradation and Failure Modes.....................383
12.1 Manufacturing-Related Problems..............................383
12.2 Storage-Related Issues 394
12.3 Issues Related to the Cell Operating Environment.................399
12.4 Battery-Specific Issues 403
12.5 Cycling and Life-Related Issues...............................405
12.6 References..413

13 Diagnostic Methods and Destructive Physical Analysis for Cells 415
13.1 Cell-Level Diagnostic Tests 415
13.2 Destructive Physical Analysis Techniques.......................434
13.3 Nickel Electrode Diagnostic Tests 438
13.4 Hydrogen Electrode Diagnostic Tests 459
13.5 Electrolyte and Separator Analysis.............................464
13.6 Expert Systems for Nickel-Hydrogen Cell Diagnostics.............469
13.7 References..475

Index...479

Preface

Nickel-hydrogen battery cells can provide one of the longest-lived and most reliable rechargeable battery systems ever developed. Not surprisingly, this battery technology was primarily developed for use in satellite and space power systems, where exceptionally long life was well worth the high cost associated with the technology. This book provides the reader with an in-depth view of nickel-hydrogen cell technology, including information about how it is developed, how and why it works, how one can realize its ultimate capability, how one can implement it, and what can go wrong if one does not properly manage it.

The technology of nickel-hydrogen cells has been developed, studied, and validated through many years of research, test, and analysis programs. All these programs have helped provide the understanding and data used in this book. The Aerospace Corporation sponsored much of this research between 1980 and 2000. During this period, these research activities elucidated many of the chemical processes that are unique to the nickel-hydrogen battery cell and developed many of the specialized analysis techniques that have been used to measure the effect of these processes on cell performance.

Between 1992 and 2000, The Aerospace Corporation also supported the development of a first-principles model of nickel-hydrogen cell performance, which ultimately included all the processes that control the performance of this cell. This model, described in chapter 8, has been successfully used as a basis for aspects of nickel-hydrogen cell technology ranging from the accurate prediction of voltage and capacity behavior over life, to the development of novel cell designs that are optimized for maximum life or minimum internal resistance. This model was ultimately successful to the point where it was used to predict "emergent behavior" in cells: performance signatures arising from the dynamic interactions between the processes occurring within the cells rather than from preprogrammed cell processes.

The life-test programs initiated in the 1980s and 1990s were critical both for validating the performance capabilities of the technology, and for identifying cell designs that could provide superior performance or life. In the early 1980s, The Aerospace Corporation was instrumental in initiating U.S. Air Force–supported life tests at the Naval Weapon Support Center (NWSC) in Crane, Indiana. These life tests, along with a major life-test program initiated by NASA during the period from 1984 to 1996, were the principal sources of the long-term performance data used throughout this book to illustrate the capability of nickel-hydrogen cells. This body of test data, along with specialized life tests and extensive destructive physical analyses of failed cells performed by The Aerospace Corporation, also provided the information needed to develop and validate the high-fidelity performance and life models described in this book.

In parallel with the maturation of the nickel-hydrogen cell technology, nickel-hydrogen batteries rapidly replaced most of the nickel-cadmium batteries used in earlier satellites, for commercial, scientific, and military satellite missions. While of great interest, the on-orbit performance of the nickel-hydrogen batteries used in these satellite programs is not discussed here, largely because it is outside the scope of this book and has potential commercial and national security sensitivities. While on-orbit failures of nickel-hydrogen batteries have occurred in a few instances, each has been attributed to cell manufacturing flaws or improper management of the batteries. Today some nickel-hydrogen batteries are approaching 20 years of successful operation in orbital satellites, and the best available models project that these batteries should be capable of many more years of reliable operation.

Nickel-hydrogen technology is an example of one of the most successful developments of technology to support the specific needs of space systems. It is ironic that in recent years nickel-hydrogen battery technology has begun to be replaced by lithium ion battery technology, a battery system originally developed in the 1990s to meet the needs of the terrestrial commercial market. Nickel-hydrogen batteries are slowly succumbing to the simple fact that they are two to three times heavier than are lithium ion batteries, and that their ultimate lifetime capability of 20–30 years is not required for most space missions. It remains to be seen how long the unique life capabilities of nickel-hydrogen batteries will continue to be used in high-reliability remote applications such as space systems.

Organization of the Book

This book is organized into three separate parts intended to be read in order; they provide a balanced picture of the development of nickel-hydrogen cell technology, the fundamental principles of operation for the key components in the cell, and the key concepts involved in the use of the cells in satellite power systems. Part I is the introduction, part II is entitled Fundamental Principles, and part III discusses the application and practice of using state-of-the-art nickel-hydrogen cells in battery power systems.

Part I provides an overview of the technology in chapter 1, followed by a historical discussion in chapter 2 of the principal developments that have shaped the design and use of nickel-hydrogen cells over the years. The final chapter in the first part of the book, chapter 3, summarizes the key performance characteristics of nickel-hydrogen cells and thus provides reference information that can be used in the design and use of these batteries in power systems.

Part II concentrates on the fundamental physical and chemical principles that control the behavior of nickel-hydrogen cells. This portion of the book begins

Preface

appropriately with chapter 4, which contains a detailed description of how the nickel electrode is built and how it functions. The nickel electrode is generally regarded as the key component that governs the performance and life of the nickel-hydrogen cell. Much has been learned about the fundamentals of this electrode from study over the past 40 years by numerous researchers. Chapter 4 introduces the basics of all the processes that have been found to influence this electrode over its lifetime in nickel-hydrogen cells.

Chapters 5 and 6 provide a description of the other key components involved in the chemistry of the nickel-hydrogen cell: the separator, the hydrogen electrode, and the electrolyte. Chapter 6, in particular, provides a good source of information on the key physical properties of the potassium hydroxide electrolyte used in these battery cells. This information was originally compiled in this work to serve first as key input for understanding the dynamics of how the components function together in the nickel-hydrogen cell environment as discussed in chapter 7, and second as required physical data for the first-principles models of the nickel-hydrogen cell that are described in chapter 8.

Chapter 8 presents a detailed discussion of the various types of performance models, life prediction models, and storage models that have been successfully used to guide the development and use of nickel-hydrogen batteries. The models that are discussed range from highly empirical regression models for predicting cycle life that are based on an extensive life-test database, to first-principles models that can accurately predict cell performance by including all the chemical and physical processes known to occur in these cells. This chapter is highly recommended as a resource for readers interested in the approaches that have been successfully applied to model nickel-hydrogen cells in the past, and it serves as a springboard for those interested in extending or improving the present modeling capability. The references included in chapter 8 serve to introduce the reader to the detailed literature that describes the methods used for state-of-the-art modeling of battery cells.

Part III presents much of the more practical knowledge involved in the usage of nickel-hydrogen cells and batteries. This part draws heavily on the fundamentals presented in part II, but does not require the reader to have an in-depth knowledge of all the underlying chemistry and physics. Chapter 9 discusses the issue of charge management, which must be properly implemented if these batteries are to provide a long and reliable cycle life. While nickel-hydrogen batteries are extremely robust in terms of tolerating abusive charge control and environmental exposure compared to most batteries, inappropriate charge management is still the most common cause for the premature failure of nickel-hydrogen batteries. Chapter 10 discusses the key issues involved with proper control of the nickel-hydrogen battery thermal environment, and it includes a discussion of the key aspects of cell and battery reliability.

Over the past 30 years a significant life-test database has been accumulated for nickel-hydrogen cells of various designs. Chapter 11 provides a critical review

Preface

and analysis of this database, which serves as a key element in understanding the life and reliability that are attainable from these cells. As a part of the extensive testing of nickel-hydrogen cells over the years, analysis of the degradation and failures that have been seen has added to our understanding of the technology. The last two chapters of the book address this important area. Chapter 12 describes the various degradation and failure modes that have been seen, and chapter 13 discusses the methods that have been developed for analyzing cells and their internal components to deduce why the cells have eventually failed. The final portion of chapter 13 describes a software package that has been developed to interactively guide the user through an analysis of why a nickel-hydrogen cell has undergone degradation or failure, thus directing a powerful array of modern analytical techniques to problem-solving for even the novice battery engineer or technician.

No overall conclusions are provided at the end of the book, because its only goal is to enable the reader to understand in detail the factors that govern the performance and chemical complexities of a battery system that has powered the satellites of the world for decades.

—*Albert H. Zimmerman*

Acknowledgments

This book is largely based on the results of research programs supported by The Aerospace Corporation between the years of 1980 and 2000. These programs were part of the Aerospace Supported Research program (known later as the Aerospace Internal Research and Development program), and the Mission-Oriented Investigation and Experimentation program. The early research efforts in these programs focused on understanding the unique chemistry in the nickel electrode that seemed to control its performance in the nickel-hydrogen cell. During the latter ten years of this period, these research efforts were largely devoted to the development of models for accurately predicting performance and life. These research efforts are acknowledged as a key effort without which the developments described in this book would not have occurred.

Numerous collaborations and discussions with colleagues at The Aerospace Corporation have also been instrumental in developing the understanding and knowledge presented here. Without the tireless support and technical input from Dr. Charles Badcock and Lawrence Thaller, the research and testing responsible for much of the contents of this book would not have been possible. Dr. Margot Wasz and Valerie Ang have also provided invaluable technical discussions that have contributed to the understanding of nickel-hydrogen battery technology described in this book. Michael Quinzio and Gloria To have contributed immensely to development of the specialized testing and the diagnostic cell analysis methods described herein.

The life-test database that has been developed for nickel-hydrogen cells has resulted from tests conducted over 20–25 years at contractors' and cell manufacturers' facilities, as well as from extensive life-test programs at the Naval test facility in Crane, Indiana. All of these organizations are acknowledged for making data from these tests available for analysis in support of this effort. In particular, Michelle Manzo and Tom Miller are thanked for providing permission to access and utilize the Crane NASA test database. Similarly, Ralph James is acknowledged for providing permission to access and utilize the U.S. Air Force nickel-hydrogen life-test database at Crane. Much of the validation of the nickel-hydrogen technology has depended on these test efforts. The personnel at Crane who have provided the life-test data, and who have enthusiastically shared in many discussions involving the test data, are also thanked: Dr. Harlan Lewis, Harry Brown, Bruce Moore, Stephen Wharton, and Evan Hand.

Finally, technical discussions with many researchers who have also been involved in developing, understanding, and modeling nickel-hydrogen cell technology are acknowledged. Such discussions of shared research interests have been

Acknowledgments

invaluable in helping develop the key concepts needed to understand how nickel-hydrogen cells work in sufficient detail to accurately model their performance based on fundamental chemical and physical principles. In particular, discussions with Dr. John Newman, Dr. Ralph White, Paul Timmerman, and Joseph Stockel, who were also involved in modeling nickel-hydrogen cells, were most helpful during the development of an accurate first-principles model of the nickel-hydrogen battery cell. The work of these and many other researchers has played a significant role in shaping the understanding of nickel-hydrogen battery technology that is presented in this book.

Sources

Certain figures in this book are based on figures that have appeared in earlier publications.

Chapter 2: Figures 2.3 and 2.6 are based on drawings from Ref. 2.2: J.D. Dunlop, G.M. Rao, and T.Y. Yi, NASA Handbook for Nickel-Hydrogen Batteries (NASA Ref., Pub. 1314, September 1993), pp. 1–54. Published courtesy of NASA.

Chapters 3 and 7: Figures 3.24 and 7.4 both appeared in Ref. 3.3, in an article cowritten by the author, "Dynamic Calorimetry for Thermal Characterization of Battery Cells," on pp. 281–286 of *Proc. of the 17th Annual Battery Conf. on Appl. and Adv.,* IEEE 02TH8576, ISBN 0-7803-7132-1 (Long Beach, CA, 2002). ©2002 IEEE. Reprinted with permission.

Chapter 4: Figure 4.28 is based on Fig. 6 from page 235 of Ref. 4.12. Reproduced with permission from the author's contribution to *Power Sources 12*, T. Keily and B.W. Baxter, eds. (Taylor and Francis, Ltd., Basingstoke, England, 1988).

Figure 4.36 is based on a figure from Ref. 4.24. Reproduced with permission from the author's article "The Role of Anionic Species in Energy Redistribution Processes in Nickel Electrodes in Nickel Hydrogen Cells," in *Hydrogen Storage Materials, Batteries, and Electrochemistry, Proc. 91-4* (The Electrochemical Society, Inc., Pennington, NJ, 1991).

Chapter 8: Figure 8.18 appeared in Ref. 8.11, in the author's "Virtual Life Testing of Battery Cells," *Proc. of the 2005 NASA Battery Workshop* (Huntsville, AL, November 2005). Published courtesy of NASA.

Figures 8.20 through 8.26 appeared in Ref. 8.15, in the author's cowritten article "Model for Predicting the Effects of Long-Term Storage and Cycling on the Life of NiH_2 Cells," *Proc. of the 2003 NASA Battery Workshop*, NASA/CP-2005-214190 (Huntsville, AL 20 November 2003). Published courtesy of NASA.

Chapter 10: Figure 10.8 is based on a figure in Ref. 10.5, the author's cowritten article "Expert System for Nickel Hydrogen Battery Cell Diagnos-

Acknowledgments

tics," NASA/CP-1999-209144, *Proc. 1998 NASA Battery Workshop* (Huntsville, AL, November 1998), pp. 297–316. Published courtesy of NASA.

Chapter 13: Figure 13.12 is based on a figure in Ref. 13.10, the author's cowritten article "Nickel Electrode Failure by Chemical De-activation of Active Material," *Proc. 1998 NASA Battery Workshop*, NASA/CP-1999-209144, pp. 317–328. Published courtesy of NASA.

Figures 13.23 through 13.25 are based on figures in Ref. 13.1, the author's cowritten article "Expert System for Nickel Hydrogen Battery Cell Diagnostics," *Proc. 1998 NASA Battery Workshop*, NASA/CP-1999-209144, pp. 297–316. Published courtesy of NASA.

Part I

Introduction to Nickel-Hydrogen Technology

1 Overview of Nickel-Hydrogen Cell Technology

The nickel-hydrogen battery cell is a rechargeable electrochemical cell that has found wide use in high-reliability space applications that require extended service life. These applications include satellites operating for 15 or more years in geosynchronous orbits, as well as orbital spacecraft (in low Earth orbits) that require many tens of thousands of charge and discharge cycles from the batteries. In these applications, the nickel-hydrogen battery system has largely replaced the sealed nickel-cadmium technology that was widely and successfully used to power space systems from the early days of spaceflight.

Nickel-hydrogen technology was developed for space use beginning in the early 1970s as a spin-off of the hydrogen-oxygen regenerative fuel cell and is a technology capable of long life and high reliability without the periodic refueling and servicing needs typical of fuel cells. The nickel-hydrogen battery cell performs essentially as a quasireversible hydrogen-oxygen electrochemical cell, with the oxygen being stored in the positive electrode as a metastable nickel oxyhydroxide and the hydrogen being stored as high-pressure gas in a pressure vessel. The positive electrode was selected to be the highly reliable and robust nickel electrode that had been used for years in the nickel-cadmium cell. The energy storage reaction at the positive electrode can be represented by the simplified reaction in Eq. (1.1).

$$Ni(OH)_2 + OH^- \xrightarrow{Charge} NiOOH + H_2O + e^- \qquad (1.1)$$

The negative electrode is a catalyst-based gas electrode that has been largely developed and utilized in the fuel cell industry. This electrode electrochemically stores energy by the reaction in Eq. (1.2) in the form of hydrogen gas, which is contained in the cell container at high pressures.

$$2H_2O + 2e^- \xrightarrow[\text{Pt catalyst}]{\text{Charge}} H_2 + 2OH^- \qquad (1.2)$$

The combination of these two electrodes began the development of a nickel-hydrogen battery technology that today is unmatched for providing highly reliable performance combined with a service life that can easily support decades of continuous operation in the most demanding applications.

After more than 30 years of development and maturation, nickel-hydrogen technology has reached a point of key historical significance. While the technology is now the dominant rechargeable energy-storage system used in the space and satellite industry, support for further nickel-hydrogen cell development, system optimization, and technology maintenance is decreasing each year. Without a critical level of technology development or maintenance support, it frequently has been the case that even the most established battery technology can be compromised by erosion of production infrastructure and manufacturing expertise. This pattern was a real factor in the replacement of nickel-cadmium technology with nickel-hydrogen in the 1980s. One of the principal goals of this work is to point out key areas where the existing nickel-hydrogen technology requires continued attention if it is to meet the future needs of space power.

This book is divided into three parts. Part I provides an introduction to nickel-hydrogen technology, Part II covers the fundamental principles of the technology, and Part III discusses the practical and applied aspects of the technology. In Part I the existing experience with nickel-hydrogen technology will be introduced, from both a historical viewpoint and a performance-capability viewpoint. A brief historical monograph will provide an understanding of how today's nickel-hydrogen cells evolved from the earliest concepts. This historical perspective will cover the development of all variations of nickel-hydrogen cell design that have been used or are available for use today. The typical electrical and thermal performance characteristics of the nickel-hydrogen cell will also be described in Part I.

Part II will deal with the fundamental principles of the nickel-hydrogen cell and its components. This discussion will start with the basic designs and key processes in the cell components: the nickel electrode, the hydrogen electrode, and the separator. Many of the key performance characteristics of the nickel-hydrogen cell are related to the dynamics of how all its components function together in the operating cell or battery. For the nickel-hydrogen cell this is probably more true than it is for most other types of battery cells, for the simple reason that gas, liquid, and solid species must all interact and function as designed for proper performance to be maintained. A detailed discussion of cell dynamics will describe the most important interactions that can affect the performance of the cell components and how the components function together to dictate cell and battery behavior.

During the past 10 to 15 years, significant effort has centered on modeling nickel-hydrogen cell performance, either from using an empirical model or from a first-principles approach. Various approaches for such modeling will be discussed in Part II, along with the validation, utility, and successes of each approach. Examples of such modeling efforts will be provided to suggest that very-high-fidelity performance modeling is made possible by including all chemical, physical, and electrochemical processes in a realistic microscopic and macroscopic model of the cell. High-fidelity models that are well validated will become an essential part of designing nickel-hydrogen cells and characterizing cell behavior in a climate where the time and funding are not always available to support testing.

Part III will discuss the important factors involved in the application of nickel-hydrogen cells and batteries, and the practices involved in properly maintaining and using them. Charge management, as for all types of battery cells, is an essential factor in realizing high reliability and long cycle life. Similarly, appropriate thermal management is critical in nickel-hydrogen cells, largely because their performance is highly sensitive to the thermal environment. Testing of cells and batteries will also be discussed, including typical acceptance, qualification, and life cycle testing. Typical life test experience for the various cell designs that have been tested will be presented and discussed.

Life test results invariably point to degradation modes, failures, and weak points in any battery cell design. Part III will include a discussion of nickel-hydrogen cell and battery degradation and failure modes in the context of test experience. This section will provide detailed information on how nickel-hydrogen cell and battery technology and the management of batteries can be improved or optimized to get the best performance possible from a given battery system. A detailed discussion of destructive physical analysis and diagnostic techniques will also be provided to list the observed performance signatures associated with the chemical and physical root causes for cell and battery degradation. Finally, expert systems that have recently been developed to link observed performance signatures to their underlying root causes will be reviewed.

1.1 Position of Nickel-Hydrogen Technology in Space Power

The development of nickel-hydrogen technology for space power use began in the early 1970s, largely to obtain a longer battery lifetime capability than could be reliably provided by nickel-cadmium batteries. Nickel-cadmium batteries were reliably providing three years in low Earth orbit applications and seven to ten years in geosynchronous satellite applications. Very early in the process of developing nickel-hydrogen technology, it became very clear that this technology was capable of exceedingly long life. In fact, early testing of nickel-hydrogen cells found it very difficult to induce failures, in some cases creating performance expectations that were not always borne out by more comprehensive and realistic testing and modeling. The cells often appeared capable of handling the stresses of inadvertent reversal or excessive overcharge with little evidence of the damage or performance degradation that had been the rule for nickel-cadmium cells.

Chapter 2 provides details on the history of nickel-hydrogen technology development. Because of the technology's robustness and excellent performance, nickel-hydrogen batteries quickly found their way into two flight tests within 10 years of the start of its development. These flight tests were the NTS-2 flight and the Air Force Flight Experiment, both of which were launched in 1977, performing well for 1.5 and 4 years respectively. As a result of these successful flight tests, along with the life test programs initiated by the United States Air Force, the National Aeronautics and Space Administration (NASA), Martin Marietta, and battery manufacturers, nickel-hydrogen battery technology became a viable space power option that began to be baselined for new space missions. However, because of the 5–10 year (or longer) period involved in the planning, design, acquisition, and launching of these missions, it was not until the early 1990s that nickel-hydrogen batteries had replaced nickel-cadmium batteries as the dominant system on newly launched spacecraft.

Figure 1.1 shows the decline of spacecraft launched with nickel-cadmium batteries along with the rise of those with nickel-hydrogen batteries for noncommercial launches in the United States from 1980 to 2000, a period when nickel-hydrogen batteries essentially replaced nickel-cadmium batteries in satellites. There is another plot similar to Fig. 1.1 reflecting the increase in commercial satellite use of nickel-hydrogen batteries at the expense of nickel-cadmium batteries; however, commercial usage has entailed large spikes in the number of launches arising from the launch of large satellite constellations such as Iridium, as well as periodic fluctuations in the commercial space market.

Today there are essentially no new spacecraft launched with nickel-cadmium batteries, with the smaller and shorter-lived missions having transitioned to lithium-ion batteries. The satellite missions that previously utilized nickel-cadmium batteries are the most likely candidates for transitioning to the emerging lithium-ion battery technology. During the past few years, another battery usage plot similar to Fig. 1.1 has been developing, showing the growth of lithium-ion battery use in space missions at the expense of nickel-hydrogen batteries.

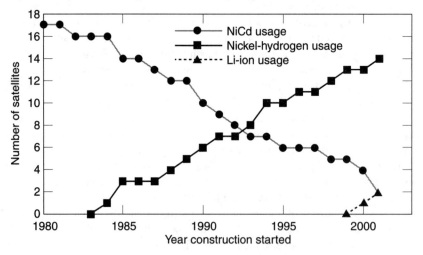

Figure 1.1. Noncommercial usage of nickel-cadmium and nickel-hydrogen batteries in satellites during the period when nickel-hydrogen largely replaced nickel-cadmium.

The issues and factors that drive the transition from one established space battery technology to another emerging one are numerous. Clearly, a successful ground and flight test history for the emerging technology is important, as was the case for nickel-hydrogen. In addition, the safety record of the emerging technology has become a very significant factor with the advent of crewed launch systems such as the space shuttle. It should be noted that with today's safety standards, if nickel-cadmium batteries were an emerging technology, it would probably be regarded as unsafe, because sealed nickel-cadmium batteries do not tolerate high-rate overcharge or reversal safely. Nickel-hydrogen batteries, on the other hand, have been repeatedly demonstrated to be safe under such conditions of abuse. Table 1.1 lists several of the most important drivers that have historically influenced new battery technologies to be either successful or unsuccessful in transitioning into space power systems.

The last factor in Table 1.1, mission requirements, is one of the most elusive. Spacecraft design practices are extremely conservative. A new battery technology may be lighter, smaller, and less costly than the technology presently in common usage, but unless these advantages actually enable a mission at the time it is being planned, the new technology will often be viewed as involving unnecessary risk. Nickel-hydrogen battery technology has been fortunate in this area. Longer-lived missions that were being planned in the late 1970s and the 1980s were actually enabled by nickel-hydrogen technology. Geosynchronous satellite missions of 15 years or longer could be considered, and low Earth orbiting satellites lasting 5 to 10 years could be designed. A good example of the advantages of nickel-hydrogen is provided by the Hubble Space Telescope, which has operated for more than 15 years on nickel-hydrogen batteries with limited battery degradation. These capabilities, when coupled with growing manufacturing and performance problems with spacecraft nickel-cadmium batteries in the mid-1980s, clearly pushed nickel-hydrogen batteries into space use.

Table 1.1. Factors Driving Space Battery Technology

Factor	Importance for Nickel-Hydrogen
Weight	Moderate
Life	High
Reliability	High
Cost	Low
Safety	Moderate
Ease of handling	Low
Size or volume	Low
Mission requirements	High

8   Overview of Nickel-Hydrogen Cell Technology

Nickel-hydrogen batteries do, however, have a number of strengths and weaknesses that have influenced and will continue to influence their use in space power systems relative to other battery technologies. Weight has historically been a significant factor in the development of technologies for use in space systems, and the technology of batteries is not an exception. The cost of launching each kilogram into low Earth orbit is high, and weight is very costly indeed to lift into geosynchronous orbit. The horizontal axis of the graph in Fig. 1.2 shows how nickel-hydrogen cell energy density compares to other rechargeable battery technologies that can be considered for space power use. Clearly, nickel-hydrogen technology would not be the obvious choice for space power systems based on weight alone. Nickel-metal-hydride technology has a slight advantage, and lithium-ion technology a significant advantage, in terms of weight. Indeed, this weight advantage may eventually enable lithium-ion battery technology to replace nickel-hydrogen in many space missions.

The vertical axis of the graph in Fig. 1.2 shows the expected life capability of rechargeable battery systems, in terms of the watt-hour throughput over typical high-reliability lifetimes. Clearly, nickel-hydrogen is one of the best technologies available based on life expectancy. The wide range that is indicated in Fig. 1.2 for lithium-ion battery cycle-life is a result of the present lack of completed life test databases for this emerging technology. The range indicated in Fig. 1.2 for lithium-ion batteries extends from demonstrated performance levels in flight-type cells at the low end, to the optimistic extrapolations of the test data at the high end. Clearly, such extrapolations do not rule out lithium-ion life as being competitive with nickel-hydrogen. Increased life capability can typically be translated into increased reliability by simply increasing the required performance margins.

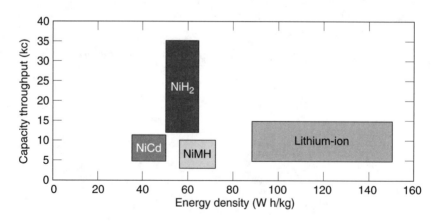

Figure 1.2. Relative life and energy density characteristics of common rechargeable battery types used in satellite applications.

It is generally possible to increase reliability at the expense of life and weight. One example of this trade is the classic choice of a depth of discharge that is consistent with the required lifetime and reliability.

While the cost of batteries is a significant driver in selecting technology for terrestrial applications, it is usually a secondary factor in selecting batteries for spacecraft systems, because of the high "paper" costs associated with testing, verification, and traceability. Nickel-hydrogen batteries are significantly more costly than any of the other technologies shown in Fig. 1.2 in terms of manufacturing costs. However, this difference is significantly reduced if similar extra costs of extensive testing and documentation are included for all the technologies, although nickel-hydrogen remains the most costly by a factor of about 50–100%. The higher cost of nickel-hydrogen has not been a major technology driver, because of the significantly enhanced life and reliability possible with it.

Nickel-hydrogen cells and batteries have historically displayed a robust safety record relative to nickel-cadmium and lithium-based batteries. They are tolerant of high rate overcharge, reversal, and short circuits, often responding to these events with no change in performance. The safety issues for nickel-hydrogen batteries are discussed in detail in chapter 3. The principal issues are associated with the possibility of hydrogen gas leaking and its flammability.

The handling and storage of nickel-hydrogen cells and batteries has historically been an area fraught with pitfalls, although to some extent this has been true of most batteries used in space power systems. Handling and storage will probably remain an area of difficulty in the future. As will be discussed in more detail in chapters 3 and 4, nickel-hydrogen cells and nickel electrodes in these cells are sensitive to precharge, have a high self-discharge rate, are very sensitive to temperature, may require active storage, and can display highly history-dependent performance. While these factors have not been major issues in selecting nickel-hydrogen over nickel-cadmium batteries in most cases, they have clearly been an inconvenience that has increased the costs associated with managing nickel-hydrogen batteries on the ground.

Nickel-hydrogen batteries are not generally chosen for an application based on their volume. The pressure vessels that contain the hydrogen gas in the cell typically occupy a large volume even in high-pressure cells that operate at 1000 psi or more. The packing density of these pressure vessels in a battery structure also generally adds more volume. It is this characteristic that makes nickel-hydrogen batteries particularly unattractive for powering small satellites that have little extra volume. It is in the microsatellites, nanosatellites, and picosatellites that lithium-ion batteries will likely be required to enable small-satellite missions, simply because there is insufficient volume for other battery types.

The combination of factors discussed above and listed in Table 1.1 has resulted in nickel-hydrogen technology largely replacing nickel-cadmium during the past 25 years in rechargeable space power applications. The principal driver in this transition has been the cycle life capability of nickel-hydrogen. As part of this transition, space power systems, their manufacturers, and their users have had to learn how to best deal with the performance, charge control, storage and handling, testing, modeling, degradation modes, and diagnostics related to nickel-hydrogen usage. This book will discuss the details of how to best deal with nickel-hydrogen batteries in all these areas.

2 The Historical Evolution of Nickel-Hydrogen Cell Designs

2.1 Background of Nickel-Hydrogen Development

The technological background of the nickel-hydrogen cell centers on a number of developments in the late 1960s. In this time period significant efforts were under way to develop truly regenerative fuel cells to power electric vehicles, support load leveling for power utilities, and support long-term power needs in high-technology areas such as spacecraft. Much of this development effort was stimulated by concerns about the continued supply of inexpensive petroleum-based fuels. For space power systems, regenerative fuel cells were envisioned that could store energy and cycle reliably in closed systems for decades. Such systems could provide power for satellites operating in low orbits, midaltitude orbits, or the higher geosynchronous orbits; for space launch systems such as the (then conceptual) space shuttle, space stations, and interplanetary probes and spacecraft; and for planetary exploration stations on the moon, Mars, or other bodies in the solar system. With extensive support from the National Aeronautics and Space Administration (NASA), a number of practical fuel cell systems were developed in the United States for space use.

However, the ultimate goal of a long-lived and reliable fully regenerative system remained very elusive. The concept of a regenerative hydrogen-oxygen fuel cell power system also attracted commercial visionaries who saw numerous terrestrial applications for a highly efficient and pollution-free energy source. The "hydrogen economy" was seen by some to be the ultimate future for world energy consumption, in some scenarios replacing the existing petroleum-based economy by the early twenty-first century. For space power, regenerative hydrogen-oxygen fuel cells were attractive because of the simple chemistry, which involved no toxic components or emissions, and the potential for integrating such systems into the life-support systems of astronauts or the fuel needs of rocket thrusters.

Two major problems have plagued these developments, problems that have not been fully resolved to this day. The first is a straightforward engineering issue that is extremely difficult to handle in long-lived space systems. The pumps, motors, and plumbing required for fuel cell operation have to be extremely reliable if these systems are to be depended on in long-term space missions. System engineering solutions to this problem can be devised in terms of high-reliability parts and redundant systems. However, such solutions significantly increase the size, weight, complexity, and cost of the system. The second problem has had no clear technical solution. The oxygen electrodes used in all these systems to regenerate the oxygen fuel are unstable relative to electrochemical oxidation. Even the best catalysts used for oxygen regeneration slowly oxidize, leading to chemical or physical changes that cause an eventual loss of the necessary catalytic activity. This problem persists for regenerative fuel cells to this very day.

However, the development activities that began in the late 1960s led ultimately to the development of robust and highly reliable primary fuel cells, which remain the electrical generation system of choice in short-duration space missions requiring more energy than can be collected from solar panels, such as the space shuttle missions. Just as important for space power, the problems in developing long-lived regenerative fuel cells led to the concept of a nickel-hydrogen battery cell as a viable alternative to the fully regenerative hydrogen-oxygen fuel cell.

The earliest concept of a nickel-hydrogen battery cell was put forward in Germany in 1970, as part of an effort to use hydrogen-based fuel cells to power electric vehicles. The concept offered a brilliant alternative to deal with the stability problems of oxygen regeneration electrodes in fuel cells. The nickel electrode, which had been used for decades in nickel-cadmium and nickel-iron battery cells, provided a surrogate oxygen regeneration electrode. Not only did the nickel electrode reversibly cycle oxygen, but it also stored the oxygen in a metastable solid matrix of nickel oxyhydroxide. (The detailed chemistry of the nickel electrode will be discussed in chapter 4.) The alkaline nickel electrode could make the oxidative power of the oxygen-rich oxyhydroxide matrix available using readily reversible reactions for many thousands of energy-storage cycles.

The nickel-cadmium cell, which also uses the nickel electrode, suffers from the relatively poor lifetime and reliability of the cadmium electrode. The nickel-hydrogen cell concept combined the highly reliable nickel electrode with the advanced hydrogen electrode concepts that came from fuel cell development programs. Because both the nickel and the hydrogen electrodes were seen as very robust, early nickel-hydrogen cell development efforts identified few failure modes that were expected to limit the life of these battery cells. Ultimately, through very long-term testing and usage, a number of key degradation and failure modes of nickel-hydrogen cells have been identified, and these are discussed in detail in chapter 12.

Unfortunately for the original commercial developers of the nickel-hydrogen cell concept, an important advantage of the fuel cell concept had been lost. The nickel-hydrogen cell now had to carry its own fuel in the form of hydrogen gas and oxygen-rich nickel hydroxides, which limited the energy density, volume, and scale-up of the nickel-hydrogen system. Recharging involved more than refueling with fresh hydrogen and oxygen. In addition, nickel-hydrogen cells used expensive catalysts and required the safe containment of large amounts of high-pressure hydrogen gas. These drawbacks made the nickel-hydrogen cell concept less than economically feasible for most commercial applications, including electric vehicles and large-scale load leveling.

However, space applications provided an almost ideal market for such a battery cell. High cost was a minor detail for a technological innovation that was capable of significantly extending the life and performance of satellites costing hundreds of millions of dollars. In the late 1960s and the 1970s, the improved power system life and reliability possible with nickel-hydrogen batteries was seen as a truly enabling technology for future generations of long-lived space power systems. It is in this technology climate of the early 1970s that the story of nickel-hydrogen batteries begins, within the space programs of the United States, the Soviet Union, Japan, and Europe.

2.2 Overview of Nickel-Hydrogen Development History

The history of nickel-hydrogen technology presented here covers the initial design period of the early 1970s, through numerous technology improvements and design changes, up to a relatively mature technology at the start of the twenty-first century. This thirty-year period covers not only the maturation of the technology, but also the entry of new cell producers and the variety of novel cell concepts, both successful and unsuccessful, that have been explored. A historical view of nickel-hydrogen cell technology is provided by the time line shown in Fig. 2.1, which will serve as the basis for the discussions in the following sections.

In Fig. 2.1, wide lines that have been shaded differently correspond to the major nickel-hydrogen cell design variants that have been used in space missions. These styles allow the evolution of the early COMSAT and AF/HAC (Air Force/Hughes Aircraft Company) cell designs to be easily followed, and the entry of later-generation cell designs, such as the EPI (EaglePicher Industries)/ManTech, to be tracked.

In general, the history of nickel-hydrogen cell development can be viewed in three phases. The 1970s saw the development, testing, and initial applications of the first-generation cell designs. The 1980s saw the development of second-generation cell designs, experimentation with numerous design variants, the proliferation of nickel-hydrogen technology in space missions, and the entry of numerous new producers into the market. The 1990s marked the maturation of the most successful technologies. During the 1990s, as nickel-hydrogen became the dominant rechargeable space battery technology, test data became available to identify the design features and cell variants that gave the best long-term performance. It is noteworthy that the 1990s were marked by a significant consolidation in the suppliers of nickel-hydrogen cells. The following sections will fill in the details within each of these three decades of nickel-hydrogen technology, and a final discussion will explore the numerous less-successful design variations that have been tried.

14   The Historical Evolution of Cell Designs

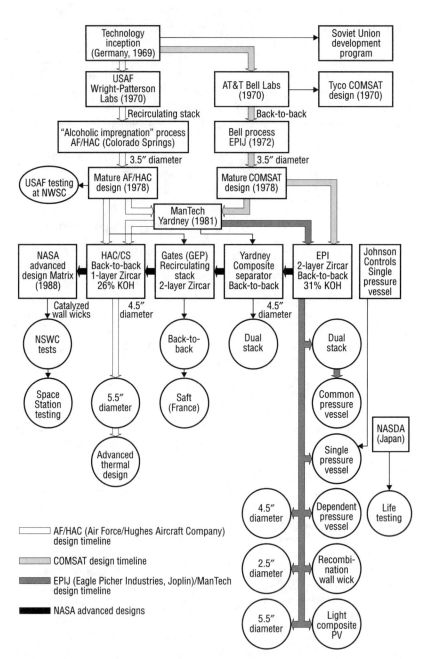

Figure 2.1. Flowchart and time line showing the development of nickel-hydrogen battery cell technology through 2000.

2.3 First-Generation Nickel-Hydrogen Cells

2.3.1 The COMSAT Cell Design

The concept of the nickel-hydrogen cell, first conceived of in Germany, was recognized in 1970 by AT&T/Bell Labs in the United States as a potential technology of key value in space power. AT&T/Bell Labs took on the role of evaluating the feasibility of the basic technology for space use, developing initial component and cell design concepts, and transitioning this technology to battery cell manufacturers and satellite contractors for evaluation. This effort also began the initial commercial nickel-hydrogen cell development for nonspace applications. A number of initial test cells were constructed and evaluated. Early in the development effort, the nickel electrode designs available at the time were identified as the life-limiting components in the cells. While historically adequate for nickel-cadmium cells, the existing nickel electrodes could not meet the more demanding performance levels required of nickel-hydrogen cells.

In response to the need for improved nickel electrodes, AT&T/Bell Labs developed the initial version of what is now referred to as the aqueous electrochemical impregnation process. This process, variants of which are still in use today, produced improved sintered nickel electrodes by electrochemically depositing the energy storage material directly within a conductive substrate. It produced significantly improved nickel electrodes when this process was combined with improved sintered nickel substrates based on those that had been used for years in nickel-cadmium cells. Testing of the early "proof of concept" cells demonstrated that nickel-hydrogen cells could be used to significant advantage in space systems as a replacement for nickel-cadmium cells.

To transition the new nickel-hydrogen technology into an actual space program, AT&T/Bell Labs initiated a joint effort with Tyco International to manufacture cells containing the new technology, and with the International Telecommunications Satellite Consortium (Intelsat) to use these cells in AT&T/Bell Labs's geosynchronous communications satellites. This general cell design, which was initially acquired and tested by COMSAT Laboratories (the technical arm of Intelsat in the United States), became known in later years as the COMSAT cell design.

The COMSAT cell, which was one of the first viable nickel-hydrogen cells for space use, contained nickel electrodes made by aqueous electrochemical impregnation of nickel/cobalt hydroxides into a 30 mil thick nickel sinter made by a slurry process. The sintered starting substrate had a porosity of about 76%, and the finished nickel electrode had a porosity of about 40–45%. This electrode design offered significant improvements in life cycle capability over the standard chemically loaded or pasted nickel electrodes that had been used for years in nickel-cadmium cells. Two of these nickel electrodes were stacked back-to-back with asbestos cloth separators on each face of the back-to-back electrodes. The asbestos separators, which were borrowed from fuel cell technology, were in contact with the hydrophilic surface of platinum-catalyst-based hydrogen electrodes, which had also been developed for fuel cell use. The hydrophobic side of these catalytic gas electrodes was in contact with a plastic screen that allowed hydrogen gas to flow freely to the catalyst sites in the hydrogen electrode. This stacking unit was then repeated, with all the nickel electrodes connected in parallel to a bus bar on one side of the cell and all the hydrogen electrodes to a similar bus bar on the other side of the cell. This stacking arrangement is illustrated in Fig. 2.2.

All the components in the COMSAT design shown in Fig. 2.2 were made to fit into a cylindrical pressure vessel having a nominal diameter of 3.5 in. The stack of components was held together by a steel bolt that passed through the center of all the stack components to tightly clamp them together. One end was bolted to a plastic end plate and a metal ring that attaches the stack to the pressure vessel, while the other end of the stack was simply bolted against a plastic end plate. Positive and negative leads connected to the bus bars on each side of the stack carried the current to positive and negative cell terminals sealed into the ends of the pressure vessel. The stack was attached to the pressure vessel by welding the metal ring at one end of the electrode stack to the wall of the pressure vessel. The pressure vessel was sealed (except for an electrolyte fill tube) by welding the two halves of the pressure vessel together. (The halves were typically hydroformed pieces of Inconel 718.) Tungsten inert gas welding was originally used for this purpose but subsequently gave way to laser and e-beam welding. A tube was also

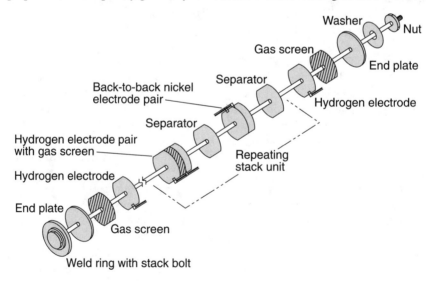

Figure 2.2. Stacking of nickel and hydrogen electrodes in the COMSAT design.

2.3 First-Generation Nickel-Hydrogen Cells

welded into one end of the pressure vessel to allow the later addition of electrolyte, a process known as cell activation.

About two years after the joint effort to transition nickel-hydrogen cell technology into space systems was started, Tyco was replaced as the cell manufacturer by EaglePicher Industries of Joplin, Missouri (EPIJ). The aqueous nickel electrode impregnation process developed by AT&T/Bell Labs was scaled up and put into place at EPIJ, and production of nickel-hydrogen cells was begun in 1974. This COMSAT cell was one of the first-generation nickel-hydrogen cell designs, as indicated in the time line in Fig. 2.1. It had some features that remain to this very day in nickel-hydrogen cells, while many of its other design characteristics have since been altered to provide cells that are more robust.

All nickel-hydrogen cell designs are essentially "electrolyte starved," meaning that there is little or no free electrolyte sloshing about in the pressure vessel. All electrolyte is contained within the porous stack components. The early COMSAT cell had a hydrophobic coating over the inner wall of the pressure vessel, the purpose of which was to force any electrolyte that was expelled from the stack during cell operation back into the stack in the zero-gravity environment of space. This hydrophobic coating has been replaced in designs that are more modern by a hydrophilic wall wick that enables electrolyte to move by capillary forces from wet to dry regions of the stack. The electrolyte used in the original COMSAT cell was potassium hydroxide, which, after a process of activation and conditioning, ended up at a typical concentration of 38–40% by weight. Most state-of-the-art cells today use 26 or 31% potassium hydroxide electrolyte. The original COMSAT cell had a hydrogen precharge of 14.7 psi of hydrogen gas, which has been supplanted by nickel precharge in modern nickel-hydrogen cell designs. Later chapters will detail the reasons for these and many other changes that have occurred throughout the years.

The basic COMSAT cell with some design variants was in production at EPIJ until about 1997. Production of this "heritage" cell ended because it became clear that this cell was not as capable of high-rate performance and long life as were the improved second-generation cell designs. Major factors in the demise of this cell line were the lack of high-quality asbestos separator and the liability issues associated with the handling and use of asbestos.

Nickel-hydrogen cells of the COMSAT design were extensively tested for many years by COMSAT Laboratories and other aerospace contractors, giving performance that was generally superior to that of nickel-cadmium cells. Some of these test data will be reviewed in chapter 11. COMSAT batteries were flown in two spaceflight experiments in 1977. The NTS-2 satellite flew nickel-hydrogen batteries successfully in geosynchronous orbit for 5 years. Nickel-hydrogen batteries also performed well for about 8 months in the low Earth orbiting Air Force Flight Experiment satellite. COMSAT batteries were subsequently flown on satellites of the Intelsat IV and Intelsat V design during the next 10–12 years.

In general, the nickel-hydrogen batteries performed well in these geosynchronous satellites for much beyond 10 years. Some issues were encountered with increasing cell impedance, which was attributed to asbestos degradation, as well as movement of electrolyte and its loss from the cell stack in response to thermal gradients. These observations were important for the future design and handling of nickel-hydrogen cells and batteries, because they showed how sensitive the performance of the cells could be to electrolyte distribution. Later cell designs have invariably included features to manage the distribution of electrolyte in the cell stack, and thermal control systems have been designed to keep the thermal gradients within an operating cell relatively small.

2.3.2 The First-Generation AF/HAC Cell Design
In parallel with the development of the COMSAT cell design, the U.S. Air Force (USAF) began supporting the development of a nickel-hydrogen cell design that would be more robust in a low Earth orbit environment, where more than 50,000 cycles could be experienced over a 10-year period. In 1971, this development began under the jurisdiction of Wright-Patterson Air Force Laboratories (WAFL). Hughes Aircraft Company (HAC) was chosen as the cell and battery manufacturer in this development effort; thus the resulting cell design is termed the AF/HAC design.

Like AT&T/Bell Labs, WAFL found that a high-quality nickel electrode was critical for obtaining a long-lived nickel-hydrogen cell. An electrochemical impregnation method was also chosen to make sintered nickel electrodes at WAFL. However, rather than using a sinter made by a slurry process, WAFL developed a process by which the substrate was made by sintering a layer of dry nickel powder to a nickel screen using no binders or additives. This is still referred to as the "dry powder" process. To reduce corrosion of the nickel sinter during the subsequent electrochemical impregnation process, the temperature of the highly acidic aqueous bath used by AT&T/Bell Labs was reduced. Because a water/ethanol mixture was used to control the reduced temperature, this process is referred to as the "alcoholic impregnation" or Pickett process. This process incorporates a cobalt additive deposited at about a 10% level with the nickel hydroxide, which is about twice the 5% cobalt additive level used in the aqueous process developed at AT&T/Bell Labs.

The alcoholic nickel electrode impregnation process was put first into pilot plant, and later into full-scale production, at the EaglePicher Industries facility in Colorado Springs, Colorado. The nickel electrodes fabricated using the early alcoholic impregnation processes were built into a new nickel-hydrogen cell design by developers at HAC with support from USAF. This cell used the "recirculating stack" design illustrated in Fig. 2.3.

In this stacking arrangement, each nickel electrode had one face in contact with a gas screen and the hydrophobic side of the hydrogen electrode, and the other face in contact with a separator that in turn was in contact with the hydrophilic

2.3 First-Generation Nickel-Hydrogen Cells   19

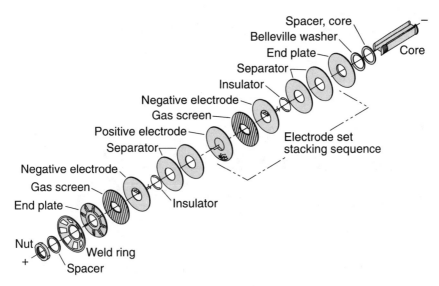

Figure 2.3. Arrangement of stack components in a recirculating-stack nickel-hydrogen cell. Reprinted courtesy of NASA.

side of the hydrogen electrode. The gas screen was an open-mesh polypropylene screen that allowed oxygen formed at the nickel electrode to freely flow across to the hydrogen electrode, where it could catalytically recombine with hydrogen gas. All ionic current flow passed only through the side of the nickel electrode in contact with the separator. Thus, the recirculating stack design tended to pump electrolyte from one end of the stack to the other, leading to a natural concentration gradient across the stack. The cell design also instituted a porous wall wick consisting of a thin flame-sprayed zirconium oxide layer coating the inner wall of the pressure vessel. The wall wick was also in contact with the edges of the separator, which protruded slightly from the stack. The function of the wall wick was to diffusively redistribute the electrolyte gradient set up by the recirculation pattern established by the stack design.

The AF/HAC cell design also used a novel stack geometry that has been referred to as the "pineapple slice" stack. As indicated in Fig. 2.4, all the stack components—nickel electrodes, separators, hydrogen electrodes, and gas screens—are shaped like thin pineapple slices. In this design, leads that run through the center hole of the pineapple slices make the positive and negative connections to the electrodes. This configuration maximizes the contact area between the pressure vessel wall and the edge of the stack to allow optimal thermal conduction to the pressure vessel wall and the thermal sleeve in which the pressure vessel is mounted.

The nickel electrodes in this design were nominally 3.5 in. in diameter and 30 mils thick. The separator material consisted of two layers of a zirconium

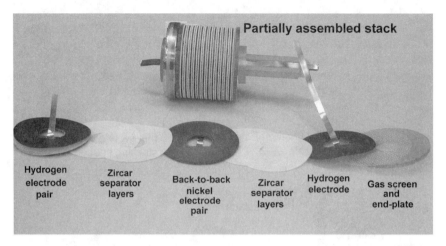

Figure 2.4. Typical pineapple-slice nickel-hydrogen cell components, with a partially assembled stack.

oxide fabric, either woven or knit. The hydrogen electrodes were fuel-cell-quality, platinum-catalyst-based electrodes with hydrophilic platinum catalyst on the side facing the separator, and the side facing the gas screen consisting of a hydrophobic porous Teflon layer. The electrolyte was aqueous potassium hydroxide, 31% by weight. The cell components were stacked on a polysulfone core that fit into the center hole of the pineapple slice components. The electrode leads were routed through compartments inside the core, and passed out the top and bottom of the core as lead bundles that were welded onto the cell terminals. The stack was compressed between polysulfone end plates using a polysulfone nut threaded onto one end of the core. Belleville washers were used with the end nut to provide a controlled force to the stack, and to allow the stack to swell. A weld ring at one end of the stack was attached to a hydroformed pressure vessel made from Inconel 718. The cell terminals were sealed in the pressure vessel dome using Teflon compression seals. A 1/8 in. electrolyte fill tube was typically provided in the pressure vessel dome for activating the cell with electrolyte after it was fully assembled. Figure 2.5 shows an example of an AF/HAC cell before being built into a battery. The finished nickel-hydrogen cell shown in this figure, like all variations of this basic design, physically appears as a sealed pressure vessel with two electrical terminals to pass current.

Cells of the AF/HAC design were originally activated using a precharge of 50 psi of hydrogen gas, thus enabling the full capacity of the nickel electrodes to be used. In the early 1980s it was found that the hydrogen precharge reacted with and degraded the nickel electrodes during storage in the fully discharged state, leading to capacity loss by a mechanism termed "hydrogen sickness." This degradation

Figure 2.5. Typical 50 A h AF/HAC nickel-hydrogen cell. The white tie-wrap holds a thermistor in place.

mode is described in detail in chapter 12. Starting in about 1984, a nickel precharge was adopted as standard in the AF/HAC design.

Early AF/HAC cells provided a robust design that was tolerant of reversal, withstood overcharge even at high rates, and provided good long-term cycle life performance. Testing in standard low Earth orbit 90 min cycling regimes typically gave 7000 to 9000 cycles at 80% depth of discharge, and more than 60,000 to 70,000 cycles at 40% depth of discharge.

2.4 The ManTech Program

In 1981, Yardney was awarded the Nickel-hydrogen Manufacturing Technology (ManTech) Program to develop a lower-cost design and improved manufacturing methods for nickel-hydrogen cells. From this program emerged what is referred to today as the ManTech cell. The ManTech cell design and manufacturing methods developed by Yardney are described in detail in Bentley and Denoncourt[2.1] and are summarized in Dunlop et al.[2.2] These changes include the following.

- The Inconel 718 pressure vessels are hydroformed to a uniform thickness, with no chemical milling.
- A machined terminal boss is e-beam welded to the pressure vessel dome.

- E-beam girth welding is used rather than tungsten inert gas welding.
- The fill tube is made an integral part of the cell terminal.
- A Teflon compression seal is baselined for the terminals; however it is not clear whether this was a major improvement over the nylon Zeigler compression seal used in the COMSAT design.
- Molded polysulfone end plates are used at each end of the stack, rather than Inconel end plates.
- A modified molded polysulfone core is used, along with dual Belleville washers to maintain stack compression on a floating core.
- A dual-layer separator consisting of one layer of Zircar and one layer of asbestos is used, with the asbestos layer facing the nickel electrode surface. This design change did not prove best compared to dual-layer Zircar in life tests.

Each of the other cell manufacturers developed their own versions of the ManTech cell design that used many of the key changes developed in the Yardney ManTech Program, but included other major improvements specific to each manufacturer. The following section describes many of these second-generation cell designs.

2.5 Second-Generation Cell Designs

In the five years following the ManTech program, a number of changes occurred both in the nickel-hydrogen cell manufacturing industry and in the cell designs that were being produced. Many of these changes were either a direct or indirect consequence of the ManTech program. These changes began the trend to replace nickel-cadmium with nickel-hydrogen batteries in satellites. Coincidentally, the rising number of performance problems that vintage nickel-cadmium batteries made by Gates Energy Products began to experience in the 1985–87 period probably spurred the transition to nickel-hydrogen batteries. One significant change was the entry of several new nickel-hydrogen cell manufacturers into the market. Yardney developed a line of nickel-hydrogen battery cells based on the results of that company's work for the ManTech program. Gates Energy Products began to produce a cell design that was similar to the AF/HAC design. Finally, Johnson Controls began efforts to market a single pressure vessel battery design that was initially targeted towards commercial applications such as electrical vehicles or load leveling, and was based on Johnson's lead-acid battery experience. In addition, changes also occurred within the existing AF/HAC cell design and in the cell designs produced by EPIJ.

The AF/HAC cell design, which was largely being produced by EaglePicher Industries in Colorado Springs by this time, underwent some major changes in the mid-1980s. The cells began to be made in a larger, 4.5 in. diameter size, primarily to support higher-power satellites having a single battery. The single-battery power system provided a somewhat higher energy density than could be achieved

2.5 Second-Generation Cell Designs

with multiple batteries employing smaller cells. This was an important advantage for the geosynchronous satellites in which these cells and batteries were primarily finding applications. In addition, the HAC cell design adopted the back-to-back stacking arrangement that was pioneered in the COMSAT design, and recommended with the pineapple slice components by the ManTech program. The arrangement of the stack units in this back-to-back configuration, which is common to many modern nickel-hydrogen cell designs, is shown in Fig. 2.6.

Other changes in the AF/HAC cell design included the adoption of a single layer of Zircar (a zirconium oxide fabric) separator and the use of 26% potassium hydroxide electrolyte. While the decreased amount of electrolyte retained by the single layer of separator clearly made the cell less robust in terms of susceptibility to dry-out, the use of 26% electrolyte compensated by making the stack less prone to dimensional changes throughout its life. The nickel electrode design and fabrication methods used in the earlier cells were retained, although the thickness of the nickel electrode was increased from 30 mils to 35 mils. Cell activation techniques that left free electrolyte in the cell tended to be the major problem throughout the long term, causing popping, occasional stack damage, and a few short-circuit failures. These problems have been dealt with by improved activation methods that assure all free electrolyte is drained from the cells. The methods

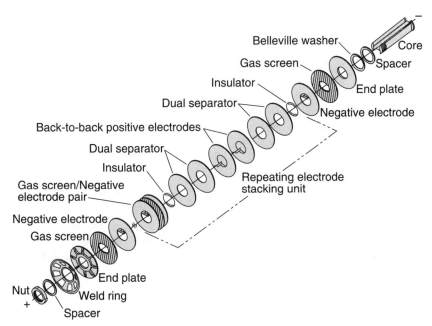

Figure 2.6. Arrangement of stack components in a typical back-to-back ManTech nickel-hydrogen cell. Reprinted courtesy of NASA.

have included better drain procedures, and have included centrifuging cells to remove excess electrolyte. This cell design, while perhaps not optimum for the more than 40,000 cycles needed in a low-orbit satellite, was perfectly appropriate for a lightweight geosynchronous satellite power system.

The second-generation HAC (later Hughes Space and Communications and Boeing Satellite Systems) cell design has been utilized extensively in geosynchronous satellites during the past 10–15 years. Beginning in the late 1990s a new 5.5 in. diameter cell design was developed by HSC for higher-power satellites, providing cells having capacities up to 350 A h. These cells were built at the Hughes Space and Communications facility in Torrance, California, as well as in Colorado Springs. Along with the larger cell capacities, improved thermal control systems were developed to help dissipate the increased heat generated in the cells. The improvements seen in these systems included carbon composite thermal sleeves that offered better thermal conductivity with lighter weight, and the use of heat pipes to provide active cooling to cells. In 2000, the Hughes Space and Communications production facilities were acquired by Boeing Satellite Systems. Since 2000, Boeing has developed a low Earth orbit version of their standard cell, which goes back to the dual-layer Zircar separator and 31% electrolyte that was used in the earliest HAC cells.

In the mid-1980s, Gates Energy Products developed a nickel-hydrogen cell line that was produced in the company's Gainesville, Florida, facility. This design was similar to that of the AF/HAC cell in that it used a recirculating stack and dry sinter in the nickel electrodes. An aqueous electrochemical impregnation process was developed using 5% cobalt additive in the nickel hydroxide. The cells used two layers of Zircar separator, 31% potassium hydroxide electrolyte, and a nickel precharge. The seals used in the pressure vessels were ceramic-to-metal seals similar to those used for years by Gates Energy Products in its sealed nickel-cadmium satellite cells. Performance problems associated with the recirculation pattern of electrolyte in the relatively long 3.5 in. diameter cells eventually led to a shift to the back-to-back design that had been adopted by the rest of the nickel-hydrogen industry by this time.

The sale of Gates Energy Products to the French company Saft in 1994 enabled Saft to offer the Gates Energy Products cell design to its customers, a design that soon replaced the existing Saft nickel-hydrogen cell designs. Saft had developed a nickel-hydrogen cell design using the nylon separators and chemically impregnated nickel electrodes used in its nickel-cadmium cell lines. While performing reasonably well, the existing Saft design had exhibited problems associated with the relatively compressible and thermally reactive nylon separator. Several ongoing research programs to develop improved separator materials, such as Zirfon, to replace nylon, lost much of their impetus after Saft obtained the Gates Energy Products cell designs. As was found by others attempting to develop improved separators, it was difficult to identify materials that worked better than Zircar.

2.5 Second-Generation Cell Designs

In parallel with its highly successful ManTech program, Yardney developed the capability to produce nickel-hydrogen cells that essentially followed the features specified by that program. Yardney developed a facility for the aqueous electrochemical impregnation of nickel electrodes, using 5% cobalt additive and a slurry-based nickel sinter. This facility is the only one that has ever made nickel electrodes individually. Each nickel electrode was impregnated within an individual cell through which solution from a common manifold was flowing. This facility was capable of producing some electrodes of very high quality, but it also suffered from variability in quality of electrodes made at different points in time, or from different production cells. The nickel electrodes were stacked in a back-to-back configuration using a novel composite separator. The separator consisted of a layer of Zircar that supported a layer of asbestos. This separator was supposed to combine the incompressible physical structure and electrolyte retention characteristics of Zircar with the ability of asbestos to channel oxygen into the gas spaces at the edge of the cell. The Yardney cell also used a Teflon compression seal to seal the terminals into the pressure vessel, and it was activated with one atmosphere of hydrogen precharge. Cells made by Yardney were placed into a number of life tests. These cells were generally marked by a significant spread in their performance. Some cells gave excellent performance, while other cells gave much poorer performance. In about 1983, Yardney ceased to produce nickel-hydrogen cells.

Following the ManTech program, EPIJ developed a ManTech cell line to go along with the standard COMSAT cell line that they already produced. This cell design used the electrochemical impregnation method developed for nickel electrodes in the COMSAT cell design, employing either slurry sinter or dry sinter. The nominal thickness of the nickel electrodes was 35 mils, as opposed to the 30-mil thickness of the electrodes used in the COMSAT cells. The cells were built with a back-to-back stacking arrangement that included two layers of Zircar separator, and they used 31% potassium hydroxide electrolyte. They were built with about a 15% nickel precharge. As in the COMSAT cell, a nylon compression seal was used to seal the terminals into the pressure vessel. A flame-sprayed zirconium oxide wall wick was included in the design to allow electrolyte gradients to equilibrate by diffusion. Extensive electrolyte draining procedures were employed to ensure that no free electrolyte existed in the cell, because free electrolyte was found to result in popping damage and internal cell short circuits.

The initial cells in the EPIJ ManTech line were in the 35–50 A h capacity range and were 3.5 in. in diameter. The length of the stack that could be supported in a pressure vessel while meeting vibration and mechanical safety margins limited the maximum cell capacity. To satisfy the need for cells in the 60 to 100 A h capacity range, EPIJ developed a dual-stack cell design that enclosed two stacks within the same pressure vessel. The two stacks were attached to a weld ring that was welded to the center of the cylindrical portion of the pressure vessel, thus providing mass balance about the point where the stacks were supported. Electrical connection of the two stacks in parallel provided twice the capacity that could be achieved from a single stack. Because the wall wick on the inner wall of the pressure vessel could not extend across the weld region at the center of the cell, Zircar wicking assemblies through the weld ring were provided to enable electrolyte diffusion between the two stacks. The dual stack design has achieved cell level energy densities up to 60 W h/kg based on capacity at 10°C, which is a record for 3.5 in. diameter nickel-hydrogen cells containing two layers of Zircar separator. The ManTech cell design containing a dual stack has performed quite well in a wide range of life tests, and in a large number of orbital satellites.

The successful dual stack nickel-hydrogen cell suggested a design alternative that could reduce the number of pressure vessels required in some power systems. Rather than connecting the two stacks in the pressure vessel in parallel, the two stacks could be connected in series to provide a unit having twice the voltage but only half the capacity. This configuration was termed a common pressure vessel, which refers specifically to two cells connected in series within the same pressure vessel. This is quite different from the single pressure vessel configuration, which involves a large number of cells that are series-connected within the same pressure vessel, or the individual pressure vessel configuration, in which all electrodes are connected in parallel.

While in principle the common pressure vessel design should perform as reliably as the individual pressure vessel cells, an additional degradation mode does exist for the common pressure vessel. Any electrolyte bridging that occurs between the two stacks will result in ionic migration (driven by the voltage difference between the cells) and the development of an electrolyte gradient. A hydrophobic barrier strip on the pressure vessel wall between the two stacks is provided to minimize this possibility. In addition, if oxygen that is generated in one stack recombines in the other stack, an electrolyte volume gradient can develop. These issues are likely to be driven by significant thermal gradients across the cells, or by excessive overcharge. These cells have not yet been tested as extensively as the standard individual pressure vessel cells; however, the test data that are available do not show premature degradation modes.

A number of other nickel-hydrogen cell design variants have been developed by EPIJ. Cells have been built in both a smaller-diameter (2.5 in.) size, as well as in larger sizes (4.5 and 5.5 in. diameter). These other cell sizes are essentially scale versions of the 3.5 in. diameter ManTech cell design. The 2.5 in. diameter cell has been built in capacities ranging from 6 to 32 A h. It is intended primarily for small satellites, or any other applications needing high reliability and long cycle life. The 2.5 in. diameter cell has been built in both the common pressure vessel as well as the individual pressure vessel configuration. Cells having a 4.5 in. diameter typically cover the capacity range from 100 to 220 A h, and the 5.5 in. diameter cells have a capacity of 250 to 350 A h.

Efforts to make lighter-weight nickel-hydrogen cells at EPIJ have led to the use of single layers of Zircar separator, as well as the use of larger cell sizes. The highest energy density reported for nickel-hydrogen cells to date is 70 W h/kg for a 4.5 in. diameter cell design using a single layer of Zircar separator. Some efforts to make lighter cells have also involved the use of lightweight composite pressure vessels. Generally, when the weight of nickel-hydrogen cells has been reduced to give energy densities above about 55–60 W h/kg, the weight reduction has been accompanied by some reduction in cell robustness. The 70 W h/kg cell cited above is not likely to be the best cell for a low Earth orbit satellite that must operate throughout 40,000 or more eclipse cycles, but is likely to be adequate for a geosynchronous satellite requiring 88 cycles per year. As will be seen in later sections of this chapter, a number of other design changes intended to reduce cell and battery weight have been explored. Almost invariably, the weight reductions are accompanied by reduced life or reduced reliability.

One innovation that has been pursued by EPIJ for improving cell performance during overcharge is the use of catalytic strips on the wall wick in the ManTech-type cell. These catalytic wall wicks have spiral strips of platinum catalyst firmly embedded in the zirconium oxide wall wick. The catalyst sites on the wall of the pressure vessel allow the direct recombination on the cell wall of any oxygen that is generated during overcharge and that finds its way to the edge of the stack. This deposits a portion of the heat from overcharge directly on the cell wall where it can be most easily dissipated by the thermal control system, allowing cells to run cooler and better tolerate overcharge. Life tests of cells containing this feature have demonstrated significantly improved cycle life when compared to cells cycled in the same test that did not have catalytic wall wicks. This design feature has only slowly been accepted in the aerospace industry, and it is now used in the battery cells that power several satellites.

A final design variant that was pioneered in nickel-hydrogen cells manufactured by EPIJ is one that uses what is known as "rabbit ear" terminals. In this design, both the positive and negative terminals are situated in the top of the cell, emerging from opposing sides of the top dome of the pressure vessel. The lead bundles from both the positive and the negative plates are run up through the core to the top dome of the cell, where they are welded to the terminals. This cell design allows cells to be efficiently wired together in series strings by positioning the negative terminal of one cell immediately adjacent to the positive terminal of the next cell in the string. These terminals can then be connected together by an extremely short length of conductor, thus allowing a significant reduction in the contribution of the intercell conductors to battery impedance. A typical ManTech cell with the rabbit-ear terminal configuration is shown in Fig. 2.7.

While the rabbit ear design has been extremely popular, it can have a negative impact on cell performance. With this design, the resistance is now greater to the electrodes at the bottom of the cell because of the longer leads, thus imposing a gradient in depth of discharge, particularly at high discharge rates. In addition, heating is greater at the top of the cell because of the greater local discharge near the top of the cell, as well as the flow of both current and heat upwards through the lead bundles. These issues make these cells somewhat more sensitive to the thermal environment and the charge control system used for maintaining cell state of charge.

2.6 NASA Advanced Design Matrix

In 1988 NASA entered the nickel-hydrogen development arena by initiating a program intended to acquire both standard and (several) advanced cell designs from each cell manufacturer for the purpose of conducting life tests on these cell designs. The manufacturers providing cells for this program were Gates Energy Products, Saft, Yardney, and EPIJ. This test program was intended to evaluate nickel-hydrogen designs for future NASA missions, including the space station (initially Space Station Freedom, SSF, and later the International Space Station), which was planned to use nickel-hydrogen batteries for secondary power. Each of the manufacturers involved in this program constructed a range of design variations for life testing,[2,3] as shown in Table 2.1. In this table, the stack types are back-to-back (BB) or recirculating (Rec), and the separator is either dual-layer Zircar (ZZ) or a Zircar/asbestos composite (ZA).

Life testing of the cells in the advanced design matrix continued many years. One somewhat surprising conclusion became evident from this test program. Of the cell designs under test, very few of the advanced designs functioned even as well as the standard designs submitted by the manufacturers. It appeared that the advanced designs were simply experiments with new features having promise of improvement. To actually realize the hoped-for performance improvements, an

Figure 2.7. ManTech 100 A h nickel-hydrogen cell with a rabbit-ear terminal configuration.

2.6 NASA Advanced Design Matrix

Table 2.1. International Space Station Nickel-Hydrogen Cell Design Matrix

		Capacity (A h)	Stack BB/Rec	Stack Single/Dual	Separator Zircar/Asbestos	Serrated Separator	Ni sinter Dry/Slurry	Catalyzed Wall	KOH conc (%)	Cell diam (in)
Eagle Picher	Standard design	65	BB	Dual	ZZ	No	Slurry	No	31	3.5
	Advanced design	65	BB	Dual	ZA	No	Slurry	Yes	31	3.5
	Advanced design	81	BB	Dual	ZA	No	Slurry	Yes	31	3.5
	KOH test	81	BB	Dual	ZZ	No	Slurry	No	31/26	3.5
	22 cell SPV	50	BB	N/A	ZZ	No	Slurry	No	31	10
Gates	Standard design	65	Rec	Single	ZZ	No	Dry	No	31	3.5
	Advanced design	81	Rec	Single	ZZ	No	Dry	No	31	3.5
	KOH test	65	Rec	Single	ZZ	No	Dry	No	31/26	3.5
Yardney	Standard design	65	Rec	Dual	ZA	No	Slurry	No	31	3.5
	Advanced design	65	BB	Dual	ZA	Yes	Slurry	Yes	26	3.5
	Advanced design	81	BB	Dual	ZA	Yes	Slurry	Yes	26	3.5
SAFT	KOH test	81	BB	Single	ZZ	No	Slurry	No	31/26	3.5

iterative redesign and test program was likely to be needed. These results drive home an important point that has been demonstrated by numerous tests of nickel-hydrogen cells. The majority of the stresses in the cells that ultimately limit performance and life arise from strong interactions between a large number of processes that affect the dynamics between all the cell components. Simply changing one component or adding a new design feature can have a nonlinear effect on the dynamics of cell operation. Design improvements must encompass the entire cell dynamic, not just the performance of isolated components.

The International Space Station cell-testing program became a significant additional test effort supported by NASA. Cells from Gates Energy Products were tested as the first generation of nickel-hydrogen cells planned for the International Space Station. Cells from EPIJ were tested as an alternative after Gates Energy Products was sold to Saft. The International Space Station program ultimately used the standard cell designs produced by these manufacturers, probably because of the NASA experience with advanced designs. The standard ManTech designs used a back-to-back stack with dual-layer Zircar separator and 31% electrolyte. The batteries eventually used on the International Space Station contained 81 A h cells made by EPIJ.

2.7 Single Pressure Vessel Cells and Batteries

Nickel-hydrogen cells have a relatively poor volumetric energy density, which partially results from the difficulty in closely packing cylindrical domed devices into a battery. The single pressure vessel nickel-hydrogen battery encloses a large number of individual cells within a single large pressure vessel. In addition to minimizing the volume of the battery, the single pressure vessel battery can have significantly reduced impedance, because the cells can be directly connected together within the pressure vessel, eliminating the need for interconnecting cell wires.

In the early 1980s, Johnson Controls began a program to develop large single pressure vessel nickel-hydrogen battery systems for terrestrial uses. This development effort was highly dependent on the battery- and cell-packaging technologies that Johnson Controls had developed for lead-acid batteries. While a few test batteries were built and tested with some success, the commercial market for nickel-hydrogen batteries never really developed. By the late 1980s Johnson Controls was targeting satellites as the most realistic market for single pressure vessel nickel-hydrogen batteries. During the next several years a number of single pressure vessel nickel-hydrogen battery variations were produced, each providing somewhat better performance than the last. The technical issues in making a single pressure vessel battery operate reliably over a long cycle life were daunting. These issues included removing heat, maintaining all electrolyte within each cell while allowing hydrogen to flow into and out of each cell, recombining all oxygen within the cell in which it was generated, handling popping in the cells, and minimizing the impacts of a leak in the single pressure vessel. By the early 1990s, Johnson Controls had developed a design that worked reasonably well and was targeted for the large satellite constellations planned for the mid-1990s, such as Iridium, Teledesic, etc.

EPIJ had also been working for some years on a single pressure vessel design, which also targeted the large-satellite-constellation market. In 1992 EaglePicher acquired the battery division of Johnson Controls and transferred all the single pressure vessel technology to the Joplin facility. The company possessed the facilities to produce the single pressure vessel batteries in the volume that would be required for large constellations of satellites, and the single pressure vessel battery system could be produced much more rapidly and cheaply than the individual pressure vessel batteries traditionally used in satellites. EaglePicher manufactured the single pressure vessel batteries that were flown in the Iridium constellation of 88 satellites (98 satellites including orbital spares). While these batteries have functioned reasonably well, some difficulties have been encountered in maintaining all the cells at a similar state of charge when they share a common hydrogen reservoir. This problem stems from subtle differences in loss rates between individual cells combined with the limited overcharge to which these batteries are subjected, resulting from concerns with oxygen recombination and movement of oxygen between the cells. However, these single pressure vessel batteries continue to perform well in orbit, and ground tests suggest they can provide 75–80% of the lifetime expected from individual pressure vessel batteries.

2.8 Nickel-Hydrogen Cell Design Variants

A number of other viable nickel-hydrogen cell designs have been produced and tested, and have given good life test results, but have not been selected for actual flight use in satellites. For a design variant to be selected for satellite use, it generally must meet some key need; otherwise it is not selected in place of standard designs for which there is a large body of experience and test data. Thus, these variants may be as capable as the more standard designs, or even better in some cases. They simply need more testing, or a satellite program that matches their capabilities.

2.8.1 The Dependent Pressure Vessel Cell Design

The first design variant is the dependent pressure vessel cell design. This cell design uses a flattened pressure vessel, giving the cell the shape of a discus. The flattened surface alone is not capable of supporting the internal cell pressure; thus multiple cells must be mechanically tied together (i.e., using tie-rods) to hold the pressure. The internal cell stack is in thermal contact with the flattened face of the cell, thus providing a large thermal dissipation surface. In addition to providing better heat dissipation, these cells can be more closely packed to give a battery better volumetric energy density. Because this cell design is dependent on the pressure in adjacent cells or a battery end plate for support, internal structural accommodation must be provided to prevent an individual cell from being crushed by its adjacent cells if it loses pressure for any reason while the other cells are charged.

Such pressure loss could occur if a cell fails by a short circuit, if a leak occurs, or if a sufficiently large imbalance in state of charge develops. Dependent pressure vessel cells have been built and tested at EPIJ, providing good performance.

2.8.2 The Dual-Anode Design

A second design variant is termed the dual-anode design. It uses a stack design that is a variation of the commonly used standard back-to-back design. In the dual-anode cell, each face of each nickel electrode is in contact with a separator layer, which in turn is in contact with a hydrogen electrode. Thus, it uses twice as many layers of separator and twice as many negative electrodes. If a single thickness of Zircar separator is used for each separator layer, then the weight of separator and electrolyte is similar to that in a back-to-back cell with two layers of Zircar separator. The extra negative electrodes make the dual-anode design somewhat larger and heavier than the standard back-to-back design.

However, the dual-anode design offers some significant performance advantages that can more than offset the extra size and weight. Both faces of each nickel electrode are fully and symmetrically used, making the superficial current density half of that in standard designs. The distance for ionic current flow and mass transport through the nickel electrode and separator is about 50% of that in standard designs, significantly decreasing internal impedance and improving oxygen recombination pathways. These changes result in significant decreases in recharge voltage and increases in discharge voltage, as well as improved capacity and reduced degradation rates. Cells of this design are being tested at The Aerospace Corporation to evaluate their capability for long-term cycling at 60% depth of discharge. If the cells can reliably operate for 60,000 cycles or more at 60% depth of discharge, their improved capability, as compared to that of standard cells, would more than compensate for the slightly higher weight. These cells may be appropriate for future high-power low Earth orbiting satellites. As of late 2008, these cells have demonstrated 40,000 cycles at 60% depth of discharge with little evidence of cell wear.

2.9 Experimental Nickel-Hydrogen Cell Designs

A number of nickel-hydrogen cell design and component design variations have been tried over the years. While many have functioned adequately, they have not been introduced into production, because they provided insufficient advantage over existing designs, or in some cases did not perform as well as existing designs. These are referred to as experimental cell designs, and they are documented here to illustrate the range of design options that have been pursued.

2.9.1 Bipolar Cells

Bipolar nickel-hydrogen cells have been explored in the context of single pressure vessel systems. In a bipolar nickel-hydrogen cell, cells are stacked up in a single pressure vessel, with a solid conduction plate providing direct electrical contact between the positive electrode of one cell and the negative electrode of the adjacent cell. The reliable management of both gas and electrolyte in these systems, while assuring that no electrolyte bridges occur between cells, is not simple. Because of these issues and assembly problems, successful long-term operation of a bipolar nickel-hydrogen cell has never been demonstrated.

2.9.2 Separators for Cells

Most cell developers have explored the development of improved separators for nickel-hydrogen cells. The benchmark separator material that has successfully been used is Zircar, a material that is relatively incompressible, is chemically stable, and holds electrolyte within relatively large pores compared to the positive and negative electrodes. Zircar is costly to make, and it is brittle and difficult to handle. Many other materials have been evaluated to replace it, but in no cases have any performed better than or even as well as Zircar. Alternative separator materials have included asbestos, asbestos composites, Zirfon (a zirconium oxide polysulfone blend), nonwoven nylon, and a range of polymer composites.

The essential reason why Zircar has performed better than all these alternatives is its incompressibility. As the nickel-hydrogen cell cycles, the nickel electrodes tend to swell until the opposing stack compression, which is set by the Belleville washer and the core elasticity, matches the compressibility of the stack components. Any separator that is much more compressible than the nickel electrode will introduce dimensional instability to the cell stack, which will reduce performance by its increased impedance as it compresses and by dry-out as the expanding nickel electrodes draw in electrolyte from the separator.

2.9.3 Catalysts for Cells

The platinum black catalyst used in the hydrogen electrode is costly. For this reason, significant effort has been devoted to finding cheaper catalysts that will function adequately for use in nickel-hydrogen cells. Standard hydrogen electrodes probably contain ten times the amount of platinum black required for adequate functionality. The most successful efforts for better catalysts have simply reduced the amount of platinum black in each negative electrode. This does increase the polarization of the hydrogen electrode, which does become an important design issue at high currents. In addition, a platinum electrode with lower platinum loading is not as robust for tolerating the decreases in active catalyst surface area that result when the platinum is exposed to an oxygen environment. This happens every time a nickel-precharged cell is fully discharged and put into storage. Because nickel precharge is standard in today's cell designs, most manufacturers have chosen to retain significant platinum loading margin. Catalysts other than platinum have not had the needed catalytic activity, or they had activity that was not stable over long-term operation, or they were chemically unstable.

2.9.4 Energy Density of Cells

The energy density of the nickel-hydrogen cell is strongly influenced by the energy density of the nickel electrode. One technique that has been explored for increasing cell energy density is to use thicker nickel electrodes. Most cells today use 35-mil-thick nickel electrodes placed back-to-back, resembling a single 70-mil-thick nickel electrode. The use of thicker nickel electrodes (45 and 60 mil) has been explored. Clearly in a back-to-back design there will be reduction in usable capacity and increase in impedance that results from the use of thicker nickel electrodes. However, the most significant drawback of thicker nickel electrodes has been the lack of sintering techniques capable of producing the thicker sinter with a uniform porosity and pore-size distribution through its thickness. Sinter having a poor uniformity is difficult to load properly with active material, and when loaded, does not give the expected utilization of its active material.

Increasing the active material loading levels in the nickel electrode has also been evaluated as a method to improve energy density. The standard loading level used in today's nickel-hydrogen cells is 1.6–1.8 g per cubic centimeter of void volume. While higher loading can increase the capacity of the nickel electrode, the stability of the capacity tends to become poorer. The nickel electrodes tend to expand more rapidly as they are cycled, their internal structure suffers more rapid breakdown, and the capacity drops with cycle life. Because most nickel-hydrogen cells are used in applications requiring very long life, decreased capacity stability throughout life has not been found a very productive trade.

Similarly, energy density can be improved by using higher-porosity substrates, whether nickel sinter, fiber, foam, or felt. With all these materials, it has been found that movement towards higher porosity tends to decrease electrode strength and life. The nickel sinter commonly used today for nickel electrodes ranges from 80% porosity (for most slurry sinters) to 84% porosity (for most dry sinters). When the porosity of either sintered or alternative substrates is made much greater than these levels, the utilization of the active material typically drops markedly and the strength of the substrate is not able to keep the electrode from undergoing excessive expansion as it is cycled.

In addition, the felt and fiber substrates tend to have a relatively high compressibility, which can also lead to cell stack dimensional changes with cycling. Although the fiber substrates have lower utilization and life than sintered substrates, they have found a place in commercial nickel-cadmium or nickel-metal-hydride battery cells using pasted nickel electrodes. The large void spaces between the fibers in these materials enable active material slurry to be physically pasted into the substrate structure. The performance of these electrodes is quite adequate for commercial cell applications.

2.9.5 Additives for Improving Cell Performance

Additives of various types have been introduced into nickel electrodes and electrolytes in attempts to improve cell performance. The principal ones are cobalt or cadmium in the nickel electrodes, and lithium in the electrolyte. Of these, cobalt is consistently used at either a 5% or 10% level in the nickel electrodes to provide significantly improved performance and life, as will be discussed in more detail in chapter 4. Cadmium additives improve the charge efficiency of the nickel electrodes by poisoning the oxygen evolution reaction that occurs in overcharge. However, cadmium additives can lead to dendritic short circuits and can poison the platinum catalyst in the hydrogen electrodes. Lithium additives in the electrolyte can increase capacity early in life by enabling the nickel electrodes to be more easily recharged to higher oxidation states. However, the capacity has been found to degrade much more rapidly throughout life with lithium additives, presumably from the greater volume changes and stresses that occur in the nickel electrodes.

2.9.6 Potassium Hydroxide Electrolyte Concentration

The concentration of potassium hydroxide electrolyte has been found to have a significant impact on the capacity and performance of nickel-hydrogen cells. Higher concentrations tend to give higher cell capacities, at the cost of significant reduction in cycle life. Today's nickel-hydrogen cells typically operate with electrolyte concentrations of either 26% or 31% by weight. When test cells of the same design activated with both these electrolyte concentrations have been cycled in the same test regime, the cells with 26% electrolyte have invariably given better cycle life. Cells tested with 35% to 40% electrolyte have given significantly lower cycle life. Cells containing 20% potassium hydroxide electrolyte have exhibited some problems supporting high discharge rates, possibly because of having insufficient ionic strength to support the required ionic current flow.

2.10 References

[2.1] J. G. Bentley and P. J. Denoncourt, *Manufacturing Technology for Nickel/Hydrogen Cells*, AFWAL-TR-87-4051, October 1987.

[2.2] J. D. Dunlop, G. M. Rao, and T. Y. Yi, *NASA Handbook for Nickel-Hydrogen Batteries* (NASA Ref., Pub. 1314, September 1993), pp. 1–54.

[2.3] B. A. Moore, H. M. Brown, and T. B. Miller, "International Space Station Nickel Hydrogen Battery Cell Testing at NAVSURFWARCENDIV Crane," *Proc. of the 32nd International Energy Conversion and Engineering Conf.*, Vol. 1 (1997), pp. 174–179.

3 Electrical and Thermal Performance

The electrical and thermal characteristics of nickel-hydrogen cells are among the most important variables to consider when designing a battery power system. This chapter discusses the general electrical and thermal responses of nickel-hydrogen cells, and it includes examples to illustrate the typical behavior of specific cells. In most cases, the examples focus on the behavior of state-of-the-art cell designs that use Zircar separator. In a few cases, the behavior of earlier COMSAT cell designs is discussed to allow the comparison of performance signatures. The performance characteristics described in this chapter reflect the beginning-of-life behavior of nickel-hydrogen cells. The gradual changes in performance seen over years of cycle life are discussed in chapter 11.

3.1 Steady-State Voltage Behavior

Nickel-hydrogen cells typically exhibit a discharge voltage plateau between 1.2 and 1.3 V, and a recharge voltage plateau between 1.4 and 1.5 V. Figure 3.1 shows the typical variation of charge and discharge voltage with state of charge for a range of currents. In most applications, nickel-hydrogen cells are discharged at maximum rates of 0.4–0.8C (the C-rate discharges the nameplate capacity of a cell in 1 hr), while peak recharge rates are generally 0.5C or less. In most practical battery systems, the peak recharge rate is reduced as the state of charge (or cell

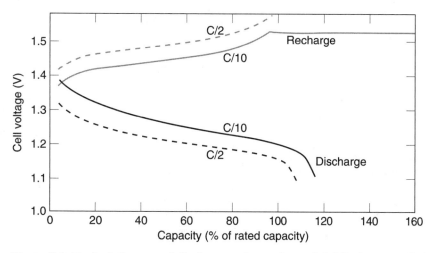

Figure 3.1. Typical charge and discharge voltages for a nickel-hydrogen cell at several rates at a temperature of 10°C.

voltage) approaches higher levels where overcharge can occur. This avoids the significant heat generation and stress that are created with high-rate overcharge.

Voltage data such as those shown in Fig. 3.1 may also be expressed in terms of an I/V (current/voltage) curve for the cell, as shown in Fig. 3.2. Figure 3.2 shows a family of I/V curves at specific states of charge. The slopes of these I/V curves, whether during charge or discharge, represent the resistance of the cell at each state of charge.

Several important signatures of nickel-hydrogen cells may be recognized in Fig. 3.2. First, there is a significant discontinuity, or hysteresis, between the discharge and the recharge I/V curves for each state of charge. This apparent discontinuity does not result from curvature of the I/V curves to allow them to connect smoothly at low currents, but remains even at very low charge or discharge currents. As discussed in chapter 4, a number of explanations have been put forward for the voltage discontinuity, which arises from the fundamental chemical processes within the nickel electrode.

The second important signature seen in Fig. 3.2 is the linearity of the I/V curves during discharge and recharge, even down to relatively low currents. This means that nickel-hydrogen cells in steady-state operation behave electrically as an ideal dc voltage source in series with a resistor. The voltage of the source and the value of the resistance may change with state of charge, temperature, or other operating conditions, but this simple equivalent circuit describes the dc performance

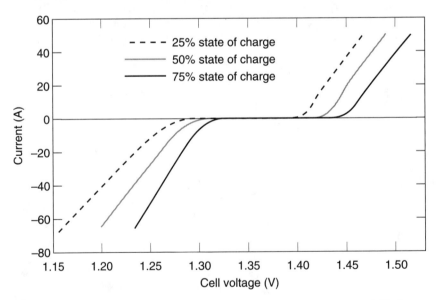

Figure 3.2. I/V curves during the recharge (positive current) and discharge (negative current) of a 100 A h nickel-hydrogen cell at a temperature of 10°C.

during recharge of the nickel-hydrogen cell quite well. While the overall resistance of the electrolyte and the electronic resistance of the cell leads are expected to give a fixed series resistance, the I/V behavior of the electrochemical reactions is expected to follow a Butler-Volmer I/V relationship, which should give an exponential increase in current with increasing voltage.

The reason for the linear behavior is that the Butler-Volmer equation predicts a linear I/V curve when the current is low relative to the exchange current density for the electrodes. In other words, high-surface-area electrodes, such as the sintered nickel and hydrogen electrodes, have an impedance low enough to keep them operating in the linear regime of the Butler-Volmer equation near zero overpotential. Even at high rates of cell charge or discharge, the electrochemical overpotential at the nickel and hydrogen electrodes typically is less than 50 mV.

The other important factor to note from Fig. 3.2 is the magnitude of the cell resistance that can be obtained from the I/V line segments. The nickel-hydrogen cell must have a low dc operating resistance, as is the case for any battery cell capable of supplying high currents. Figure 3.3 shows the steady-state cell resistance as a function of cell state of charge during discharge and recharge. This dc resistance, which is typically in the range of 1–1.5 mΩ, includes the ionic resistance of the electrolyte, the electronic resistance of the metal current conductors, and the electrochemical polarization of the electrodes.

The relatively simple description here of how the steady-state cell voltage depends on state of charge and current is, nonetheless, a response that can be

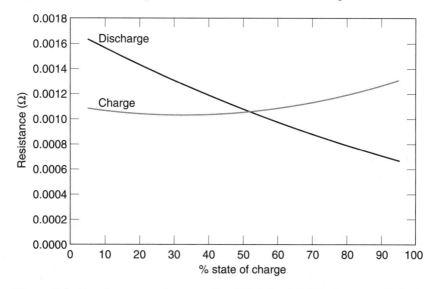

Figure 3.3. Steady-state resistance of a 100 A h nickel-hydrogen cell during C/10 charge and C/2 discharge at a temperature of 10°C.

changed by several complicating factors. First, the formation of γ-NiOOH can occur in the nickel electrode during overcharge to result in a lower discharge voltage, in spite of the increased state of charge caused by generation of this higher-capacity phase. Because the γ-NiOOH phase is only formed while the cell is in overcharge, it is often difficult to fully separate its effects on cell behavior from those of the overcharge processes (oxygen evolution and heat generation). Figure 3.4 illustrates how γ-NiOOH formation affects the charge voltage behavior of the nickel-hydrogen cell at low temperatures, where there is the most separation in voltage between the formation of γ-NiOOH and the evolution of oxygen.

While the formation of γ-NiOOH can complicate the voltage response of a cell during recharge, the discharge of the γ-NiOOH phase during cell discharge can also complicate cell voltage response. Figure 3.5 shows the discharge voltage of a nickel-hydrogen cell that contains significant γ-NiOOH. Clearly there is a transition from the higher β-NiOOH discharge voltage plateau (represented by the I/V lines in Fig. 3.2) to a γ-NiOOH discharge plateau that is lower and displays a different variation of I/V behavior with state of charge. Thus, a full understanding of the I/V response of a nickel-hydrogen cell during discharge depends on the amount of γ-NiOOH present in the nickel electrode.

An added complexity in the I/V curves of nickel-hydrogen cells during recharge can be caused by α-Ni(OH)$_2$, which is the discharge product of γ-NiOOH.

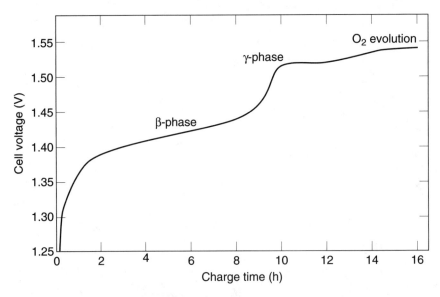

Figure 3.4. Phases generated in each of the recharge voltage regions typically seen for a nickel-hydrogen cell. Overcharge (oxygen evolution) begins when the voltage rises above about 1.50 V.

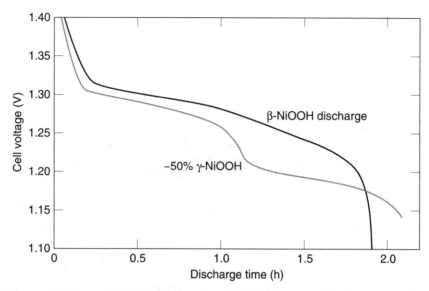

Figure 3.5. Normal C/2 discharge of a nickel-hydrogen cell compared to discharge following extensive cycling that converted ~50% of the active material to γ-NiOOH.

The α-Ni(OH)$_2$ phase is not thermodynamically stable in the nickel-hydrogen cell environment, and it will crystallize over time to the β-Ni(OH)$_2$ phase represented by the I/V curves of Fig. 3.2. However, if recharge is done before all the α-phase has recrystallized, the recharge will occur at a significantly reduced voltage level.

Typically, the lower recharge voltage response is only significant if recharge is done before recrystallization occurs (i.e., within several hours following the discharge of significant amounts of γ-NiOOH). Therefore this lower recharge voltage behavior is not often evident during recharge following typical battery reconditioning discharges. It is clear from these discussions that the complex I/V behavior of the nickel-hydrogen cell is best predicted either from an empirical database that includes all the operational parameters or from a model that accurately captures all the fundamental processes described here.

No discussion of the voltage behavior of the nickel-hydrogen cell would be complete without including the effects of fundamental electrode thermodynamics and kinetics on voltage. These effects are described by the Nernst equation, which predicts how the reversible cell voltage depends on both temperature and hydrogen pressure, and the Butler-Volmer equation, which predicts how cell overpotential depends on current, temperature, and pressure. The thermodynamic temperature dependence arises from the variation of standard electrode potentials with temperature, an effect driven by the entropy change for the overall electrochemical cell reaction. For the nickel-hydrogen cell reaction, which has a negative overall entropy change, this thermodynamic effect shifts the voltage upward as temperature decreases. This thermodynamic effect combines with the kinetic effect, which causes an upward shift in charge voltage predicted by the Butler-Volmer equation as temperature decreases. The overall combination of these two effects makes the cell recharge voltage increase 1–2 mV for every 1°C drop in cell temperature.

Figure 3.6 shows the typical increase in charge voltage seen with decreasing temperature. Most of the changes in recharge voltage occur in the overcharge region and stem from the overcharge, or oxygen evolution reaction. The recharge voltage during the plateau between 1.4 and 1.45 V is influenced just as much by the temperature-dependent kinetics of the relaxation of α-Ni(OH)$_2$ to β-Ni(OH)$_2$ before recharge is completed, as it is by the temperature dependence of the recharge reactions.

While the variation of recharge voltage with temperature is important for understanding cell performance, it is also important in charge management to understand how cell overcharge voltage changes with temperature. This temperature dependence is primarily driven by the thermodynamics and the kinetics of the oxygen evolution reaction, as opposed to the nickel-electrode charge reaction. The voltage of the oxygen evolution reaction is much more sensitive to temperature than is that of the nickel-electrode recharge reaction, as indicated in Fig. 3.6. The

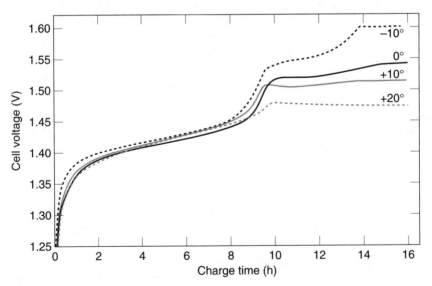

Figure 3.6. Variation in typical nickel-hydrogen cell recharge voltage with temperature at a C/10 charge rate.

difference in temperature coefficient for these reactions, which increases the voltage separation between the recharge and overcharge processes, is one reason why nickel-hydrogen cells have significantly improved charge efficiency at reduced temperature.

The effect of temperature on discharge voltage at a C/2 discharge rate is shown in Fig. 3.7, and is not as pronounced as that shown in Fig. 3.6 for recharge voltage. The discharge plateau voltage level is roughly centered at 1.25 V for all temperatures. As temperature decreases, the internal resistance of the electrolyte increases while that of the metal conductors decreases. The most pronounced effect of temperature is the significant decrease in capacity that is obtained at the higher temperatures. This is a result of the decrease in charge efficiency at higher temperatures in nickel-hydrogen cells.

Changes in the cell pressure also change cell voltage by influencing the reversible potential and the kinetics of the hydrogen reaction. These voltage changes are again governed by the Nernst and Butler-Volmer equations, respectively. For the current densities and pressures customarily used in nickel-hydrogen cells, the influence of pressure on charge and discharge voltage is not large unless hydrogen gas depletion occurs (i.e., at the end of discharge when a cell has nickel precharge). Figure 3.8 shows how pressure changes and charge or discharge current affect cell voltage for typical recharge and discharge.

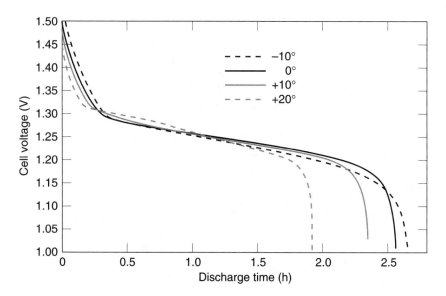

Figure 3.7. Variation in typical nickel-hydrogen cell discharge voltage with temperature at a C/2 discharge rate.

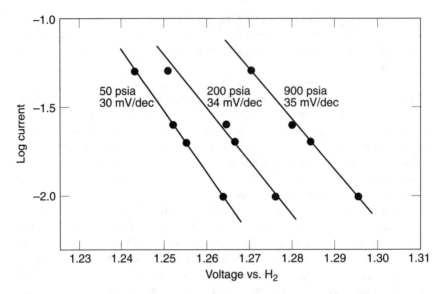

Figure 3.8. Dependence of the voltage of nickel-hydrogen cells (in mV/decade) on hydrogen pressure and the logarithm of the current density (as a fraction of the C-rating for the cell).

3.2 Cell Power Response

The power response of a nickel-hydrogen cell during discharge depends directly on the internal resistance of the cell, as indicated by the slopes of the discharge I/V curves illustrated in Fig. 3.2. This assumes, of course, that the high power demand occurs while the cell is discharging in the lower left quadrant of Fig. 3.2. If a high-power discharge pulse is required while a cell is recharging, a more complex capacitive transient will occur as the voltage characteristic of the cell shifts from the upper-right quadrant to the lower-left quadrant. If high power is demanded during discharge, specific power-density curves such as the ones in Fig. 3.9 are obtained for a typical nickel-hydrogen cell (55 W h/kg).

It is possible to design a nickel-hydrogen cell having higher specific power capabilities. Such a cell design should have a significantly reduced internal resistance. The dual-anode stack design allows cells to be built with approximately 50–60% of the internal resistance of a more standard back-to-back cell stack design. The reduced resistance is realized by cutting the current density on the electrodes in half, decreasing the electrolyte resistance nearly in half, and decreasing the lead resistance to 50–75% of that in the more standard cell designs. With these design changes, the cell is about 15% heavier than a standard cell of the same capacity but is capable of significantly improved specific power, as illustrated in Fig. 3.9. High-power nickel-hydrogen cells have been built in pressure vessels

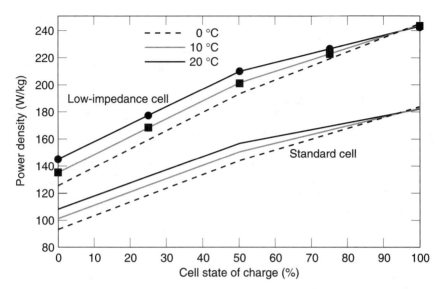

Figure 3.9. Power density as a function of cell state of charge and operating temperature for a standard nickel-hydrogen cell (55 W h/kg) and a low-impedance cell design (47.8 W h/kg).

that were not optimized for weight, and they were tested to provide the performance response curves shown in Fig. 3.9.

3.3 Cell Capacity Behavior

In a discussion of the capacity of nickel-hydrogen cells, it is important to define the standard recharge/discharge cycle used to measure capacity. Nickel-hydrogen cells, if adequately cooled, can be pumped up to remarkably high capacities[3.1] by the application of high-rate overcharge for a long enough period. The drawbacks of such artificially pumped-up capacity are that first, it is not attainable in any realistic power system, because of high heat evolution, and second, the high rate and amount of overcharge can rapidly degrade the cell cycle lifetime. For these reasons, standard-capacity measurement methods have been developed and used by nickel-hydrogen cell manufacturers. The most common include C/10 recharge for either 12 or 16 h, typically at temperatures of –10, 0, 10, and 20°C. The standard discharge almost universally involves a C/2 discharge to a cell voltage of 1.0 V, although sometimes capacity is also reported at higher voltage limits, such as 1.10 V. A variation of the standard recharge cycle has been used for COMSAT cells to reduce the overcharge rate, and involves recharge at C/10 to 90% of rated cell capacity, followed by recharge at C/20 to 160% of rated cell capacity.

The capacity of nickel-hydrogen cells is probably most strongly influenced by cell temperature. Reduced temperature allows a higher cell capacity to be attained for a number of reasons. First, reduced temperature increases the separation between the recharge voltage and the overcharge voltage (see Fig. 3.6), enabling charge efficiency to be much better at lower temperature. Second, the decreased self-discharge rate in nickel-hydrogen cells at low temperature enables more of the charge to be retained until discharge occurs. Third, reduced temperature enables more facile formation of the higher-capacity γ-NiOOH phase. Finally, the resistance of the metallic conductors within the cell decreases as temperature is reduced. The only factors that conspire against improved capacity at reduced temperature are the increased electrolyte resistance, reduced diffusion rate, and increased electrochemical overpotential. When capacity is plotted as a function of temperature, the behavior shown in Fig. 3.10 is typically seen for the lower-impedance cells with dual-layer Zircar separator, and for the higher-impedance cells with asbestos separator.

For ManTech cells utilizing Zircar separator, the capacity generally levels off at low temperatures, but does not drop until the temperature falls below –10°C. For COMSAT cells, which utilize asbestos separator, the higher impedance of the separator and the electrolyte cause a capacity drop-off at temperatures below 0°C. However, the COMSAT cell design holds its capacity slightly better at higher temperatures because of the higher electrolyte concentration and reduced-rate

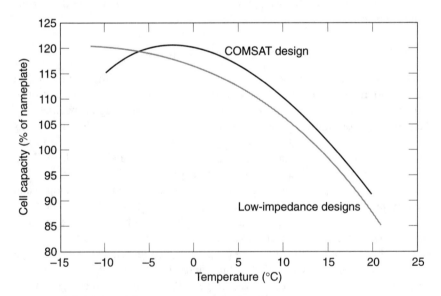

Figure 3.10. Variation of nickel-hydrogen cell capacity with temperature for two design types when discharged at a C/2 rate.

oxygen gas flow through the less permeable asbestos separator. An additional difference between these two cell designs is that recharge of the COMSAT cell design shown in Fig. 3.10 requires about a 50% longer recharge time, because of its lower tolerance to overcharge at a normal C/10 rate.

As this discussion has implied, a higher electrolyte concentration can give improved cell capacity. For example, cells with 38% potassium hydroxide electrolyte can attain about 10–15% higher capacity than can cells with 31% electrolyte in a standard-capacity cycle, largely because the higher concentration facilitates the formation of γ-NiOOH in the nickel electrodes. In spite of allowing a higher cell capacity, the higher electrolyte concentration is not typically preferred because it results in accelerated nickel electrode degradation and significantly reduced cycle life. State-of-the-art nickel-hydrogen cells use either 26% or 31% potassium hydroxide electrolyte. Figure 3.11 illustrates the typical trade between improved beginning-of-life capacity and reduced cycle life that occurs when using higher electrolyte concentration. The 100% utilization in Fig. 3.11 refers to storage of one electron per nickel site in the nickel electrode active material; thus utilization of 100% or more implies that significant amounts of the higher-capacity γ-NiOOH phase have been formed.

The prior cycling, thermal, or open-circuit history of the cell can also influence cell capacity behavior. For this reason, it is very important to use a standardized sequence of capacity and performance evaluation cycles whenever characterizing nickel-hydrogen cells, evaluating cell lot acceptability, or matching

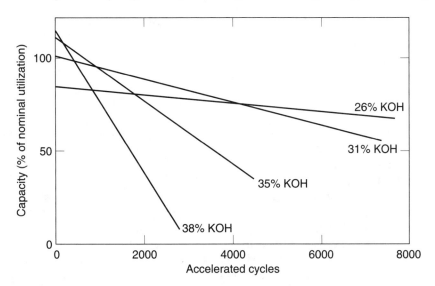

Figure 3.11. Effect of electrolyte concentration on the beginning-of-life capacity and the accelerated cycle life of nickel-hydrogen cells.

cells into a battery. Nickel-hydrogen cells will exhibit different capacity at warm temperatures for several cycles after being operated cold, and will display cold-temperature capacity performance that is altered for several cycles just after cells are cooled from a warmer environment.

Another factor that can dramatically affect cell capacity is the recent open-circuit history of the cell. Open-circuit periods in the fully discharged state will result in a temporarily reduced capacity along with a slightly elevated recharge voltage. This is one reason favoring a "wake-up" cycle after any significant period of open-circuit stand. Active material in the nickel electrode that has not recently undergone full recharge and discharge tends to be more difficult to charge, thus providing reduced capacity. A wake-up cycle that fully recharges and then discharges the cell will eliminate this condition. Significant open-circuit periods in the charged state can also change cell capacity behavior by allowing the charged phases to redistribute into layered structures in the nickel electrode active material, altering self-discharge behavior and cell capacity characteristics. Again, a full charge/discharge cycle will largely eliminate these types of "charge memory" effects.

3.4 Impedance Behavior

The impedance behavior of a nickel-hydrogen cell consists of several contributions. First is a steady-state resistance within the cell that is independent of frequency. This resistance dictates the slopes of the steady-state I/V curves shown in Fig. 3.2. The resistance is composed of the electronic resistance of the metallic current-conducting components of the cell, in series with the ionic resistance of the electrolyte. While the pure cell resistance can be influenced by the development of any significant electrolyte concentration gradients during high-rate operation or if separator dry-out occurs, the frequency-independent cell resistance typically does not change a large amount during normal charge or discharge.

The cell resistance does, however, change with temperature according to the expected variation of the resistance of the metallic and ionic conductors with temperature. For standard cell designs that have resistances from 1.0 to 1.2 mΩ at 20°C, this leads to the variation of cell resistance with temperature illustrated in Fig. 3.12. As temperature is reduced, the metallic conductors decrease in resistance, while the electrolyte resistance increases. Therefore, the variation in the resistance with temperature for any particular cell design will depend on the relative contributions of the metallic conductors and electrolyte-containing ionic conductors to the total cell resistance. In a standard cell the electrolyte resistance typically contributes most significantly to the cell resistance at normal operating temperatures. Low-resistance cell designs have been built in which the electrolyte and the metallic conductors contribute more equally to the overall cell resistance,

3.4 Impedance Behavior   49

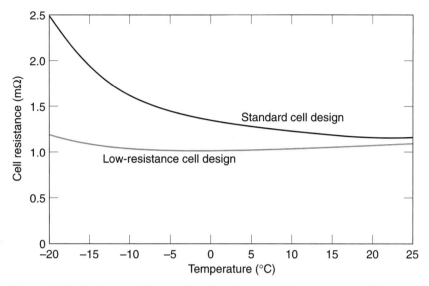

Figure 3.12. Variation of internal cell resistance with temperature for a standard nickel-hydrogen cell design and a low-resistance cell design.

giving the lower curve in Fig. 3.12, which displays very little resistance variability over the typical −10 to +10°C operating temperature range of nickel-hydrogen cells.

The ac impedance of nickel-hydrogen cells includes contributions from the cell inductance as well as the capacitive and resistive contributions of the hydrogen and the nickel electrodes. Figure 3.13 shows the typical ac impedance of a nickel-hydrogen cell over a range of frequencies. The cell inductance dominates the impedance at frequencies greater than 1 kHz, while the cell is purely resistive at about 500 Hz. The hydrogen electrode impedance causes the capacitive impedance loop at 13.4 Hz, and the low-frequency impedance loop at 0.0011 Hz results from diffusional polarization impedance in the nickel electrode.

At low states of charge, depletion of the nickel electrode (in a hydrogen-precharged cell) will increase the magnitude of the nickel electrode impedance loop, while depletion of hydrogen (in a nickel-precharged cell) will increase the size of the hydrogen electrode impedance loop. The ac impedance of the open-circuit nickel electrode is dominated by the diffusional polarization of a dipole layer at the interface of the active material with the electrolyte. It is the polarization of this layer that enables the reaction of solid-state protons within the nickel-electrode active material with hydroxide ions in the electrolyte, and that also accounts for the hysteresis between the charge and discharge I/V curves.

The impedance behavior of a nickel-hydrogen cell can also be defined by the transient voltage response of the cell to a step change in cell current. Such a transient

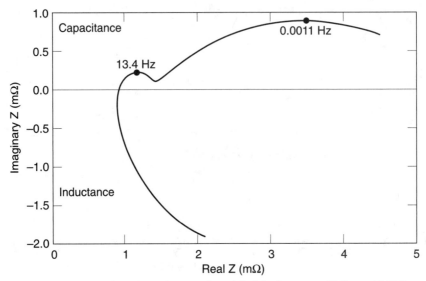

Figure 3.13. Typical ac impedance of a nickel-hydrogen cell from 10 kHz to 0.0005 Hz.

response during discharge is shown in Fig. 3.14. The rapid voltage dip seen on the inset millisecond time scale results from the cell inductance, followed by a rapid voltage rise as the inductance saturates. The voltage rises to a level commensurate with the dc resistance of the cell, and then rapidly decays with an initial behavior that is nearly exponential as the hydrogen electrode polarizes. At longer times, the voltage drops according to the square root of time as the proton diffusion processes in the nickel electrode adjust to the new current. At the longest times, the voltage approaches the steady-state (nearly linear) change associated with the changing state of charge of the cell as discharge continues. The linear change in voltage resulting from the changing state of charge in the cell at long times has been removed from the data shown in Fig. 3.14.

If the cell is in an open-circuit state rather than charging or discharging, the same processes that have been discussed here control the cell impedance. However, the time constants and magnitude of the diffusion impedance associated with the nickel electrode increase significantly. This is because the impedance of the nickel electrode in the open-circuit state is controlled by the polarization or depolarization of a dipolar layer at the interface between the active material and the alkaline electrolyte, as shown in Fig. 3.15.

This diffusional polarization of an interfacial layer enables current to flow across the interface by electrostatically allowing hydroxide ions to move to the surface layer, where they react with protons in the active material. The polarization layer is responsible for the voltage hysteresis between the charge and discharge I/V curves

3.4 Impedance Behavior 51

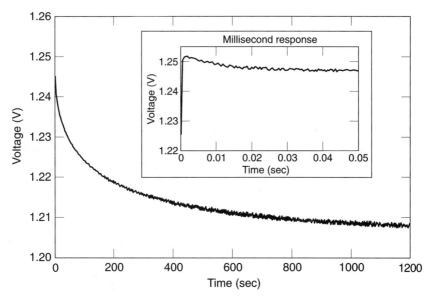

Figure 3.14. Transient voltage response of a nickel-hydrogen cell to a step change in the discharge current. The inset shows the inductive transient seen on the millisecond time scale.

of the nickel electrode shown in Fig. 3.2, and it does not contribute to the impedance when steady-state current is flowing, because the dipolar layer becomes fully polarized by the steady-state current following its characteristic polarization time.

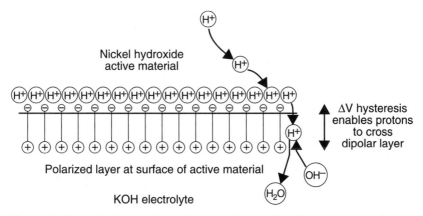

Figure 3.15. A polarization layer forms at the interface between the active material and the electrolyte, which enables protons from the active material to jump across the interface to be neutralized by hydroxide ions.

If the cell transitions from a state of current flow to an open-circuit condition, the dipolar layer will thermally depolarize completely over a period of several hours.

3.5 Self-Discharge Behavior

Nickel-hydrogen cells have a self-discharge rate that is significantly higher than is typical for most other types of battery cells. As discussed later in chapter 4 on nickel electrodes, self-discharge can occur by three processes:

- evolution of oxygen
- direct reaction of hydrogen with the charged nickel electrode
- electrocatalytic discharge of hydrogen on the metallic surfaces in the nickel electrode

Each process depends on the hydrogen pressure in the cell, the state of charge of the nickel electrode, and the temperature. Figure 3.16 indicates the typical variation in self-discharge rate seen in nickel-hydrogen cells with increasing temperature and cell state of charge. The data in this figure were obtained for a nickel-hydrogen cell that operated at about 600 psi of pressure when fully charged. The self-discharge rate is somewhat greater for higher-pressure cells and lower for cells that operate in a lower pressure range. Figure 3.16 illustrates why nickel-hydrogen cells are normally operated at low temperatures, and why their capacity drops

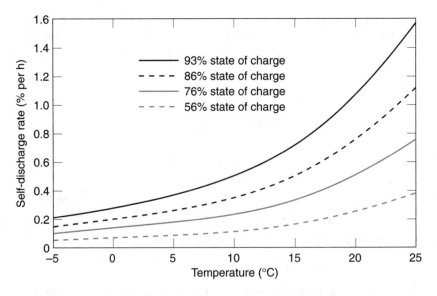

Figure 3.16. Variation in self-discharge rate for typical nickel-hydrogen cells with temperature for different states of charge.

rapidly at temperatures greater than about 10°C. It is primarily the high internal loss rates of these cells at elevated temperature that makes the low-temperature operating environment so attractive for nickel-hydrogen batteries.

The rapid rise in self-discharge rate at the higher temperatures and at the higher states of charge is primarily from the evolution of oxygen in the cells, which is the dominant loss process. The data in Fig. 3.16 show that a C/100 trickle-charge rate should be able to maintain about a 93% state of charge in a typical cell at 20°C, while a C/200 trickle-charge rate should only hold a 76% state of charge at 20°C. If the temperature is dropped to 10°C, a C/200 trickle-charge rate should be able to hold a 93% state of charge in a typical cell. While some variation in self-discharge rate occurs for different cells in a production lot, the data in Fig. 3.16 are a good representation of the average behavior of a battery.

3.6 Secondary-Plateau Discharge Behavior

As described in chapter 4, the nickel electrode exhibits a secondary-discharge plateau at a voltage about 0.8 to 0.9 V greater than the hydrogen electrode voltage. This secondary plateau arises from the formation of a semiconducting depletion layer at the surfaces where the active material contacts the nickel metal in the nickel electrode. In a nickel-hydrogen cell, a wide range of secondary-plateau characteristics can be seen during discharge, depending on the cell precharge and history of storage. Nickel-hydrogen cells that contain more than about 10–12% nickel precharge, and that have been exposed to limited storage, typically exhibit essentially no secondary-discharge plateau. This is because hydrogen gas depletion causes the cell voltage to drop off at the end of discharge from hydrogen-electrode polarization, rather than nickel-electrode voltage polarization. As the cell is operated or stored for many years, the nickel precharge is eventually reduced to essentially zero, allowing cell capacity to increase and the secondary-discharge plateau to be seen, as shown in Fig. 3.17.

As cells continue to age, they eventually build up significant hydrogen precharge, and then will exhibit an extended secondary-discharge plateau. As long as the nickel electrodes are not degraded from either improper storage or long-term cycling, the capacity discharged on the upper-voltage plateau increases or remains intact, while the capacity discharged on the secondary plateau increases. However, as the nickel electrodes degrade from age or cycling, the capacity discharged at the lower plateau can increase at the expense of the upper-plateau capacity. Such a shift in capacity provides an indication of chemical and structural degradation of the nickel electrodes as a result of how they have been operated or stored in the cells over the battery lifetime.

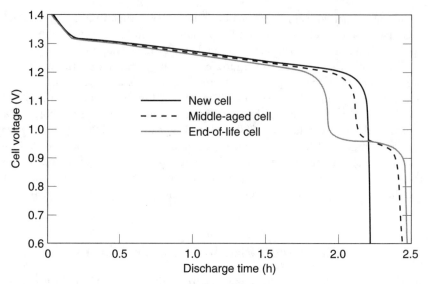

Figure 3.17. Evolution of the lower voltage plateau during discharge. Comparison of voltage for a new nickel-precharged cell with voltage for one that is near end of life reveals loss of significant capacity to the lower plateau.

3.7 Charge Efficiency Behavior

The charge efficiency of nickel-hydrogen cells is a strong function of temperature. This is the primary reason why this type of battery cell is normally operated at temperatures lower than 10°C. The charge efficiency drops off at higher temperatures because the self-discharge rate increases sharply at higher temperatures, and because the rate of oxygen evolution also increases relative to the rate of the recharge reactions in the nickel electrode. Typical coulombic charge efficiency characteristics of nickel-hydrogen cells are shown in Fig. 3.18 as a function of state of charge and temperature for a C/10 recharge rate.

The charge efficiency at reduced state of charge in Fig. 3.18 is an indication of the self-discharge rate relative to the recharge rate, because at less than 50% state of charge the evolution of oxygen is minimal and charge inefficiency is largely controlled by self-discharge. As the state of charge approaches 100% the charge efficiency drops off dramatically as the oxygen evolution rate increases at the nickel electrode. Interestingly, the charge efficiency does not drop to zero at 100% state of charge. Charge can continue to be stored in the cell, albeit at a low efficiency, at up to more than 160% state of charge, particularly at lower operating temperatures. This is a result of the gradual (and relatively inefficient) formation of γ-NiOOH when the cell is overcharged.

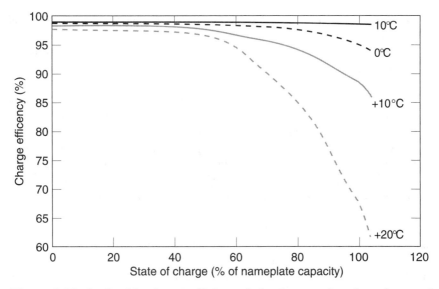

Figure 3.18. Coulombic charge efficiency behavior as a function of state of charge and temperature for a nickel-hydrogen cell using a C/10 recharge rate.

When nickel-hydrogen cells are recharged at rates higher than C/10 the charge efficiency is greater at a given state of charge. This is because the rates for the processes responsible for efficiency loss, self-discharge, and oxygen evolution increase more gradually with increasing current than do the charge storage reactions in the nickel electrodes. However, recharge of nickel-hydrogen cells at high rates (e.g., C/2) must be done carefully. In spite of the improvement in charge efficiency at high recharge rates, significant overcharge heating can occur and can damage cells if the recharge voltage is allowed to increase to a level greater than the voltage level associated with a C/10 steady-state overcharge condition. Normally, it is desirable to minimize overcharge during high-rate cell recharge by reducing the recharge rate when the cell voltage reaches a prescribed level greater than 1.47 to 1.52 V (depending on the temperature). This reduction in charge rate may be done either by switching to a lower recharge rate, or by transitioning to a charge mode where the cell or battery voltage is held constant and the current is allowed to taper down to whatever level is needed to hold the prescribed constant voltage.

3.8 Pressure Behavior

A unique feature of nickel-hydrogen cells is that the internal hydrogen gas pressure provides an excellent indication of the state of charge of the cell. Nickel-hydrogen cell pressure vessels are often fitted with strain gauges that can provide a quantitative pressure measurement based on the flexing of the pressure vessel domes as hydrogen pressure rises or falls in the pressure vessel during cell operation. These strain-gauge pressure monitors are often temperature-compensated to eliminate the effects of temperature variation on the strain-gauge device itself. However, the pressure indications must still be corrected for the effect of temperature changes on the hydrogen gas within the cell. Hydrogen gas is relatively ideal, but as its pressure increases, deviations from ideal gas behavior can become significant. At a pressure of 1000 psia (~68 atm), hydrogen deviates by more than 4% from the ideal gas relationship between pressure and volume. For this reason, the gas law $PV = znRT$ is customarily used to describe and correct pressure data from nickel-hydrogen cells, where z is the compressibility of hydrogen. The compressibility factor z is a linear function of pressure as indicated in Fig. 3.19 for the operating pressure and temperature range of nickel-hydrogen cells. The compressibility becomes greater at lower temperatures, increasing from about 1.063 at 20°C to more than 1.067 at −10°C at a hydrogen pressure of 100 atm.

The typical pressure behavior of a nickel-hydrogen cell during recharge and discharge is shown in Fig. 3.20. It is clear from this behavior that pressure can provide an accurate indication of cell state of charge. However, a number of factors can perturb the highly precise pressure vs. state-of-charge behavior shown in Fig. 3.20. At high rates of charge or discharge, significant heating can occur within the cell to produce thermal gradients. Such thermal gradients can make

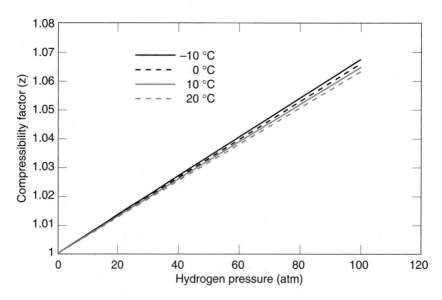

Figure 3.19. Compressibility factor for hydrogen gas as a function of pressure for typical operating temperatures of nickel-hydrogen cells.

3.8 Pressure Behavior   57

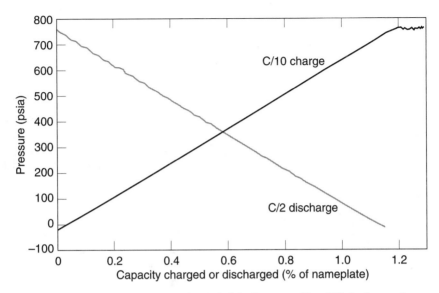

Figure 3.20. Pressure behavior of a nickel-hydrogen cell at 0°C during recharge and discharge as a function of capacity charged or discharged. No corrections were made to the data for temperature variations.

both measurement of the true average temperature of the gas within the cell and the resulting pressure correction somewhat uncertain.

A second effect that can perturb cell pressure measurements during recharge is the evolution of oxygen at moderate to high rates of charge. As oxygen evolution commences when a nickel-hydrogen cell enters overcharge, the oxygen gas will displace some hydrogen within the pores of the nickel electrode, resulting in a pressure rise that is not associated with an increase in state of charge. Such a pressure response from oxygen displacement of hydrogen is shown during a C/10 constant current recharge in Fig. 3.21.

The pressure response in Fig. 3.21 during recharge shows about a 14 psi pressure greater than that expected if the cell accepts charge with a 100% efficiency, in spite of entering overcharge after about 10 h (36,000 sec) of recharge. Because the charge efficiency is almost certainly lower than 100% for the final 3 h of this recharge, the excess pressure indicated in Fig. 3.21 is likely to underestimate the actual pressure contribution from oxygen accumulation in the stack components.

Other secondary pressure corrections include the small pressure-vessel volume changes that can occur in response to the ~1000 psi pressure change as a cell is cycled, and the small active-material volume changes in the nickel electrode from the phase changes that occur during charge and discharge.

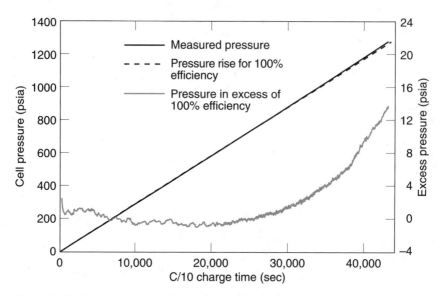

Figure 3.21. Excess pressure from oxygen displacement of hydrogen as a cell at 0°C enters overcharge after about 36,000 sec of recharge at C/10.

Another factor that must be considered in interpreting nickel-hydrogen cell pressure data is the possibility of long-term drift in the actual internal cell pressure, or in the strain-gauge devices themselves. Drift in cell pressure does occur during years of cell operation as additional hydrogen is produced by slow corrosion of the nickel metal in the nickel electrodes. This drift is always toward higher pressure, as indicated in Fig. 3.22 by the pressure from two cells in nickel-hydrogen cell life-test pack 3214E.

The gradual pressure rise results largely from corrosion of the nickel electrodes. In life tests of this kind, sometimes a drop-off in pressure at the end of life is seen, which reflects the formation of soft short circuits, which are high-impedance leakage paths through the separator that cause the cell state of charge to drop to the point where the cell fails. Clearly, it is possible to see real drifts in internal pressure of several hundred psi, making it essential to periodically recalibrate the pressure reading that corresponds to either 100% or 0% state of charge in the cell. The 100% state-of-charge point can be recognized by the observation of an appropriate temperature rise at full charge as the cell enters overcharge. Reconditioning, which involves fully discharging the cell at an appropriately low rate, can be used to determine the 0% state-of-charge point.

Long-term drift in the strain gauges themselves has been observed, and in several instances has been found comparable to the drift in internal hydrogen pressure.[3.2] Strain-gauge drift can result from relaxation of the adhesives that bond

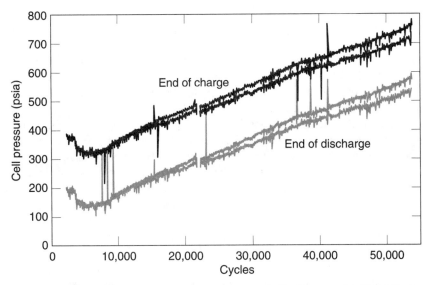

Figure 3.22. Upward drift of the pressure in two cells from pack 3214E during life-testing (40% depth of discharge and 10°C operation).

the strain-gauge element to the pressure vessel and that are influenced by bonding methods and bond cure procedures, as well as by thermal fluctuations and humidity during operation. Strain-gauge drift to both higher and lower pressures has been observed. Fortunately, periodic calibration of the cell pressure reading against a known cell or battery state-of-charge point can enable compensation for strain-gauge drift as well as for real drift in the internal hydrogen pressure.

3.9 Heat Generation and Thermal Behavior of Cells

The generation of heat in nickel-hydrogen cells has been measured over a range of rates (up to 4C), temperatures (−10 to +20°C), and states of charge using dynamic calorimetry.[3.3] Heat generation has been found to depend on the enthalpy of reaction for the energy storage/discharge processes, the charge efficiency, and the work associated with compression or expansion of hydrogen as the pressure in the constant volume pressure vessel changes. In addition, there are a number of transient (or non–steady state) thermal effects that may be encountered during nickel-hydrogen cell operation.

The primary heat dissipation processes in the nickel-hydrogen cell result from the enthalpy changes associated with the energy storage and discharge processes, in association with the charge efficiency, as indicated by Eq. (3.1):

$$Q = I\eta(V - E_{th}) + IV(1-\eta) + \frac{nF}{RT} \ln \frac{dP}{dt} \tag{3.1}$$

Equation (3.1) is based on several assumptions. The first term represents the electrochemical heat associated with the nickel and hydrogen electrode energy-storage processes that take place with coulombic efficiency η. The assumption implicit in this term is that the enthalpic voltage E_{th} for all the energy-storage reactions in the nickel electrode is constant over the range of operating temperatures and state of charge. The second term represents the overcharge reaction, including the evolution and recombination of oxygen in the cell, and the assumption implicit in this term is that the oxygen recombines and generates heat as rapidly as it is generated (i.e., there is no oxygen accumulation). The third term in Eq. (3.1) reflects the heat associated with the work of compressing the hydrogen gas in the constant volume pressure vessel during recharge, and conversely the equivalent cooling associated with expansion of the hydrogen gas during discharge.

Dynamic calorimetry has been used to evaluate the enthalpic voltage E_{th} as a function of temperature and state of charge. This technique involves measurement of the heat Q in Eq. (3.1), which is then used to calculate E_{th} based on the known voltage, current, and pressure changes. Figure 3.23 indicates how E_{th} has been found to change with state of charge and temperature in a typical nickel-hydrogen cell based on calorimetry. The large excursions at the highest states of charge for the lines in Fig. 3.23 result from heat released by the recombination of entrained oxygen at the start of discharge, and thus do not result from the thermodynamics of nickel electrode discharge.

The ideal value for E_{th} that may be calculated from thermodynamic parameters for the cell reactions in the nickel-hydrogen cell is about 1.52 V.[3,4] This value assumes zero cobalt additive concentration in the nickel electrodes, and does not differentiate between the β – β and the β – γ reactions in the nickel electrodes. The measurements shown in Fig. 3.23 are during discharge for a cell having 10% cobalt additive in its nickel electrodes, and the change in slope at low states of charge correlates with the point where the γ-NiOOH material begins to discharge to α-phase.

During cell recharge, a transition to a higher E_{th} at high states of charge for the lower temperature is seen in the calorimetry data, and is associated with the β – γ reaction during recharge. During recharge at higher temperatures a drop-off in E_{th} at high states of charge is generally seen, and is caused by oxygen evolution and the heat that it produces during the recharge process. The data in Fig. 3.23 indicate that E_{th} varies slightly with state of charge, and averages ~1.50 V for the β – β reaction in the nickel electrode with or without cobalt additive. E_{th} is somewhat higher (~1.54 V) for the β – γ reaction, and lower (~1.475 V) for the γ – α reaction. Its value does not seem to vary greatly with temperature. Calorimetry

3.9 Heat Generation and Thermal Behavior of Cells

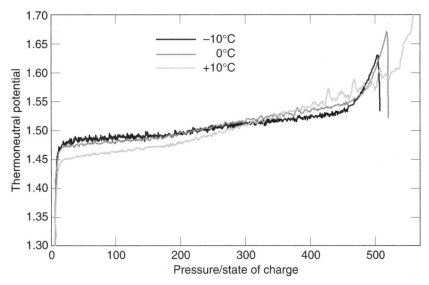

Fig. 3.23. Calorimetry measurements of the thermoneutral potential of a nickel-hydrogen cell during discharge at a C/10 rate at various temperatures.

measurements for discharge rates up to 4C also indicate that E_{th} does not change noticeably at high discharge rates.

The charge efficiency term η in Eq. (3.1) assumes that all the heat associated with oxygen evolution appears instantly (i.e., the oxygen recombines quickly with hydrogen). This is an approximation, of course. As a nickel-hydrogen cell enters overcharge, there is a buildup of oxygen gas in the pores of the cell components. This buildup constitutes storage of latent heat in the cell as oxygen gas, heat that will be released at a later time when the oxygen eventually undergoes recombination with hydrogen. The oxygen accumulation causes a small increase in cell pressure, as well as lower-than-anticipated heat generation from the cell at the point when it goes into overcharge. The reverse of the process, which is the release of latent heat from stored oxygen, is seen at the point when the cell current is switched from recharge to high-rate discharge. At this time, the hydrogen pressure begins dropping rapidly, which drives a burst of high-pressure oxygen gas out of the nickel electrodes and separators to undergo rapid recombination and generation of heat at the hydrogen electrode. Figure 3.24 shows the typical heat dissipation response to suddenly switching a nickel-hydrogen cell from C/10 recharge to C/10 discharge.

The thermal response shown in Fig. 3.24 can generate a transient flux of oxygen gas from the nickel electrode that is high enough to actually cause popping damage to the hydrogen electrode as the oxygen recombines. Short circuits have

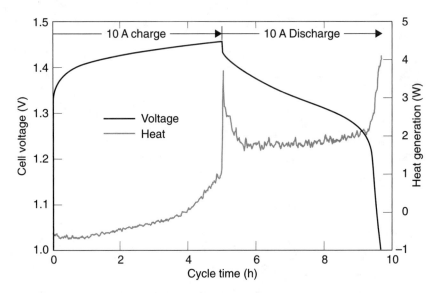

Figure 3.24. Burst of heat from oxygen recombination at the start of discharge following a period of overcharge. ©2002 IEEE. Reprinted with permission.

occurred in nickel-hydrogen cells during the first few minutes of high-rate discharge as the result of such damage. This is particularly true if the cell is switched from overcharge to high-rate discharge without allowing an equilibration period of low-rate trickle-charge or open circuit during which accumulated oxygen gas can slowly recombine.

Another dissipation source that is often overlooked for nickel-hydrogen cells is the work associated with expansion and compression of the hydrogen gas during discharge and recharge, as represented by the third term in Eq. (3.1). This work is negative as the hydrogen in the domes of the pressure vessel expands during discharge, thus cooling the cell, and it is a heating effect during recharge. Over a full round-trip charge/discharge cycle where the cell is returned to its starting pressure, the sum of the pressure/volume (PV) heating during recharge and cooling during discharge is zero. However, up to 20% of the discharge heating expected during cell operation can be shifted to the recharge portion of the cycle by the effects of PV work on the hydrogen. Because the PV work is primarily done on the gas stored in the domes of the pressure vessel, temperature sensors on or near the domes can be significantly perturbed by this contribution to cell heating. Figure 3.25, for example, shows the thermal response recorded on a cell dome as the pressure in the cell rapidly begins to drop as a C/2 discharge begins. The sensor on the dome records a significant drop in temperature as the result of gas PV cooling from decreasing gas pressure, while the overall cell produces significant discharge heating.

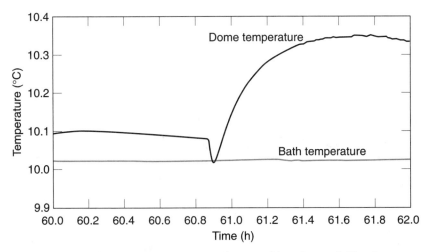

Figure 3.25. Cell dome temperature at the transition from trickle-charge to a C/2 discharge (10°C constant bath temperature).

The role of PV cooling is also quite sensitive to the precharge in the nickel-hydrogen cell. If the cell has a nickel precharge, the hydrogen pressure at the end of a high-rate discharge will drop rapidly towards zero. Because the total cooling from the third term in Eq. (3.1) remains essentially constant as the pressure drops to zero, the effect of the cooling on the gas temperature in the cell domes increases dramatically as the density of gas falls to zero. This effect can be seen in the rapidly changing dome temperature at the end of the high-rate discharge for a nickel-precharged cell. As Fig. 3.26 indicates, in a nickel-precharged cell the pressure vessel dome temperature can cool precipitously at the end of a high-rate discharge when the overall rate of cell heat generation is high.

The examples discussed here illustrate the thermal complexity present in the nickel-hydrogen cell. The details of these thermal processes must be understood to properly design, monitor, and manage the thermal control system associated with nickel-hydrogen batteries.

3.10 Precharge

The precharge in a nickel-hydrogen cell is the amount of excess charge that remains in either the nickel electrode, or the hydrogen electrode (as hydrogen gas), when the cell is completely discharged. In the fully discharged state either the nickel electrode or the hydrogen electrode will be fully discharged, while the other electrode will have some remaining charge. Nickel-precharged cells have excess charge remaining in the nickel electrode after the hydrogen gas in the cell has been completely consumed. Hydrogen-precharged cells have residual hydrogen

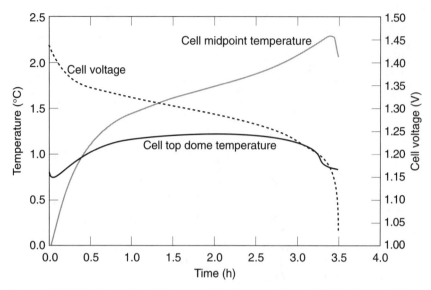

Figure 3.26. Cooling of cell dome as discharge voltage falls and overall heat generation rises.

gas remaining after the nickel electrode has been completely discharged. In principle, it is also possible to have a perfectly balanced cell in which both electrodes simultaneously become depleted; however, such a condition will not persist for long. Generally, as a cell ages, the amount of active nickel precharge will slowly decrease or the amount of hydrogen precharge will slowly increase.

The type and amount of precharge has a significant effect on the electrical behavior of nickel-hydrogen cells. During discharged storage, for example, hydrogen-precharged cells maintain a hydrogen potential on the fully discharged nickel electrode. At the hydrogen potential, the nickel electrode active material is thermodynamically unstable and will chemically degrade over time, eventually exhibiting significant capacity loss. Additionally, fully discharged nickel electrodes will cause cells to exhibit an elevated recharge voltage and reduced charge efficiency during the first recharge following a complete discharge of the cell. Nickel-precharged cells, when fully discharged, maintain a chemically stable environment at the nickel electrode, but produce an oxidizing environment at the hydrogen electrode that is capable of slowly oxidizing the platinum catalyst in this electrode. Plating of the oxidized platinum catalyst at a later time back onto the negative electrode, or onto the nickel electrode, can have a significant effect on charge efficiency, self-discharge, and heat-generation rate. Precharge also affects the hydrogen pressure in a nickel-hydrogen cell, thereby affecting its voltage slightly through the Nernstian dependence of the hydrogen electrode voltage on hydrogen activity. Nickel-precharged cells tend to have hydrogen pressure several hundred psi lower than do equivalently hydrogen-precharged cells, thus having slightly lower voltage and lower self-discharge rates.

The behavior of cells during reversal events is also influenced by precharge. Nickel-precharged cells produce oxygen from the platinum catalyst electrode when reversed, and can exhibit negative cell voltages between -0.4 and -1.6 V, depending on amount of nickel precharge and the extent of reversal. Recharge of nickel-precharged cells following a period of reversal will result in a recharge voltage plateau at about 0.4 V as the oxygen gas in the cell undergoes recombination. When the oxygen gas is exhausted the recharge voltage will rise to more than 1.3 V and the normal cell recharge reaction will commence.

Hydrogen-precharged cells exhibit a very different signature during reversal. The voltage of a hydrogen-precharged cell drops to 0 to -0.1 V and remains in this range as long as the reversal continues. During reversal, hydrogen is generated from the depleted nickel electrode and undergoes recombination at the platinum catalyst electrode, a zero-volt process. Heat production is the only net cell energy change during reversal in this situation. During recharge following reversal, the potential of a hydrogen-precharged cell rapidly rises to more than 1.3 V, where normal recharge processes occur.

Modern nickel-hydrogen cells are exclusively built with nickel precharge at beginning of life, largely to take advantage of the stability that this confers on the nickel electrode during discharged storage. Over extended periods of storage or cycling, these cells can lose their active nickel precharge and become hydrogen-precharged. When this occurs, cells must be kept at operating voltages (> 1.2 V) during storage to maintain stable cell performance. This type of storage, termed "active storage," is discussed in considerably more detail in chapter 9.

3.11 Cycle-Life Behavior

The cycle-life behavior of nickel-hydrogen cells is described in detail in chapter 11. However, nickel-hydrogen cells in general are capable of extremely long cycle life if optimally managed to minimize the stresses of cycling, thermal degradation, and overcharge. While in well-designed cells, it is always the nickel electrode that limits life, the degradation that limits life can be accelerated by electrolyte and gas management issues within the cell. Well-designed nickel-hydrogen cells are intrinsically capable of more than 60,000 cycles in rapid cycling applications such as low Earth orbit satellite applications, or more than 30 yr in geosynchronous satellite applications that have many fewer charge/discharge cycles but an extremely long calendar life.

3.12 Acceptance and Qualification Test Procedures

While acceptance and qualification test procedures vary somewhat depending on cell manufacturer, intended cell application, and detailed cell design, there are a number of tests common to nearly all nickel-hydrogen cell acceptance and qualification tests. Acceptance testing typically begins following cell activation (which may include some burn-in cycles to ensure capacity stability), setting of precharge, and pressure vessel pinch-off and sealing. The purposes of acceptance testing are first, to ensure that each cell falls within the normal distribution of performance characteristics for that design, and second, to gather data indicating how the cells are to be grouped into batteries. Acceptance tests are normally run at controlled cell temperatures, because cell performance is highly sensitive to temperature. Acceptance tests also normally involve monitoring of cell current, temperature, voltage, and pressure for cells with strain gauges.

Capacity tests are always performed during acceptance testing. Capacity tests are typically done at temperatures of –10, 0, 10, or 20°C. Acceptance test procedures for different manufacturers may not include all these temperatures, and specific applications for the cells may dictate that the standard capacity tests focus on either the low- or the high-temperature regions. In a few isolated cases, standard capacity tests at 30°C have been performed in spite of the relatively low capacity normally obtained from nickel-hydrogen cells at this elevated temperature.

Acceptance test procedures for nickel-hydrogen cells also always include a charge-retention test that is intended to verify that self-discharge rates for each cell are normal. Standard charge-retention tests involve full recharge of the cell, using the same recharge procedure as for a standard capacity test, followed by an open-circuit stand period (36–72 h have been used, with the longer periods giving a better measure of self-discharge), and finally a high-rate discharge at the same rate as in the standard capacity test. The cell capacity in this charge-retention test is compared to its capacity in the equivalent standard capacity test that did not include the open-circuit period, and the percentage of the standard capacity retained over the open-circuit period is calculated. For a 72 h open-circuit period, 85% charge retention is typical at 10°C. At 20°C, charge retention of ~65–70% is more typical.

An important result from the charge-retention test is that all the cells within each test group should be in family, and that no cells should be unusually low, either in their capacity retained or in their voltage at the end of the open-circuit period. Low results in either of these measurements would suggest that that cell may have a "soft" short, which is a high-impedance leakage path that makes the cell run down more rapidly than normal. This test does not distinguish whether such soft shorts result from unusually high self-discharge, or from high-resistance electronic leakage paths through or around the separators that isolate the positive and negative electrodes. The charge-retention test should not be the final cycle in the acceptance test procedure, because the charged open-circuit stand can leave some active material temporarily isolated, a condition that can be corrected by following the charge-retention test with a standard capacity test cycle.

Acceptance tests also often include a pulse discharge test that allows the resistance of each cell to be determined from its voltage response to the change in current, and frequently are done midway through the cell discharge. Such pulse discharge tests typically look at the voltage change after 10 sec with the pulse on. Acceptance tests also may involve vibration of the cells at acceptance levels selected to detect any workmanship problems in the cell assembly, but at levels below where the cell components could be damaged or degraded by the vibration. Other routine acceptance tests include electrical insulation tests between the cell mounting sleeve or cell case and the positive and negative cell terminals, as well as phenolphthalein tests of the pressure vessel seals for electrolyte leakage, and sniffer tests for hydrogen gas leakage.

Cell-level qualification tests typically involve standard capacity and charge-retention tests as in acceptance testing, but they include tests with added high and low temperature margins, typically a 10°C margin at the upper and lower limits relative to the extremes of temperature expected in actual operation. Similarly, qualification-level vibration involves vibration at levels significantly higher than acceptance levels, and is designed to identify components and interfaces in the cell that have inadequate margin for mechanical fracture. Qualification may also involve thermal cycling of the inactive cell between worst-case upper or lower temperature limits to evaluate the robustness of seals and thermal interfaces, although such qualification testing is frequently done at the battery level.

Cell qualification may also involve conducting a life test to verify the expected performance over life. Initial qualification for a given type of application should utilize realistic charge and discharge profiles, realistic thermal profiles, and realistic charge control. Follow-on life tests frequently validate cell cycle life using highly accelerated conditions, for which a new cell design or production lot is compared to an earlier cell design run through the identical accelerated conditions. The requirement for passing such an accelerated life test is that the new design that is to be qualified should perform as well as or better than the previous design. Such accelerated life tests may take anywhere from 2 to 5 yr or more to complete, and therefore are often conducted in parallel with other system procurement activities. The incrementally increasing amount of data from the life test can be used for risk reduction over time as the life test is completed in parallel with the other procurement activities.

3.13 References

[3.1] R. Barnard and C. F. Randell, *J. Applied Electrochem.* **13**, 97 (1983).

[3.2] L. H. Thaller and A. H. Zimmerman, *Nickel-Hydrogen Life Cycle Testing* (The Aerospace Press, El Segundo, CA, 2003), p. 118.

[3.3] M. V. Quinzio and A. H. Zimmerman, "Dynamic Calorimetry for Thermal Characterization of Battery Cells," *Proc. of the 17th Annual Battery Conf. on Appl. and Adv.*, IEEE 02TH8576, ISBN 0-7803-7132-1 (Long Beach, CA, 2002), pp. 281–286.

3.4 A. H. Zimmerman, "Calculation of the Thermoneutral Potential of NiCd and NiH$_2$ Cells," NASA Conf. Pub.3254, *Proc. of the 1993 NASA Battery Workshop* (Huntsville, AL, 1994), pp. 289–294.

Part II

Fundamental Principles

4 The Nickel Electrode

Thomas Edison originally invented the alkaline nickel electrode in 1901. As conceived by him, the design of all nickel electrodes to this day still involves an active material (nickel hydroxide) maintained in close contact with an electrically conductive substrate, or current collector. The function of this substrate is to provide structural support as well as a stable conductive surface for electron transfer to the active material. Early nickel electrodes used metal pockets or tubes for the structural support and often mixed the active material with metal powders or flakes to improve its electrical conductivity. Nickel-hydrogen cells have exclusively used a sintered type of nickel electrode, which was originally developed in the early 1940s and is still used in many nickel-cadmium and nickel-metal-hydride cell designs. The sintered nickel electrodes used in nickel-hydrogen cells, however, have been optimized for use in space systems, where low weight, long life, and high reliability are essential.

Following a discussion of the design and construction of nickel electrodes for use in nickel-hydrogen cells, this chapter describes the basic chemical, thermodynamic, and physical processes involved in nickel electrode operation.

4.1 Construction and Design of Nickel Electrodes

The sintered nickel electrode employed in nickel-hydrogen cells uses a current-collecting substrate consisting of a nickel metal screen onto which is sintered a 30–40 mil thick layer of nickel metal powder. The sintered substrate, commonly referred to as plaque, has an overall porosity of 75–85% depending on how it is made; the pores between the sintered particles are typically 20–30 μm in average size. The nickel hydroxide active material is loaded into the pores to fill approximately 50% of their volume by a process referred to as impregnation.

Before the electrodes can be used in cells, they must first go through a "formation" process that stabilizes their electrochemical activity, and then be cleaned of all particulate surface or edge materials that could come loose in a cell. The typical nickel electrode used in nickel-hydrogen cells is about 40% void volume, which mostly becomes filled by electrolyte in an operating cell. The active material makes up approximately 50% of the dry electrode weight, and the nickel metal in the plaque makes up the other 50%.

Each manufacturer has specific processes and procedures for producing its own nickel electrodes. While different manufacturers' electrodes have subtle differences in behavior in cells, all have been optimized based on years of experience to give reliable long-term cell performance.

The following sections outline the generic methods that have been successfully used to fabricate plaque, impregnate the plaque, form the electrodes, and verify proper electrode performance.

4.1.1 Nickel Plaque

All high-quality nickel electrodes require the use of a strong and uniformly porous sintered plaque that is produced using well-controlled manufacturing processes. The nickel plaque is fabricated around a nickel grid or screen substrate that provides mechanical support and distributes electrical current to the sinter structure. The substrate is coated with a nickel powder that is sintered to the substrate to form a porous nickel plaque that ranges in thickness from 0.030 to 0.036 in. for state-of-the-art nickel electrodes. The nickel plaque serves as a porous and electrically conductive supporting structure for the nickel hydroxide active material.

Two methods are commonly used for making nickel plaque. The "slurry" sinter method involves mixing nickel powder into slurry having appropriate viscosity and shear properties, then coating the slurry onto the substrate, where it is dried and sintered. The "dry powder" method involves laying a smoothed deposit of the dry nickel powder over the substrate in a form of the desired thickness and dimensions, followed by directly sintering the powder and the substrate in their preshaped form.

The method for making dry powder sinter is conceptually the most simple. However, to make a plaque with reproducible thickness and porosity, it is necessary to carefully control the size and shape of the nickel powder particles used. Particle shape and size control the porosity characteristics of the plaque by dictating how the powder settles into the form and how it packs around the substrate.

After the form has been uniformly filled with a smooth layer of nickel powder, the unsintered plaque is placed on a ceramic sheet and run through a sintering furnace. The sintering furnace contains a reducing atmosphere (typically hydrogen-based), which protects the nickel particles from oxidation that would interfere with the sintering process. The sintering process itself is carried out at a temperature just below the melting point of nickel metal, and it should melt together all the surface protrusions on the metal powder particles to produce a sturdy and well-bonded layer of sintered nickel plaque of the desired thickness.

The slurry sinter production process differs from the dry powder process in that the nickel powder is mixed into either aqueous or alcohol-based slurry before it is coated onto the substrate. In addition to the water or ethanol used to liquefy it, the slurry contains appropriate organic pore-forming materials. The solvent and pore-formers are removed during drying and sintering, leaving behind the void spaces needed to ensure the desired sinter porosity. The slurry sinter is made by drawing the substrate sheet through the slurry mixture, and then between a pair of doctor blades that are spaced for the desired plaque thickness. These blades size the slurry coating on the substrate to the required thickness and remove all excess slurry mix. The coated substrate is then passed through a drying oven that removes the solvent; then it is passed through a sintering furnace that contains a reducing atmosphere. Finally, the organic pore-formers are oxidized to provide the finished sintered sheet.

4.1 Construction and Design of Nickel Electrodes

The slurry sintering process is typically performed as a continuous operation that feeds substrate screen into the process from a roll and continuously draws the coated substrate through the various steps of the process before it is cut into individual sheets of plaque at the end. In contrast, the dry powder method involves the fabrication and processing of individual sheets of plaque.

To produce a high-quality slurry sinter, it is critical to maintain the slurry composition, slurry mix uniformity, and slurry viscosity during the entire production run. This requires that after the slurry is properly mixed, it must be stirred sufficiently to keep it uniform. However, excessive agitation must be avoided, because that can result in the entrapment of gas bubbles, which results in voids within the sinter. The viscosity of the slurry must also be well controlled. It must be protected from evaporative loss of solvent to the ambient atmosphere during mixing and processing, because this will affect the viscosity and shear properties of the slurry. If the viscosity is too high, the slurry mix can shear from the substrate surfaces as the substrate is drawn through the slurry. This will result in voids adjacent to the substrate wires where the slurry is pulled away from the wire surfaces. If the coated substrate is dried too rapidly or nonuniformly before being sintered, it can develop cracks that can significantly reduce the strength of the plaque.

Because slurry plaque is sintered at temperatures near the melting point of nickel metal in the presence of organic pore-formers, a low level of carbon is always incorporated into the nickel sinter. The carbon content results in a slight reduction in the melting point of the nickel and also makes it slightly more resistant to oxidation in the nickel-hydrogen cell environment.

For both of these plaque production processes, properly sintered plaque has the points of contact between all the nickel particles and between the particles and the substrate melted together, but the bulk sinter particles have not been melted together to fill in any of the void spaces between the particles. Each manufacturer has developed a sintering rate and furnace-temperature profile that produces this structure. Inadequate sintering as a result of insufficient sintering time, too low a temperature, or excessive particle size will produce a weakened sinter structure that can easily fracture and shed fragments during subsequent production steps or during operation in a cell. It is also possible to have excessive particle melting in the sintering process resulting from too high a temperature or excessive time in the furnace. Excessive sintering can produce plaque with inadequate internal surface area and internal pores that are larger than desired, resulting in poor utilization of the active material loaded into the sinter.

A properly made sintered plaque, whether by the dry powder or the slurry process, must have adequate strength. Plaque strength is typically verified by testing the bend strength of plaque samples periodically removed from the production line.

Another key property used to control the production of plaque is its average porosity, which is easily monitored during production based on the thickness and the weight of the plaque. Dry powder plaque typically has an average porosity of about 80%, and slurry plaque is typically about 76% porous. Slurry plaque production processes that provide an average porosity of about 80% have been developed, and have seen limited use in nickel-hydrogen cells. In each type of plaque, the sintered regions are about 4% more porous than the average plaque porosity; the difference is a result of the solid nickel substrate.

The bulk characteristics of strength and average porosity are relatively easy to monitor and maintain during plaque production; however, the internal microstructure of the plaque, which is more difficult to monitor during production, is also important to the performance of the plaque in nickel electrodes. The internal microstructure should ideally consist of a relatively uniform distribution of pore sizes throughout its thickness.

Some deviations from this idealized pore structure are normally seen. For dry powder sinter, some settling of the powder always occurs before it is sintered, making one side of the sinter slightly denser than the other. The gravitational settling can also make the substrate settle toward the more dense side of the plaque, resulting in a substrate that is not centered in the plaque structure.

The slurry plaque, on the other hand, typically has the substrate centered within the plaque structure. However, the slurry plaque typically has a densified skin of lower porosity sinter on each surface of the plaque. The skin of lower porosity results from the surface tension characteristics of the slurry, which draws a high density of nickel particles to the skin of the slurry before it is dried and sintered.

Figure 4.1 shows the typical sinter porosity variation seen through the cross section of plaques made by either the dry powder or the slurry process.

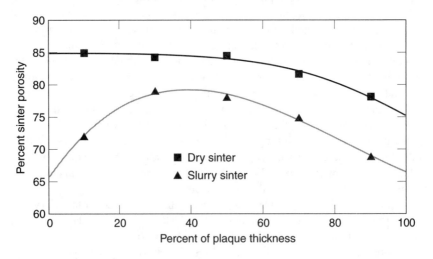

Figure 4.1. Typical variation of porosity through the thickness of slurry and dry sinters.

This figure clearly shows the gradient in porosity across the thickness of the dry powder plaque, as well as the lower porosity at the surfaces of the slurry sinter.

A number of process controls must be in place to ensure that the porosity of the sinter in plaques made by any process is relatively uniform. Nonuniform sinter can result from particulate contaminants in the nickel powder, from undispersed clumps of nickel powder that form dense regions within the sinter, or from voids within the sinter. All these microstructural nonuniformities are undesirable, because they will degrade the performance of the plaque in nickel electrodes.

Figure 4.2 illustrates some examples of these kinds of microstructural defects. Foreign particles of zirconium oxide are sometimes found in nickel powders, and they can result in sinter occlusions, as shown in Fig. 4.2(a). These foreign particles add weight, decrease conductivity, and occupy volume that could otherwise be filled with active material for greater energy storage. If foreign

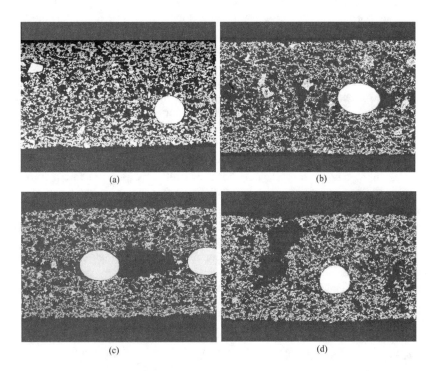

Figure 4.2. Common defects in cross sections of nickel sinter used in nickel electrodes: (a) foreign particle; white particle at upper left is a zirconium oxide fragment; (b) regions of dense sinter from clumps of nickel powder; (c) large void associated with separation of slurry from the grid; (d) large void from entrained bubbles or oxidizable particles.

particles are reactive with potassium hydroxide (KOH) electrolyte or if they are oxidizable, they can also introduce contaminants that could affect cell performance.

Also shown, in Fig. 4.2(b), is the microstructure that results when the nickel particles are allowed to clump together before sintering, either in the dry powder or the slurry process. What results is a region of very dense sinter, which adds weight without leaving the needed volume for active material. The most damaging effect of these clumps is that, if the desired average plaque porosity is to be realized, there will be regions of equivalently high porosity to make up for the dense regions. The resulting porosity gradient between the high- and low-density regions can produce internal stresses that promote cracking, blistering, or delamination of the sinter from the substrate during nickel electrode operation.

Another microstructural feature of general concern, shown in Figs. 4.2(c) and (d), is the presence of large voids in the sinter. These voids can occur in dry powder sinter if the powder does not pack uniformly in its form before sintering, or if there are particles in the powder that burn away during sintering. Typically the voids within dry sinter are likely to occur at any location through the plaque cross section. In slurry sinter, voids are much more likely to occur near the center of the plaque, and they can arise from poor mixing of the slurry, entrained bubbles of gas, or shearing of the slurry from the substrate as it is drawn through the doctor blades. For sintered nickel electrodes that contain a high incidence of large voids, such as those shown in Figs. 4.2(c) and (d), utilization of the active material can be reduced by up to 15–20% relative to plaque having a uniform sinter microstructure. Large voids that are filled with active material do not allow the charge stored in that active material to be effectively discharged at useful rates. In addition, active-material impregnation processes may have difficulty loading these internal voids with active material, resulting in overloading of other regions in the sinter to get the needed average loading. Voids within the sinter also provide points where the internal stresses from expansion of the active material during cycling are concentrated. Most cracks and blisters that occur in nickel electrodes during extended cycling operation are initiated at the location of voids in the sinter for this reason.

Following the production of sintered nickel plaque by either a dry powder or a slurry process, the plaque may be passivated by an oxidation process that produces a compact protective layer of nickel oxide on the surfaces of the nickel particles within the sinter. The primary purpose of this passivation process is to reduce the rate of sinter corrosion during the electrochemical impregnation process, which involves exposure of the plaque to a highly acidic impregnation solution.

The proper distribution of pore sizes within nickel plaque is critical to obtain the desired uniform impregnation of the pore volume with active material, as well as proper utilization of that active material for energy storage. Ideally, a uniform distribution of pore sizes in the 10–40 μm range is desired. A well-controlled production process for either the dry powder or the slurry methods, as shown in Fig. 4.3, can produce a relatively uniform pore size distribution in this size range,

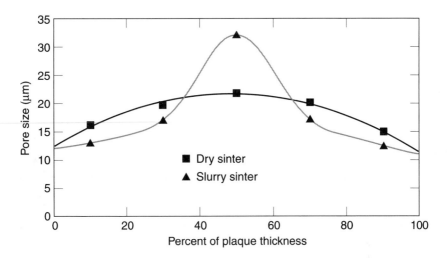

Figure 4.3. Typical pore size variations through the thickness of dry powder and slurry sinters.

although it is not unusual for slurry sinter to have somewhat larger pores in the center of the plaque.

Because it is difficult to continuously monitor the internal pore structure and pore size distributions during plaque production, plaque is sometimes produced with less-than-optimum pore size distribution characteristics. Measurements of pore size distributions[4.1] on samples of plaque taken from each production lot can ensure that any unexpected changes in sinter microstructure will not change nickel electrode performance from one cell lot to another. This allows the incidence of the most common microstructural defects in sinter to be measured and tracked throughout the production history, and can prevent unexpected changes in performance. For example, Fig. 4.4 shows the pore size distribution that can result when numerous large voids such as those in Figs. 4.2(c) and (d) are present within the plaque. These voids are the most commonly observed departure from the ideal sinter pore size distributions shown in Fig. 4.3.

4.1.2 Impregnation of Nickel Electrodes

Impregnation refers to the process of loading nickel hydroxide active material into the pores of the sintered nickel plaque. Many methods for doing this have been developed over nearly sixty years. Here, discussion is limited to the processes used to make the sintered nickel electrodes that have been used widely in nickel-hydrogen cells since about 1970. These processes are exclusively electrochemical impregnation procedures that cathodically precipitate nickel hydroxide directly within the pores of the sinter from a nickel nitrate solution. Other processes,

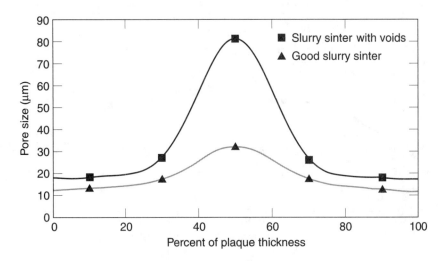

Figure 4.4. Pore sizes through the thickness of slurry sinter having numerous large voids.

including chemical precipitation methods and physical pasting methods (processes still widely used for loading active material into nickel electrodes for use in nickel-cadmium and nickel metal hydride cells), will not be discussed here.

There are two electrochemical processes that have been used to load active material into sintered nickel plaques for use in nickel-hydrogen cells. They are similar in terms of directly precipitating nickel hydroxide within the pores of the sinter by cathodically producing hydroxide ions within the pores. The two processes differ in the make-up of the solutions in which the cathodic deposition is carried out, and in the conditions used during deposition. They are referred to as the "aqueous" and the "alcoholic" impregnation processes and should not be confused with the aqueous and the alcoholic slurry sinter production processes. As their names suggest, the aqueous impregnation process[4.2-4.4] uses a several-molar aqueous solution of cobalt and nickel nitrates, while the alcoholic process[4.5,4.6] uses a mixture of ethanol and water as the solvent for a similarly concentrated solution of nickel and cobalt nitrates.

Each of these processes exposes a large number of parallel-connected sheets of nickel plaque to its respective processing solutions in a large vat, while passing the large current required to simultaneously deposit active material in the pores of all the plaques. The metal nitrate solutions in both processes are highly acidified so that any hydroxides precipitated on the outer surfaces of the plaques, which are exposed to the bulk vat solution, will be dissolved. If this surface material is not dissolved as it forms, it will rapidly block the transport of metal ions into the pores and prevent the proper loading of the sinter with active material.

4.1 Construction and Design of Nickel Electrodes

The ethanol used in the solvent for the alcoholic process enables this process to be carried out at temperatures well below 100°C, while the aqueous process is carried out at significantly higher temperatures. For this reason, corrosion of the nickel sinter by the acidic impregnation solution tends to be somewhat more of an issue that requires careful process controls for the aqueous process.

In both processes, corrosion is minimized by cathodic protection of the plaques (i.e., they are held at a potential below the corrosion potential of nickel metal) while they are immersed in the impregnation vat. However, after the plaques are removed from the vat, corrosion can take place during the time required to rinse the impregnation solution from the sinter. Typical sinter corrosion levels are less than about 5% of the nickel initially present in the sinter structure.

The cobalt hydroxide additive in the active material is codeposited with the nickel hydroxide as a solid solution by using a mixture of nickel and cobalt nitrates of the desired composition in the impregnation solution. The alcoholic impregnation process typically uses a cobalt level that is 10% of the nickel concentration, while the aqueous process uses a 5% cobalt additive level. The actual levels of cobalt in the active material can differ somewhat from these nominal levels, because there are differences in the hydroxide deposition chemistry of these two metals.[4.7] Figure 4.5 shows the range of cobalt additive compositions that can be measured through the cross section of several sintered nickel electrodes. It is possible to get either depleted cobalt levels near the surface of the sinter or enhanced cobalt levels, depending on the local impregnation conditions of pH, solution agitation, current density, and temperature.

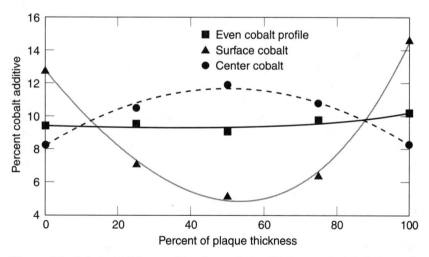

Figure 4.5. Cobalt additive profiles through the thickness of nickel electrodes, showing a uniform concentration, as well as surface-enhanced and surface-depleted cobalt levels.

For long-cycle-life nickel electrode operation, which is required for nickel-hydrogen cells, active material must be loaded into the pores of the sinter at a relatively well-controlled density, which is typically in the range of 1.6 to 1.8 g/cm^3 of void volume. If the loading level falls significantly below this range, the specific capacity of the electrode will suffer, and performance can actually degrade as a result of reduced contact area between the interior pore surface area and the active material. If the loading level exceeds about 1.8 g/cm^3, the cycle life of the electrode tends to be reduced because there is insufficient free volume within the pores to accommodate the cyclic expansion and contraction of the active material without exerting large stresses on the sinter structure. These volume changes will result in accelerated electrode swelling and sinter fracturing during long-term cycling. With high active-material loading levels, the capacity of a nickel electrode may be very high at the beginning of its life, but the capacity will degrade rapidly, resulting in much lower capacity and premature failure after many years of cycling. Because nickel-hydrogen cells nearly always are required to provide a long cycle life, excessive loading levels are carefully avoided.

The uniformity of active-material impregnation is important in producing high-quality nickel electrodes. The loading of sintered plaque is generally controlled based on the average loading level of an entire nickel electrode or an entire sheet of plaque (from which up to nine electrodes may be cut). The loading level is monitored by the gain in weight as the active material is deposited. The assumption normally made is that the average loading is the same as the local loading at various locations on a sheet of plaque, or through the thickness of the plaque. If this assumption of loading uniformity is not adequately realized in an impregnation process, significant variability in the performance of individual nickel electrodes may be seen.

As an example of this type of loading nonuniformity, Fig. 4.6 shows the range of loading levels that have been measured through the cross section of electrodes of a particular design. It is clear that significant variations in loading level are possible from the surface of an electrode into its interior. In the measurements of Fig. 4.6 the average loading level of all the electrodes is close to the specified requirement of 1.70 g/cm^3 of void volume. However, in some of the electrodes, local loading—either at the surface or in the interior of the sintered plaque—significantly exceeds or falls below the specification. Of course, because the average loading can be accurately controlled based on the weight pick-up of the plaque during impregnation, the layers of high loading are typically compensated for by other layers with very low loading density.

The electrodes in Fig. 4.6 that showed large variations and layered structures in the loading level were also the electrodes that had unexpectedly low capacity and utilization. During long-term cycling of electrodes having internal layers of highly dense loading, there is likely to be a gradual movement of active material from the highly loaded layers to fill in the layers of low loading. The stress that

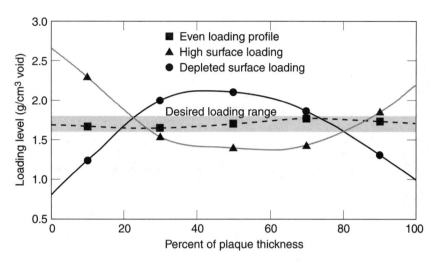

Figure 4.6. Loading-level variations through the thickness of nickel electrodes showing uniform, surface-enhanced, and surface-depleted loading levels.

this movement exerts on the sinter structure may cause premature physical degradation of the nickel electrode and is likely to significantly degrade its cycle life. If the high-density layer is at the surface of the sinter, significant extrusion of active material out of the pores of the sinter and into the porous separator is likely. Because the particles of active material extruded from nickel electrodes during cell operation are typically charged, they have significant electrical conductivity and can contribute to low-level electrical short circuits through the separator if their density in the separator becomes sufficiently high.

Another type of nonuniformity that can arise in the nickel electrode impregnation process is the variability in loading from one location in the vat to another location. This variability is influenced by the patterns of solution flow through the vat and the amount of agitation, thermal gradients, pH gradients, and variations in current density. Large variability of this type is undesirable, because it is often the weakest electrode in a cell that can initiate accelerated degradation and premature cell wear-out. Impregnation processes that have a history of consistently producing good nickel electrodes typically exhibit less than about 10% variability in loading and corrosion levels through the impregnation vat. Because loading level and utilization are typically measured for a number of sample electrodes from each production run, this type of variability can be statistically tracked to provide assurance that the impregnation process is operating as expected.

The uniformity for the loading level in a typical impregnation vat is shown in Fig. 4.7, which gives a top view of a vat containing dozens of sheets of nickel plaque. Loading levels and corrosion levels were measured for a large number

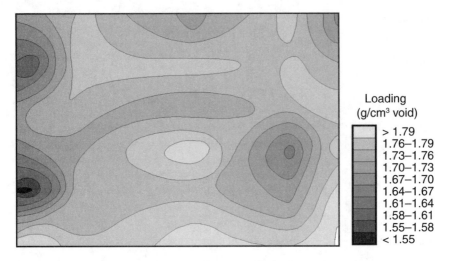

Figure 4.7. Loading-level profiles across a typical electrochemical impregnation vat.

of samples from this impregnation vat, and they typically indicate about a ±10% variation in loading level for different plaque locations. Figure 4.8 shows a similar view of the amount of sinter corrosion during the impregnation process, which is strongly influenced by the local pH and temperature. The maximum amount of sinter corrosion is about 10%, and the average is about 5%. The regions of greatest

Figure 4.8. Nickel corrosion levels measured throughout a typical impregnation vat.

loading and corrosion variability in these figures appear to follow the solution flow patterns, and they suggest that regions of reduced solution-mixing may exist near the corners of the vat. Measurements such as those in Figs. 4.7 and 4.8 can be used to both characterize and optimize the nickel electrodes produced by any particular impregnation process.

4.1.3 Formation of Nickel Electrodes

The active material that is deposited within the pores of sintered plaques by electrochemical impregnation is precipitated as relatively high-density layers on the surfaces in the pores of the nickel sinter. Figure 4.9(a) shows the typical structure of the active material "as deposited" by the impregnation process. While this compact, crystalline deposit may exhibit good initial capacity for a few cycles, it is physically unstable. Repeated charge and discharge will rapidly fracture the compact active-material deposit into a powder of finely divided crystallites that will eventually fill nearly all the internal volume in each pore of the sinter.

This process of physically stabilizing the active material in the pores of the nickel sinter is referred to as the formation process. The process typically involves repeated high-rate cycling of the nickel electrodes, which is often accompanied by electrode reversal and overcharge to generate gassing from the electrodes. The formation process will fully break up the active-material crystallites into a physically stable powder because of the volume changes associated with high-rate cycling, and the gassing will help force the powdered crystallites to uniformly fill each pore. A fully formed sintered electrode is shown in Fig. 4.9(b), which shows a uniform active-material deposit that now fully fills essentially all the space within the pores, except for the cracking that occurred in the active material when the electrode was dried.

The consequences of putting incompletely formed nickel electrodes into nickel-hydrogen cells are decreased charge efficiency and premature evolution of

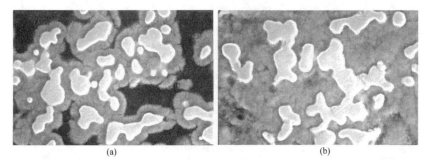

Figure 4.9. Cross sections of sintered nickel electrodes as deposited (a), and after complete formation (b). The sinter is white, the active material is gray, and the void regions are black.

oxygen. However, continued cycling of the cell will eventually result in complete formation of such electrodes within the cell. The inset in Fig. 4.10 shows the cross section of a partially formed nickel electrode, where numerous fragments of the compact "as deposited" active material still exist in poor contact with the sinter and many unfilled void spaces exist in the pores. The figure shows the performance of a partially formed nickel electrode when cycled at high rate with an 80% depth of discharge and a limited recharge ratio (102.5%).

Figure 4.10 also shows the performance of a fully formed nickel electrode for comparison. The partially formed electrode, because of its degraded charge efficiency, runs down in capacity for the first ~30–40 cycles, after which it begins to recover capacity. In this type of test, many electrodes that are not well formed will actually become fully depleted because of their poor charge efficiency and will fail to maintain adequate voltage. After about 100 cycles, the capacity has fully recovered, because the electrode in Fig. 4.10 that was initially only partly formed has completed the formation process, and the active material has become physically stabilized. Thereafter, the electrode that was initially partially formed displays charge efficiency, capacity, and cycle life performance that is equivalent to the electrode that was initially fully formed.

In a nickel-hydrogen cell, if only one or two nickel electrodes are not fully formed, the cell can fail; while if most of the nickel electrodes are not well formed,

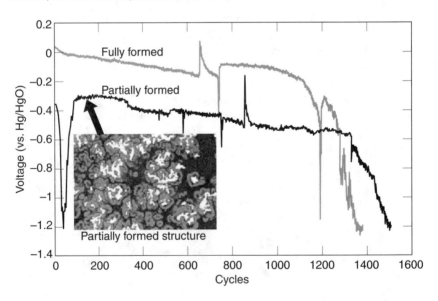

Figure 4.10. Stress test performance of a fully formed nickel electrode and a partially formed electrode. The inset shows the internal active-material structure of the partially formed electrode.

the cell will just display low capacity. The reason for this behavior is that a single poorly formed nickel electrode in a cell will begin to evolve oxygen at significantly lower voltages and will go into overcharge at quite a high rate well before the other electrodes, particularly at low temperatures. The very high rate of oxygen evolution from this one poorly formed nickel electrode can initiate explosive oxygen/hydrogen recombination events whenever the oxygen streaming from the nickel electrode impinges on the platinum catalyst in the negative electrode. The damage from these events can break up the nickel electrode and the separator at either the nickel electrode surface or at the edges of the nickel electrode. This damage can potentially result in a cell short circuit early in life before the poorly formed electrodes can undergo "in-cell" formation.

Interestingly, the worst-case scenario occurs when only a single poorly formed nickel electrode is put into a cell, because in such a case, oxygen evolution is prematurely concentrated on that one electrode to the maximum possible extent. A cell that has most or all of its nickel electrodes in a poorly formed state would be unlikely to fail as a result of a short circuit, because all electrodes would prematurely go into overcharge together and share the evolution of oxygen relatively well. However, this cell could fail as a result of low initial capacity because of the poor charge efficiency associated with nickel electrodes that are not fully formed.

Because nickel electrodes can actually undergo improvement in capacity for many cycles, it is not unusual to subject cells to a number of "burn-in" cycles after they are activated. The purpose of these burn-in cycles is to fully stabilize the capacity of all cells in a lot so that they can be acceptance-tested and matched into batteries based on a completely stabilized capacity.

4.1.4 Verification Testing of Nickel Electrodes

Because the nickel electrode is typically the component whose degradation eventually limits the life of nickel-hydrogen cells, significant effort is usually devoted to tests and analyses capable of verifying that the nickel electrodes in a cell will not degrade and wear out prematurely. Detailed discussions of such testing are provided in chapter 13. However, the key tests typically used to verify the quality of sintered nickel electrodes are described here.

The first type of verification test involves ascertaining that the nickel electrodes have acceptable electrical performance. Several types of electrical tests are customarily applied to samples from each production lot for verification purposes. Stress tests are performed on several sample electrodes to show that the electrodes do not degrade more rapidly than normal when subjected to the stresses of extremely-high-rate cycling and overcharge. Utilization tests are performed to demonstrate that the material loaded into the plaque during impregnation has the expected ability to store charge and to deliver that charge at usable voltages. The second type of test for verifying the quality of finished nickel electrodes uses chemical analysis to confirm the correct chemical composition of the electrodes.

One of the most common methods for verifying the quality of nickel electrodes is a stress test. This test exposes a sample electrode to a series of high-rate charge and discharge cycles to verify the ability of the finished electrode structure to hold up to the rigors of cycling without developing unexpected damage. Test samples of nickel electrodes are typically cycled in an electrolyte-flooded cell using an extremely high charge and discharge rate (typically 10C). At least several hundred of these 100% depth-of-discharge cycles are usually run. This test causes large and rapid volume changes in the active material, as well as significant stress levels from gassing during overcharge. These stresses will make electrodes that have weak sinter, sinter voids, or excessive loading crack, swell, or blister significantly more than is normal. The criteria for passing this test are, typically, swelling of the electrode thickness of less than several mils, and a capacity that does not drop abnormally before the end of the stress cycles. In addition, there is generally a requirement that surface blistering of the nickel electrodes after this test be minimal. This test has historically been highly effective for screening out nickel electrode batches that have problems with the sinter structure or the active-material loading.

Utilization tests are typically performed on several electrode samples from each impregnation batch. Utilization testing involves charging the electrodes in a flooded test cell at a high rate (typically C/2) for sufficient time to bring the electrodes to full charge. Flooded test cells typically contain 31% potassium hydroxide electrolyte and are operated at room temperature. The electrodes are discharged at a C/2 rate (20–30 mA/cm^2) to measure the amount of capacity that they can deliver. The utilization is typically determined as a percentage of the theoretical capacity associated with the transfer of one electron per nickel or cobalt hydroxide molecule in the active material. The theoretical capacity is based on the weight of active material picked up by each electrode in the impregnation process.

Typical utilization levels range from 110 to 130% of the theoretical capacity, depending on the electrode design and the current densities used in the utilization test. The utilization measurements allow statistical variability and trends in the electrical performance to be tracked over time, and they enable performance changes in electrodes to be detected and traced back to the structural characteristics of the plaque and its loading before the electrodes are built into cells. Utilization is directly influenced by the sinter porosity, cobalt level, loading level, and uniformity. Figure 4.11 shows a typical correlation between utilization and loading level.

Chemical analyses are typically performed on several samples that are selected from each production batch of finished electrodes. These analyses generally measure the loading level in the electrode, the amount of cobalt additive, and the sinter porosity. Each of these measurements should be within the range expected based on the specification for the electrode design. Analyses are also usually performed to determine the levels of common contaminants of concern in

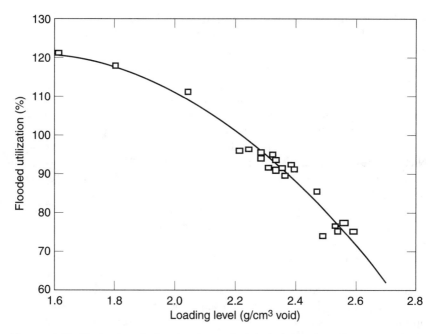

Figure 4.11. Typical variation in sintered nickel electrode utilization with the local loading level. Charge rate is 2 mA/cm^2 and charge input is 160% of rated capacity. Discharge rate is 10 mA/cm^2.

nickel electrodes. These include iron, copper, silicon, calcium, magnesium, zinc, sulfur, carbonate, chloride, fluoride, nitrate, lead, aluminum, and other potential trace contaminants that could indicate problems with handling or processing the electrodes.

4.2 Chemistry of the Nickel Electrode

A wide range of reactions can occur in the nickel electrode, resulting in chemical and electrochemical behavior in nickel-hydrogen cells that is quite complex and not fully understood to this day. The rich chemistry of the nickel electrode stems from interactions involving many structures and processes: a number of multiple-phase structures of varying stoichiometry, spatial layering of phases within the microscopic active-material deposits, charge diffusion processes, and crystalline disruption and ripening processes. Additional reactions occur involving oxygen gas, chemisorbed surface species, hydrogen gas, cobalt additives, and platinum oxidation products (from the hydrogen electrode)—and all of this makes a complete theoretical understanding of this electrode extremely difficult. The following subsections elaborate the range of reactions that control the behavior of nickel electrodes in the nickel-hydrogen cell.

4.2.1 Charge and Discharge Reactions in the Nickel Electrode

The phases formed during charge and discharge of the nickel electrode are significantly more complex than what has often been simplistically written about in this area—ideas suggested by the oxidation reaction of $Ni(OH)_2$ to $NiOOH$. The rectangular interconversion diagram developed by Bode et al.[4.8] (Fig. 4.12) provides a more complete description of the nickel-containing phases that can form during charge and discharge. However, as will be discussed later, even the Bode reaction scheme is a simplification of the true chemistry of these nickel phases.

The fully discharged active material normally exists as the hexagonal β-$Ni(OH)_2$ phase, which is the only phase shown in Fig. 4.12 that is thermodynamically stable. The β-$Ni(OH)_2$ phase can be charged to the hexagonal β-$NiOOH$ phase in an apparently reversible reaction entailing only a small decrease in the unit cell volume. However, when closely examining the potential and kinetics of this reaction, one finds that it clearly is not microscopically reversible, as is schematically indicated. A hysteresis of about 40 mV is observed between the voltage of the charge reaction and the voltage of the discharge reaction, even when the current is sufficiently low that kinetically induced overpotential is negligible. In addition, there is a very large capacitance associated with charge or discharge of the active material; the apparent open-circuit potential depends on the state of charge; and the voltage of the charging reaction can vary significantly depending on prior electrode history. All these observations suggest that Fig. 4.12 does not fully capture the true mechanism of the β-$Ni(OH)_2$ ↔ β-$NiOOH$ reaction.

Examination of the microscopic sequence of events that must occur during the charge and discharge of a nickel electrode shows why the β-$Ni(OH)_2$ ↔ β-$NiOOH$ reaction is not a simple reversible electrochemical process. The physical structure of typical active material consists of small nickel oxyhydroxide crystallites embedded within the pores of the sintered plaque. Figure 4.13 shows the active material, along with the sequence of five processes involved in the charge or discharge of a particular site within a crystallite of active material.

The five processes that are involved in the charge or discharge of the nickel electrode are indicated by Eqs. (4.1) through (4.5). The first process involves electronic transfer between the nickel current-collecting surfaces and the active-material particles that are in contact with the metal surfaces, as expressed by Eq. (4.1):

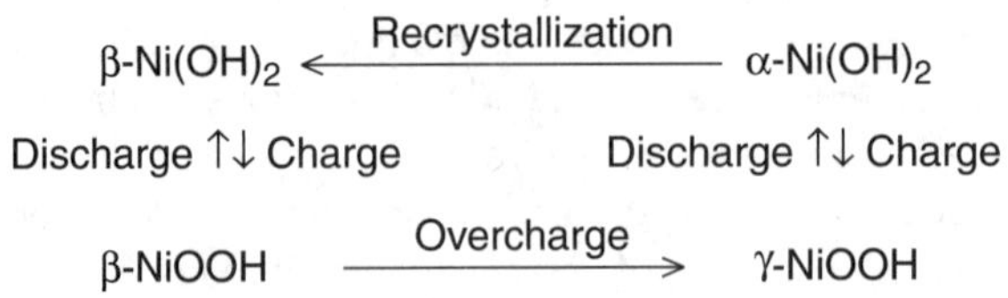

$$\begin{array}{ccc}
\beta\text{-Ni(OH)}_2 & \xleftarrow{\text{Recrystallization}} & \alpha\text{-Ni(OH)}_2 \\
\text{Discharge} \updownarrow \text{Charge} & & \text{Discharge} \updownarrow \text{Charge} \\
\beta\text{-NiOOH} & \xrightarrow{\text{Overcharge}} & \gamma\text{-NiOOH}
\end{array}$$

Figure 4.12. Bode reaction scheme for phases in the nickel electrode.

4.2 Chemistry of the Nickel Electrode

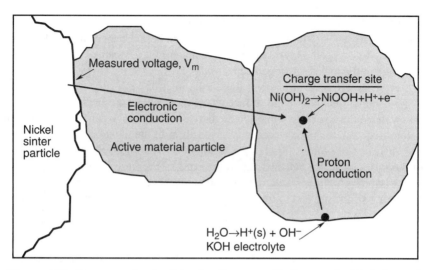

Figure 4.13. Processes involved in the charge or discharge of nickel electrode active sites.

$$e^-(Ni) \rightarrow e^-(AM) \quad (4.1)$$

While the transfer indicated by Eq. (4.1) may be as simple as the conduction of an electron from the metal to a hole in the active material, it may also include electron insertion from catalytic reactions or surface sites at the metal interface. Inert or nonconductive layers at the nickel interface can prevent the movement of electronic charge into the active material, thus deactivating the electrode capacity for high-rate discharge.

The second process involves conduction of electronic charge through the active-material crystallites, and it occurs by the hopping of electron-deficient lattice sites (holes) through the structure rather than by movement of electrons. Electrons undergo conduction to a distant charge-transfer site in the active material by being injected into a hole at the metal interface, then moving through the potential gradients in the lattice to a site (s) where electrochemical charge transfer can occur. The conduction process is expressed by Eq. (4.2):

$$e^-(AM) \rightarrow e^-(s) \quad (4.2)$$

The electronic conductivity of the active material depends strongly on the state of charge of the active material because the density of filled electron vacancies (holes) in the lattice increases as the state of charge increases. At low states of charge, where most holes are vacant, movement of an electron requires it to traverse a large number of lattice sites. At high states of charge, where most holes are filled, the addition of an electron at one point in the lattice can force an electron at a distant site to hop into the nearest vacant hole. Changes in the electronic conductivity in specific layers of the active material is a principal cause of capacity losses in the nickel-hydrogen cell.

For charge transfer to be possible in the interior of the crystallites of active material rather than just on the surface where they contact the electrolyte, it is necessary that ionic conduction through the lattice also occurs to deliver the chemical species required for the charge-transfer reaction. In the nickel oxyhydroxide lattice, diffusion of protons through the solid lattice from the surface of a crystallite to the charge-transfer site fills this role. Eq. (4.3) represents the third process, which is the splitting of water molecules that must occur at the surface of the crystallites at surface site (ss) to form lattice-stabilized protons and hydroxide ions.

$$H_2O \xleftrightarrow{\text{site (ss)}} H^+(ss) + OH^- \qquad (4.3)$$

Protons at the surface site (ss) can then diffuse by the fourth process into the interior of a crystallite as indicated by Eq. (4.4). The interlamellar spacing between the hexagonal sheets in the nickel hydroxide lattice provides relatively facile solid-state proton diffusion.

$$H^+(ss) \longrightarrow H^+(s) \qquad (4.4)$$

The balance between the proton diffusion rate and the electronic conductivity dictates the preferred site (s) where electrochemical reaction, which is the fifth process, will occur by charge transfer, as indicated in Eq. (4.5).

$$Ni(OH)_2 \xrightarrow{\text{site (s)}} NiOOH + H^+(s) + e^-(s) \qquad (4.5)$$

While the charged material in Eq. (4.5) and the other charged materials discussed later in this section all contain hydrogen atoms, these hydrogen atoms are delocalized as mobile protons within the lattice. This has been confirmed by infrared spectroscopic studies, which have shown no evidence of O-H bonds, and by neutron-scattering studies, which have not detected any hydrogen atoms localized within the charged structures.

If the electronic conductivity is high and the proton diffusion rate is also high (as would be the case with small crystallites and a high state of charge), the active material will be oxidized (charged) or reduced (discharged) uniformly throughout the thickness of the particulate deposit. If proton diffusion is low (as with large crystallites), the active material will charge or discharge preferentially at the interface of the active-material crystallites with the electrolyte. If electronic conductivity is low (as it would be with a low state of charge), charge and discharge will tend to be localized at the interface with the nickel metal current collector.

The sequence of processes detailed in Eqs. (4.1)–(4.5) can result in widely varying electrochemical behavior, particularly when coupled with the range of reactions shown in Fig. 4.12, the structural variability possible in the active material, and the overcharge processes in the nickel-hydrogen cell. The reaction on the left side of Fig. 4.12 involving the β-Ni(OH)$_2$ ↔ β-NiOOH couple is the most desirable for long-term cycling stability, because it involves the least change in active volume during charge and discharge. The stoichiometry of this reaction is reasonably well represented by Eqs. (4.1)–(4.5), involving a one-electron charge-transfer reaction between divalent and trivalent nickel at the local site (s), as indicated in Eq. (4.6).

$$\beta-\text{Ni(OH)}_2 + \text{OH}^- \xrightarrow{\text{charge}} \beta-\text{NiOOH} + \text{H}_2\text{O} + \text{e}^- \quad E_o = 0.49 \text{ volts} \tag{4.6}$$

However, it must be kept in mind that the overall active-material deposit includes a wide range of physically and chemically varying nickel sites. It is impossible to reduce all these sites to the divalent state because of conductivity limitations at low states of charge; however, with sufficient charging, virtually all sites may be oxidized. The β-Ni(OH)$_2$ ↔ β-NiOOH couple typically cycles between average nickel oxidation states of about 2.2 and 3.0 at high rates, and average oxidation states as low as 2.03 can be reached at discharge rates less than C/100.

As suggested by the reaction at the bottom of Fig. 4.12, oxidation of the nickel active material does not stop with the formation of the β-NiOOH phase. Continued charging, often occurring in conjunction with overcharge and oxygen evolution, can result in further oxidation to what is usually termed the γ-NiOOH phase for simplicity. The actual chemical formula for this phase is written as NiOOH(NiO$_2$)$_2$·KOH.

This phase, which has an average oxidation state of 3.67, involves a rhombohedral structure that extends over three of the hexagonal nickel sheets initially present in the β-NiOOH starting phase. These sheets undergo lateral distortion from the hexagonal β-phase stacking arrangement, along with some interlayer expansion and incorporation of one potassium atom per unit lattice cell. Because potassium cannot move through the lattice with the same facility as protons, this compound only forms slowly at relatively high potentials, in spite of its being about 40 mV more thermodynamically stable than the β-NiOOH phase.

There is a significant change in volume associated with the formation of γ-NiOOH as the hexagonal lattice undergoes restructuring. While the formation of this phase potentially enables 67% more capacity to be stored in the nickel electrode, it is not a phase that should be frequently charged and discharged, because the sizable volume changes can rapidly degrade nickel electrode structure and performance. The formation of the γ-NiOOH phase from β-NiOOH involves a two-electron oxidation process that includes restructuring of three adjacent nickel-containing layers, as indicated in Eq. (4.7):

$$3\beta-NiOOH + 2OH^- + KOH \xrightarrow{charge} NiOOH(NiO_2)_2 \cdot KOH + 2H_2O + 2e^-$$
(4.7)

Alternatively, Eq. (4.7) may be written as an oxidative removal of two of the three delocalized protons from the β-NiOOH material, leaving a single proton delocalized over the rhombohedral unit cell, as indicated in Eq. (4.8). Because the protons are delocalized and can diffuse through the lattice, there is no discrete intermediate corresponding to the removal of only one of the protons indicated in Eq. (4.8).

$$3\beta-NiOOH + KOH \xrightarrow{charge} NiOOH(NiO_2)_2 \cdot KOH + 2H^+(s) + 2e^-$$
(4.8)

The overall reaction for formation of $NiOOH(NiO_2)_2 \cdot KOH$ (aka γ-NiOOH) from β-NiOOH is not reversible, because γ-NiOOH is the more stable structure, although kinetically it is slow to form. Likewise, there is no clear evidence that it is possible to electrochemically form γ-NiOOH directly from the β-Ni(OH)$_2$ structure. However, the higher-potential β-NiOOH can undergo a slow chemical conversion to the lower-potential γ-NiOOH phase. Because potassium hydroxide incorporation from the electrolyte is required to stabilize the γ-NiOOH structure, the formation of γ-NiOOH is much more facile at higher electrolyte concentration. In 26% potassium hydroxide electrolyte, the formation of γ-NiOOH is very slow; in 31% potassium hydroxide, moderate; and in 38% potassium hydroxide, the formation of γ-NiOOH is relatively facile. This is one of the key reasons why nickel electrodes not only can provide much higher capacity at greater electrolyte concentration but also undergo faster degradation and reduced cycle life.

The γ-NiOOH phase cannot discharge to the less stable β-NiOOH phase; however, it can undergo a reversible reaction to the α-Ni(OH)$_2$ phase, as indicated by the reaction on the right side of Fig. 4.12. The α-Ni(OH)$_2$ phase has a rhombohedral structure that has an expanded lattice relative to γ-NiOOH, and significantly greater molar volume. This significant difference in volume between these two phases is the primary reason why it is not desirable to cycle nickel electrodes between them. Because the oxidation state of nickel in α-Ni(OH)$_2$ is essentially 2.0, full discharge of the γ-NiOOH unit cell must involve the transfer of five electrons to the three nickel atoms in the unit cell. Because the reaction occurs at a single well-defined potential and is reasonably reversible, one may infer that there are five energetically similar sites for mobile protons within the rhombohedral lattice of the γ-NiOOH phases that must become occupied before the lattice restructures to the α-Ni(OH)$_2$ phase. The reaction for this overall process is expressed by Eq. (4.9).

$$\text{NiOOH(NiO}_2)_2 \cdot \text{KOH} + 5\text{H}_2\text{O} + 5e^- \xrightarrow[E_o=0.45\text{volt}]{\text{discharge}} 3\alpha\text{-Ni(OH)}_2 + \text{KOH} + 5\text{OH}^- \quad (4.9)$$

While the reaction in Eq. (4.9) is reversible, it is not generally possible to repeatedly charge and discharge between the α-Ni(OH)$_2$ and the γ-NiOOH phases in a nickel electrode. The reason for this is that the α-Ni(OH)$_2$ phase is unstable in potassium hydroxide electrolyte, normally recrystallizing back into the original and more stable β-Ni(OH)$_2$ phase before recharge can significantly occur by the reverse of Eq. (4.9).

The process at the top of Fig. 4.12 represents the recrystallization process. The recrystallization can occur because the divalent nickel hydroxide phases have some solubility in the potassium hydroxide electrolyte. It is important to note that recrystallization results in a β*-Ni(OH)$_2$ that exists as larger, more well-ordered crystallites than the same phase formed electrochemically by the reverse of Eq. (4.6). This recrystallization process, expressed by Eq. (4.10), has important electrochemical implications that will be discussed later.

$$\alpha - \text{Ni(OH)}_2 \xrightarrow{\text{recrystallization}} \beta * - \text{Ni(OH)}_2 \quad (4.10)$$

Stabilization of the α-Ni(OH)$_2$ structure so that one can reversibly charge and discharge it by Eq. (4.9) could result in up to 67% more capacity than can be provided by Eq. (4.6). While it is possible to stabilize the α-Ni(OH)$_2$ structure in potassium hydroxide electrolyte by the incorporation of anionic species such as nitrate or chloride, repeated cycling results in loss of the stabilizing anions and the gradual conversion of Eq. (4.10). Nickel hydroxide has been modified by adding iron to form the pyroaurite structure, which can undergo reversible cycling as in Eq. (4.9). However, the large volume changes that accompany cycling between nickel oxidation states of 2.0 and 3.67 make these electrodes degrade quickly in comparison with standard nickel electrodes.

4.2.2 Voltage Hysteresis in the Nickel Electrode

The voltage values of the charge and discharge reactions in the nickel electrode exhibit a phenomenon known as hysteresis, which means that there is a voltage offset between the recharge voltage and the discharge voltage. This voltage offset (or hysteresis) appears to have a thermodynamic origin, because it exists even if the charge and discharge voltages are extrapolated to zero current, as shown in Fig. 4.14.

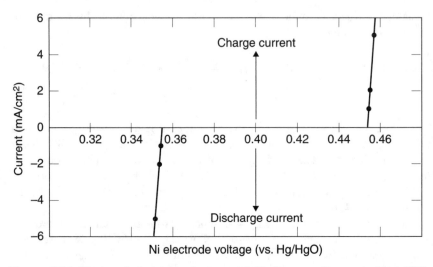

Figure 4.14. Hysteresis between charge and discharge voltage at about 50% state of charge for a nickel electrode that was charged to β-NiOOH, then discharged to β-Ni(OH)$_2$.

The magnitude of the hysteresis can be nearly 100 mV, and the hysteresis loop is seen irrespective of the magnitude of the state-of-charge range covered by the charge/discharge cycle. The origin of the hysteresis is not fully certain, although a number of possibilities exist. Hysteresis appears to be present for all the reversible processes in the nickel electrode, although it is generally observed for the β-β reaction, as a result of the instability of the α-nickel hydroxide. The reversible potentials for the electrochemical reactions in the nickel electrode are generally assumed to be halfway between the charge and discharge portions of the hysteresis loop.

The voltage that is measured for a nickel electrode is the potential difference between the conduction band in the metal current collector and the electrolyte that is in contact with the active material and the reference electrode (assuming no uncompensated ohmic resistance between the nickel electrode and the reference electrode). As indicated in Fig. 4.13, a number of interfaces exist and processes take place between the current collector and the electrolyte. The origin of the hysteresis must lie in the chemistry shown schematically in Fig. 4.13.

One possible explanation for the hysteresis is that it arises from a small energy barrier that exists for protons to hop from one lattice site to another. Significant current cannot flow until a potential gradient sufficient to overcome the energy barrier has been generated. Once the extra potential of about 20 mV is present, protons can move freely through the lattice. A variation on this theory involves a similar energy barrier to the formation and neutralization of lattice protons by the alkaline electrolyte at the surfaces of the active-material crystallites.

Another potential explanation for hysteresis is that it results from the lack of microscopic reversibility for the charge and the discharge reactions that is implicit in Fig. 4.13. Because of the interplay that can occur between electronic and proton diffusion, it is possible that the location in the active-material layer where recharge occurs is not the same as the location where discharge occurs. For example, recharge may be constrained by reduced conductivity to occur heterogeneously by the growth of a layer of charged material outward from the current collector, while subsequent discharge of the conductive charged material can occur homogeneously throughout the active-material layer. Again, the difference of about 40 mV corresponds to the gradient in reversible potential that exists between the active materials in these different physical locations.

Other explanations for hysteresis are based on surface states that produce interfacial barriers to conduction between the individual crystallites in the active material, or between the active material and the metal current collector. However, an explanation for the hysteresis that has been found to consistently fit the observed behavior of nickel electrodes involves a process whereby water molecules from the electrolyte interact with the active-material surface to produce lattice-stabilized protons and hydroxide ions. This process, indicated in Eq. (4.3), must involve the binding of a water molecule to the surface site, the splitting of the water bonds to produce an adsorbed hydroxide ion and a lattice-stabilized proton, and finally, de-adsorption of the hydroxide ion. For this process to occur, there must be a dipolar layer at the active-material surface having the correct polarity to attract the water molecule, then repel the negatively charged hydroxide ion. This requires a polarized layer of surface species wherever the electrolyte is in contact with the active material, and this dipolar layer must be polarized in the correct direction to allow the facile passage of charged species across the interface between the active material and the electrolyte.

The hysteresis results because the dipolar layer requires the opposite polarization for charge current than it requires for allowing discharge current through the interface. The mechanism for switching between charge and discharge current therefore involves flipping the polarization of this surface layer before current can flow. The polarization of the surface layer is a source for the large surface capacitance always seen in nickel electrodes, a capacitance that is critical to enabling current to flow between the electrolyte and the solid active material. The other consequence of the polarization flip between charge and discharge is that a voltage hysteresis will exist between charge and discharge that is equal to the voltage difference between a positively polarized surface layer and a negatively polarized one. This voltage hysteresis is very similar to the hysteresis seen when magnetic domains flip during the polarization of ferromagnetic films.

4.2.3 Proton Diffusion Behavior of the Nickel Electrode

Divalent nickel hydroxide has little ability to conduct charge by either ionic or electronic conduction mechanisms. However, the active material in the nickel electrode is never completely reduced to divalent nickel, and thus there always are some hydrogen vacancies in the discharged hydroxide lattice. As the active material is recharged, hydrogen species are removed from the lattice. When the active materials are fully charged, each nickel site has a lattice vacancy that is capable of being occupied by a hydrogen atom (or in the case of γ-NiOOH, an average of 1.67 vacancies per nickel site). It is the process of filling and emptying these vacancies that is involved in the discharge or recharge of the active material in the nickel electrode. As noted earlier, hydrogen species exist in the lattice vacancies as protons that can move easily through the active material from one lattice vacancy to the next. This ionic conduction process within the active-material lattice is referred to as proton diffusion.

As will be discussed in the next section, proton conductivity enables the charge and discharge of active material that has relatively poor electronic conductivity. Proton conductivity also allows electrode capacity to recover after discharging to depletion at high rate, simply by providing an open-circuit recovery period during which protons can diffuse into depleted regions to replenish their capacity.

The proton diffusion coefficient determines the rate of discharge or charge that can be supported by proton diffusion. It is expected that the proton diffusion coefficient will depend on the density of proton vacancies in the lattice and thus will increase as the active-material state of charge increases. This does appear to be the case; however, the active materials used in nickel-hydrogen cell electrodes always contain cobalt additive at a level of either 5% or 10%. The cobalt additive exists as CoOOH sites within the lattice, and it thus provides proton-deficient sites that support proton diffusion independent of state of charge.

The proton diffusion coefficient in the active material of the nickel electrode has been measured by a variety of kinetic and impedance measurements. Proton diffusion coefficients have been measured between about 1×10^{-9} and 1×10^{-11} cm^2/sec for sintered nickel electrodes of the general type used in nickel-hydrogen cells. Examination of Fick's diffusion equation indicates why this range has been measured for sintered nickel electrodes.

The current that can be supported by diffusion is $I = nFA\ D\Delta c/L$, where n is the number of equivalents per mole (assumed to be 1), F is Faraday's constant, A is the area throughout which diffusion occurs, D is the diffusion coefficient, and Δc is the concentration difference across the diffusing layer of thickness L. In a 1 cm^2 area of sintered nickel electrode there are typically about 100 cm^2 of actual sinter area in contact with the active material that is capable of supporting proton diffusion. L is on the order of 5 μm, because the active-material layers within the pores of a sintered electrode are typically less than about 10 μm thick.

If one assumes a density for the active material of 3.6 g/cm^3, the diffusion current in a sintered nickel electrode should be $I \approx 37.8 \times 10^6\ D\Delta c^*$ mA/cm^2, where Δc^* is the normalized concentration difference across the diffusing layer (i.e., 1.0

for a fully diffusion-limited situation). Thus, a diffusion coefficient of 10^{-9} cm^2/sec can fully support a maximum rate somewhat higher than the typical C-rate for a sintered nickel electrode (~30 mA/cm^2). Diffusion coefficients that are several orders of magnitude lower (as low as 10^{-11} cm^2/sec) can also be very effective for supporting electrode operation with diffusion layers less than 1 μm in thickness when combined with good electronic conductivity in the active material.

4.2.4 Electronic Conductivity of Active Material

The electronic conductivity of the active material allows electrons to move to active-material sites distant from the current-collecting surfaces, where they may engage in charge-transfer reactions. This process is shown schematically in Fig. 4.13 and described by Eq. (4.2). If the active material does not maintain adequate electronic conductivity, all charge transfer is forced to occur at the interface between the active material and the current collector, thus forcing all charge movement within the active material to occur by solid-state ionic conduction by mobile protons.

The electronic conductivity depends both on the state of charge of the active material and whether it was homogeneously or heterogeneously charged or discharged. Homogeneous charging involves bringing the state of charge of the entire active-material layer uniformly to the desired level with essentially no boundary layers existing between charged and discharged material within the active-material particles. Heterogeneous charging, which probably occurs more commonly at the rates at which electrodes are normally operated, involves propagation of a moving boundary between charged and discharged material as the state of charge changes.

Conductivity measurements have been performed on active material that was purified from sintered electrodes that had been brought to varying states of charge or discharge. These measurements are typically made by grinding up the rinsed and dried electrode after bringing it to the desired state of charge. The active material is then magnetically separated from the metal particles. A sample of active material is compressed at a standardized force into a pellet between two flat nickel contacts, and the resistance is measured across the active-material pellet. When resistance measurements of this kind are taken using a sintered nickel electrode that has 10% cobalt additive and that was charged at 2 mA/cm^2 to various states of charge, the charge behavior shown in Fig. 4.15 is seen.

The active material retains the high resistance associated with fully discharged material until the state of charge reaches about 50% (nickel oxidation state of ~2.5), after which the resistance abruptly falls to a level that is many orders of magnitude lower. When the nickel oxidation state increases to more than about 3.1, another drop of about an order of magnitude occurs. These resistance levels correspond to the conductivity of the homogeneous β-Ni(OH)$_2$, β-NiOOH, and γ-NiOOH phases. X-ray diffraction measurements of the active material in

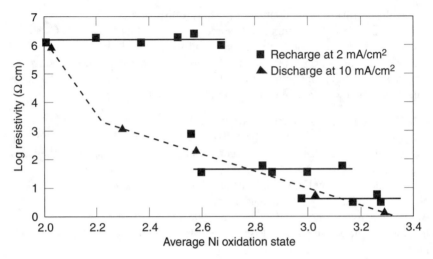

Figure 4.15. Resistivity changes resulting for nickel-electrode active material as a result of charge or discharge as a function of nickel oxidation state.

each resistance region show diffraction patterns for essentially only the one phase structure that has been homogeneously charged. When the density of protons within a given lattice reaches the critical level where the lattice becomes unstable, it suddenly restructures to a new phase. There is significant overlap between the resistance of the β-NiOOH and γ-NiOOH phases because it is very difficult to homogeneously generate the γ-NiOOH phase.

If, after being fully recharged, the 10% cobalt-containing electrode is discharged at 10 mA/cm^2, or if it is initially recharged at 10 mA/cm^2, the lower sloping resistance line is obtained as a function of state of charge. The sloping resistance relationship is indicative of heterogeneous charge or discharge, where the active material of any specific average oxidation state is actually composed of a mixture of charge and discharge phases. As the cobalt-additive level is decreased to 5% or less, it becomes more difficult to homogeneously charge the active material, and lower rates must be used.

The resistivity behavior shown in Fig. 4.15 provides some interesting insights as to how charge moves through the active materials. The ability to oxidize a poorly conducting nickel hydroxide phase to a stability limit where half of the nickel sites in the hydroxide lattice are trivalent is a testimony to the ability of protons to efficiently distribute charge through the nickel hydroxide lattice. It must be remembered that pure nickel hydroxide is not a good proton conductor. The proton conductivity requires that a significant density of proton vacancies exist in the hydroxide lattice, vacancies that are always present in active material because it is never fully discharged completely to divalent nickel. It is also likely that the cobalt additive provides trivalent metal sites in the lattice that maintain good proton conductivity for even completely discharged active material.

The principal performance consequence of the changes in active-material conductivity is premature depletion of the nickel electrode, leaving significant residual capacity that cannot be effectively discharged at a high rate and at a high potential. At high discharge rates, the state of charge drops until eventually the active material at the interface with the current collector transitions to the high-resistance nickel hydroxide phase. At this state of charge, which is typically in the 10–20% range, proton diffusion will generally not be sufficient to support the discharge rate, thus forcing all charge transfer into the thin layer at the current-collector surface. This, of course, forces the rapid depletion of a layer of active material at the current collector and results in collapse of the electrode voltage and apparent depletion of the electrode capacity. However, the 10–20% active-material capacity that remains more distantly situated from the current collector can move by proton diffusion to replenish the depleted layer, thus allowing further discharge at a high rate after an open-circuit recovery time. The replenishing of the depleted layer also allows significant further discharge of residual capacity at lower rates, for which proton diffusion can better support continued discharge. While 15% residual capacity may remain in a typical nickel electrode after a C/2-rate discharge to 1 V (*vs.* hydrogen), after a subsequent C/100 discharge only 3–5% residual capacity may remain undischarged.

The data provided in Fig. 4.15 can be effectively used to model nickel electrode behavior at several levels. If a model is being used that does not keep track of the amounts of each phase throughout the active-material layer, the sloping resistivity curve should be used as a function of state of charge. If the model does keep track of the amount of each phase at all locations within the active-material deposit and the moving phase boundaries, it may compute the composite resistance at any location from the resistivities of the pure phases.

4.2.5 Active-Material Volume Changes

The volume changes involved as the active material is charged or discharged in the nickel electrode are a key factor that influences the gradual degradation of the electrode as a result of repeated cycling. Each time the electrode is cycled, the expansion of the active material puts a force on the sinter structure as well as the surrounding active-material particles. Repeated stress cycles result in expansion and possibly cracking of the sinter structure, as well as extrusion of active material from the interior of the electrode onto its surface or into the separator. Cycling between phases that involve larger volume changes will accelerate these types of electrode degradation.

The density of the active material in a sintered electrode can be measured as a function of its charge state using a traditional Archimedean method. An electrode that contains a known amount of nickel hydroxide is brought to the desired state of charge in alkaline electrolyte. The electrode is exposed to vacuum to draw all oxygen from its pores, then reimmersed and accurately weighed as a function of time. The fully immersed weight, extrapolated to the time the electrode was evacuated of gas, provides the active-material volume and density when combined with the weight of the rinsed and dried electrode.

The density of the electrolyte must also be precisely measured (by weighing a 50 cm^3 volumetric aliquot) after each measurement, because the vacuum exposure will slightly change it. The electrode may then be returned to the electrolyte and allowed to age before the measurement is repeated, thus allowing density changes that occur during extended open-circuit stand to be measured.

A number of factors influence the volume of the phases present in the nickel electrode, including overcharge and open-circuit ripening periods. The density of pure nickel hydroxide is 4.14 g/cm^3. However, the density of the discharged active material in nickel electrodes is always significantly lower than this, and can be variable as a result of residual charge, the high density of lattice defects, and grain boundaries between the active-material particles.

Figure 4.16 shows the stabilized density of active material in a sintered nickel electrode as a function of state of charge. The β phases have significantly higher density than the γ-NiOOH or α-Ni(OH)$_2$ phases. The discharged phases have a lower density than do the charged phases for both the β-β and the α-γ couples. If either the charged or discharged phases are allowed to stand open-circuited, the

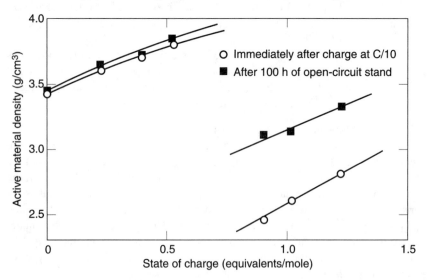

Figure 4.16. Change in density with state of charge for chemically impregnated nickel-electrode active material containing 10% cobalt hydroxide after charging in 31% potassium hydroxide electrolyte.

density will gradually increase over time as the lattice disorder diminishes. There is also a very large change in volume that accompanies both the transition from the hexagonal β-NiOOH to the rhombohedral γ-NiOOH, as well as the onset of oxygen evolution. This decrease in density is likely to result from fracturing of crystallites by lattice expansion and the pressure of expanding oxygen bubbles within the capillary structure between the active-material particles. These inter-particle pores are extremely fine, and they can retain electrolyte with internal capillary pressures of up to 1000 psi.

The uptake of potassium that stabilizes the γ-NiOOH structure can be readily measured during the density measurements described here from the increase in the weight of the dry active material when the nickel valence is increased to more than 3.0. This behavior is illustrated in Fig. 4.17, which shows that the rinsed and dried active material increases its weight with a slope that corresponds to almost precisely 0.33 potassium atoms per nickel atom. This extra weight, which can be attributed to potassium from chemical analysis, cannot be rinsed from the charged γ-NiOOH material. However, discharge of the material immediately leads to loss of the extra active-material weight, as well as loss of the potassium bonded to the active material.

4.2.6 Overcharge Reactions in the Nickel Electrode

All the charged phases discussed in the previous section are thermodynamically unstable relative to the evolution of oxygen, and indeed, oxygen evolution is the

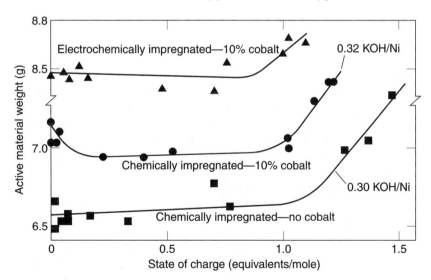

Figure 4.17. Weight of the active material in different nickel electrodes as a function of state of charge.

dominant self-discharge process in nickel electrodes when they are at high states of charge. In fact, the only reason the nickel electrode functions at all is that the process of oxygen evolution from the surfaces of the charged active-material crystallites is not a facile reaction. As indicated in Eq. (4.11), the reversible potential for oxygen evolution is nearly 90 mV less than that for charging nickel hydroxide, yet significant oxygen evolution does not actually occur until the voltage is 50 mV or more greater than the potential where nickel hydroxide charges.

$$4OH^- \xrightarrow{NiOOH} O_2(g) + 2H_2O + 4e^- \qquad (4.11)$$

As indicated in Eq. (4.11), the charged nickel oxyhydroxide phases provide the catalytic surfaces from which the oxygen gas is evolved. The oxygen evolution voltage from the β-NiOOH phase typically occurs at a slightly higher potential than does that from the γ-NiOOH. This often results in a drop in nickel electrode voltage by up to 10–20 mV once some γ-NiOOH has begun to form. Because γ-NiOOH usually begins to form in nickel-hydrogen cells while oxygen is also being evolved, it is often difficult to separate this voltage effect from the effects of rising temperature as the oxygen recombines with hydrogen gas.

The kinetics of the oxygen evolution reaction at the nickel electrode follows the Tafel kinetics expected of an essentially irreversible electrochemical process, as indicated in Eq. (4.12):

$$i = i_o A_{NiOOH} e^{-(E-E_o)/nRT} \qquad (4.12)$$

The exchange current density i_o is typically about 10^{-5} mA/cm^2. The area of charged material from which oxygen is being evolved, A_{NiOOH}, is explicitly included, because this area will vary with the amount of each charged phase present in the electrode as it is recharged. The logarithmic slope, which is nRT in Eq. (4.12), does not correspond to $n = 4$ as suggested by Eq. (4.11), because the oxygen evolution process is not a simple one-step charge-transfer process. The reaction involves a series of transport, adsorption, and charge-transfer steps that can produce a range of n values from $n = 2$ (30 mV/decade at 25°C) to $n = 1$ (59.6 mV/decade) at high current densities where transport influences the kinetics, as shown in Fig. 4.18.

The large activation energy or overpotential for oxygen evolution makes this process much more sensitive to changes in temperature than are the normal charging reactions in the nickel electrode. Because it is the voltage separation between the recharge reaction and oxygen evolution that gives good charge efficiency, the charge efficiency of the nickel electrode is quite sensitive to temperature. As the temperature is raised above about 15 to 20°C, the oxygen evolution voltage falls

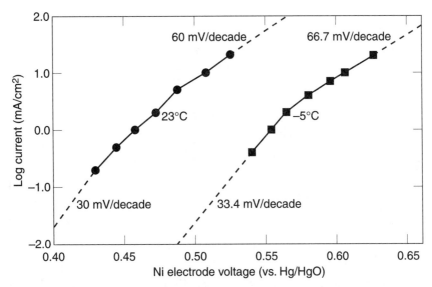

Figure 4.18. Tafel plots of oxygen evolution rate as a function of electrode potential for nickel electrodes at 23 and –5°C, illustrating both high and low current slopes.

to a level low enough to compete effectively with the charge processes. Conversely, at low temperatures (those below 0°C) the charge efficiency of the nickel electrode is sufficiently high to readily convert a significant portion of the active material to γ-NiOOH. Under these conditions it is not uncommon to attain an average nickel oxidation state of 3.1 to 3.3 or more.

The dominant self-discharge process in the nickel-hydrogen cell at high states of charge is the evolution of oxygen from the nickel electrode. Even with no external current flow to drive the oxygen evolution reaction, oxygen continues to evolve at the rate predicted by Eq. (4.12) from the open-circuit electrode voltage. When a highly charged nickel electrode is placed in electrolyte, it is normal to see bubbles of oxygen coming out of the electrode in the open-circuit condition. Direct measurements of oxygen evolution rates for open-circuited electrodes have been used to confirm that Eq. (4.12) applies to self-discharge as well as to overcharge-driven oxygen evolution.

The evolution of oxygen gas from the nickel electrode appears to be accompanied by a significant amount of capacity that is stored in high-potential surface states. These are the very surface states that are involved as adsorbed intermediates in the oxygen evolution process, and they are significant because of the extremely large surface area associated with the fragmented crystallites of oxygen-evolving active material. This high-voltage capacity, which can amount to up

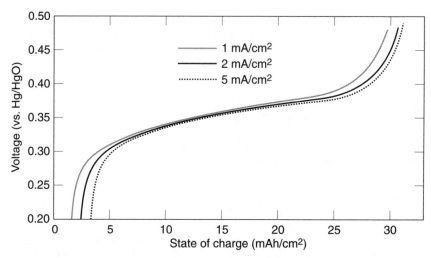

Figure 4.19. Nickel electrode discharge at varying rates showing high voltage capacitive behavior at the start of discharge. Recharge was at 13°C in 31% potassium hydroxide electrolyte.

to 10% of the total electrode capacity, is termed pseudocapacitance. Discharge of the charged surface-state pseudocapacitance gives a high-potential capacitive discharge region during initial electrode discharge, as is shown in Fig. 4.19. It is not surprising that these high-potential surface states are initially lost to self-discharge at an accelerated rate as predicted by Eq. (4.12).

4.2.7 The Role of Crystalline Disorder in the Nickel Electrode

As Eq. (4.10) suggested, there is a significant difference between the behavior of a highly ordered β*-Ni(OH)$_2$ phase formed by recrystallization and that of a highly disordered β-Ni(OH)$_2$ material formed by electrochemical discharge. The degree of crystalline order or disorder plays a major role in the performance of the phases present in the nickel electrode. It is clear that large active-material crystallites will expose less surface area to the electrolyte and to contact with other crystallites or the current collector, which can significantly shift the rate-limiting processes as discussed in relation to Fig. 4.13. In addition, the reversible potential of the electrochemical reactions depends on the density of lattice sites that are filled with mobile lattice protons, which in turn decreases as the crystal domains become more fragmented. The solid-state lattice acts essentially as a solid solution in which the mobile lattice protons produce a Nernstian dependence of reversible potential on concentration. Lattice defects or cobalt additive sites that stabilize lattice protons can therefore reduce the reversible potential significantly.

The recrystallization of α-Ni(OH)$_2$ to a deactivated β*-Ni(OH)$_2$ phase was discussed in the context of Eq. (4.10). Phases marked with an asterisk should be

4.2 Chemistry of the Nickel Electrode

regarded as highly ordered phases consisting of large crystallites; they are electrochemically deactivated phases that will exhibit higher reversible potentials and higher overpotential. While the recrystallization of Eq. (4.10) can rapidly form deactivated nickel hydroxide, a slower deactivation process also applies to deactivation of β-Ni(OH)$_2$ by Ostwald ripening.

$$\beta - Ni(OH)_2 \xrightarrow{\text{ripening}} \beta*-Ni(OH)_2 \qquad (4.13)$$

Ostwald ripening involves the local dissolution of surface lattice sites, which can migrate over crystallite surfaces to eventually recrystallize to minimize surface energy. For the active nickel hydroxides, this process will result in the fragmented crystallites (~200–300 Å) formed by electrochemical cycling ripening to ~1200–1400 Å in size before surface stabilization is achieved. The ripening process, if allowed to proceed fully (which may require many weeks), can result in an increase of about 20 mV in the recharge voltage. Partial ripening can result in either two distinct recharge voltages separated by about 20 mV, caused by two different populations of crystallites, or alternatively a single broader-voltage population that simply increases in voltage as it is allowed to fully ripen. The higher recharge voltages associated with the deactivated phases always show reduced charge efficiency and increased self-discharge. It is clear that these types of ripening or recrystallization processes can make the performance of the nickel electrode strongly dependent on the details of its previous history.

Once the nickel hydroxide has become either fully or partially deactivated, recharging it to a potential (or state of charge) high enough to fragment the crystallites can reactivate it as a result of electrochemically induced volume changes. Reactivation tends to occur at an average nickel oxidation state just above 3.0. Once reactivated, the fully charged materials remain in the fragmented and highly active state until discharged. Once discharged, the reactivated materials begin to deactivate and may become significantly deactivated before recharge is completed. The deactivation process is highly temperature-dependent because the solubility of the nickel hydroxides increases with increasing temperature. It is not unusual to see significant deactivation effects increasing the voltage during recharge at high temperatures, while at low temperatures these effects are not normally seen during recharge.

The deactivated phases of active material that can undergo electrochemical processes have distinct voltage signatures differentiating them from the highly active α, β, and γ phases commonly involved in repetitive cycling. To include the chemistry of the deactivated phases, the simplified reaction scheme of Fig. 4.12 must be modified as indicated in Fig. 4.20. The recrystallization processes at the top of Fig. 4.20 result in the formation of deactivated $\beta*$-Ni(OH)$_2$ as discussed for Eqs. (4.10) and (4.13). However, as indicated on the left side of Fig. 4.20, the

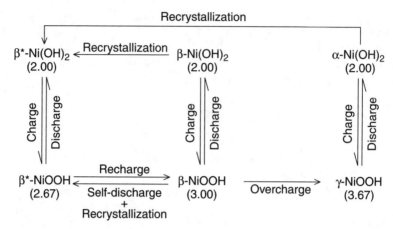

Figure 4.20. Chemistry of the nickel electrode, showing both activated and deactivated phases. The average nickel oxidation state for each phase is shown under each formula.

deactivated β*-Ni(OH)$_2$ can undergo several electrochemical processes before lattice volume changes caused by oxidation result in formation of the more disordered β-NiOOH structure. The deactivated β*-Ni(OH)$_2$ can first undergo reversible oxidation to a highly ordered β*-NiOOH structure having an average nickel oxidation state of 2.67. With continued oxidation, the β*-NiOOH phase is converted to the more disordered β-NiOOH structure having a nickel oxidation state of about 3.0. The disordered β structure can revert directly back to the disordered structure as a result of gradual self-discharge and recrystallization.

The electrochemical reactions involving β*-NiOOH in Fig. 4.20 can lead to relatively complex charge and discharge voltage behavior for nickel electrodes, particularly after extended periods of open-circuit stand in either the charged or discharged states. For example, recharge after a period of fully discharged stand can result in multiple recharge voltage regions corresponding first to the formation of the β*-NiOOH structure and then the β-NiOOH structure. The electrochemical voltage spectroscopy[4,9] signatures in Fig. 4.21 show two well-defined charge voltage peaks corresponding to these two sequential reactions. Only one such peak is seen during subsequent cycles that do not involve a rest period in the fully discharged state.

Similarly, if a charged nickel electrode is allowed to stand open-circuited and undergo self-discharge, a well-defined discharge peak emerges as β*-NiOOH gradually forms, as shown in Fig. 4.22. The β*-NiOOH phase appears to have a discharge potential that is intermediate between that of β-NiOOH and γ-NiOOH, which implies that it is less stable than γ-NiOOH but more stable than β-NiOOH. As indicated by the very long stand times in Fig. 4.22 that are required to restructure the β-NiOOH to a more ordered lattice structure, the reordering process is quite slow.

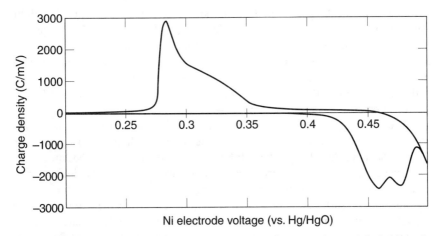

Figure 4.21. Electrochemical voltage spectroscopy spectra for a nickel electrode exhibiting two recharge peaks after long-term discharged storage.

4.2.8 Second Plateau Discharge

At the end of the discharge of a nickel electrode, the voltage can exhibit a discharge plateau at about 0.8 V (*vs.* hydrogen) that does not coincide with the potential of

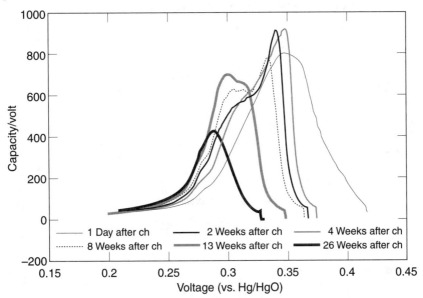

Figure 4.22. Electrochemical voltage spectroscopy spectra showing voltage behavior during discharge of β-NiOOH containing nickel electrodes after varying periods of charged stand.

any electrochemical reactions in the electrode. Nickel electrodes typically exhibit this behavior at rates of C/10 or lower. However, in the presence of hydrogen gas in the nickel-hydrogen cell, the duration of the lower voltage discharge plateau is typically significantly longer, and it can be significant at rates as high as C/2. In addition, the level of the second plateau voltage is observed to depend on the hydrogen pressure in the cell, increasing to nearly 1.0 V at several hundred psi of hydrogen pressure, as illustrated in Fig. 4.23.

The second plateau has frequently been attributed to discharge of oxygen within the nickel-hydrogen cell, because it has also been found to disappear if all oxygen gas is removed from the cell before the cell can drop to the lower voltage level toward the end of discharge. Oxygen, which has a reduction potential of 1.31 V *vs.* hydrogen, can potentially be reduced with a large overpotential on the nickel metal surfaces in the sintered electrode. However, the oxygen reaction cannot account for the complete inability to observe the second plateau at voltages greater than 1.0 V (even at very low rates), as well as the fact that up to 50% of the nickel electrode capacity has been observed to discharge on the second plateau. The nickel-hydrogen cell cannot contain more oxygen than about 2% of the hydrogen pressure for any length of time without initiating (on the platinum catalyst surfaces) a chain reaction with hydrogen gas that will rapidly burn the oxygen. However, it is clearly possible for the nickel sinter to act as a poor catalyst and reduce oxygen during electrode discharge at very low rates.

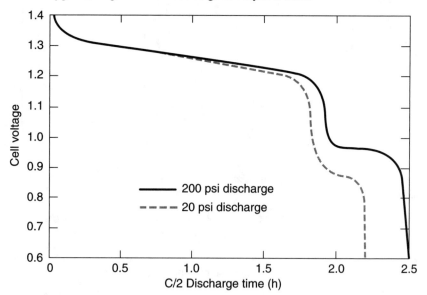

Figure 4.23. Second plateau discharge for a nickel electrode at different hydrogen pressures.

Other suggestions for the cause of the second plateau have associated it with reduction of carbonate species, or a direct electrochemical reduction of γ-NiOOH directly to β-Ni(OH)$_2$, a process that has never been substantiated to occur at a significant rate. The second plateau, however, has been shown to actually arise from the normal electrochemical discharge reactions, as demonstrated by an identical enthalpy of reaction measured by calorimetry on both the upper and lower discharge plateaus. The second plateau results from an interesting interaction between the electronic and electrochemical behavior of the active material at the interface between the sinter nickel metal and the active material as the active-material potential plummets at the end of discharge. Figure 4.24 illustrates this condition.

As the nickel electrode discharges, the state of charge of the active material adjacent to the current collector decreases, and the resistivity of the active material at this interface increases. When the resistivity of the active material reaches a critical level, the voltage drop through the interfacial region forces further discharge to become concentrated in a thin film at the interface with the current collector. The thin film of active material at the current-collector interface rapidly becomes depleted in this situation, and the discharge voltage of the electrode drops sharply as the interfacial region undergoes depletion. If no other processes took place, further discharge of the electrode would be impossible without some open-circuit recovery time to allow charge to diffuse back into the depleted film.

However, the depleted active material does not act as a simple resistive layer. Depleted active material is a p-type semiconductor that will begin to transport charge by hole conduction when a potential of about 0.4 V is applied across the p-n junction formed where the depleted active material contacts the current collector.

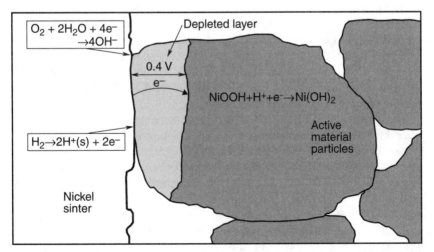

Figure 4.24. Depletion layer that forms at the interface of the active material and the current collector to cause second-plateau discharge.

This situation corresponds to the formation of a Schottky barrier diode across the interfacial layer, which exhibits good electronic conduction when biased sufficiently. Thus, the potential of the depleted active material at the junction only falls to a level 0.3–0.4 V below the normal charge-transfer potential of the active material; then the depleted active-material layer becomes capable of efficiently conducting charge to remaining charged material located some distance from the current collector. Thus, a mechanism exists to continue discharging isolated residual charged material, as long as some additional surface process exists to pin the potential at the current collector at a lower voltage level.

There are several surface processes in the nickel-hydrogen cell that are capable of supporting charge-transfer processes at the very low rates needed to pin the potential of the active material while further discharge proceeds via barrier-layer conduction to isolated charge. The reduction of oxygen gas on the nickel metal surface, while kinetically slow, can occur at the $\sim 10^{-5}$ mA/cm^2 rate required to support the electrode potential during low-rate discharge. This leads to the second-plateau discharge that is commonly seen for nickel electrodes during low-rate discharge of their residual capacity, and that is sensitive to the presence of oxygen gas in the nickel electrode. However, the kinetics of the oxygen reduction process are slow, and this reaction is not generally able to support the electrode potential at a second voltage plateau during discharge of sintered electrodes at rates greater than about C/10.

In the nickel-hydrogen cell, however, there is also a significant pressure of hydrogen gas that is in contact with the nickel electrode. Although the oxide-covered nickel current collector does not provide a highly active catalytic surface for hydrogen reduction, it is normally more than 1 V above the hydrogen potential, and it is capable of reducing hydrogen gas at a low rate. As will be discussed later, this is in fact one mechanism for self-discharge of the nickel electrode in a nickel-hydrogen cell and can typically occur at a rate of microamperes per square centimeter. This process can very effectively pin the potential of the nickel current collector at the voltage where the barrier layer of depleted active material begins to become electronically conductive.

The hydrogen reduction process is capable of pinning the potential at much higher discharge rates than is the oxygen reduction process, giving a lower discharge plateau in nickel-hydrogen cells at about 0.9 V, even at rates as high as C/2 or more. The level of the lower plateau is also dependent on hydrogen pressure, rising to more than 1.0 V at pressures of several hundred psi, which are frequently found in well-cycled cells. The hydrogen reduction process is also capable of improving the conductivity of the depleted layer by injecting protons into the depleted active-material layer where it contacts the current collector. This catalytic hydrogen reduction process can enable up to about 50% of the active material to be discharged at a lower voltage plateau for well-cycled and degraded nickel electrodes.

4.2 Chemistry of the Nickel Electrode 111

The thickness of a barrier layer that can be biased enough to maintain conductivity sufficient to support a C/2 discharge therefore appears to be limited to about 5 μm, or approximately 50% of the typical 10 μm active-material layers found in sintered nickel electrodes. At low discharge rates, barrier-layer conduction can give highly efficient second-plateau discharge, which can actually result in reduction of much of the active-material layer to a green nickel hydroxide material that is relatively free of proton vacancies.

Nickel-hydrogen cells are often constructed with nickel precharge, and thus complete cell discharge involves depletion of all the hydrogen gas in the cell, leaving behind 10–15% remaining charge in the nickel electrode. Therefore, cells built with nickel precharge typically do not exhibit any second plateau when new. After extended cycling the nickel electrode will have undergone degradation, and additional hydrogen gas will have gradually built up by corrosion of the nickel sinter. These changes will result in the observation of increasing capacity in the lower voltage plateau as the nickel-hydrogen cell ages. If the cell eventually fails by loss of usable capacity at end of life, a quite large fraction of the cell capacity may be discharged at the lower voltage plateau.

4.2.9 Active-Material Layering, Deconditioning, and Reconditioning

The layer of depleted active material illustrated in Fig. 4.24 plays a key role in the voltage behavior of the nickel electrode at high depths of discharge, and it can be the major contributor to loss of usable capacity. The formation of layers of the charged active-material phases can influence performance, as can the formation of gradients in the cobalt additive typically present in the nickel electrode. (The role of additive gradients will be discussed later.) A range of nickel-hydrogen cell performance signatures results from layers of charged and discharged materials that can form during long-term cycling or during charged open-circuited stand periods.

The nickel electrode in a nickel-hydrogen cell is referred to as "reconditioned" if it has been fully discharged at low rate ($< $ C/50) to a voltage below 0.5 V $vs.$ hydrogen. This reconditioned state involves a relatively uniform active-material layer of relatively ordered β-Ni(OH)$_2$ in the nickel electrode. Recharge results in a relatively uniform layer of disordered β-NiOOH.

If the cell is overcharged, a layer of γ-NiOOH will begin to grow out from the surface of the current collector, because this is the location of highest electrochemical potential. Subsequent discharge will preferentially discharge the layer of β-NiOOH that is distant from the current collector, because the discharge potential of β-NiOOH is about 40 mV higher than that of γ-NiOOH. Partial discharge, which is how nickel-hydrogen cells are typically cycled, will thus result in cycling of the β-NiOOH layer while the γ-NiOOH layer is not cycled unless the cell is discharged to a sufficiently low state of charge. During each cycle, the small amount of overcharge applied to the cell will result in a small amount of additional γ-NiOOH formation until the β-NiOOH discharged exactly balances that formed during recharge. This results in the layered structure illustrated in Fig. 4.25, in which essentially all the charged active material that is not repeatedly cycled exists as a layer of γ-NiOOH phase while the remaining active material exists as β-NiOOH and β-Ni(OH)$_2$ layers.

The type of electrode illustrated in Fig. 4.25 is often referred to as a "deconditioned" nickel electrode. Large amounts of overcharge will hasten the development of the deconditioned state, because more γ-NiOOH will be generated during each cycle. Typically, 300–500 cycles will result in the complete deconditioning of the nickel electrode with normal amounts of overcharge. The electrical signatures that accompany deconditioning are a drop of up to 40 mV in the end-of-discharge voltage, irrespective of the state of charge that corresponds to the end of discharge, and a shift in overall state of charge. The complete discharge voltage curve (Fig. 4.26) shows two distinct voltage plateaus. The higher one, which has a distinct slope with state of charge, is caused by discharge of the β-NiOOH layer, and the lower plateau is caused by discharge of the γ-NiOOH layer. For a fully deconditioned electrode the transition between these two discharge plateaus occurs at the maximum depth of discharge to which the cell has been recently cycled.

Deconditioned nickel electrodes cycle with a somewhat diminished energy efficiency because of the lower discharge voltage at high depths of discharge. The deconditioned state can be completely eliminated by reconditioning, which involves a complete discharge of the nickel electrode at a low rate. As discussed, the reconditioned state involves regenerating a relatively uniform active-material layer of relatively ordered β-Ni(OH)$_2$ in the nickel electrode. While there is no evidence that reconditioning nickel electrodes in nickel-hydrogen cells can reverse

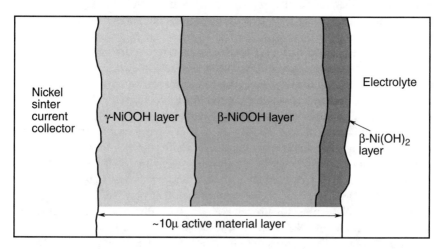

Figure 4.25. Active-material layering in a "deconditioned" nickel electrode.

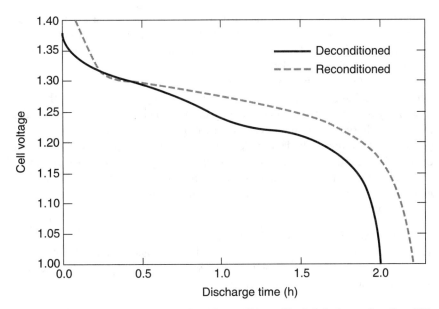

Figure 4.26. Discharge voltage of a "deconditioned" nickel electrode after 500 cycles at 60% depth of discharge in a nickel-hydrogen cell, compared to the same cell after reconditioning.

any of the chemical or structural degradation processes that can occur during long-term operation, reconditioning can provide a 300–500 cycle improvement in discharge voltage and energy efficiency. Reconditioning followed by complete recharge is capable of eliminating most of the complex layers of phases that can result from periodic partial discharge and recharge, with occasional periods of electrical inactivity.

A specific layered structure of interest can result when a nickel electrode containing some γ-NiOOH and some β-NiOOH, as illustrated in Fig. 4.25, is allowed to stand in the open-circuited condition. Self-discharge of the active material will slowly convert the β-NiOOH layer to β-Ni(OH)$_2$ and the γ-NiOOH layer to α-Ni(OH)$_2$. However, the remaining β-NiOOH has a potential high enough to initiate a local couple between the layers that will recharge the α-Ni(OH)$_2$ from self-discharge back to γ-NiOOH. The consequence is that the higher-potential β-NiOOH will undergo preferential self-discharge, and the self-discharge products will accumulate in the region between the γ-NiOOH and the β-NiOOH layers.

When the electrode is completely discharged, the layer of self-discharge products will completely cut off the discharge of all residual charge remaining in the β-NiOOH layer once a depletion layer develops at the interface between the current collector and the active material. Because the layer of self-discharge products is not in contact with the current collector, it is unable to form a conductive p-n junction that will support continued discharge of isolated capacity in the β-NiOOH layer. This process can result in complete isolation of up to 15% of the capacity in a nickel electrode if it is allowed to stand open-circuited in the fully charged state for three days or more. Fortunately a full recharge, which will recharge all intermediate depletion layers, will fully restore the ability to fully discharge all residual capacity in the nickel electrode. However, the temporary charge isolation that can occur following charged open-circuit stand periods could temporarily deactivate all the nickel precharge in a cell, enabling degradation reactions to occur with residual hydrogen gas during subsequent storage. The use of appropriate procedures that properly prepare nickel-hydrogen cells for storage will eliminate this potential source of nickel electrode degradation.

4.2.10 Thermal Stability of Active Material

The primary mechanism for decomposition of the charged nickel electrode in the absence of hydrogen gas is evolution of oxygen as the charged active material decomposes, as indicated in Eq. (4.14):

$$4NiOOH + 2H_2O \xrightarrow{thermal} 4Ni(OH)_2 + O_2 \quad (4.14)$$

Increased temperature accelerates the rate of this decomposition process. Because significant amounts of heat are released by the self-discharge, if the charged nickel electrode is in a nearly adiabatic environment it can go into thermal runaway and rapidly release all its energy if the temperature gets up to 85–90°C. Between 180 and 450°C the nickel hydroxides undergo thermal decomposition to nickel oxide and water.

The thermal stability of the discharged α-$Ni(OH)_2$ and β-$Ni(OH)_2$ phases can be measured by thermogravimetric analysis. Evidence has been found for several types of bound water in the active-material structure. The water is both adsorbed onto the active material, and intercalated into the structure. Water bound by these interactions is typically lost below 150°C. The hydration of the active materials is quite important to their performance in battery cells. Temperatures above about 60°C appear to result in loss of weakly bound water, which causes some changes in nickel electrode performance. All water of hydration seems to be lost at temperatures of 150°C, as well as some of the activity of the active material.

4.2.11 Corrosion in the Nickel Electrode

The sintered nickel substrates used in nickel electrode plaques are thermodynamically unstable at the highly oxidizing potentials at which this electrode operates, corroding to produce nickel hydroxide and hydrogen gas.

4.2 Chemistry of the Nickel Electrode

$$\text{Ni}(m) + 2\text{H}_2\text{O} \xrightarrow{\text{corrosion}} \text{Ni(OH)}_2 + \text{H}_2 \qquad (4.15)$$

The sintered nickel metal surfaces of the substrate passivate in alkaline solution to give a seminoble substrate. However, during years of operation, corrosion of the sinter does slowly occur in the nickel electrode. The corrosion rate of nickel substrates seems to vary considerably for different cell designs and for different cell lots of the same design, and it is likely influenced by trace levels of dissolved salts that may either accelerate corrosion or protect the nickel surfaces from attack. However, it is clear that the rate of corrosion is accelerated by cycling activity of the nickel electrode, as well as by allowing the electrode to remain at the high potentials associated with oxygen evolution or cell overcharge.

Typically the rate of sinter corrosion during cell operation is sufficiently slow that many years of operation are needed before the corrosion significantly alters the electrode performance. For nickel-hydrogen cells, which normally have cycle-life requirements of many years, substrate corrosion eventually is a significant factor in causing performance degradation and ultimate electrode failure. Failure typically occurs when 30–50% (depending on discharge rate) of the sinter has corroded, which can occur after about 40,000 cycles at 80% depth of discharge. In nickel-hydrogen cells, substrate corrosion will cause an increase in hydrogen gas pressure in the cell, thus providing an indication for the extent of corrosion, but at the same time reducing the effectiveness of pressure as a simple state-of-charge indicator without needing periodic recalibration.

The corrosion of nickel electrode substrate leads to the formation of nickel hydroxide as the corrosion product, as indicated in Eq. (4.15). The corrosion process consumes water, thus leading to an increase in the electrolyte concentration. The nickel hydroxide product of corrosion is capable of being electrochemically cycled; however, because it does not contain any cobalt additive, it can result in a layer between the current collector and the active material that prevents full discharge. Typically in high-rate cycling, the expansion and contraction of the active materials during cycling will provide some mixing of corrosion products with the active material. Thus, it is not unusual to see corrosion initially result in an increased nickel electrode capacity. Continued corrosion will eventually decrease both the area and the electrical continuity of the current-collecting substrate.

The increase in electrolyte concentration resulting from corrosion will increase the stresses of cycling by making the γ-NiOOH phase more readily formed during recharge. In addition, the buildup of corrosion products in the nickel electrode will begin to plug the pores and the increased volume of these products will force swelling of the entire substrate structure. Loss of electrical continuity from substrate depletion and swelling damage, as well as pore filling, will eventually cause a significant drop-off in active-material discharge efficiency, thus causing low electrode capacity. Significant swelling of the substrate structure can cause either this structure, or active material forced from the structure, to penetrate the cell separation, resulting in a short circuit in the cell. Fortunately, corrosion of nickel is slow enough that electrode failure from corrosion takes 10 yr or more of continuous high depth-of-discharge cycling, or more than 20 yr of low depth-of-discharge or intermittent cycling.

4.2.12 Pore Structure of the Nickel Electrode

As illustrated in Fig. 4.13, the nickel electrode active material consists of a sintered nickel structure filled with particles of active material. These particles are very small, ranging from 300 Å for freshly charged or discharged material to 1500 Å for highly ordered material that has had the opportunity to fully ripen. The nickel electrode thus contains pore volume in several different sizes of pores. Relatively large pores consisting of large cracks in the active-material deposit or unfilled regions of the sinter are on the order of 10 μm in size. Some pore volume is also often seen from small cracks in the active material at 0.5 to 1.0 μm. However, most of the volume within the nickel electrode is contained in the spaces between the particles of active material, which are typically smaller than 500 Å.

A typical pore size distribution measured by mercury intrusion porosimetry is shown in Fig. 4.27. The pore structure of the nickel electrode is important in controlling the proportion of electrolyte that typically fills the nickel electrode in the operating nickel-hydrogen cell.

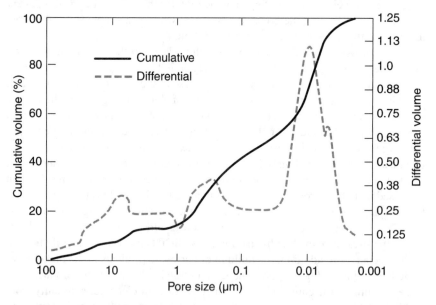

Figure 4.27. Typical distribution of pore sizes within a sintered nickel electrode designed for use in a nickel-hydrogen cell. Mercury intrusion porosimetry was used for this measurement.

The pores within the nickel electrode are sufficiently small compared to the other components of the cell that they are nearly fully filled with electrolyte at all times. In the rest state, the nickel electrode typically has substantially more than 95% of its internal void volume filled with electrolyte. The smaller pores between the active-material particles have capillary pressures of hundreds of psi that maintain their electrolyte fill. The larger pores also tend to remain fully filled with electrolyte except during overcharge, when oxygen gas generated in the nickel electrode can displace a significant amount of electrolyte from the larger pores. The oxygen generated by overcharge tends to clear the largest pores of electrolyte when its pressure exceeds the pore capillary pressure, then streams through the cleared channels to the surfaces of the electrode during continued overcharge.

4.2.13 Thermodynamics of the Nickel Electrode

The thermodynamics of the nickel electrode are represented by the composite free energy change ΔG, enthalpy change ΔH, and entropy change ΔS for all of the electrochemical reactions that are occurring at a specific time in the electrode. For example, during recharge 80% of the reaction current may involve the conversion β^*-Ni(OH)$_2$ $\leftrightarrow$ β^*-NiOOH and the other 20% may involve the conversion β^*-NiOOH $\leftrightarrow$ β-NiOOH. For each reaction i, the following relationship exists between these thermodynamic parameters:

$$\Delta H_i = \Delta G_i + T\Delta S_i \qquad (4.16)$$

For electrochemical reactions, these thermodynamic parameters are typically expressed as potentials: $\Delta G = -nFE_r$, $\Delta S = dE_r/dT$, and $\Delta H = -nFE_{th}$. In these relationships n refers to the number of electrons involved in the electrochemical charge transfer reaction, F is Faraday's constant, E_r is the reversible potential, T is the absolute temperature, and E_{th} is the enthalpy of reaction expressed in terms of potential (otherwise known as the thermoneutral potential). As indicated in Eq. (4.17), the thermoneutral potential for each reaction may be expressed in terms of the reversible potential and its temperature derivative:

$$E_{th_i} = E_{r_i} - T\left(\frac{dE_{r_i}}{dT}\right) \qquad (4.17)$$

The heat Q produced by the nickel electrode as a result of current flow is then obtained from Eq. (4.18) as a function of the partial current I_i involved in reaction i.

$$Q = \sum_i I_i (E - E_{th_i}) \qquad (4.18)$$

118   The Nickel Electrode

Thus, when the electrode potential E equals the thermoneutral potential for a given reaction, no net heat is produced by that reaction. During recharge the nickel electrode absorbs heat as the recharge voltage is often less than the thermoneutral potential, and during discharge the electrode typically produces heat because the discharge voltage is below the thermoneutral potential. The oxygen evolution reaction customarily is viewed as producing heat in a battery cell as a result of overcharge. However, the overcharge heating does not arise from the nickel electrode, but from subsequent recombination of the oxygen gas on the negative electrode (hydrogen electrode in a nickel-hydrogen cell). Clearly this creates the possibility for latent thermal effects involving the production of oxygen at the nickel electrode followed by accumulation of the oxygen, and the subsequent release of its heat at a later time when it undergoes eventual recombination.

The thermoneutral potential for any given electrochemical reaction in the nickel electrode may be calculated from Eq. (4.17) using the Nernst equation for the reversible potential of that reaction. For example, Eq. (4.19) describes the reversible potential for the β-Ni(OH)$_2$ ↔ β-NiOOH process of Eq. (4.6), where the bracketed quantities refer to the activity of each chemical species relative to the standard state of unit activity.

$$E_r = E_o + \frac{RT}{F} \ln \frac{[H_2O][NiOOH]}{[OH^-]} \qquad (4.19)$$

The result, which is indicated in Eq. (4.20), is independent of temperature (to first order) because of cancellation of the temperature-dependent term in the reversible potential by a term with similar temperature dependence from the entropy. Thus, the thermoneutral potential of nickel electrode processes depends only on the standard potential and the entropy change (expressed as the temperature derivative of the standard potential).

$$E_{th} = E_o - T \frac{dE_o}{dT} \qquad (4.20)$$

Because each of the two terms on the right side of Eq. (4.20) have identical dependences on temperature, all temperature dependence disappears in the resulting thermoneutral potential, giving Eq. (4.21):

$$E_{th} = E_o^{T_s} - T_s \left(\frac{dE_o}{dT} \right)_{T_s} \qquad (4.21)$$

Thus, to first order, the thermoneutral potential is only a function of the standard potential at a given temperature T_s and the change in entropy at that same temperature. The values for the standard potential, entropy, and thermoneutral potential for the reversible reactions indicated in Fig. 4.20 are listed in Table 4.1 with different levels of cobalt added to the nickel active material (discussed in detail in the next section).

The standard potentials in Table 4.1 are based on unit activity for the reagents involved in the electrode reactions, and are relative to the standard hydrogen electrode. These standard potentials were determined using voltammetry as a function of cobalt level in the active material for the β and α reactions,[4.10] and are estimated for the β* reaction. To obtain the thermoneutral potential for a battery cell that uses the nickel electrode, the thermoneutral potential of the nickel electrode must be subtracted from that of the negative electrode. For example, the thermoneutral potential of the cadmium electrode is –0.507 V and that of the nickel electrode with 5% cobalt is 0.947 V, giving a thermoneutral potential for the nickel-cadmium cell of 1.454 V, which is in agreement with the best reported calorimetric measurement.[4.11]

Table 4.1. Thermodynamic Parameters for Nickel Electrode Reactions

Reaction	Reversible Potential (31% KOH)	Standard Potential	Entropy (mV/°C)	Thermoneutral potential (V)
β*-Ni(OH)$_2$ ↔ β*-NiOOH (0% cobalt)	0.554	0.508	–1.48	0.949
β-Ni(OH)$_2$ ↔ β-NiOOH (0% cobalt)	0.533	0.487	–1.581	0.958
α-Ni(OH)$_2$ ↔ γ-NiOOH (0% cobalt)	0.515	0.469	–1.52	0.921
β-Ni(OH)$_2$ ↔ β-NiOOH (5% cobalt)	0.530	0.484	–1.554	0.947
α-Ni(OH)$_2$ ↔ γ-NiOOH (5% cobalt)	0.510	0.464	–1.51	0.914
β*-Ni(OH)$_2$ ↔ β*-NiOOH (10% Co)	0.531	0.485	–1.43	0.911
β-Ni(OH)$_2$ ↔ β-NiOOH (10% cobalt)	0.527	0.481	–1.460	0.916
α-Ni(OH)$_2$ ↔ γ-NiOOH (10% cobalt)	0.504	0.458	–1.503	0.906

All the reactions in Table 4.1 have relatively similar thermoneutral potentials, within the uncertainty that is always present in these measurements. The values that are known most accurately are those for the β reactions and the α reaction at 10% Co, which tend to be the principal reactions that occur in nickel-cadmium and nickel-hydrogen cells. The uncertainty determined for the best-known thermoneutral potentials is typically ±0.01 V.[4.11] Other values are likely to have somewhat greater uncertainties.

4.3 Role of Additives and Contaminants in the Nickel Electrode

A wide range of additives has been used throughout the years to modify and improve the performance of the nickel electrode. Many have never been used in electrodes for nickel-hydrogen cells, and will not be discussed here, but have been reviewed previously. The additives that have been used in nickel-hydrogen cell positives include cobalt, cadmium, lithium, and platinum. Of these, cobalt is almost universally used at concentrations that are either 5% or 10% of the nickel in the active material. Platinum is an additive that is not intentionally added to the nickel electrode but migrates as platinate ions into it during the storage of nickel-hydrogen cells that have a nickel precharge. Thus, most modern nickel-hydrogen cells contain some level of platinum oxide in the active material. Two additives that can produce improvements in performance (but are not commonly used in nickel-hydrogen cell positive electrodes) are cadmium and lithium.

In addition, a number of contaminants can adversely affect electrode performance when present in the nickel electrode at levels that exceed specific concentration thresholds. These include silicates, iron, chlorides, fluorides, and nitrates.

4.3.1 Cadmium Additives

Cadmium is typically introduced into the nickel electrode in the form of cadmium hydroxide. The cadmium hydroxide can be introduced either by gradual incorporation into the nickel electrode active material from the electrolyte, or by the addition of some solid cadmium hydroxide. In either case the general mechanism for cadmium incorporation seems to be adsorption onto the surfaces of the nickel oxyhydroxide crystallites. This adsorption process can result in 2–3% cadmium by weight. While it occurs naturally over time in nickel-cadmium cells, in nickel-hydrogen cells the cadmium hydroxide must be added to the active material and stabilized into an adsorbed state before the nickel electrodes are built into a cell.

Cadmium hydroxide provides a clear beneficial effect when present in nickel electrodes. Treatment of the nickel electrode with cadmium poisons the oxygen evolution reaction that occurs on the surfaces of the active material in nickel electrodes as they go into overcharge.[4.12] This significantly increases the oxygen overpotential, as shown in Fig. 4.28.

4.3 Role of Additives and Contaminants in the Nickel Electrode

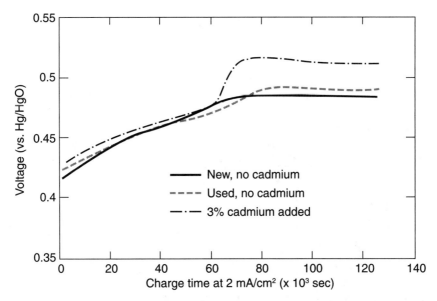

Figure 4.28. Effect of cadmium additive on the recharge voltage of nickel electrodes at 20°C in 31% potassium hydroxide electrolyte. The used electrode was removed from a nickel-cadmium cell.

At 20°C a cadmium level of 3% produces an increase of about 30 mV in the overcharge voltage and has very little effect on either the recharge voltage prior to overcharge or the discharge voltage.

The increase in oxygen evolution overpotential indicated in Fig. 4.28 enables the nickel electrode to be recharged with much-improved efficiency and reduces the rate of self-discharge by oxygen evolution at all states of charge. Because of its relatively high self-discharge rate, the presence of cadmium can significantly improve cell performance, particularly at elevated temperatures. However, if the recharge rate is relatively high, or if the temperature is substantially less than 10°C, the benefits of the cadmium additive become relatively small simply because the charge efficiency is quite good even without the cadmium additive under these conditions.

Cadmium also has a secondary effect of impeding the formation of γ-NiOOH from β-NiOOH, in spite of the higher electrode potential that the cadmium additive induces during overcharge. While this prevents nickel electrodes containing cadmium from readily enjoying the higher capacity offered by γ-NiOOH, it can significantly improve cycle life by minimizing the relatively large active-material volume change associated with the formation of γ-NiOOH.

However, cadmium additive has not been commonly used in nickel-hydrogen cells because of a concern that some cadmium hydroxide would detach from the surfaces within the nickel electrode, dissolve in the electrolyte, and migrate to the hydrogen electrode. At the hydrogen electrode, cadmium can act as a poison to impede the hydrogen reaction on the platinum catalyst, thus degrading cell performance by significantly raising the hydrogen overpotential and causing higher cell recharge voltages and lower cell discharge voltages. In addition, if the nickel electrode potential is allowed to fall to the hydrogen potential, it is possible to grow dendrites of cadmium metal from the nickel electrode to the hydrogen electrode by reducing the cadmium hydroxide in the nickel electrode. However, in the few instances where nickel-hydrogen cells have been made using cadmium-treated nickel electrodes, careful management of the cells has generally provided good cycle life.

4.3.2 Lithium

Lithium is typically added to nickel electrodes by introducing 2–3 weight percent lithium hydroxide into the potassium hydroxide electrolyte used in the cell. Lithium acts to both increase the oxygen overpotential and decrease the recharge voltage of the nickel electrode before it goes into overcharge. The combination of these two effects produces an increase in charge efficiency and a reduction in the self-discharge rate that is similar to that obtained with cadmium additives. Indeed, the effects of these two additives have been often combined in nickel-cadmium cells to significantly improve both performance and life.

Figure 4.29 shows the effect of lithium hydroxide as an electrolyte additive on the open-circuit voltage and the charge-transfer resistance of the nickel electrode. The lithium ions in the electrolyte undergo intercalation into the active-material lattice, and thus the additive changes the potential at which the active material is charged and discharged. Such intercalation can alter the propensity of the active material to convert to the γ-NiOOH phase during overcharge and prevent potassium ions from being incorporated into the active material.[4.13,4.14] These effects, along with modifications in the amount of interstitial water,[4.15] are likely to be responsible for the increase in oxygen evolution voltage typically seen when lithium is present. It is clear that the increased oxygen evolution voltage gives many of the same benefits from improved charge efficiency that are provided by the cadmium additive, albeit by a different mechanism.

The lithium additive also appears to reduce the charge-transfer resistance at low states of charge. Simply replacing the electrolyte with lithium-free potassium hydroxide can largely eliminate the effect of the lithium additive, and then the effect may be restored by again adding lithium. Thus, the interaction of lithium ions with the active-material lattice appears to be a reversible intercalation process that reduces the potential by about 35 mV. Examination of Fig. 4.29 also indicates that lithium exerts an irreversible residual effect on the depletion-layer resistance in the region of the second plateau below 0 V vs. Hg/HgO. This irreversible effect suggests that some lithium remains adsorbed on the surfaces of the active-

4.3 Role of Additives and Contaminants in the Nickel Electrode

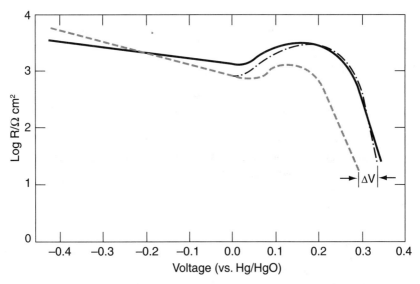

Figure 4.29. Charge-transfer resistance as a function of voltage during a C/100 discharge of residual capacity in a nickel electrode in 31% potassium hydroxide electrolyte: (a) without additives; (b) after addition of 3% lithium to the electrolyte; and (c) after removal of lithiated electrolyte. The indicated ΔV is the shift in open-circuit potential with lithium at a fractional state of charge of about 0.2.

material crystallites to influence the resistance where the depleted active material contacts the nickel metal substrate.

Lithium has not been used as an additive in nickel-hydrogen cells, except for experimental purposes. The potential drawback of the lithium additive is that it can degrade long-term cycle-life performance, perhaps by the formation of a lithium nickelate structure[4.16] that involves significant stresses from active-material volume changes. The severity of this effect seems to be variable,[4.17] depending on electrode and active-material structure and composition, the electrolyte concentration in the cell, and charge/discharge conditions. The reduced cycle life makes this additive somewhat undesirable in cells that have long cycle-life applications, such as nickel-hydrogen cells.

4.3.3 Cobalt

Cobalt is universally used as an additive in nickel electrodes that are found in nickel-hydrogen cells. It is present in the active material as a cobalt oxyhydroxide at either 5% or 10% of the nickel hydroxide content, depending on the process by which the nickel electrodes are manufactured. A number of mechanisms have been identified by which cobalt additive improves the performance of nickel electrodes. Cobalt significantly improves charge efficiency by lowering the active-material recharge voltage and by decreasing the rate of oxygen evolution at a given electrode voltage. Cobalt additives also increase the conductivity of the active material, enabling a more complete discharge of the capacity stored within it. In addition, cobalt additives can influence the tendency to form γ-NiOOH during charge by reducing the voltage at which it is formed relative to the oxygen evolution voltage, as well as improving the mobility of protons in the active-material lattice. These cobalt-additive–induced effects can significantly change the tendency of nickel electrodes to discharge a portion of their capacity on the second-plateau voltage.

The electrochemical and chemical processes involved in codeposition of 5–10% cobalt in nickel hydroxide active material leave the cobalt additive as a solid solution in the nickel oxyhydroxide lattice that fills the pores of sintered nickel electrodes. In a properly made nickel electrode there is no chemical or physical evidence of isolated cobalt compounds in the active-material lattice, because the cobalt-containing sites are incorporated into the nickel hydroxide lattice. X-ray diffraction patterns of the active material in sintered nickel electrodes can be used to confirm that the cobalt additive is completely incorporated into the nickel oxyhydroxide lattice.

One key effect that the cobalt additive has on the electrochemistry of nickel hydroxide is that it decreases the reversible potentials for the charge and discharge reactions of the active phases, as is indicated in Fig. 4.30, where the regions of stability for the various nickel-cobalt oxyhydroxides are shown. While the decrease

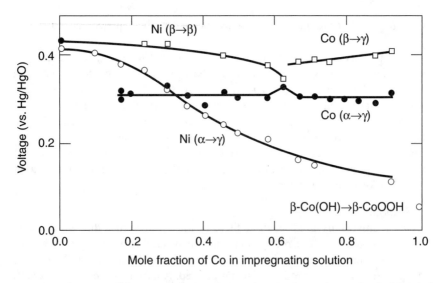

Figure 4.30. Reversible potentials for electrochemical reactions of cobalt/nickel oxyhydroxides as a function of composition.

4.3 Role of Additives and Contaminants in the Nickel Electrode   125

in reversible potential is only 10–20 mV for the 5–10% cobalt composition range used in nickel electrodes, this voltage shift has a significant influence on electrode performance. Because the reversible potentials of the β- and γ-oxyhydroxide phases are shifted downward, and because cobalt additives also shift the oxygen evolution potential upward, the nickel electrode charges with much less parasitic oxygen evolution when cobalt additives are present. This enables a higher state of charge to be attained by a nickel electrode containing cobalt, and thus a higher specific energy for the cell. Well-made nickel electrodes can gain up to 20% capacity as a consequence of having the cobalt additive.

The decrease in the reversible potential resulting from cobalt additive is caused by the presence of cobalt defect sites in the active-material crystalline lattice, defects that have reduced electronic energy levels that lower the electronic band structure of the lattice as a whole. These defect sites generally remain in a local trivalent state (CoOOH) and thus produce a decrease in the lattice redox potential. As indicated in Fig. 4.30, the reduction in reversible potential becomes progressively greater as the level of cobalt is increased. At cobalt levels greater than about 20%, however, the cobalt-containing lattice sites do not remain completely dissolved in the active-material lattice during long-term charge/discharge cycling. The result is that some of the cobalt additive can separate from the lattice to form discrete crystallites of the phases indicated in Fig. 4.30 that are stable at higher cobalt levels. These cobalt-rich phases may have reduced electrochemical activity, or may be electrochemically inert, and in some cases can be electronic insulators. Separation of cobalt phases from the active material invariably leads to degraded nickel electrode performance. Such phase separation can be prevented if the level of cobalt is kept less than about 15%, and if the nickel electrode potential is maintained higher than the reversible potential of the trivalent cobalt sites in the active-material lattice (greater than about 1.05 V *vs.* hydrogen).

The β-nickel hydroxide lattice can be restructured into a stable superlattice containing one cobalt site for every nine overall lattice sites,[4.18] as illustrated in Fig. 4.31. This superlattice structure thus has an 11.1% cobalt level. The cobalt superlattice has a significantly larger unit cell than does the normal β-nickel hydroxide lattice, and it displays a significantly lower charge and discharge voltage, as shown in Fig. 4.31.

The nickel electrode performance shown in Fig. 4.31 corresponds to a situation where an electrode having nominal cobalt content of 10% ended up with about 20–25% of its active material structured into a nickel/cobalt superlattice. It should be noted that in this particular electrode the actual average cobalt level was close to 11%. The slightly higher-than-nominal cobalt level is responsible for the significant domains of the superlattice that were formed during the active-material deposition.

Special processing would likely be required to produce a nickel electrode in which the majority of the active-material crystallite domains consist of the super

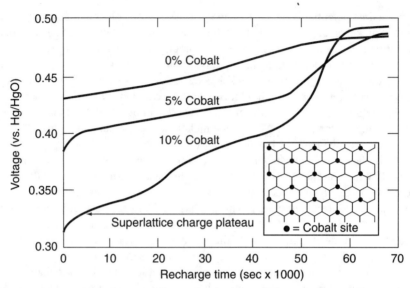

Figure 4.31. Lower charge-voltage plateau caused by cobalt superlattice in a 10% cobalt nickel electrode.

lattice shown in Fig. 4.31. However, those electrodes in which a significant fraction of the active material is found in the form of a nickel/cobalt superlattice have tended to exhibit exceptional capacity and performance stability during long-term cycling. Analyses of nickel electrodes have not been performed after long-term cycling in a nickel-hydrogen cell to determine how well the superlattice holds up throughout years of charge/discharge operation.

Cobalt additives in sintered nickel electrodes also influence the oxygen evolution behavior of the electrode. Cobalt tends to increase the potential at which oxygen is evolved, thus enabling the electrochemical reactions involved in charging the electrode to occur more completely before becoming dominated by oxygen evolution. Figure 4.32 shows the effect of cobalt on the oxygen evolution voltage level.

The increase in charge efficiency resulting from a 10–20 mV increase in oxygen evolution potential is significant. However, by coupling the higher oxygen evolution voltage with lower redox potentials for the active material, cobalt additives can enable a significant fraction of the active material to be efficiently converted to γ-NiOOH during recharge. This is particularly true at low temperatures, which also tend to preferentially increase the oxygen evolution potential more than the active-material recharge potential. While the facile formation of large amounts of γ-NiOOH can leave the nickel electrode with a very high state of charge (up to 167% utilization is possible), frequent charge and discharge of the capacity stored in the γ-NiOOH phase can induce large active-material volume changes, which will tend to reduce cycle life. An ideal use of the γ-NiOOH capacity

4.3 Role of Additives and Contaminants in the Nickel Electrode 127

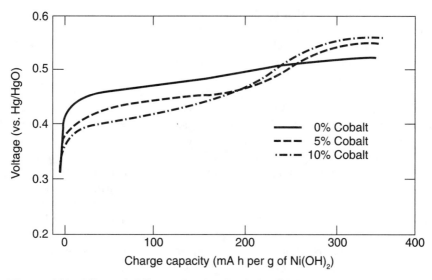

Figure 4.32. Effect of different levels of cobalt additive on the recharge and oxygen evolution voltage levels in nickel electrodes. The temperature is 20°C.

is as a reservoir of capacity that is maintained in the battery cells for discharge only in an emergency situation, while the capacity that is frequently cycled involves only charge and discharge of the β-NiOOH phase.

Because cobalt additives provide an abundance of trivalent defect sites in the nickel oxyhydroxide lattice, the electronic conductivity of the active material can be maintained more effectively during discharge. The stable trivalent defect sites in the lattice act to maintain the electronic degeneracy of the semiconducting lattice until essentially all the charged nickel sites have been discharged. Without the cobalt defects, up to 10–15% of the trivalent nickel sites would have to remain charged to allow the lattice to maintain its electronic conductivity. The trivalent cobalt sites in the lattice also provide a source of mobile ionic holes (proton vacancies) that act to maintain the ionic conductivity of the solid-state lattice as it reaches low states of charge.

The combination of improved electronic conductivity and improved ionic conductivity enables a significantly greater fraction of the capacity stored in the nickel electrode to be discharged at usable voltages and currents when cobalt is present. Figure 4.15 shows the role that improved proton-hole conductivity can play in enabling the effective recharge or discharge of a nickel electrode in spite of having very poor electronic conductivity. The interplay between the electronic and the ionic conductivity of the layers of charged and discharged β and γ phases typically present in a cycled nickel electrode can result in a wide range of performance characteristics as the operating conditions of the electrode are changed.

An additional role that cobalt additives play in many commercial nickel electrodes used in nickel-cadmium and nickel metal hydride cells involves the deposition of conductive CoOOH on the surfaces of the active-material crystallites. Cobalt-containing deposits of this kind improve electrode performance by maintaining an electronically conducting network throughout the active-material particles as the electrode is discharged. This function of cobalt additives is not operative in the sintered nickel electrodes used in nickel-hydrogen cells, largely because the nickel sinter matrix has been found to provide a conductive matrix having superior long-term physical and electrical stability.

The principal drawbacks of cobalt additives are caused by reactions that can strip the cobalt from the solid solution in the active material, or by corrosion processes that can gradually cause accumulation of layers of cobalt-depleted active material near corroding nickel metal surfaces. The cobalt additives in sintered nickel electrodes are stable as long as the electrode potential is maintained above the approximately 1.0 V (vs. H_2) potential where Co(III) is reduced to Co(II). Below this potential Co(II) can slowly form (this reaction is very slow because the active material loses all its electronic conductivity when the Co(III) defects are lost), as indicated in Eq. (4.22):

$$CoOOH + H_2O + e^- \longleftrightarrow Co(OH)_2 + OH^- \qquad E_o = 1.00 \text{ (vs. } H_2) \qquad (4.22)$$

Because Co(II) hydroxide is slightly soluble in potassium hydroxide electrolyte, forming the dicobaltite ion by reaction (4.23), the cobalt will tend to dissolve from the regions where it has been reduced to the divalent state, leaving a cobalt-depleted layer.

$$Co(OH)_2 + OH^- \longleftrightarrow HCoO_2^- + H_2O \qquad (4.23)$$

In practice, the cobalt-depleted layer resulting from Co(III) being reduced to Co(II) is so thin that it has little impact on performance, except in situations where the nickel electrode is allowed to fall below 1.0 V (vs. H_2) for very long periods (i.e., years).

A more troublesome process that can very rapidly degrade cobalt additives and electrode performance[4.19] occurs when the nickel electrode is held below about 0.10 V (vs. H_2). Below this potential range, Ni(II) and Co(II) lattice sites are reduced to the metallic state, leaving a very finely divided and highly reactive layer of metal particles. These processes are indicated by Eqs. (4.24) and (4.25):

$$Ni(OH)_2 + 2e^- \longleftrightarrow Ni(m) + 2OH^- \qquad E_o = 0.09 \text{ (vs. } H_2) \qquad (4.24)$$

4.3 Role of Additives and Contaminants in the Nickel Electrode 129

$$Co(OH)_2 + 2e^- \longleftrightarrow Co(m) + 2OH^- \quad E_o = 0.10 \ (vs. \ H_2) \quad (4.25)$$

Because the particulate products are metallic, they have good electronic conductivity and therefore enable facile reaction access to additional lattice sites within the active material. Thus, this reaction can propagate into the active material fairly rapidly to generate a relatively thick layer of modified active material. Layers with a thickness on the order of several microns have been observed to form in several months' time in electrodes held at 0 V vs. H_2. Subsequent electrochemical oxidation (during electrode recharge) of the reactive metal particulates to nickel and cobalt hydroxides can allow the cobalt hydroxide to dissolve in the potassium hydroxide electrolyte, and leave behind a thick layer of cobalt-depleted active material. This process has been observed in nickel-hydrogen cells to result in the loss of significant capacity in as little as two weeks' time. Because the cobalt-depleted layer is relatively thick (several microns or more), it can have a major effect on electrode capacity, producing up to a 30% capacity loss. Capacity loss resulting from this type of thick cobalt-depleted region is generally difficult to recover. While repeated high-rate cycling has, in some cases, restored capacity (probably by physically remixing the layers), the long-term performance of such electrodes has been found to be significantly degraded in subsequent life tests.[4.20]

The capacity loss processes described here result when layers of cobalt-depleted active material occur in the region that directly contacts the current-collecting sinter. It is actually possible to turn the physics responsible for this capacity loss into a performance improvement of the nickel electrode. The active material can be intentionally deposited containing cobalt additive such that the level of cobalt at the interface with the current collector is significantly higher than the level found at locations far from the current collector. Nickel electrodes built in this way are referred to as "gradient nickel electrodes," and they can exhibit significant improvements in capacity.[4.21,4.22] This concept is discussed in more detail in a later section on advanced nickel electrodes.

4.3.4 Platinum

Platinum is not an additive to the nickel electrode in the strict sense. It is a material that is used as a catalyst in the hydrogen electrode, and in some situations, it can oxidize and migrate to the nickel electrode, where a range of chemical processes can occur to modify the performance of the nickel electrode. At the normal operating potential of the nickel electrode in a nickel-hydrogen cell that is above a cell voltage of 1.1 V, the platinum catalyst in the hydrogen electrode is completely stable because of the presence of hydrogen gas in the cell. However, if the hydrogen gas is depleted, and the hydrogen electrode is allowed to rise close to the potential of the partially charged nickel electrode, the platinum catalyst will undergo slow oxidation to form platinum oxide on the surfaces of the catalyst particles, as indicated by Eq. (4.26):

$$Pt(m) + 2OH^- \longleftrightarrow PtO + H_2O + 2e^- \quad E_o = 0.98 \; (vs. \; H_2) \tag{4.26}$$

Platinum oxide is slightly soluble in the potassium hydroxide electrolyte, as indicated in Eq. (4.27); thus platinum oxidation will continue until the electrolyte becomes saturated with platinate ions, $Pt(OH)_6^=$:

$$PtO + H_2O + 4OH^- \longleftrightarrow Pt(OH)_6^= \tag{4.27}$$

This situation typically occurs when a nickel-precharged nickel-hydrogen cell is fully discharged, enabling the hydrogen electrode to consume all the hydrogen gas in the cell and ultimately be forced by traces of oxygen gas to a potential within 0.2 V of the partially charged nickel electrode. In this condition, the open-circuited cell potential is typically 0.2–0.3 V. During long-term storage in this condition, the platinate-saturated electrolyte is free to chemically interact with the active material in the nickel electrode.

After partially charged nickel electrodes experience long-term exposure to platinate-saturated potassium hydroxide electrolyte, a compound that is not normally present in the nickel electrode is frequently detected.[4.23] This compound appears to form by incorporation of platinum as a hydroxide into the γ-NiOOH at sites occupied by cobalt additives. The formation of this compound is shown in Eq. (4.28):

$$Pt(OH)_6^= + CoOOH + NiOOH \longrightarrow Pt(OH)_6[CoO][NiO] + 2OH^- \tag{4.28}$$

This compound is not formed in nickel electrodes that contain no cobalt additive. This Ni-Co-Pt oxyhydroxide compound is easily detected by its characteristic reduction reaction at about 1.09 V (vs. H_2, or 0.16 V vs. Hg/HgO), as indicated in Fig. 4.33 and Eq. (4.29):

$$Pt(OH)_6[CoO][NiO] + H_2O + e^- \longleftrightarrow Pt(OH)_6[CoOOH][Ni(OH)_2] + OH^- \tag{4.29}$$

The Ni-Co-Pt complex has been found to contain platinum at levels up to several percent of the weight of the active material in the nickel electrode. This complex appears to be indefinitely stable in its oxidized form, which is how it normally forms in the nickel electrode. If the Ni-Co-Pt complex is reduced by complete discharge of the nickel electrode below 1.09 V (vs. H_2) by the reaction described in Eq. (4.29), it becomes unstable and begins to slowly disproportionate.

4.3 Role of Additives and Contaminants in the Nickel Electrode

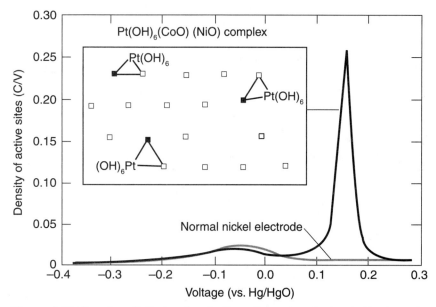

Figure 4.33. Density of electrochemically active states during reduction of a normal nickel electrode and an electrode containing a platinum complex. The inset shows a possible structure for the platinum complex.

It releases the platinum back into the electrolyte. This release of the platinum typically takes 12–24 h to fully occur. If the nickel electrode is recharged before the platinum complex can disproportionate, an oxidation reaction corresponding to the reverse of reaction (4.29) can be seen just above 1.09 V. This oxidation reaction corresponds to the electrochemical regeneration of the stable oxidized form of the Ni-Co-Pt complex from the reduced complex that has not decomposed.

The presence of a stable platinum-containing complex in the nickel electrode active material can have several effects on the performance of the nickel electrode. First, the presence of the platinum-containing complex appears to increase the capacity of the nickel electrode by 3–5%, as shown in Fig. 4.34. The capacity increase disappears if the nickel electrode is fully discharged, because complete discharge generally causes the platinum-containing complex to dissociate. The effect of platinum on the capacity is typically manifested by measurement of a higher-than-normal cell capacity during the first capacity cycle following long-term storage of a nickel-precharged nickel-hydrogen cell. Repeated capacity cycles will show a reduced (and more normal) capacity following the first cycle, as indicated in Fig. 4.34.

A secondary effect that can occur in nickel electrodes with significant amounts of platinum-containing complex can be initiated by the release of that

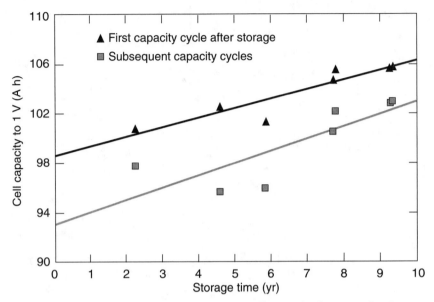

Figure 4.34. Difference in nickel-hydrogen cell capacity between the first capacity cycle following long-term storage and subsequent capacity cycles.

platinum back into the electrolyte when the nickel electrode is fully discharged. If the nickel electrode is discharged below ~1.0 V *vs.* H_2 with platinate ions present in the electrolyte, platinum metal can be plated onto the nickel electrode. The platinum metal on the nickel electrode will remain during subsequent operation until it is gradually oxidized, redissolves in the electrolyte, and is replated back onto the hydrogen electrode as platinum metal.

This oxidation process can take several weeks of nickel electrode operation at normal nickel-hydrogen cell potentials. During this period of time the platinum metal on the nickel electrode can cause a significantly increased self-discharge rate, poor charge efficiency, and significant reduction in cell capacity. This situation can be avoided by refraining from discharge of the nickel electrode below 1.10 V *vs.* H_2. If this situation does occur in a nickel-hydrogen cell, the cure is to simply hold the charged cell on trickle charge at a low rate for several weeks, which allows the platinum to be oxidized and removed from the nickel electrode.

4.3.5 Nickel Electrode Contaminants

A number of contaminants have been found to influence the performance of nickel electrodes in nickel-hydrogen cells. These contaminants fall into several classes. The first general category consists of materials that influence the oxygen evolution behavior of the nickel electrode, and typically includes metallic contaminants such as iron and copper. The second category of contaminants involves materials

4.3 Role of Additives and Contaminants in the Nickel Electrode 133

that accelerate the rate of sinter corrosion during nickel-hydrogen cell operation, and includes anionic contaminants such as chloride, sulfate, and nitrate. The third category involves organic materials and silicates that can permeate the porous structure of the active material within the nickel plaque, thereby blocking its facile charge and discharge.

4.3.5.1 Contaminants That Influence the Nickel Electrode's Oxygen Evolution
Contaminants such as copper and iron typically reside in the nickel electrode as oxides that may be either dispersed uniformly on the surfaces of the active-material crystallites or concentrated in localized regions as contaminating particles within the active material, depending on the source of the contaminant. These oxides are generally not directly incorporated into the lattice structure of the nickel oxyhydroxide active materials, owing to the mismatch between the lattice characteristics of these metallic oxides and those of the stable nickel oxyhydroxide phases. The contaminating oxides thus exist as particulates in the active material, or as oxide domains adsorbed onto the surfaces of the active nickel oxyhydroxide crystallites.

These surface metal oxide domains can act as catalytic oxygen evolution sites that can significantly reduce the oxygen evolution potential of the active material if they are widely present at the interface between the active material and the electrolyte. Bulk levels of 0.1% or more can be sufficient to significantly affect the oxygen evolution potential. Iron, in particular, is quite effective in decreasing the oxygen potential. Iron levels greater than 1% in the nickel electrode active material can reduce the oxygen potential 30–40 mV, resulting in large reductions in charge efficiency and nickel electrode capacity.

Copper is a contaminant that typically gets into nickel-hydrogen cells and nickel electrodes from copper wires or electrodes used in electrode or cell production processes, and therefore it is often initially present within the cell in metallic form. When the copper is in contact with the charged nickel electrode in potassium hydroxide electrolyte, it is oxidized to copper oxide. This oxide material can react with the electrolyte to form copper hydroxide. Copper hydroxide has some solubility in the electrolyte, thus enabling the copper to migrate to and be dendritically plated onto the negative electrode of the nickel-hydrogen cell. Such plating can lead to cell short circuits. Copper oxides that remain within the nickel electrode can cause a significant reduction in the oxygen potential, even at levels of several hundred ppm or less.

4.3.5.2 Materials That Accelerate the Rate of Sinter Corrosion
A second type of contaminant consists of corrosion-enhancing materials such as chloride, nitrate, or sulfate species. These materials can get into nickel electrodes from the impregnation solutions, from a contaminated or dirty production environment, or from other cell components such as asbestos separator (which frequently contains a mix of minerals having a wide range of composition).

Chloride, fluoride, or nitrate contaminants, in particular, can significantly increase the rate of sinter corrosion in the nickel electrode, as indicated in Fig. 4.35. These contaminants can lead to the formation of green corrosion spots on nickel electrodes during long-term storage.

The slow corrosion of the nickel sinter used in nickel electrodes is a normal degradation process that cannot be prevented. While this process is quite slow, eventually it can limit the life of well-managed nickel-hydrogen cells after more than 20 yr of operation. Thus, a 25–50% increase in the rate of corrosion can result in a 5–10 yr reduction in the ultimate cell-life capability. For this reason, close attention to minimizing the level of all contaminants that can increase the rate of corrosion is most important for nickel electrodes in nickel-hydrogen cells.

Sulfate contaminants often arise from sulfate-containing minerals from dirt particles or asbestos separator, or from reaction of the nickel hydroxide active material with sulfur dioxide air pollutants during storage. Sulfate species not only act to increase the rate of sinter corrosion, but also can cause the slow formation of a cobalt sulfate complex that is slightly soluble in potassium hydroxide electrolyte. Throughout long-term periods of nickel electrode operation in test cells that contain significant sulfate levels, significant loss of cobalt from the nickel electrode has been seen to occur as a result of complexation with sulfate.[4.24] Figure 4.36 shows the effect that this process can have on the performance of nickel electrodes.

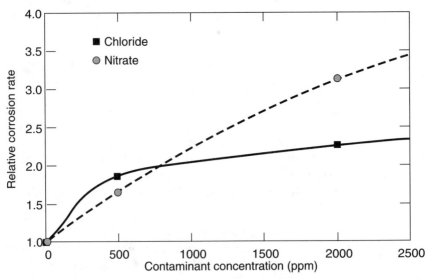

Figure 4.35. Relative corrosion rate of nickel sinter in 31% potassium hydroxide as a function of chloride and nitrate contamination levels.

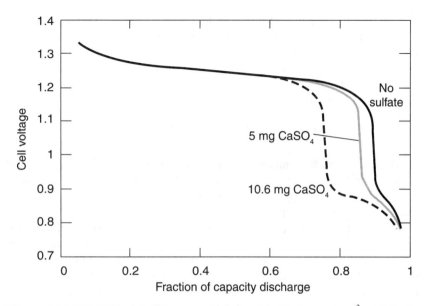

Figure 4.36. The effect of different sulfate levels added to a 4 cm^2 nickel electrode in a starved nickel-hydrogen cell containing 31% potassium hydroxide electrolyte.

4.3.5.3 Organic Materials and Silicates That Permeate the Active Material's Porous Structure

The final class of contaminants includes silicates, as well as organic species such as oils, greases, and polymers that can be deposited on the surfaces of the nickel electrode or in the active material and block the facile access of electrolyte to the active material. This effect includes both physical isolation of active-material crystallites, and decreased wettability and electrolytic conductivity to the charge-transfer sites.

The common contributing factors for such contaminants include oil-laden aerosols and handling in the electrode production and cell assembly environment, as well as poorly controlled electrode-processing procedures. For example, a polymer coating is often applied to the edges of nickel electrodes to prevent particulate flaking, which could lead to short circuits within a cell. Excessive edge coating or improper polymer viscosity can lead to deactivation of significant amounts of electrode active material.

While many organic contaminants are inert in the nickel electrode, some can be slowly oxidized at the operating potential of the nickel electrode. Any organic contaminants that are gradually oxidized on the nickel electrode in a nickel-hydrogen cell will cause a reduction in the level of nickel precharge in the cell.

One specific contaminant that fits into this final class of materials is soluble silicate, which can leach from a wide variety of silica-containing contaminating minerals in potassium hydroxide electrolyte. Most silicate-containing minerals react slowly with the hydroxide ions in the electrolyte, and this results in buildup of the silicate ion concentration, while the minerals are converted to the more stable hydroxides or oxides as indicated in Eq. (4.30):

$$MgSiO_3 + 2OH^- \longrightarrow Mg(OH)_6 + SiO_3^= \qquad (4.30)$$

These contaminating minerals often arise from accumulations of dust or dirt on the electrodes or other cell components. It is typically difficult to avoid low levels (less than 50–100 ppm) of silicate contaminants without carrying out electrode and cell processing operations in a clean-room environment.

Fortunately, silicate levels do not have a severe effect on nickel electrode performance until the concentration in the alkaline electrolyte exceeds about 0.5% by weight. When the concentration approaches or exceeds this level, it becomes possible to form high-impedance silicate deposits in the nickel electrode during high-rate discharge, particularly when the temperature is low (< 5°C). Silicate ions appear to adsorb onto the surfaces of the active-material crystallites, where their limited mobility and their affinity for water can effectively block the water molecules that are involved in the charge-transfer reactions from facile movement between the active-material surface and the electrolyte. The signature produced by high levels of silicate is a severely depressed discharge voltage, typically combined with a higher-than-normal cell impedance, as shown in Fig. 4.37.

The source of silicate contamination that has historically been the most significant in nickel-hydrogen cells is asbestos separator, which was used in a number of early cell designs. Asbestos is a mixture of various (largely magnesium) silicate minerals that can have a somewhat variable composition if not reprocessed or purified. All of these silicate minerals (more than 200 are known in asbestos) react with alkaline electrolyte to produce soluble silicate in the electrolyte and magnesium hydroxide. Some of them react quickly, while others react very slowly. The most desirable minerals for asbestos separator are those that react very slowly, in some cases requiring decades of exposure to the electrolyte before significant amounts of reaction have occurred. However, in some instances, more reactive minerals have been present in asbestos, leading to significant silicate buildup in the electrolyte in a period of several years or less. It must also be recognized that the asbestos decomposition process will lead to a hydroxide powder-containing separator that has much greater impedance than the more desirable silicate fiber structure of asbestos.

Potassium silicate levels of 0.5 to 1% by weight in the electrolyte will significantly degrade nickel electrode performance. The highest level that has been

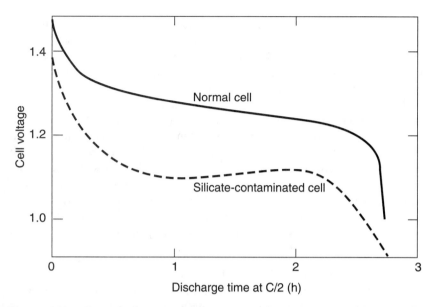

Figure 4.37. Effect of silicate contamination on the discharge voltage of a nickel electrode in a nickel-hydrogen cell at 0°C.

measured in the electrolyte in an asbestos-containing nickel-hydrogen cell is about 4% potassium silicate by weight, which was measured after cell failure as a result of high resistance following 3.5 yr of operation in a life test. This level corresponds to about 15% of the potassium hydroxide in the cell having reacted with silicate minerals in the asbestos. If nickel electrodes from such a cell are removed from the cell, rinsed well to remove the soluble silicate, and then operated in an uncontaminated potassium hydroxide electrolyte, they will return to normal performance.

4.4. Interactions of Hydrogen Gas with the Nickel Electrode

The self-discharge of nickel electrodes in nickel-hydrogen cells is significantly faster than in battery cells that do not contain hydrogen gas at high pressure, such as nickel-cadmium or nickel metal hydride cells. The reason for this difference is the significant effect that reactions involving hydrogen gas have on the behavior of the nickel electrode. In general, there are four types of reactions whereby hydrogen gas can influence the behavior of the nickel electrode.

4.4.1 Recombination of Evolved Oxygen with Hydrogen Gas

The first of these reactions is the relatively rapid recombination of oxygen evolved from the nickel electrode with hydrogen gas within the nickel-hydrogen cell. Because the oxygen evolution process is relatively irreversible at the nickel electrode, this process does not significantly increase the oxygen evolution rate or decrease the charge efficiency of the nickel electrode. However, it does ensure that all the heat that can be produced as a result of oxygen evolution (overcharge) will be generated quickly, typically with a time constant of about 15 min or less.

This is not the case in nickel-cadmium cells, where significant latent heat from overcharge can be stored in the form of oxygen gas, which can build up to high pressures and then recombine during many hours to release the heat from overcharge. In nickel metal hydride cells, the high hydrogen pressure and highly active catalyst available in nickel-hydrogen cells are not present to ensure the rapid release of all overcharge heat. The rate of the oxygen evolution process increases exponentially with nickel electrode voltage, and the current follows Tafel dependence on the oxygen overpotential. The oxygen overpotential is the electrode voltage in excess of the standard oxygen potential, which is ~1.30 V vs. H_2.

4.4.2 Direct Reaction of Hydrogen with Charged Active Material

The second type of reaction in the nickel electrode that is influenced by hydrogen gas is generally referred to as the "direct" reaction of hydrogen with the charged active material. This process is best thought of as a direct chemical self-discharge reaction between hydrogen (whether gaseous or dissolved in the electrolyte) and the charged sites in the active material. It does not require any electronic conduction by the current-collecting structures within the sintered nickel electrode. This is actually the dominant self-discharge process at intermediate states of charge ranging from about 10% up to 80%, as indicated in Fig. 4.38. Above about 80% state of charge, self-discharge by oxygen evolution from the charged active material tends to be dominant.

The direct self-discharge process is responsible for the linear increase in the self-discharge rate with increasing nickel electrode state of charge shown in Fig. 4.38. The linear dependence arises from the linear increase in the number of charged reaction sites in the active material as the state of charge increases. Because hydrogen gas can react with charged cobalt sites in the active material as well as charged nickel sites, the state-of-charge controlling rate of this reaction includes all active-material sites, nickel and cobalt, that are above the divalent oxidation state. In addition, the rate of this direct self-discharge reaction is directly proportional to the hydrogen pressure for typical operating conditions in nickel-hydrogen cells.

The direct self-discharge process increases in rate as the temperature is increased, as shown in the Arrhenius plot of Fig. 4.39. This temperature dependence is the principal reason for the high sensitivity of the nickel-hydrogen cell capacity to temperature. The activation energy for this process is 4.90 kcal/mole, a value consistent with a chemical recombination process that is rate-limited by diffusion

4.4. Interactions of Hydrogen Gas with the Nickel Electrode

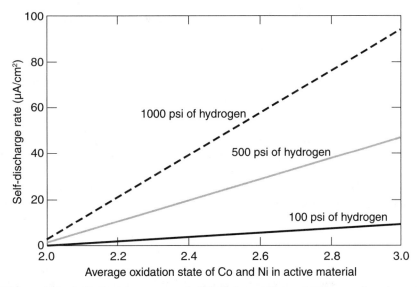

Figure 4.38. Self-discharge rate of nickel electrodes at 10°C as a function of active-material oxidation state at different constant hydrogen pressures (pressure is not normally constant in a nickel-hydrogen cell).

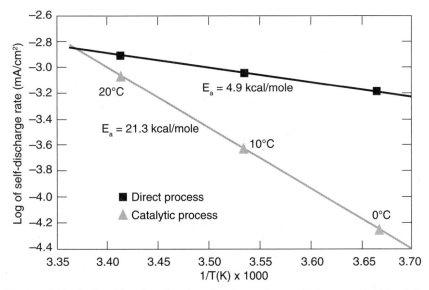

Figure 4.39. Arrhenius plot showing the dependence of the rates for the "direct" and "catalytic" self-discharge processes on temperature at an average active-material oxidation state of 2.10 and at 100 psi of hydrogen pressure.

140   The Nickel Electrode

of dissolved hydrogen to the charged sites in the active material. At hydrogen pressures greater than 500 psi and temperatures above 20°C, the self-discharge rate resulting from the direct process has been observed to reach a limiting rate that does not increase as the pressure is further increased.[4.25] This limiting behavior is likely to occur when the hydrogen diffusion rate becomes great enough that it ceases to limit the rate of hydrogen reaction with the charged sites.

4.4.3 Catalytic Self-Discharge Process
The third type of self-discharge process is indicated by the nonzero self-discharge rate when the rate shown in Fig. 4.38 is extrapolated to a fully discharged active-material oxidation state. There is a self-discharge process that occurs at a low rate that is independent of the amount of charged active material. This is referred to as the "catalytic" self-discharge process. It arises from the low catalytic activity of the oxide-covered nickel sinter surfaces, which can act as a very poor hydrogen electrode. While the rate of this catalytic process is quite low relative to that of activated (unoxidized) nickel particles, it is an important component of self-discharge in the nickel electrode at low states of charge if there is a significant hydrogen pressure present. It is this catalytic process that causes nickel electrodes to self-discharge completely to the hydrogen potential (0 V vs. H_2) in nickel-hydrogen cells containing excess hydrogen gas (hydrogen-precharged cells).

The catalytic self-discharge process involves the oxidation of hydrogen atoms formed at catalytic sites on the nickel surface from hydrogen gas, and it has an activation energy of 21.3 kcal/mole. This oxidation process injects a proton into a vacancy in the nickel oxyhydroxide lattice and leaves an electron in the conduction band of the metal current collector. Subsequent recombination of the electron with a protonated lattice site results in the discharge (reduction) of that site to the divalent oxidation state. If the electron cannot find its way to a trivalent protonated site, as will eventually occur when the active material in contact with the sinter becomes fully discharged, the current-collector potential will be forced down to the hydrogen potential and the proton will begin to reduce the oxide and hydroxide materials that are in contact with the current collector to the metallic state.

4.4.4 Reduction of Divalent Oxides and Hydroxides
The reduction to the metallic state of the divalent oxides and hydroxides in contact with the nickel sinter surfaces constitutes the fourth type of reaction in the nickel electrode involving hydrogen gas. These processes first reduce the compact protective oxide passivation layer on the nickel sinter by the reverse of the reaction that is Eq. (4.31), then reduce the more porous hydroxide active materials where they contact the nickel sinter by reaction (4.24).

4.4. Interactions of Hydrogen Gas with the Nickel Electrode

$$Ni(m) + 2OH^- \longleftrightarrow NiO + H_2O + 2e^- \qquad E_o = 0.28 \text{ v } (vs.\ H_2)$$
(4.31)

Because this process exposes catalytic sites on the nickel sinter to hydrogen gas by removing the passivating oxide and hydroxide layers, it increases the catalytic reduction rate significantly, as indicated in Fig. 4.40.

One consequence of the process by which hydrogen gas can catalytically reduce active materials in nickel electrodes to finely divided metal particulates is that if this reaction is allowed to proceed far enough, a Raney nickel electrode will be produced that has very high catalytic activity. This electrode will function as a hydrogen electrode rather than as a nickel electrode when attempts are made to recharge it. In a nickel-hydrogen cell, this will result in an apparent chemical short circuit that is caused when the hydrogen generated at the negative cell electrode will be electrochemically recombined at the Raney nickel electrode as fast as it is formed.

When a chemical short circuit of this type forms, the only way to restore normal nickel electrode operation is to increase the recharge current high enough to

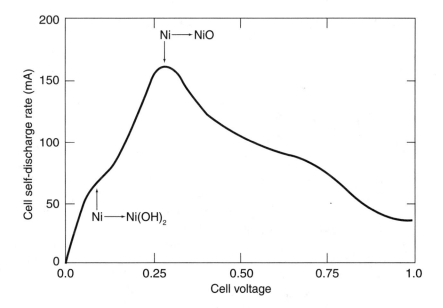

Figure 4.40. Passivation of the hydrogen catalytic activity of nickel electrodes in hydrogen-containing nickel-hydrogen cells below 0.28 V at room temperature. As the voltage is increased, nickel hydroxide is first formed at 0.1 V, then a compact passivation layer of nickel oxide at 0.28 V.

raise the overpotential above the threshold for passivating the nickel surfaces. As indicated in Fig. 4.40, the threshold for passivation is 0.2 to 0.28 V *vs.* H_2. This sequence of reduction and reoxidation processes has been verified[4.25] by reducing a nickel electrode in 1000 psi of hydrogen for 21 days. After this reduction time, the nickel electrode could be continuously recharged at a C/50 rate for days without having its voltage rise over 0.15 V *vs.* H_2. However, increasing the recharge rate to C/25 brought the potential to 0.28 V in less than 5 min, allowed the nickel sinter to be fully repassivated within several hours, and enabled the electrode to subsequently charge and discharge at its normal operating potential.

4.5. Storage of Nickel Electrodes

Nickel electrodes can be stored either in a dry state, where they are not in a cell that has been activated with electrolyte, or in an activated cell. In a dry, unactivated condition, the chemical processes that can occur within the electrodes during storage are limited. If the electrodes are partially charged, they can slowly evolve oxygen and self-discharge until the state of charge of the active material falls to a level where it is stable and ceases to evolve oxygen. This self-discharge process, however, requires that some water be present. If the electrodes are well dried and are stored in a desiccated environment, even the self-discharge as a result of oxygen evolution is largely arrested. In addition, the presence of some moisture during storage can initiate corrosion processes when coupled with trace amounts of nitrate species that often remain from the impregnation processes. For these reasons, nickel electrodes before activation in a cell are typically stored in a well-dried and sealed environment, often an inert gas, to ensure that they are not exposed to any gasborne contaminants. In some cases, nickel electrodes are stored within the unactivated nickel-hydrogen cell pressure vessels after completion of the cell assembly up to the point of adding electrolyte.

Another form of degradation during storage can arise from the reaction of airborne hydrogen sulfide (a common air pollutant) with the nickel hydroxide in stored nickel electrodes. This can occur whenever the electrodes are not protected from air exposure; the reaction converts the nickel hydroxide to nickel sulfide, as shown in Eq. (4.32)—most heavily at the surfaces where the electrodes are exposed to the atmosphere, in general.

$$Ni(OH)_2 + H_2S \longrightarrow NiS + 2H_2O \tag{4.32}$$

Nickel sulfide is electrochemically inert and an insulating material, and thus it will degrade both the capacity of the electrode and the conductivity of the active material. Similar degradation can be caused by exposure to sulfur dioxide, another common atmospheric contaminant.

4.5. Storage of Nickel Electrodes

The only chemical degradation process that has been seen to occur in well-dried nickel electrode active material is dehydration to form an electrochemically inert compound, Ni_2O_3H:

$$NiOOH + Ni(OH)_2 \longrightarrow Ni_2O_3H + H_2O \qquad (4.33)$$

Outside nickel-hydrogen cells, this reaction has only been seen[4.26] in instances where the continuous, long-term exposure to a desiccant maintained a very low humidity, which is expected to accelerate Eq. (4.33). The reaction is accelerated significantly when platinum oxides are present in the nickel electrode active material, presumably catalyzing or mediating the dehydration process. This process has also been seen in nickel-hydrogen cells that have been severely overcharged at high temperatures. Nickel electrodes that have experienced this form of dehydration have always been removed from an activated nickel-hydrogen cell, rinsed, and stored for periods from months to years in a desiccator. The dehydration process in reaction (4.33) has never been seen in new, unactivated nickel electrodes.

Once nickel electrodes are exposed to alkaline electrolyte in a nickel-hydrogen cell, a number of chemical processes can occur during long-term storage. The processes that are possible depend on the potential of the nickel electrode during storage, and thus on the type of precharge built into the cell. A nickel precharge, which is customary in today's nickel-hydrogen cells, keeps the nickel electrodes partially charged during cell storage, and thus at their normal operating potential. A hydrogen precharge, on the other hand, allows the nickel electrodes to fully discharge over time and eventually fall to the hydrogen potential, where a completely different set of chemical reactions are possible. Hydrogen precharge is not recommended because of the reduction mechanisms initiated by the hydrogen gas, and the subsequent oxidation processes when the cell is recharged (see section 4.4).

When an activated nickel electrode is stored in a nickel-hydrogen cell, it is at a highly oxidizing potential, where the following chemical processes will occur. (In addition, because the partially charged nickel electrode maintains an oxygen pressure within the cell, the negative cell electrode will also rise to a level near the oxidizing potential of the nickel electrode.) As the parenthetical references indicate, these processes are discussed in detail in earlier sections of this chapter:

1. corrosion of nickel sinter (as described in section 4.3)
2. oxidation of platinum metal, followed by solubilization of platinate ions in the electrolyte, and subsequent incorporation of platinum oxides into the nickel active material (section 4.3)
3. slow recrystallization of the partially charged active material to the most stable lattice structure, typically a well-ordered and deactivated form of γ-NiOOH (section 4.2)

These processes place some real constraints on the duration of storage for activated nickel-hydrogen cells, as well as the selection of the optimum storage method. These points are detailed in the discussion of nickel-hydrogen cell storage provided in chapter 11 as part of the nickel-hydrogen cell test experience.

4.6 Advanced Nickel Electrode Concepts

A number of advanced concepts have been developed throughout the years for the purpose of decreasing nickel electrode weight, increasing specific capacity, improving efficiency, improving power, or optimizing cycle life. While many of these concepts have never been widely used in the design of nickel-hydrogen cells, they are discussed here simply because they offer promise of cell performance improvements in areas that may be critically important for some applications.

4.6.1 Lightweight Nickel Electrodes

A number of concepts have been explored throughout the years as alternatives to the standard sintered nickel electrode for the purpose of reducing electrode and cell weight. The most common approach is to increase the porosity of the metallic structure that distributes current to the active material and provides structural support to hold the electrode together. This metallic structure, which consists of the sinter and the current collector, can constitute up to 50% of the weight of standard sintered nickel electrodes.

One approach to reduce the electrode weight is to increase the porosity of the sinter. Sinter porosities as high as 95% have been evaluated.[4.27] These high-porosity electrodes do work. However, they never provide as long a cycle life or as high a utilization of the active material as do standard sintered nickel electrodes. Cycle life is limited by increased electrode swelling rate, and increased sinter fracturing occurs as the sinter corrodes and weakens over time. The decreased current-collector area that is in contact with the active material, and the greater average distance between the current collector and the active material, typically reduces active-material utilization.

While these reductions in life and performance can be minimized by careful choice of the design details, there has always been a compromise associated with nickel electrodes containing sinter with porosity greater than about 85%. Because most applications of nickel-hydrogen cells choose this cell type precisely because of its extremely long cycle-life capability, there have not been any strong incentives to compromise cycle life and actually use lightweight sinter in real-world nickel-hydrogen cell applications. Today, the prospect of doing so has become even less likely, with the advent of lithium-ion cells as a lightweight alternative to nickel-hydrogen cells where weight is critical and the life requirements are not demanding.

A number of other lightweight structures have been developed as alternatives to nickel sinter for use in nickel electrodes. These include nickel felts,[4.28] nickel foam materials,[4.29] and nickel-coated polymer fiber mats,[4.30] as well as combinations of these kinds of structures with surface area and conductivity enhancers such as nickel flake or powder. All of these lightweight structures have been found to provide less cycle life and generally lower utilization than do standard sintered nickel electrodes. However, they are lighter and do work, and they can be used in battery cells that have less-demanding cycle-life requirements. These lightweight electrode concepts have found wide application in commercial cells. Electrodes using mats of nickel-plated polymer fibers have been successfully used in sealed nickel-cadmium cells. Lightweight nickel felt and foam substrates have found wide use in pasted nickel electrodes used in commercial nickel-cadmium and nickel metal hydride cells.

4.6.2 The Gradient Nickel Electrode
The gradient nickel electrode concept was developed[4.21,4.22] to make optimum use of the way cobalt additives influence the electrochemistry of sintered nickel electrodes, and it can increase the usable specific capacity of standard nickel electrodes by up to 20% through increasing the utilization of the active material. The gradient electrode concept is based on the fact that the concentration of cobalt additive alters the reversible potential of the active material in a regular and predictable way. In the gradient electrode, the active material is intentionally deposited so that high cobalt-additive levels are found adjacent to the current collector, with cobalt levels decreasing as the active material is deposited increasingly farther from the current-collector surface. The result is a gradient in the cobalt additive, which is found in percentages as high as 15% in material at the current-collector surface, and at 5% or lower in material far from the current collector. The average cobalt level remains in the range of 5–10%, as it is in standard nickel electrodes.

The cobalt-additive gradient will impose a potential gradient that actively prevents the formation of a poorly conducting depletion layer in the active material where it contacts the current collector. This allows capacity that is normally only available at low rate or at the lower discharge voltage plateau to be discharged at useful voltages. The cobalt gradient thus forces the active material to charge with a well-defined boundary layer that gradually moves outward from the current collector, and then to discharge in an orderly sequence with a boundary layer moving gradually back toward the current collector. This concept improves charge efficiency and allows use of the 10–20% residual capacity that normally cannot be discharged at a usable voltage level at a high rate.

The slow corrosion of the nickel sinter during many years of cycling can build up a layer of cobalt-free oxidation product (nickel hydroxide) at the surface of the current collector, causing loss of the high cobalt level at the current collector and preventing the cobalt gradient from functioning as effectively. The cobalt gradient can be maintained in this situation by plating the nickel sinter with a layer of cobalt metal or nickel/cobalt alloy,[4.31] thus maintaining the cobalt additive level within the active material as the sinter slowly oxidizes during the life of the electrode. The gradient electrode has not been routinely used in nickel electrodes, largely because of the added cost associated with adequately controlling the active-material deposition, as well as the cost of adding some cobalt to the sinter.

4.6.3 Chemically Modified Nickel Electrodes

Standard sintered nickel electrodes are typically capable of long-term cycling of the active material across a state-of-charge range corresponding to the transfer of one electron per nickel atom. In principle, the active material can be charged to the γ-NiOOH phase with sufficient overcharge, enabling a maximum nickel oxidation state of 3.67 and a theoretical capacity corresponding to transfer of 1.67 electrons per nickel atom. However, the volume changes and the heat dissipation that result when attempts are made to use the full capacity of the γ-NiOOH phase make this high capacity impossible to maintain during long-term cycling of nickel electrodes.

The goal of many advanced nickel electrode concepts has been to realize a practical capacity of up to two electrons per nickel atom in the active-material lattice. While such efforts have not yielded nickel electrodes having this high specific capacity, concepts have been developed that seek to stabilize the γ-NiOOH phase so that it can be efficiently and reversibly charged and discharged. While additives such as cobalt do influence the stability of the γ-NiOOH material, they do not keep this phase from reverting back to the β-NiOOH phase when discharged.

The development of a new type of nickel active material has had some success in realizing reversible capacity of more than one electron per nickel atom. The use of a pyroaurite active-material structure[4.32] enables cycling at capacity levels similar to those of γ-NiOOH. This material is an iron-stabilized nickel oxyhydroxide that contains 20% iron to stabilize the pyroaurite brucite structure so that it does not restructure during charge and discharge. Although this material has shown promise in nickel electrodes, it has not been used in nickel electrodes for nickel-hydrogen cells.

4.6.4 Microporous Nickel Electrodes

In recent years, nickel foam materials that have an extremely fine pore structure have become available. The pores can be substantially less than 10 μm in average size, and they offer a potentially improved conductive matrix for containing the active material used in nickel electrodes. However, these materials have not been successful in nickel electrodes for two reasons: They tend to exhibit increased rates of electrode swelling from charge/discharge cycling, and they tend to have relatively poor charge efficiency as a result of their extremely high internal nickel surface area, which seems to increase the rate of oxygen evolution relative to the rate of the active-material charging reactions.

4.7 References

[4.1] A. H. Zimmerman, G. A. To, and M. V. Quinzio, "Scanning Porosimetry for Characterization of Porous Electrode Structures," *Proc. of the 17th Annual Battery Conf. on Appl. and Adv.*, IEEE 02TH8576 (2002), pp. 293–298.

[4.2] L. Kandler, U. S. Patent No. 3,214,355 (1965).

[4.3] M. B. Pell and R. W. Blossom, U. S. Patent No. 3,507,699 (1970).

[4.4] R. L. Beauchamp, U. S. Patent No. 3,653,967 (1972).

[4.5] D. F. Pickett, U. S. Patent No. 3,827,911 (1974).

[4.6] V. J. Puglisi, H. N. Seiger, and D. F. Pickett, "Fabrication and Investigation of Nickel-Alkaline Cells, Part II. Analysis of Ethanolic Metal Nitrate Solutions Used in Fabrication of Nickel Hydroxide Electrodes," *Proc. of the 1974 IECEC* (1974), p. 873.

[4.7] C. H. Ho, M. Murthy, and J. W. Van Zee, "Studies of the Co-deposition of Cobalt Hydroxide and Nickel Hydroxide," *Proc. of the 1996 NASA Battery Workshop*, NASA Conf. Pub. 3347 (1997), p. 289.

[4.8] H. Bode, K. Dehmelt, and J. Witte, Zur Kenntnis de Nickel-hydroxidelectrode-1. Uber das Nickel (II) Hydroxidehydrat, *Electrochemica Acta* **11**, 1079–1086 (1966).

[4.9] J. H. Kauffman, T. C. Chung, and A. J. Heeger, "Fundamental Electrochemical Studies of Polyacetylene," *J. Electrochem. Soc.* **131**, 2847–2856 (1984).

[4.10] A. H. Zimmerman and V. E. Johnson, Electrochemical Reactions of Mixed Cobalt/Nickel Oxide Hydroxides in Alkaline Electrolyte, The Aerospace Corporation Report No. ATR-86(9561)-4 (15 July 1987).

[4.11] A. H. Zimmerman, "Calculation of the Thermoneutral Potential of NiCd and NiH2 Cells," *Proc. of the 1993 NASA Battery Workshop*, NASA Conf. Pub. 3254 (1994), pp. 289–294.

[4.12] A. H. Zimmerman, *Power Sources 12*, T. Keily and B. W. Baxter, eds. (Taylor and Francis, Ltd., Basingstoke, England, 1988), p. 235.

[4.13] R. D. Armstrong, A. K. Sood, and M. A. Moore, *J. Appl. Electrochem.* **15**, 603 (1985).

[4.14] R. D. Armstrong, G. W. D. Briggs, and M. A. Moore, *Electrochim. Acta* **31**, 25 (1986).

[4.15] V. A. Volynskii and Y. Chernykh, *Elektrokhim.* **13**, 1874 (1970).

[4.16] E. A. Kaminskaya, N. Yu. Uflyand, and S. A. Rozentsveig, *Soviet Electrochemistry* **7**, 1776 (1971).

[4.17] H. S. Lim and S. A. Verzwyvelt, *Proc. of the 23rd IECEC, American Society of Mechanical Engineers* (New York, 1988), p. 457.

[4.18] A. H. Zimmerman and A. H. Phan, "Role of Cobalt Superlattices in Nickel Electrodes," in *Hydrogen and Metal Hydride Batteries, Proc. 94-27*, P. D. Bennett and T. Sakai, eds. (The Electrochemical Soc. Inc., Pennington, NJ, 1994), pp. 341–352.

[4.19] A. H. Zimmerman and R. Seaver, *J. Electrochem. Soc.* **137**, 2662 (1990).

[4.20] J. Matsumoto, T. Poston, A. Prater, T. Barrera, and A. Zimmerman, *Proc. 34th Int. Power Sources Symp.*, The Inst. of Electrical and Electronics Engineers, Inc. (1990), p. 246.

4.21 A. H. Zimmerman and P. K. Effa, "Nickel Gradient Electrode," *Extended Abstracts of the Spring Meeting of The Electrochemical Society*, 83-1 (The Electrochemical Soc. Inc., Pennington, NJ, 1983), p. 62.

4.22 M. Oonishi, M. Watada, and M. Oshitani, Japanese Patent No. 02,265,165 (1990).

4.23 A. H. Zimmerman, "Effects of Platinum on Nickel Electrodes in Nickel Hydrogen Cells," *J. Power Sources* **36**, 253 (1991).

4.24 A. H. Zimmerman, "The Role of Anionic Species in Energy Redistribution Processes in Nickel Electrodes in Nickel Hydrogen Cells," in *Hydrogen Storage Materials, Batteries, and Electrochemistry, Proc. 91-4* (The Electrochem. Soc. Inc., Pennington, NJ, 1991).

4.25 A. H. Zimmerman, "The Interaction of Hydrogen with Nickel Electrodes," *Proc. of the Symp. on Nickel Hydroxide Electrodes*, 90-4, D. A. Corrigan and A. H. Zimmerman, eds. (The Electrochemical Soc. Inc., Pennington, NJ, 1990), p. 311.

4.26 A. H. Zimmerman, M. Quinzio, G. To, P. Adams, and L. Thaller, "Nickel Electrode Failure by Chemical De-activation of Active Material," *Proc. 1998 NASA Battery Workshop*, NASA/CP-1999-209144, pp. 317–328.

4.27 M. W. Earl, T. P. Remmel, and A. Dunnet, "An Evaluation of Sinter Nickel Plaques," *Proc. 31st Power Sources Symposium* (The Electrochemical Soc. Inc., Pennington, NJ, 1984), p. 136.

4.28 D. L. Britton, *Proc. of the Symp. on Nickel Hydroxide Electrodes 90-4*, D.A. Corrigan and A. H. Zimmerman, eds. (The Electrochemical Soc. Inc., Pennington, NJ), p. 233.

4.29 H. S. Lim and S. A. Verzwyvelt, in *Proc. of the 1984 IECEC, American Nuclear Society* (1984), 1, p. 312.

4.30 W. W. Lee, R. A. Sutula, C. R. Crowe, and W. A. Ferrando, "Electrochemical Behavior of Thin Nickel Electrode with Composite Substrate," *Proceedings of the Symposium on the Nickel Electrode 82-4*, R. G. Gunther and S. Gross, eds. (The Electrochemical Soc. Inc., Pennington, NJ, 1982), p. 243.

4.31 J. Kuklinski and P. Russell, U. S. Patent No. 4,975,035 (1990).

4.32 O. Glemser, J. Bauer, D. Buss, H.-J. Harms, and H. Low, *Power Sources 12*, T. Keily and B. W. Baxter, eds. (Taylor and Francis, Ltd., Basingstoke, England, 1988), p. 165.

5 The Hydrogen Electrode

The hydrogen electrode that is used as the negative electrode, or anode, in nickel-hydrogen cells was originally developed for use in hydrogen/oxygen fuel cells. It allows a high surface area of platinum catalyst to be exposed to both hydrogen gas and alkaline electrolyte, thus enabling the catalytic electrochemical generation and recombination of hydrogen gas during cell charge and discharge. The hydrogen gas, which is the active material for the negative electrode, is stored within the cell pressure vessel at pressures up to about 1000 psi when a cell is fully recharged. Because the catalyst in the hydrogen electrode is highly active, the charge and discharge reactions at this electrode occur with high round-trip energy efficiency. Essentially all the coulombic inefficiencies associated with the hydrogen electrode arise from recombination of oxygen gas generated at the nickel electrode, and the self-discharge reactions of hydrogen with the nickel electrode. The hydrogen electrode has historically been one of the most robust components in the nickel-hydrogen cell, rarely contributing to problems with cell performance or cycle life. This chapter discusses the design, construction, chemistry, and functioning of the hydrogen electrode in the nickel-hydrogen cell.

5.1 Design and Function of the Hydrogen Electrode

The hydrogen electrode is designed as a classical gas electrode that uses hydrogen as its active material and functions by providing the active catalytic sites with the gaseous and aqueous species needed for oxidation and reduction of the hydrogen gas. The electrochemical oxidation and reduction process that occurs during charge and discharge of the alkaline hydrogen electrode is indicated in the reaction shown in Eq. (5.1):

$$2H_2O + 2e^- \longleftrightarrow H_2 + 2OH^- \quad E_o = -0.828 \text{ V} \quad (5.1)$$

Hydrogen gas is generated along with hydroxide ions, while water is consumed during cell charging. The reverse reaction consumes hydrogen and generates water during discharge.

For this reaction to occur reversibly at a high rate during both charge and discharge, the platinum catalyst must be simultaneously maintained in contact with a source of hydrogen gas as well as a reservoir of alkaline electrolyte. This is done by coating a layer of high-surface-area platinum black catalyst on one side of a nickel grid, while bonding a porous Teflon layer to the other side of the nickel grid. This structure is schematically shown in Fig. 5.1 (which is not drawn to scale). Each of the layers illustrated is typically several mils thick, resulting in a hydrogen electrode with an overall thickness of 0.005–0.006 in.

150　The Hydrogen Electrode

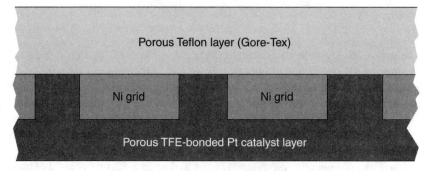

Figure 5.1. Structure of the hydrogen electrode. The top layer, porous Teflon, is hydrophobic, while the bottom, catalyst-containing layer is hydrophilic and absorbs some electrolyte from the separator.

The porous Teflon layer is generally composed of a hydrophobic material (such as Gore-Tex) that will maintain open and unflooded pores during cell operation, and it thus enables hydrogen gas to readily flow into or out of the underlying catalyst layer. This constitutes the hydrophobic side of the hydrogen electrode, while the catalyst-containing layer is much more hydrophilic and is readily wetted by contact with the electrolyte-containing separator.

The flow of hydrogen to the hydrophobic Teflon side of the electrode is ensured by the presence of a gas screen that maintains large open spaces in a 10–12 mil thick layer against the hydrogen electrode. The gas screen typically consists of a polypropylene mesh, with a mesh size large enough to prevent wetting, but small enough to maintain an open gas screen layer that is 10–12 mils thick, and through which hydrogen gas can readily flow.

In back-to-back or dual-anode cell stack designs, the overall negative electrode unit in the electrode stack, such as the one illustrated in Fig. 5.1, consists of two hydrogen electrodes placed with the Gore-Tex layers together, separated by a single gas screen that allows hydrogen to flow into the porous Teflon backing of both electrodes. In recirculating stack designs, the gas screen is placed against the porous Teflon backing and faces the next nickel electrode in the cell stack, and it serves to channel oxygen gas flow from the nickel electrode (as well as hydrogen gas) into the porous backing on the hydrogen electrode.

5.2. Hydrogen Electrode Fabrication

Hydrogen electrodes are made by sintering a layer of platinum black catalyst and Teflon binder and the Gore-Tex backing layer onto a nickel grid structure. The nickel grid controls the size and shape of the hydrogen electrode, and it typically consists of both radial and circumferential interconnected current-carrying nickel wires. The nickel grid structure has a nickel tab connected to it, either in the center of the electrode (for a pineapple-slice electrode design) or at one of the flattened edges (for designs using an edge-connected bus bar).

The platinum black catalyst particles are mixed with a Teflon emulsion to form the coating on the catalyst side of the electrode. The Teflon in the mixture serves as a binder to hold the catalyst particles together in a thin layer that is in close contact with the nickel grid structure. The catalyst mixture is typically silk-screened onto the grid, which then is pressed onto the Gore-Tex backing sheet. The wet electrode structure is then heated in a sintering oven or furnace to drive off the organic constituents of the Teflon emulsion, and to bond the Teflon particles together with the catalyst particles. The Teflon particles in the emulsion also bond the catalyst layer and the nickel screen to the Gore-Tex backing. The sintering temperature is chosen so that it is high enough to soften the Teflon particles and enable them to form good interfacial bonds, as well as to drive off the organic constituents (solvent and surfactants). However, the sintering temperature must be kept below the threshold where the Teflon will decompose or excessively coat the platinum black particles (and thus mask off much of their catalytically active surface area).

Once the hydrogen electrode structure described here has been bonded in the sintering furnace, it is cleaned of any surface residue or particulates, checked for an appropriate weight of platinum catalyst, and verified to have the correct thickness. If these physical characteristics are as required, samples of electrodes from each production lot are typically checked for proper electrochemical performance. Such electrochemical tests typically include measurement of the polarization during high-rate hydrogen evolution and recombination, as well as checking the capability of sample electrodes to recombine oxygen at an adequate rate. The platinum catalyst is generally loaded into the hydrogen electrodes at a level of 6–10 mg per square centimeter. This loading level produces an electrode that should have quite low polarization potentials and thus is not typically the performance-limiting electrode seen in nickel-hydrogen cells.

Because there is generally more platinum catalyst in the standard hydrogen electrode than is necessary to meet minimum performance needs in nickel-hydrogen cells, electrodes have been produced for some cells that have about half of the loading level described here. While such electrodes, if made properly, can provide adequate functionality at a lower cost, they have less performance margin to tolerate some of the storage-related degradation that can occur in nickel-hydrogen cells, and are thus not as desirable in high-reliability applications.

5.3. Gas and Electrolyte Management

The proper simultaneous management of both gas and electrolyte is the key requirement for proper operation of the hydrogen electrode in a nickel-hydrogen cell. The platinum catalyst layer in Fig. 5.1 must absorb electrolyte so that reaction (5.1) will not be limited by the available supply of water and hydroxide ions. Hydroxide ions must diffuse through, and liquid electrolyte must be available to flow between the separator and the platinum catalyst layer. If the catalyst layer is too hydrophobic, there may be too little electrolyte present to support the electrochemical reactions. If the platinum catalyst layer is too hydrophilic and wets completely, hydrogen gas will not be able to access much of the embedded catalyst surface area, and the hydrogen recombination rate during cell discharge will be inadequate. These types of problems are generally associated with a diffusion-limited current, whether it is electrolyte diffusion from the separator, or gas transport from the gas screen that is the problem.

The most important test of the gas and electrolyte management capability of a hydrogen electrode is at the lowest temperatures of operation. It is at the lowest temperatures that the gas and the electrolyte transport rates are lowest, and the electrolyte viscosity is the greatest. Additionally, if high-rate discharge is performed at temperatures significantly below the freezing point of water, it is possible for the water formed by discharge reactions at the catalyst to freeze before it can diffuse into the electrolyte-containing pores.

For these reasons, the best test of hydrogen electrode performance in nickel-hydrogen cells involves high-rate discharge at a temperature that has been stabilized well below the lowest temperature at which the cell is expected to operate. For example, if a cell can operate at temperatures as low as –10°C, its performance should be checked at the maximum discharge rate for a temperature stabilized at least 10°C below the minimum temperature. If all cells perform well in tests of this type, it is likely that hydrogen electrode performance will not be a factor limiting the cell performance. A stress test of this type also assesses performance margin, which can make processes in the nickel electrode that can temporarily reduce cell electrolyte concentration[5.1] less likely to influence hydrogen electrode performance.

Another key requirement for proper gas and electrolyte management in the hydrogen electrode is that the porous Gore-Tex backing and the gas screen that contacts it (see Fig. 5.1) must remain free of liquid electrolyte. The function of these porous components is to allow the free access of hydrogen gas from the cell pressure vessel to the platinum catalyst layer. If the gas screen floods, or if the Gore-Tex layer wets, the facile flow of hydrogen gas will be interrupted, and the hydrogen electrode will become unable to discharge in a cell at any significant rate. Conditions that can make gas screens susceptible to retaining too much liquid electrolyte include excessive electrolyte in a cell (inadequate draining during activation); particulates, surfactants, or contaminants that make the gas screen wettable; and too fine a gas screen weave. The Gore-Tex layer similarly must keep its pores free of liquid to function properly. This layer is sometimes referred to as a "raincoat" on the hydrogen electrode, because it protects the critical gas-flow paths from being flooded by the water that is generated on the catalyst during electrode operation. For this raincoat to function properly, it must be hydrophobic. The presence of surfactants or organic contaminants, or significant compression of its pore structure, can seriously compromise the hydrophobicity of this protective raincoat.

A final requirement of the macroscopic negative electrode/gas screen structure used in nickel-hydrogen cells is that it should enable the facile access of hydrogen gas to the hydrogen electrode. As discussed, if the gas screen fills with liquid, the hydrogen electrode cannot function properly. A more subtle manifestation of this problem can occur if a nickel-hydrogen cell contains more than several psi of an inert gas. During high-rate discharge the inert gas is swept into the gas screen along with the flowing hydrogen, and when the hydrogen is consumed the inert gas is left behind in the gas screen. Eventually, this process will completely fill the gas screen with a sufficient backpressure of inert gas so that it will block the continued flow of more hydrogen into the gas screen. When this occurs, the cell will fail to continue supporting discharge until the inert gas is given sufficient time to diffuse out of the gas screen. The solution to this problem, of course, is to keep the level of inert gases (such as nitrogen, argon, or helium) low in nickel-hydrogen cells.

5.4. Chemistry of the Hydrogen Electrode

The primary reversible electrochemical process occurring in the hydrogen electrode is that of reaction (5.1), and it involves the evolution and recombination of hydrogen gas on platinum catalyst particles. The kinetics of this reaction follows the expected Butler-Volmer dependence on overpotential.[*] Because the overpotential at the hydrogen electrode is generally quite low (< 50 mV) during high-rate nickel-hydrogen cell operation and the exchange current is relatively high, the current is nearly linearly dependent on overpotential at low currents, but becomes nonlinear at higher currents, as is indicated in Fig. 5.2. At high current densities the anticipated Tafel behavior is obtained, where current increases exponentially with increasing overpotential. However, at sufficiently high currents there is a current density threshold above which the processes controlling electrolyte or gas transport can limit the rate of electrode operation.

A second reaction that is significant in the hydrogen electrode during nickel-hydrogen cell operation is the essentially irreversible recombination of oxygen gas during cell overcharge. Oxygen recombination can occur by either of two mechanisms. The first is the electrochemical recombination of oxygen at the platinum catalyst sites, and it is indicated as reaction (5.2):

---

[*]The term "overpotential" refers to the difference between the operating potential of an electrode and the reversible potential of the electrochemical reaction occurring at the electrode.

154   The Hydrogen Electrode

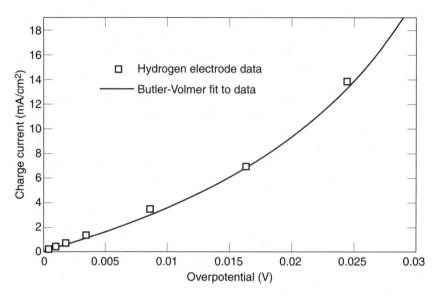

Figure 5.2. Current/voltage behavior of a typical cycled hydrogen electrode in 50 psi of hydrogen gas at 23°C, along with the best Butler-Volmer equation fit to the data.

$$O_2 + 2H_2O + 4e^- \xleftarrow{\text{Pt-catalyst}} 4OH^- \quad E_o = 0.401 \text{ V} \quad (5.2)$$

This reaction can occur in parallel with reaction (5.1), and it will reduce the net amount of hydrogen evolution by the amount of current that goes into the electrochemical oxygen recombination process. During nickel-hydrogen cell overcharge, there is an overpotential of about 1.5 V for the electrochemical oxygen recombination process of reaction (5.1). Because of this large overpotential, the rate of this process is limited by the rate of oxygen transport to the catalytic sites in the hydrogen electrode, and is thus not dependent on the voltage of the hydrogen electrode. Because of the high overpotential driving this reaction, it is essentially electrochemically irreversible under normal cell operating conditions.

The second oxygen recombination process is the direct gas-phase recombination with hydrogen through a classical chain-reaction mechanism, as indicated in the steps shown for reaction (5.3):

Chain initiation: $\quad H_2 \xrightarrow{\text{Pt-catalyst}} 2H\bullet$

5.4. Chemistry of the Hydrogen Electrode

Chain propagation: $H\bullet + O_2 \longrightarrow HOO\bullet$

$HOO\bullet + H_2 \longrightarrow HOOH + H\bullet \longrightarrow 2OH\bullet + H\bullet$

$OH\bullet + H_2 \longrightarrow H_2O + H\bullet$

Chain termination: $2H\bullet \xrightarrow{\text{surfaces}} H_2$

$OH\bullet + H\bullet \xrightarrow{\text{surfaces}} H_2O$ (5.3)

The chain initiation step involves adsorption and splitting of hydrogen molecules into radicals on the surface of the platinum catalyst. Whenever an oxygen molecule impinges on the surface of the platinum catalyst, the chain propagation steps begin, first with the formation of the hydroperoxy radical, which can then react with a hydrogen molecule in the gas phase to form hydrogen peroxide and another hydrogen radical. Because hydrogen peroxide is unstable in the gas phase, it decomposes to two hydroxy radicals that can react with an additional hydrogen molecule to form water plus another hydrogen radical. A chain reaction occurs because more free radicals are produced in the gas phase than are consumed. This chain reaction will proceed explosively until either the reactants are exhausted or the chain termination steps quench it.

The chain termination steps will occur whenever a gas-phase free radical collides with a solid surface, where the radicals will either react with the wetted surface materials or await deactivation by collision with another radical. Thus, the gas-phase recombination process will not occur significantly (or explosively) unless there are at least small pockets of gas space near the platinum catalyst. In a nickel-hydrogen cell, the chain reaction can propagate through sufficiently large gas-filled pores that contain an adequate partial pressure of oxygen. The overall rate of this gas-phase recombination process depends on the partial pressures of oxygen and hydrogen, as well as the proximity of pore surfaces that can quench the free radicals that propagate the chain reaction. Typically, this direct recombination reaction becomes most important where voids of significant volume allow relatively high fluences of oxygen gas to contact the hydrogen electrode, such as at the edges of the electrode stack. The direct reaction is also irreversible as a result of the large activation energy and enthalpy increase associated with the reverse process of splitting water molecules.

When nickel-hydrogen cells are held or operated at voltages outside the normal range of about 1.0 to 1.55 V, other reactions can become significant at the hydrogen electrode. The electrochemical oxygen reaction at the negative electrode discussed earlier, Eq. (5.2), while essentially irreversible at normal operating potentials, can become important in terms of both evolution and recombination of oxygen when a nickel-hydrogen cell with nickel precharge is allowed to discharge to near zero volts or into reversal. Under these conditions, oxygen gas can be electrochemically generated from the platinum catalyst electrode by the reverse of reaction (5.2). During nickel-hydrogen cell reversal this process typically occurs at a cell voltage plateau of about –0.3 V (as long as there is significant nickel precharge), as indicated in Fig. 5.3. The oxygen that is evolved within the cell will then undergo recombination by reaction (5.2) when the cell is recharged, with a voltage plateau at about 0.3 V. The overpotential of about 0.3 V in either the forward or reverse direction for reaction (5.2) is the direct consequence of an extremely low exchange current density for this reaction. However, the catalytic efficiency of the platinum catalyst in the negative electrode enables this process to occur readily in either the forward or reverse direction at an appropriate cell potential.

During either the storage of nickel-hydrogen cells at low potential or during cell reversal (when cells have nickel precharge), another reaction can occur in addition to the evolution of oxygen. This process is the electrochemical oxidation of platinum metal in the negative electrode, as indicated by reaction (5.4):

$$\text{Pt(metal)} + 2\text{OH}^- \longleftrightarrow \text{PtO} + \text{H}_2\text{O} + 2\text{e}^- \quad E_o = 0.98 \text{ V } (vs. \text{ H}_2) \quad (5.4)$$

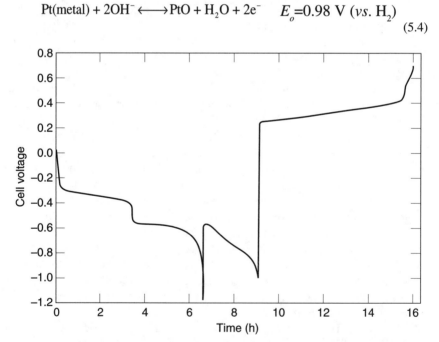

Figure 5.3. Reversal of a nickel-precharged nickel-hydrogen cell at 20°C at C/50 and then C/200 until the nickel precharge has been depleted, followed by recharge at C/50. The oxygen evolved during discharge is recombined electrochemically during recharge.

This process begins to occur when the cell potential falls below 1.0 V, assuming that the drop in cell voltage is almost entirely caused by polarization of the platinum catalyst electrode. Clearly, the hydrogen gas within the cell must be fully depleted before the platinum catalyst begins to oxidize significantly. It is the platinum oxidation process that is essentially responsible for initiating the loss of catalytic activity that can gradually occur in regenerative fuel cells using this type of gas electrode.

Once the platinum metal surface has oxidized to platinum oxide by Eq. (5.4), the platinum oxide can hydrate and slowly dissolve into the alkaline electrolyte in the cell by reaction (5.5) to form a platinate complex ion:

$$PtO + H_2O + 4OH^- \longleftrightarrow Pt(OH)_6^{-4} \qquad (5.5)$$

This dissolution process is not rapid, and at an electrolyte concentration of 31% the electrolyte becomes saturated with platinate ions at a concentration of about 6×10^{-3} M. However, repeatedly oxidizing the platinum catalyst surface, allowing some dissolution, and then reducing the dissolved platinum back to the unoxidized state can actually cause a large loss of active catalyst surface area, accompanied by an increase in negative electrode overpotential during cell operation. The loss of catalyst surface area occurs because the platinum metal that is plated back onto the catalyst by the reverse of reaction (5.4) is essentially in the form of a low-surface-area foil rather than the high-surface-area catalyst that was originally dissolved. This is the principal reason that the amount of platinum catalyst in a high-reliability hydrogen electrode should be significantly greater than is required at the beginning of life to provide a sufficiently low overpotential for the required cell discharge rate.

5.5. Negative Electrode Thermodynamics

The thermodynamics of the negative electrode reactions that are important during nickel-hydrogen cell operation involve the reversible hydrogen reaction, Eq. (5.1), as well the electrochemical oxygen reaction, Eq. (5.2). The standard potentials for these processes, the enthalpies of reaction, and the entropies of reaction are tabulated in Table 5.1.

Table 5.1. Standard Thermodynamic Properties of the Hydrogen and Oxygen Reactions at the Negative Electrode (298 K)

Reaction	Standard potential (V)	Enthalpy change (V)	Entropy change (V/°K)
Hydrogen, (5.1)	–0.828	–1.077	-8.34×10^{-4}
Oxygen, (5.2)	0.401	–0.100	-1.68×10^{-3}

When these thermodynamic characteristics are combined with those of the nickel electrode, it becomes possible to define a thermoneutral potential for the cell, which indicates the potential where heat evolution during charge or discharge is just balanced by entropic cooling.

It should be pointed out that, when using the properties of Table 5.1 to determine cell thermal dissipation, one finds that the oxygen recombination process at the negative electrode nominally releases an amount of heat proportional to the cell voltage times the cell current. However, the actual heat evolution is more complicated. Oxygen evolution first occurs at the nickel electrode based on the local nickel electrode potential and environment, and it can involve significant entropic cooling. The oxygen gas will eventually find its way to the negative electrode after some time delay for transport, where it will recombine with hydrogen. The thermodynamics of the recombination process are now dictated by the local pressure and environment at the negative electrode at a time that can be much later than when the oxygen was generated. Between these two times there can be significant changes in oxygen and hydrogen partial pressure, voltage, and electrolyte concentration, which can produce significant thermal effects from pressure-volume work and from changes in the reversible potential. The only ultimate thermodynamic requirement for the cell thermal cycle is that the cumulative heat produced by the cell as the result of oxygen evolution and recombination must equal the total electrical watt-hours that went into the oxygen evolution after all the evolved oxygen has been allowed to recombine.

5.6. Negative Electrode Contaminants

Because the hydrogen electrode utilizes a catalyst that depends on an active catalytic surface, its performance can be altered by any contaminant that can block or passivate the active catalytic sites. For this reason it is critical that inorganic contaminants in general be kept at very low levels when producing, storing, or handling hydrogen electrodes. This means that levels of trace metals such as lead, cadmium, iron, calcium, magnesium, aluminum, and others, should be kept well below 50 ppm.

It is the level of iron (which can be pervasive in the manufacturing environment) that has been observed to directly correlate with the overpotential for hydrogen evolution and recombination at hydrogen electrodes. This correlation has been noted even at very low levels of iron where the additional overpotential from the iron contaminant is small enough to not have a large effect on cell performance. At levels of more than 50 ppm, iron contaminants can clearly degrade cell performance enough to be significant.

Contaminants such as copper, which is electroactive in potassium hydroxide electrolyte, can affect cell operation by an alternative mechanism. Copper can dissolve in the cell electrolyte, and then be plated as copper metal dendrites onto the hydrogen electrode during cell recharge. These dendrites can ultimately cause short circuits in a nickel-hydrogen cell if they become extensive enough.

The platinum catalyst electrode in a nickel-precharged nickel-hydrogen cell can serve as a very sensitive sensing device capable of detecting any electroactive constituents in the electrolyte of a nickel-hydrogen cell. When a nickel-precharged cell exhausts its supply of hydrogen gas at the end of discharge, the negative electrode will shift its potential over about a 1.3 to 1.4 V potential window extending upward from the hydrogen potential all the way to the oxygen evolution potential. Throughout this potential window, the platinum catalyst offers an active surface for the oxidation or reduction of a wide range of electroactive contaminants that may be present within the cell. Such measurements can typically be performed using a controlled coulometric or amperometric scan over the voltage window. The materials normally present in the negative electrode that are seen to undergo redox reactions within this window are nickel metal and platinum metal. With appropriately controlled coulometric scanning, it is possible to detect ppm levels of other electroactive materials such as copper, lead, and arsenic.

5.7. Pore Size Distributions

The distribution of pore sizes in hydrogen electrodes is critical to the proper functioning of these electrodes in nickel-hydrogen cells. The combination of the capillary forces within the pores and the wettability of the pore surfaces controls the distribution of liquid electrolyte in the electrode, thereby dictating how hydrogen gas and electrolyte are transported through the hydrogen electrode structure. The distribution of pore sizes, as measured by mercury intrusion porosimetry, is illustrated in Fig. 5.4. This pore size distribution is distinctly bimodal, consisting of one distribution of larger pores greater than 25 µm and a second distribution of smaller pores less than 1 µm. The larger pores are primarily in the porous Teflon backing and should remain free of liquid electrolyte at all times.

The smaller-pore distribution in the hydrogen electrode, which involves pores that are primarily within the bonded catalyst layer, consists of two subdistributions. The larger one (with pores about 0.2 µm in average diameter) consists of larger channels between the catalyst particles and the Teflon binder, while the smaller subdistribution (with pores about 0.04 µm in average diameter) consists primarily of the fine channels within and between the catalyst particles. This smaller-pore subdistribution remains essentially filled with liquid electrolyte at all times. The larger of these two pore subdistributions remains partially filled with electrolyte, thus allowing both electrolyte and gas to access the high surface area of the catalyst particles. The pore substructure within the catalyst layer of the hydrogen electrode is illustrated in the cross section shown in Fig. 5.5 for a typical electrode.

If the hydrogen electrode manufacturing process does not appropriately control the pore sizes, the transport of either hydrogen gas or electrolyte may be impeded to the point where electrode polarization will be higher than normal. Excessively large pores within the catalyst layer can result in reduced diffusion

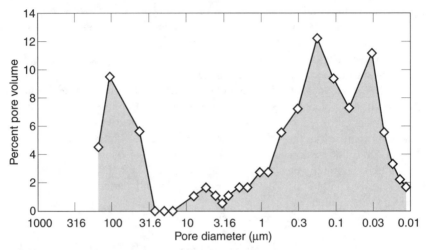

Figure 5.4. Distribution of pore sizes in a platinum catalyst hydrogen electrode as measured by mercury intrusion porosimetry.

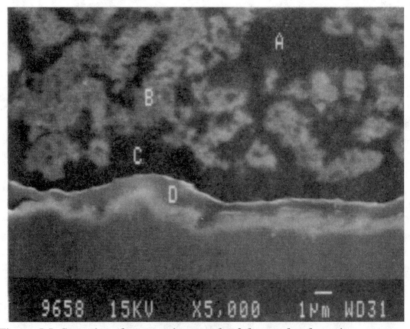

Figure 5.5. Scanning electron micrograph of the catalyst layer in a cross-sectioned hydrogen electrode showing (a) the larger channels within the catalyst layer; (b) bonded catalyst particles; (c) the channel at the boundary of the nickel grid and the catalyst layer; and (d) the surface of the nickel grid.

of electrolyte from the catalyst, and thus large gradients in electrolyte concentration as a result of the water produced or consumed during electrode operation. If the pores in the catalyst layer are excessively small, they will tend to flood with electrolyte, and gas transport into the catalyst layer will be severely impeded. In this situation, electrode polarization will be significantly increased during cell discharge, particularly at high current densities where diffusion of hydrogen (gaseous or dissolved) may not occur rapidly enough to support the discharge rate.

5.8 References

[5.1] A. H. Zimmerman, "Low Temperature Failure Mode for Nickel Hydrogen Cells," *Proceedings of the 2006 Space Power Workshop* (Redondo Beach, CA, 29 April 2006).

6 Separators and Electrolyte

Discussion of the separator in a nickel-hydrogen cell cannot be complete without an accompanying discussion of the cell electrolyte. The ways in which these cell components function are considered together because the separator, in addition to its obvious function of electrically separating the nickel and hydrogen electrodes, also serves as an electrolyte reservoir that buffers the cell electrodes from either drying out or flooding during cell operation. This electrolyte reservoir role played by the separator is common to most battery cell designs that are electrolyte starved. Electrolyte-starved designs do not allow macroscopic flooding of the cell with electrolyte. The separator can also function to control and direct gas transport in some nickel-hydrogen cell designs. This chapter discusses these separator functions, as well as the characteristics of the electrolyte typically used in nickel-hydrogen cells.

6.1 Roles of the Separator in a Nickel-Hydrogen Cell

The separator in any battery cell provides the obvious function of separating the positive and negative electrodes from any direct electrical contact. To properly serve in this role, the separator must remain physically intact to provide an electronically insulating layer between anode and cathode throughout the lifetime of the cell. Any breakdown in this insulating capability will result in cell degradation or failure as the result of a short circuit. Such breakdown can occur in a nickel-hydrogen cell either as the result of physical damage to the separator from the pressures exerted on it during operation, or as the result of accumulation of electronically conducting particles from the electrodes within the separator structure. Different separator materials have differing susceptibilities to these degradation modes, as will be discussed later in the context of specific types of separator.

A second key function of the separator is to provide ionic conductivity and allow electrolyte movement between the anode and the cathode of the cell. The charge transfer processes in the anode and the cathode produce hydroxide ions as a product along with water. In the nickel-hydrogen cell, the electrochemical production of hydroxide ions at one electrode is balanced by the consumption of hydroxide ions at the other electrode.* If the ions cannot undergo conductive migration between the anode and cathode, the battery cell will simply become an electrolytic capacitor where ionic charge storage controls cell capacity. The use of porous separators that allow ionic conduction and diffusion through electrolyte-

---

*An exception to the balance between the production and consumption of hydroxide ions occurs when the γ-NiOOH phase is formed in the nickel electrode. This phase takes one equivalent of hydroxide from the electrolyte for every three nickel atoms in the lattice.

filled pores permits the cell to operate as a true battery, with the capacity being controlled by the amount of anode or cathode active material. To keep ionic resistance low, the separator must be relatively thin, have a relatively high porosity, and hold a large volume fraction of electrolyte in a nontortuous pore structure or reservoir. The separator can also allow convective movement of electrolyte in response to pressure gradients between the anode and cathode, thus preventing the expulsion of electrolyte from the electrode edges or into the gas screens.

Typical separators operate in nickel-hydrogen cells that contain concentrated potassium hydroxide electrolyte to ensure high ionic conductivity. Different cell designs have successfully used aqueous electrolyte ranging from 26% to 38% potassium hydroxide by weight, although the most common designs in use today have electrolyte concentrations of 26% or 31%. These electrolyte concentrations ensure that, even at the highest charge or discharge rates, the concentration gradient that develops between the anode and cathode is only about 5% or less by weight. If the separators experience significant dry-out (loss of electrolyte fill fraction), this concentration gradient can grow to the point where it limits cell current. If this "diffusion-limited" current is reached, cell failure will occur unless the cell current demand is immediately reduced.

6.2 Zircar Separator

Zircar material is almost exclusively used as the separator in modern nickel-hydrogen cells. This material consists of bundles of zirconium oxide** ceramic fibers, which are woven into a fabriclike material suitable to be used as sheets of separator. Figure 6.1 illustrates the fiber structure of Zircar separator material.

The thickness of the separator sheets is typically 10–15 mils. This material is relatively incompressible, because the individual ceramic fibers are quite hard. Although they have reasonable flexibility, the fibers and sheets of Zircar separator will fracture if they are bent too sharply. Thus, cell separators made from Zircar must be maintained by the stack compression in a flattened and compressed state during operation in a cell. Cell designs may call for either a single layer or a double layer of this separator material. The single layer has the advantage of reduced ionic resistance as a result of being half as thick, but the dual layer has the advantages of providing an electrolyte reservoir with twice the holding capacity as well as having much better robustness against physical breakdown of its electronic resistance. Both single- and dual-layer Zircar separator cell designs have been, and continue to be, successfully used with potassium hydroxide electrolyte in the 26% to 31% concentration range.

**The zirconium oxide used in separator material for nickel-hydrogen cells is stabilized with yttrium oxide additive, which may be present at concentrations of up to 15%.

Figure 6.1. The fiber structure of Zircar separator, including woven bundles of individual striated fiber strands.

Zircar separator is a material that has a very open pore structure, with approximately 80% of the overall separator volume consisting of void space. As shown in the inset of Fig. 6.1, the individual fibers in the woven bundles of the separator have a striated surface that ensures that the fibers will remain wetted with electrolyte. The regions between the fibers in each bundle also tend to normally remain wet with electrolyte, while the larger spaces between the fiber bundles may be filled with electrolyte as required to serve as an electrolyte reservoir. The spaces between the fiber bundles serve as an excellent electrolyte reservoir in the nickel-hydrogen cell, releasing electrolyte to the other cell components as needed to keep them wetted with electrolyte.

When a nickel-hydrogen cell is newly activated with electrolyte (i.e., at beginning of life), the reservoir volume in the Zircar separator is typically 80–90% filled with electrolyte. Typically the amount of electrolyte within the Zircar separator drops as a cell ages, because the nickel electrodes expand and draw electrolyte from the separator to fill the increased nickel electrode volume. After a cell has been cycled for many years, only 40% or less of the separator's void volume may be filled with electrolyte. If the fraction of separator void volume filled with electrolyte falls below 20–25% (depending somewhat on electrolyte concentration), the probability of a cell's failure as a result of its current being limited by electrolyte diffusion ("diffusion-limited current") becomes significant.

Zircar is what is loosely referred to as a "low bubble pressure" separator material. This means that a low pressure differential (one that is less than several psi) is all that is needed to force gas through the relatively open weave of the electrolyte-wetted material. The implication for the operating nickel-hydrogen cell is that the convective flow of oxygen, hydrogen, and electrolyte can readily occur through the separator structure in response to even small pressure differentials that may develop between the anode and the cathode surfaces. The relative flow of liquid and gaseous phases is governed by the relative viscosity of each phase, as well as the separator capillary forces that act on the liquid as a function of the electrolyte fill fraction in the separator. The low bubble pressure allows oxygen that is generated in the nickel electrode during overcharge to easily flow from the surface of the nickel electrode into and through the separator to impinge upon the surface of the hydrogen electrode, where it will catalytically recombine with hydrogen to release heat. The fiber structure of the Zircar will thus allow facile flow of oxygen for overcharge gas management in the nickel-hydrogen cell, without allowing large pressure differentials that can damage the components. The Zircar separator also disperses the oxygen flow to prevent or control explosive gas-phase oxygen/hydrogen recombination (popping) that can blow holes in the surface of the hydrogen electrode or damage other cell components. The flow of liquid electrolyte out of the Zircar separator and into other wettable cell components is opposed only by its relatively low capillary pressure, which depends on the liquid fill fraction, as is indicated in Fig. 6.2 for the typical fiber texture, size, and spacing in Zircar.

6.3 Asbestos Separator

Asbestos was used as the separator material in the COMSAT cell design and variations on this design from the earliest nickel-hydrogen cells that were built, through the 1990s. Issues related to processing and handling, as well as the health hazards associated with asbestos, led to the discontinuation of the use of this separator material in the late 1990s. Asbestos is quite different from Zircar in terms of the details of how it functions as a separator in nickel-hydrogen cells. Asbestos separator consists of a mat of extremely fine magnesium silicate fibers that are formed into a sheet that is approximately 10–15 mils thick when uncompressed. The typical structure of this separator material is illustrated in Fig. 6.3, which shows masses of fine fibers as well as compacted assemblages of these fine fibers. The fiber mat is relatively compressible, with only small void spaces left between the fibers when the mat is compressed between the electrodes in the nickel-hydrogen cell stack. Therefore the compressed asbestos material has almost all of its internal void volume filled with electrolyte at all times in a nickel-hydrogen cell. In fact, if the asbestos is flooded with excess electrolyte, it can soften to the

6.3 Asbestos Separator

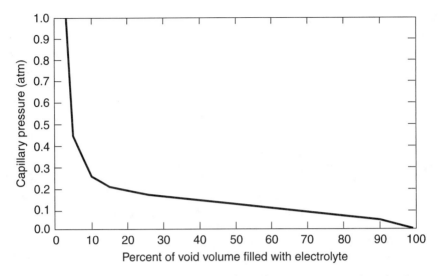

Figure 6.2. Capillary pressure of Zircar separator vs. amount of void volume filled with electrolyte.

point where it can completely lose its mechanical integrity. For this reason, it is important to avoid operating nickel-hydrogen cells containing asbestos separator using an electrolyte-flooded condition during cell activation.

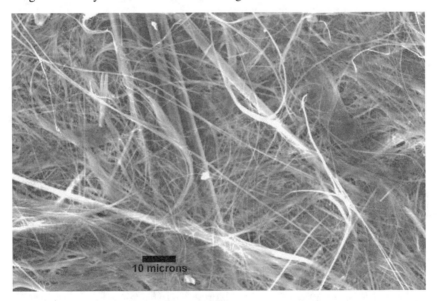

Figure 6.3. Fiber structure of asbestos separator material.

When it is wetted with electrolyte, asbestos is known as a "high bubble pressure" separator material. Because there are essentially no void gas channels through the wetted separator, it does not allow the flow of gas even when a very high gas pressure is applied across the separator. In addition, a high gas pressure will not tend to force electrolyte out of the asbestos pore structure, because of the extremely high capillary forces that draw electrolyte into and hold it within the structure. Figure 6.4 indicates the capillary pressure of asbestos that is only slightly compressed as a function of the percentage of the separator void space that is filled with electrolyte.

When compressed in a nickel-hydrogen cell stack, asbestos separator acts essentially as a gasket that blocks any flow of oxygen or hydrogen gas through the separator. This means that during overcharge essentially all undissolved oxygen generated at the nickel electrode is forced to flow out of the edges of the electrode stack and will eventually undergo recombination with hydrogen, primarily at the outer edges of the hydrogen electrodes. In the edge region of the stack there are sizeable gas spaces, which can promote explosive gas-phase oxygen recombination (edge popping). For this reason, cells with asbestos separator cannot generally tolerate significant overcharge rates, which can cause significant edge damage that can ultimately cause short circuits. If the oxygen from overcharge cannot be channeled rapidly enough to the edges of the stack, it can build up enough pressure across the separator to permanently deform the negative electrode, or force the asbestos to migrate and leave a hole in the separator. Such a hole results in catastrophic cell failure from a short circuit.

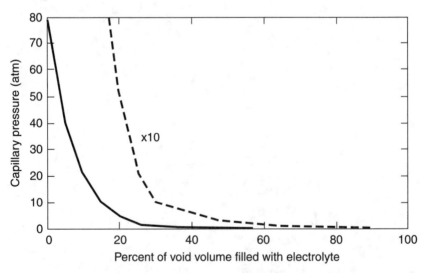

Figure 6.4. Capillary pressure of uncompressed asbestos separator vs. amount of void volume filled with electrolyte.

6.3 Asbestos Separator

While asbestos separator prevents all gaseous oxygen from passing through its fully wetted structure, it maintains an ionically conductive mat of fibers that is not subject to problems associated with dry-out as long as the fiber structure remains intact. However, the ionic resistance of the compressed mat of fine fibers is somewhat greater than that of the more open Zircar structure. As nickel electrodes expand throughout the life of a nickel-hydrogen cell, the compression of the asbestos can grow to further increase the ionic resistance. However, as long as the magnesium silicate fibers that compose the asbestos remain intact, this structure will generally maintain adequate ionic conductivity.

The magnesium silicates that make up asbestos fibers can include up to several hundred different mineral phases, each of which has a somewhat different resistance to reaction with the potassium hydroxide electrolyte. However, all these minerals react with aqueous potassium hydroxide, if given enough time, to form potassium silicate and magnesium hydroxide by the reaction shown in Eq. (6.1):

$$MgSiO_4 + 2KOH \longrightarrow Mg(OH)_2 + K_2SiO_4 \qquad (6.1)$$

The concentration of dissolved silicate in the electrolyte can thus build up over time to potentially affect cell performance, and the silicate fibers gradually turn into a highly compressed mat of magnesium hydroxide particles. This particle mat, which is highly compressible, can be squeezed to the point where its ionic conductivity becomes quite poor, leading to eventual problems with poor cell performance as a result of the high resistance.

Appropriate reprocessing of asbestos[***] has traditionally been used to eliminate its mineral constituents that react most rapidly with potassium hydroxide electrolyte, thus providing a separator that only very slowly experiences the structural damage associated with reaction (6.1). Such reprocessed asbestos has performed reasonably well in nickel-hydrogen cells for more than 15 years in some cases.

However, in some situations, typically when the reprocessing of the asbestos was not performed, cell impedance problems have been experienced quite early in life. As a general rule, cell problems from high impedance become very likely after about 10% of the potassium hydroxide in the cell has reacted with the separator by reaction (6.1). For this reason, a potassium hydroxide concentration of 38% has commonly been used in asbestos-separated nickel-hydrogen cell designs. While cell designs with 31–34% potassium hydroxide have also been made, they have sometimes experienced earlier problems from increased cell impedance.

[***]Typical reprocessing of asbestos has included a rinse in hot potassium hydroxide solution to dissolve out the more soluble magnesium silicate phases, followed by reconstitution of the fiber mat from the rinsed fibers.

Asbestos has also been a significant vector for carrying contaminants into nickel-hydrogen cells. The fibers of asbestos, in addition to consisting of naturally occurring minerals having a variety of potential mineral contaminants, present an extremely high surface area that can trap and adsorb a variety of airborne gaseous and particulate contaminants. Analysis of asbestos has commonly found clay particles from dust, organic materials from airborne emissions (pollutants), and a variety of other particulate contaminants specific to the battery factory itself. All of these contaminants serve as potential poisons to the hydrogen electrode catalyst, as materials that can degrade nickel electrode charge efficiency, or simply as sources of organic gases, carbonates, or carbon dioxide that can impede the utilization of hydrogen gas in the negative electrode. A high-reliability cell design should require verification that all such contaminants, whether originating from asbestos or some other cell material, are minimized.

6.4 Cell Electrolyte Properties

The electrolyte used in nickel-hydrogen cells, as indicated earlier in this chapter, is exclusively an aqueous solution of potassium hydroxide that can be from 26% to 38% by weight, depending on the cell design. The most commonly used concentrations in today's nickel-hydrogen cells are 26% and 31%. The electrolyte properties that are key to the performance of nickel-hydrogen cells are ionic conductivity, activity coefficient, and viscosity. The rate of hydroxide ion diffusion (diffusion coefficient D) can be determined from the viscosity η of the electrolyte by the Stokes-Einstein equation,

$$D = \frac{kT}{6\pi\eta r} \qquad (6.2)$$

where k is the Boltzmann constant, T is absolute temperature, and r is the radius of the diffusing molecule.

The key electrolyte properties to consider in the operation of nickel-hydrogen cells thus become the concentration, density, activity coefficient, ionic conductivity, and the viscosity. Figure 6.5 indicates how aqueous potassium hydroxide concentration (in moles/L) and activity coefficient are related to the weight percent concentration in the electrolyte, while Fig. 6.6 indicates the variation in electrolyte density with concentration. Figure 6.7 indicates how ionic resistance varies with concentration and temperature, while Fig. 6.8 shows how electrolyte viscosity changes with concentration and temperature.

While the molarity of the electrolyte increases nearly linearly with increasing weight percent (see Fig. 6.5), the activity coefficient increases exponentially. The density also exhibits a nearly linear increase with increasing weight percent concentration (see Fig. 6.6). The conductivity behavior shown in Fig. 6.7 is somewhat

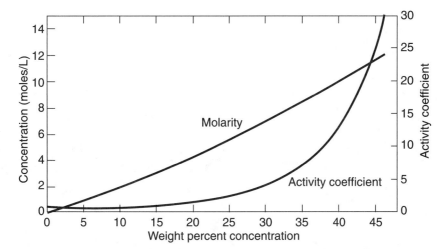

Figure 6.5. Variation in the molarity and the activity coefficient of aqueous potassium hydroxide electrolyte with weight percent concentration.

more complex, exhibiting a peak near 28%, and decreasing at lower or higher concentrations. The conductivity also increases significantly as temperature increases. The viscosity behavior shown in Fig. 6.8 shows viscosity increasing exponentially as concentration increases, and increasing significantly as temperature decreases.

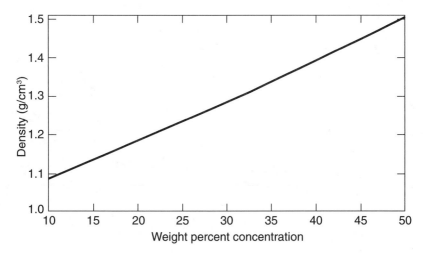

Figure 6.6. Variation in density of aqueous potassium hydroxide electrolyte with weight percent concentration.

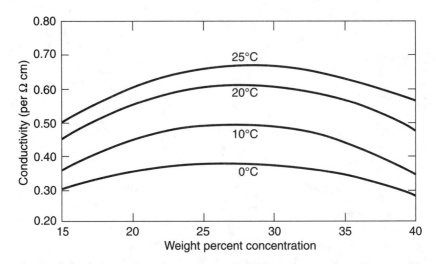

Figure 6.7. Variation in the conductivity of aqueous potassium hydroxide electrolyte with weight percent concentration throughout a range of temperatures.

It is not entirely unexpected that the most common electrolyte concentrations used in nickel-hydrogen cells (26% and 31%) bracket the concentration of maximum ionic conductivity at about 28%. As electrolyte concentrations become either very low or very high, poor ionic conductivity becomes a real issue in nickel-

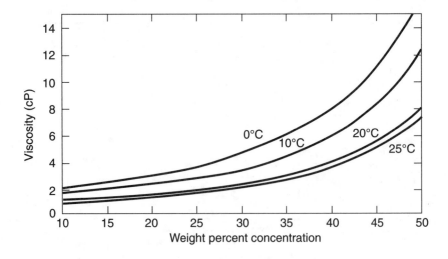

Figure 6.8. Variation in the viscosity of aqueous potassium hydroxide electrolyte with weight percent concentration throughout a range of temperatures.

hydrogen cells. Although the freezing point of aqueous potassium hydroxide electrolyte (see Fig. 6.9) is normally significantly less than the operating temperatures of nickel-hydrogen cells, at high concentrations and low temperatures electrolyte freezing can become an issue if localized electrolyte concentrations become sufficiently low. Figure 6.9 also shows why cells containing 31% potassium hydroxide are the least likely to experience problems with electrolyte freezing: because this concentration has the minimum freezing point. In addition, high electrolyte viscosity and poor diffusion can also become issues affecting cell performance at excessively high electrolyte concentration.

The vapor pressure of aqueous potassium hydroxide is a key property of the electrolyte in nickel-hydrogen cells because the composition of the gas phase within the cell is a key part of the dynamics of cell operation. A number of measurements of electrolyte vapor pressure have been reported as a function of the temperature and the concentration of the electrolyte. These measurements are summarized in Fig. 6.10 over the range of concentrations of electrolyte used in cells, and the temperatures typically used for cell operation.[6.1–6.4]

The data shown in Fig. 6.10 show some variability in the measurements of the vapor pressure of potassium hydroxide, depending on the source of the data. Data from three different sources are plotted at a temperature of 20°C. The differences between the various measurements appear to be greatest at the higher concentrations. The curve through each set of data points represents the best polynomial fit to the data points.

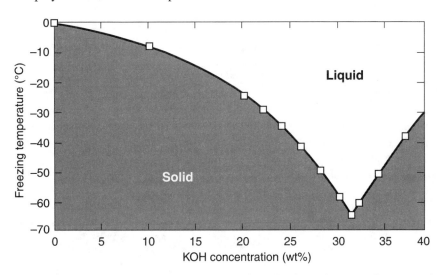

Figure 6.9. Freezing point of potassium hydroxide electrolyte as a function of concentration.

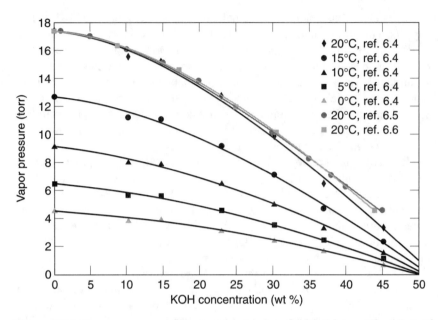

Figure 6.10. Vapor pressure of aqueous potassium hydroxide as a function of concentration for the range of temperature where nickel-hydrogen cells typically operate.

The total quantity and the distribution of electrolyte in a nickel-hydrogen cell are quite important. While hydroxide ions are not consumed during the course of the overall cell reaction, it is possible through extended overcharge to produce large amounts of γ-NiOOH in the nickel electrode. This phase incorporates 1/3 potassium hydroxide per nickel atom in the active material lattice,[****] and thus can partially deplete the electrolyte of potassium hydroxide if the cell contains an inadequate margin in its overall amount of electrolyte. Typical electrolyte fill levels are in the range of 2.8 to 3.5 g of electrolyte per A h of cell capacity. The range indicated here in the typical amount of electrolyte is dictated by the cell design; those designs that have thicker separators (dual-layer rather than single-layer) contain greater amounts of electrolyte.

In general, the appropriate amount of electrolyte in any cell design corresponds to filling 90–98% of the total void volume within the cell stack. Fill levels over 99–100% increase the risk of popping damage as the result of flooding at the

---

[****]An exception to the balance between the production and consumption of hydroxide ions occurs when the γ-NiOOH is formed in the nickel electrode. This phase takes one equivalent of hydroxide from the electrolyte for every three nickel atoms in the lattice.

edges of the stack. Fill levels lower than 90% can leave a separator that is too dry to accommodate electrolyte conductivity toward the end of the cell lifetime.

For every specific cell design, there is a probability of cell failure as the result of electrolyte issues that varies with the electrolyte fill and the age of the cell. At low fill levels, the dry-out of the separator increases the probability of cell failure, and at high electrolyte fill levels, the accumulated popping damage increases the chances of cell failure. Figure 6.11 shows an example of such a probability plot as a function of electrolyte fill for a nickel-hydrogen cell design containing 26% potassium hydroxide electrolyte and a single layer of Zircar separator. For nickel-hydrogen cell designs in general, one should avoid an excessively high electrolyte fill as well as an extremely low electrolyte fill, because these increase the risk of premature failure. Typically, a cell designer would desire an optimum electrolyte fill, which for the example of Fig. 6.11 is in the range of 2.75 to 2.9 g per A h of cell capacity. Interestingly enough, this same optimum fill level is generally found to be in the range where 90–98% of the void volume within the cell stack is filled with electrolyte.

Various electrolyte additives have been explored in attempts to improve nickel-hydrogen cell performance, but none have been found to provide significant long-term advantages. Lithium hydroxide additives have been tried, and they can provide some short-term performance benefit in terms of improved capacity. However, the lithium hydroxide additive has been found to stabilize the γ-NiOOH to the point where this phase can participate actively in the charge/discharge

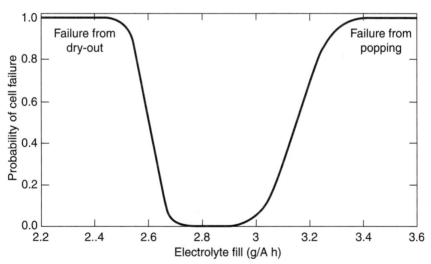

Figure 6.11. Probability of premature cell failure as a function of electrolyte fill for a design containing 26% potassium hydroxide and having a single layer of Zircar separator.

cycling of the cell. Because cycling this phase involves significant active-material expansion and contraction, the nickel electrodes can be physically stressed more than when the lithium hydroxide additive is not present. This added physical stress in the nickel electrodes has been found to reduce the cycling lifetime of the nickel-hydrogen cell. Because exceptionally long lifetime is a benchmark capability for the nickel-hydrogen cell, lithium hydroxide additives are generally not acceptable.

Additional electrolyte issues can involve contaminants. Of these, the anionic contaminants such as Cl^-, NO_3^-, $SO_4^=$, BO_4^{-3}, and $SiO_3^=$ have been found to affect cell performance in a number of ways, and they should be kept out of the cell as much as possible. Fortunately these contaminants do not appear to have large effects on cell performance until their levels exceed 200 to 1000 ppm. At levels of several hundred ppm, Cl^- and NO_3^- can accelerate nickel corrosion in the nickel electrode. At similar levels, sulfate ions can leach cobalt additives from the nickel electrode active material. Silicate and borate ions can significantly degrade electrolyte conductivity when present at a concentration greater than 1% by weight in the electrolyte. These relatively high levels are not likely to occur as the result of chance contamination, but can occur from long-term degradation of mineral separators by reactions with the electrolyte such as that indicated in reaction (6.1).

6.5 References

[6.1] P. Bro and H. Y. Kang, "The Low-Temperature Activity of Water in Concentrated KOH Solutions," *J. Electrochem. Soc.* **118**, 1430 (1971).

[6.2] A. Himy, *Silver-Zinc Battery Phenomena and Design Principles* (Vantage Press, New York, 1986), Appendix 2: KOH data.

[6.3] *International Critical Tables*, Vol. 3 (McGraw-Hill, New York, 1928), p. 298.

[6.4] R. A. Robinson and R. H. Stokes, *Trans. Faraday Soc.* **45**, 612 (1949).

7 Cell Dynamics

The individual components in a properly constructed nickel-hydrogen cell (i.e., electrodes, electrolyte, and separator) can operate together to provide a battery cell capable of exceptionally stable performance throughout many years of operation. As is the case for many types of battery cells, each of these components is often required to operate under conditions where it is thermodynamically unstable. However, the ability of a battery cell to operate without degrading under conditions of thermodynamic instability is essential for long-term operational use.

In an operating nickel-hydrogen cell, the charged active materials are unstable relative to self-discharge processes, and the nickel supporting structure in the nickel electrode is unstable relative to oxidation by the active material. If the cell voltage is allowed to drop below normal operating potentials, the active materials in the nickel electrode and the platinum catalyst in the negative electrode can become unstable. It is only through understanding these regions of instability, and how they are influenced by the interactions between the cell components, that one can properly design and optimally operate a nickel-hydrogen cell.

This chapter discusses the dynamics that allow thermodynamically unstable components to work together in a nickel-hydrogen cell. The dynamics of interactions between components in the cell are discussed in the context of how such interactions control the cell performance characteristics and degradation rates.

7.1 Charge Efficiency

For most nickel electrode–based battery cells, charge efficiency is dictated by the competition between the parallel nickel electrode energy-storage reactions and the oxygen evolution reaction during the recharge process. While this is clearly true of the nickel-hydrogen cell as well, the presence of hydrogen gas in direct contact with the charged nickel electrode active material enables additional processes that contribute to self-discharge and reduced charge efficiency. Although the charged nickel electrode is thermodynamically unstable relative to the evolution of oxygen by about 0.1 to 0.2 V, it is unstable by about 1.4 V relative to reaction with and self-discharge against hydrogen gas. This thermodynamic instability relative to hydrogen places some real constraints on how nickel-hydrogen cells can be successfully stored and operated over the long term.

7.1.1 Nickel Electrode Self-Discharge

Self-discharge normally occurs in all nickel electrodes as the less stable, charged active material decomposes to form the more stable, discharged active material and oxygen gas. This decomposition is an electrochemical process having a rate that increases exponentially with the voltage of the nickel electrode, and it thus tends to dominate the self-discharge in nickel-hydrogen cells when they are highly charged (i.e., at 80% state of charge or higher). However, when the cell state of charge falls to lower levels, self-discharge processes involving hydrogen gas can dominate. The rate of the oxygen evolution self-discharge process drops to zero at a nickel electrode state of charge of approximately 10%, which is the point where the reversible potential of the nickel active material equals the oxygen potential (dependent on the oxygen partial pressure in the nickel electrode).

Hydrogen can react with the nickel electrode to give self-discharge by two distinct processes. The first is referred to as the "direct" self-discharge process, because it involves the direct reaction of hydrogen gas molecules with the surface of the charged active-material particles. The rate of this process is directly proportional to both the state of charge of the nickel electrode and the activity of hydrogen gas in the cell. The second process is referred to as the "catalytic" self-discharge process, because it involves catalytic oxidation of hydrogen gas on the nickel sinter surfaces to produce electrons and protons that can reduce charged active-material sites that are close to the nickel sinter surfaces. The rate of this process is directly proportional to the activity of the hydrogen gas (hydrogen pressure to first order), but it is independent of the state of charge of the nickel electrode. It is this catalytic process that can bring the potential of a discharged nickel electrode all the way down to hydrogen potential.

The activation energies for these hydrogen-induced self-discharge processes make their rates highly dependent on temperature, and thus at temperatures greater than 10°C they can become major contributors to cell self-discharge. These self-discharge processes, which are described in more detail in chapter 4, are the principal factors limiting nickel-hydrogen cell charge retention time, as well as the usable capacity at elevated temperatures.

7.1.2 Parasitic Hydrogen Reactions

Because the charged nickel electrode is unstable relative to hydrogen by about 1.4 V, a number of parasitic reactions are possible in addition to self-discharge as a result of interactions between hydrogen gas and the nickel electrode. At potentials below about 0.98 V *vs.* H_2, hydrogen can react with the Co(III) additive sites in the active material, causing them to be reduced to Co(II) hydroxide. While Co(II) hydroxide is soluble in alkaline electrolyte, the cobalt additive typically exists as defect sites within the nickel hydroxide lattice, thus limiting its solubility when reduced in this way to the divalent state.

When the potential of the nickel electrode falls below 0.28 V *vs.* H_2, the nickel oxides that passivate the nickel sinter surfaces begin to be reduced slowly back to nickel metal. This process depassivates the nickel metal surfaces and makes them much more catalytically reactive to hydrogen gas. As the nickel electrode potential falls further, to less than 0.10 V *vs.* H_2, the nickel hydroxide active material in contact with the nickel sinter surfaces begins to be reduced to nickel metal. This series of reduction processes leaves the nickel sinter as a relatively active catalytic surface that can serve as a hydrogen-evolving electrode when its potential is reduced to 0.0 V or less *vs.* H_2. Evolution of hydrogen gas becomes the dominant reaction at the discharged nickel electrode when a nickel-hydrogen cell is driven into reversal (below a cell voltage of zero).

7.1.3 Platinum Dynamics

Platinum is the noble metal used in the hydrogen electrode to catalyze the hydrogen evolution and recombination processes that occur at the negative electrode during nickel-hydrogen cell charge and discharge. Because platinum metal is noble, it is not reactive in the cell environment under most conditions. However, when a nickel-hydrogen cell has nickel precharge and it is fully discharged, which is a typical storage condition for nickel-hydrogen cells, the platinum metal in the hydrogen electrode is brought up to a potential near the potential of the partially charged nickel electrode and held at that potential during storage by the partial pressure of oxygen within the cell. Under these conditions, platinum metal will develop an oxide coating consisting of PtO and $Pt(OH)_2$. These materials are slightly soluble in alkaline electrolyte, dissolving to form the platinate complex ion, $Pt(OH)_4^=$. The $Pt(OH)_4^=$ ions can then migrate to the partially charged nickel electrode, where they can slowly react with charged nickel and cobalt sites in the active material to form stable Pt/Ni/Co oxyhydroxide sites within the active-material lattice. The formation of this compound can be readily noted after years of cell storage by a characteristic reduction peak seen at about 1.09 V (*vs.* H_2), as detailed in chapter 4. This platinum-containing compound is stable as long as it remains in the oxidized state, but dissociates to release its platinum back into the alkaline electrolyte if it is electrochemically reduced.

Therefore, if a nickel-precharged nickel-hydrogen cell is stored for many years, it can accumulate a large amount of oxidized platinum in the nickel electrodes. Because there is a large excess of platinum in the hydrogen electrodes beyond the minimum amount needed, the hydrogen electrodes continue to function well. However, when the cell is removed from storage and subsequently cycled, there will come a point in its lifetime when the active nickel precharge has largely disappeared. At this point, a complete cell discharge below 1.09 V will reduce the platinum complexes that have built up in the nickel electrodes, and begin the process of releasing the platinum back into the electrolyte as the platinate ion, $Pt(OH)_4^=$. This release is quite slow as a result of slow diffusion of the large platinate ion out of the active material, as well as a relatively low solubility in the electrolyte. For these reasons, when the cell is recharged promptly, only a partial release of the platinum back into solution typically occurs.

If discharge is continued to drive the potential of the cell below 0.98 V, the electrochemical reduction of $Pt(OH)_4^=$ ions from the electrolyte to platinum metal on the conductive surfaces of the nickel electrodes will begin to compete with discharge of residual nickel electrode capacity. If the cell is allowed to discharge for a long period below 0.98 V, such as occurs during a lengthy resistive letdown, significant amounts of platinum metal can be plated onto the nickel electrodes. This plating occurs as a thin platinum foil on the conductive metal surfaces of the nickel electrode, rather than dendritically, and therefore will not normally result in a metallic short circuit between the electrodes in the cell.

If a cell that has platinum metal plated onto the nickel electrodes is recharged, a number of processes will begin that are not a part of the normal chemistry in the nickel-hydrogen cell. First, the platinum on the nickel electrode will form a protective platinum oxide coating that will slowly dissolve in the electrolyte. Second, as long as recharge current is flowing through the cell, dissolved platinate ions will be gradually reduced back to platinum metal at the hydrogen electrode. The combination of these two processes will act to strip the platinum from the nickel electrode throughout a period of several weeks of cell trickle charge. However, until the platinum is fully stripped from the nickel electrode, the charge efficiency of the nickel electrode will be low. The presence of platinum metal on the nickel electrode can make the nickel-hydrogen cell self-discharge more rapidly than normal, basically by acting as an oxide-covered hydrogen electrode that is in direct contact with a charged nickel electrode.

Fortunately, this dynamic interplay between the platinum and the nickel electrode is reversible. Maintaining a trickle charge current through the charged cell for several weeks will strip all platinum from the nickel electrodes and plate it all back onto the hydrogen electrodes. While the replated platinum certainly does not have the extremely high surface area that it originally had in the hydrogen electrode, it is restored back into a chemically stable environment. However, these dynamics may be repeated many times before the reservoir of oxidized platinum in the nickel electrode active material is completely reduced and dissolved back into the electrolyte of the cell. A full knowledge of these dynamics is crucial to understanding and controlling the behavior that can occur in nickel-hydrogen cells after lengthy periods of storage.

7.2 Hydrogen Gas Management

The dynamics of gas management are central to the proper operation of nickel-hydrogen cells. The hydrogen electrode utilizes hydrogen gas as its active material, involving gas pressures ranging from near 0, at low states of charge, up to 1000 psia or more in the fully charged state. For the hydrogen electrode to function properly at high charge or discharge rates, an adequate hydrogen flow rate to and from the catalytic surfaces must be assured.

The nickel electrode is also involved in the gas-management picture, because it can evolve oxygen gas at a significant rate during periods of overcharge. The oxygen gas can flow from the nickel electrode by both convective and diffusive flow processes, to eventually electrochemically react with hydrogen gas at the catalytic surfaces in the hydrogen electrode. The gas-management dynamics within the nickel-hydrogen cell are closely linked to managing the flow of oxygen and hydrogen gases within the cell and between the electrodes.

7.2.1 Hydrogen Flow Dynamics

Hydrogen gas is generated at the platinum catalyst sites in the hydrogen electrode during nickel-hydrogen cell recharge by electrochemical reactions that involve the consumption of water and the production of hydroxide ions. This leads to an increase in electrolyte concentration in the porous hydrogen electrode during recharge, which is accompanied by a decrease in electrolyte concentration within the nickel electrode during recharge. The hydrogen gas that is formed during recharge can dissolve in the electrolyte, where it can move by either convective or diffusive flow of the electrolyte, or it can move in the gas phase through the void regions of the negative electrode and flow through the porous Teflon backing on the hydrogen electrode into the gas screen on the back side of each hydrogen electrode. Once in the gas screen, the hydrogen gas will flow out of the edges of the gas screens, along the edges of the electrode stack, and finally into the gas reservoir regions in the domes of the cell pressure vessel.

During discharge, the hydrogen gas flow processes are reversed in response to diminished hydrogen activity within the hydrogen electrode, as hydrogen gas is electrochemically oxidized (or recombined) to form water and consume hydroxide ions. The convective flow of hydrogen between the electrode stack and the gas storage spaces of the pressure vessel will transport water vapor to and from the electrodes, as well as transport heat between the electrode stack and the pressure vessel walls.

Any process that impedes the hydrogen gas flow dynamics within the nickel-hydrogen cell can significantly degrade the performance of the cell. Generally the evolution of hydrogen at the hydrogen electrode during recharge of the cell, followed by the subsequent movement of that hydrogen into the gas storage spaces in the cell, occurs without being impeded by factors such as excessive electrolyte. In fact the evolution of hydrogen gas during recharge, as well as oxygen gas at the nickel electrode during overcharge, will tend to displace any excess electrolyte and force it out of the electrode stack. However, during discharge there are several processes that can block the required flow of hydrogen gas into the negative electrodes, thus preventing the cell from facile high-rate discharge. Excessive electrolyte in the gas screens can block hydrogen flow from the gas spaces of the pressure vessel into the catalyst-containing structures in the hydrogen electrode. Similarly, the presence of significant quantities of inert gas, such as nitrogen, helium, or argon, can cause the accumulation of a backpressure of inert gas within the gas screen during high-rate discharge, and thus block the flow of hydrogen gas.

7.2.1.1 Effect of Liquid Electrolyte on Hydrogen Flow

If the gas screens that serve as gas conduits between the hydrogen electrode and the gas reservoirs in a nickel-hydrogen cell become blocked or flooded with excessive electrolyte, the hydrogen electrodes will lose the ability to discharge at high rate, and the cell performance will be severely degraded. Discharge at very low rates may still be possible as a result of the diffusion of dissolved hydrogen through the liquid electrolyte that blocks the gas flow paths, particularly if intermittent discharge is applied, which gives the concentration of dissolved oxygen an opportunity to recover by diffusional transport. If a significant portion of the gas screens becomes blocked with electrolyte, it will typically become increasingly difficult to support high-rate discharge as the hydrogen pressure within the cell drops. Thus, the signature for this problem is increasing cell polarization as the pressure drops with no loss of open-circuit voltage, until eventually a pressure is reached below which discharge cannot be further supported without allowing a period of open-circuit recovery.

One mechanism for flooding of gas screens occurs as a result of contaminants in the cell that increase the wettability of the gas screens. Gas screens are typically made of polypropylene, and they have a mesh size that is large enough so that they do not trap free electrolyte. During normal cell activation, most free electrolyte within the gas screens will flow out as the cell is drained under gravitational forces. However, if the mesh size is too small, or if surfactants are present in the cell, large amounts of electrolyte can remain trapped in the gas screens.

The gas screen is in direct contact with the porous Teflon backing of the hydrogen electrode. The Teflon layer is normally hydrophobic so that its pores remain open to accommodate hydrogen flow; however, surfactants can also result in flooding of the pores in the Teflon backing with electrolyte. Inspection of gas screens removed from nickel-hydrogen cells for excess electrolyte can detect a problem of this kind. Inspection of the Teflon backing on the hydrogen electrodes can also indicate whether the porous Teflon layer has become wetted. Normally, the Teflon backing is white, indicating that its pores are not filled with electrolyte. If the pores become wetted with electrolyte, the appearance of the backing changes to a translucent one that is easily discerned by inspection of hydrogen electrodes taken from a nickel-hydrogen cell.

A second mechanism for flooding of gas screens can occur when there is too much electrolyte in the electrode stack. During discharge, water is produced in the hydrogen electrode by the electrochemical reaction of hydrogen gas with the electrolyte. If this water is not efficiently transported into the separator because it is already saturated with electrolyte, the liquid can weep through the porous Teflon backing and into the gas screen, and thus block gas flow into the hydrogen electrode. This problem is prevented by ensuring that the separator is not fully saturated with electrolyte, and by making sure that the distribution of pore sizes in the active catalyst layer of the hydrogen electrode (which is in contact with the separator) is appropriate to support facile wicking of the water that is produced during discharge back into the separator.

7.2 Hydrogen Gas Management

7.2.1.2 Effect of Inert Gases on Hydrogen Flow

The presence of inert gases in the pressure vessel of a nickel-hydrogen cell at pressures of several atmospheres or less would not be expected to degrade the performance of the electrodes or the cell, because these gases are electrochemically inert. Such inert gases could include nitrogen, the presence of which can result from contamination of the cell with atmospheric gases during activation; helium that was used in leak-checking the pressure vessel; and unpurged argon, if the cell was stored in the dry state prior to activation. Of these gaseous contaminants, nitrogen is probably the most commonly seen.

These gases, however, can have a significant physical impact on the dynamics of the hydrogen electrode when it is discharged at high rates, and they can significantly reduce the high-rate discharge capacity of the cell. During normal discharge, hydrogen gas will convectively flow from the pressure vessel domes where it is stored into the gas screens and the hydrogen electrode, where it will participate in the electrochemical discharge reaction. However, if another inert gas is mixed with the hydrogen, the inert gas will also be convectively carried into the gas screen and the hydrogen electrode. The inert gas will not be consumed in any electrochemical processes, but it will remain in the gas screen and the hydrogen electrode. As discharge continues, the concentration of inert gas will build up in the gas screen and the pores of the hydrogen electrode, until eventually its local partial pressure will be very close to the remaining hydrogen pressure in the pressure vessel. At this point in the discharge, it will become impossible to continue discharge because the inert gas will be blocking the convective flow of hydrogen into the hydrogen electrode. For example, a 12 psi partial pressure of nitrogen in a nickel-hydrogen cell (which is the amount introduced by typical contamination with atmospheric gas) will as a rule result in loss of 5–10% of the typical cell capacity when discharged at the C/2 rate.

Discharge can be continued at a rate consistent with the rate of diffusion of inert gas back into the pressure vessel; however this usually occurs only at a relatively low rate. Discharge may also be continued after a period of rest, during which the inert gas becomes intermixed back into the remaining hydrogen by gaseous diffusion. However, as the hydrogen becomes depleted relative to the inert gas, such diffusional mixing becomes increasingly difficult. This capacity loss mechanism will thus exhibit a characteristic diffusional signature for capacity recovery following open-circuit rest periods. However, the least ambiguous method for detecting this condition is an analysis of the amount and composition of the gas in a suspect cell after it has been fully discharged of all residual capacity at low rate.

7.2.1.3 Effect of Icing on Hydrogen Flow

When a nickel-hydrogen cell is operated at temperatures significantly below 0°C, its internal components may be at temperatures below the freezing point of water. Because the potassium hydroxide electrolyte, which is typically at 26–31%

concentration by weight, freezes at temperatures below –30°C, this normally does not present any problems with cell operation. However, if a fully charged nickel-hydrogen cell is held at temperatures of about –10°C or less while it is being trickle-charged or open-circuited, and then it suddenly begins to discharge at a high rate, ice can be temporarily formed in the negative electrodes. The formation of ice can temporarily increase the impedance of the negative electrodes and drive the discharge voltage to anomalously low levels.

When high rate discharge is initiated, water production commences within the hydrogen electrodes, along with consumption of hydroxide ions. This leads to a large concentration gradient, with nearly pure water being exposed to the cold surfaces within the hydrogen electrodes. Additionally the hydrogen oxidation process is endothermic, which further drops the local temperature of the platinum catalyst material within the hydrogen electrode. The result is that the water initially produced in the hydrogen electrode during discharge can freeze, forming a high-resistance barrier of ice in the hydrogen electrode that blocks hydrogen gas flow and hydroxide ion diffusion to the catalyst surfaces. The formation of ice occurs whenever the rate of diffusion of hydroxide ions between the hydrogen electrode and the separator is not so great as to keep the electrolyte within the hydrogen electrode concentrated enough to avoid freezing.

The process of ice formation results in an anomalous transient drop in cell voltage at the start of discharge, as illustrated in Fig. 7.1. The drop in cell voltage arises from ice formation blocking hydrogen flow as well as hydroxide ion diffusion at catalyst sites, thus starving the reaction sites for the reagents required by the discharge process in the hydrogen electrode. The high resistance produces heat resulting from the ohmic thermal dissipation that accompanies the anomalous voltage drop in Fig. 7.1. Heat is also copiously produced during discharge in the nickel electrode, and this heat will also diffuse into the hydrogen electrode. These sources of heat will typically melt the ice in the hydrogen electrodes after several minutes, allowing a complete recovery of the voltage to a normal discharge level and enabling cell discharge to be completed as shown in Fig. 7.1. The transient voltage signature of Fig. 7.1 provides a characteristic indication that a nickel-hydrogen cell is being operated at a temperature below the level where diffusion processes in the hydrogen electrode prevent the formation of ice.

The solution to the problem of degraded cell performance during discharge as a result of ice formation is either to raise the cell temperature before beginning discharge or to increase the rate of diffusion in the hydrogen electrode. For a temperature increase to be effective, the temperature must be increased to a level where diffusion processes can maintain the hydroxide ion concentration in the water produced during discharge, thus guaranteeing that it will not freeze. This temperature level is cell-design–dependent but is typically lower than –10°C for cells with 26% potassium hydroxide electrolyte, and lower than about –15°C for cells with 31% potassium hydroxide electrolyte.

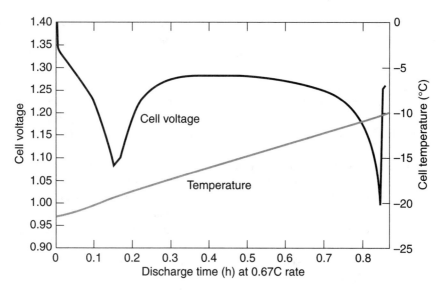

Figure 7.1. Transient voltage drop at the start of a 0.67 C discharge when a nickel-hydrogen cell is initially at about −20°C, followed by voltage recovery after the cell warms up during discharge.

It is also possible to modify a cell design to increase the rate of diffusion in the hydrogen electrode, and thus improve low-temperature cell performance. A simple way to do this is to use a dual-anode design rather than a back-to-back design. The dual-anode design, by placing a negative electrode on each side of each negative electrode, decreases the current density at the hydrogen electrode by a factor of two and increases the surface area available to support diffusion processes.

7.3 Oxygen Gas Management

During normal operation, the nickel-hydrogen cell contains hydrogen gas at pressures up to about 1000 psia. However, whenever a nickel-hydrogen cell is recharged to high states of charge, the nickel electrodes will evolve oxygen gas into this hydrogen-containing environment. When the nickel electrodes near the point of full charge, essentially all the recharge current at the nickel electrode goes into the evolution of oxygen gas. The oxygen gas from the nickel electrode, combined with the hydrogen-containing atmosphere in the cell, produces a potentially unstable mixture, particularly if the oxygen comes into contact with the platinum catalyst. Any oxygen that contacts the catalyst surfaces will undergo rapid catalytic reaction with the hydrogen to produce water and heat. If the oxygen is present in the gas phase at a concentration greater than about 2% of the hydrogen, it can initiate a catalytic chain reaction in the hydrogen to propagate an explosive recombination reaction. The shock wave from this type of event produces an audible sound, which is referred to as a "popping" event in a nickel-hydrogen cell. (Popping events are discussed in detail in subsection 7.3.3.) The shock waves from these popping events can damage nearby electrodes and separators, and in severe cases can result in cell short circuits. Because of the potential for releasing large amounts of energy in these popping events, it is critical for a successful nickel-hydrogen cell design to properly manage the evolution of oxygen gas, allowing it to recombine gracefully without causing resultant damage to other cell components.

Two somewhat differing designs have been used in nickel-hydrogen cells throughout the years to manage the dynamics of oxygen flow within the cells. The first uses a low bubble pressure separator (such as Zircar or a gas screen), which passes gas at a low differential pressure and can channel oxygen gas to the hydrogen electrode for recombination at relatively high rates. The second approach uses a high bubble pressure separator (such as asbestos), which acts essentially as a gas-impermeable gasket, forcing most oxygen out the edges of the electrode stack, where recombination must occur with hydrogen.

7.3.1 Oxygen Flow Dynamics with a Low Bubble Pressure Separator

As the state of charge in the nickel electrodes increases during cell recharge, oxygen gas is evolved within the porous nickel electrodes at an increasing rate. This oxygen gas will displace hydrogen gas from the pores of the nickel electrodes, and it will build up in the gas phase as well as in the form of dissolved oxygen in the electrolyte within the nickel electrodes. The concentration of oxygen in the porous nickel electrodes will thus be controlled by the balance between the rate of oxygen evolution and the rates of diffusive and convective flow of oxygen from the nickel electrode. When the nickel-hydrogen cell is in overcharge at rates of about C/10, models of the oxygen flow dynamics within nickel electrodes indicate, as a rule, that the nickel electrodes typically predominantly contain oxygen, as illustrated in Fig. 7.2.

As indicated in Fig. 7.2, when a cell is in overcharge, the oxygen concentration drops rapidly once the oxygen leaves the nickel electrode and flows into the separator. The oxygen concentration drops rapidly through the thickness of the separator, until it falls to near 0 at the catalytic surfaces of the hydrogen electrode, where the oxygen reacts rapidly with hydrogen on the catalyst surfaces. Figure 7.2 also shows that a crack on the surface of the nickel electrode can increase the level of oxygen that reaches the hydrogen electrode near the crack site.

As the rate of oxygen evolution increases, the concentration of oxygen in the separator increases until eventually it can reach concentrations above the explosion limit in hydrogen (2 mole percent). At this concentration, it becomes possible for hydrogen free radicals formed at the catalyst surface to ignite a chain reaction. The chain reaction can propagate throughout any gas-filled voids, as long as they

7.3 Oxygen Gas Management

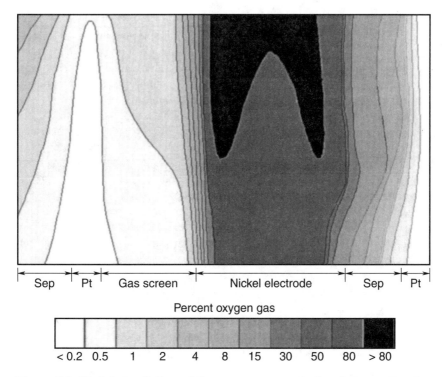

Figure 7.2. Model simulation of the oxygen concentration between the electrodes in a recirculating stack nickel-hydrogen cell while in C/10 overcharge. The top of the figure is adjacent to the pressure vessel wall, and the core is at the bottom. A small crack was modeled in the surface of the nickel electrode on the side toward the separator, which caused an oxygen level of about 6% to impinge on the platinum in the hydrogen electrode.

contain more than 2% oxygen and as long as the hydrogen free radicals do not become quenched by collision with a solid or liquid surface. If the oxygen-filled voids within the separator are sufficiently small, such oxygen-recombination chain reactions are quenched rapidly enough that they do not release enough energy and heat to do any damage to the nearby electrodes and separator. The keys to managing oxygen flow and recombination in this way are to have a separator thick enough to keep high oxygen concentrations away from the hydrogen electrode, to maintain enough electrolyte in the separator to prevent large oxygen-filled voids, and to maintain sufficient void space that gas flow will not be unduly restricted. The bundles of fibers present in Zircar separator have proven capable of providing an excellent balance between the gas flow rate and the amount of electrolyte entrained in the separator, particularly when two layers of Zircar separator are utilized.

7.3.2 Oxygen Flow Dynamics with a High Bubble Pressure Separator

When a nickel-hydrogen cell design utilizes a high bubble pressure separator, such as asbestos, oxygen gas generated in the nickel electrode during overcharge is not able to move through the separator to recombine with hydrogen at the hydrogen electrode. In this case, the oxygen gas is channeled out of the nickel electrodes at their edges. The oxygen gas from overcharge will build up to whatever pressure is needed to force its way out to the edges, including compression of the separator to open up gas flow paths if required. To minimize separator deformation, it is desirable for most of the oxygen gas to flow to the edges of the stack out the backside of the nickel electrodes in back-to-back designs. Once at the edges of the electrode stack, the streams of oxygen gas will contact the hydrogen electrodes and recombine with hydrogen at the outer edges of the stack.

Cell designs with high bubble pressure separators are generally not tolerant of extensive overcharge, even at only moderate rates. The oxygen that is evolved can build up differential pressures that can deform and damage the separators and the hydrogen electrodes. When the oxygen does find its way to the edge of the electrodes, excessive rates of oxygen evolution can lead to "edge-popping" and damage to the edges of the separator and electrodes. However, with adequate controls on overcharge rate, high bubble pressure separators such as asbestos have provided adequate oxygen gas management in nickel-hydrogen cells.

High bubble pressure separators must be mechanically able to hold together under the deformation caused by oxygen being evolved from the nickel electrode. If there is a weak spot in a high bubble pressure separator, if the asbestos is overly wet, or if the overcharge rate is too high, the separator can come apart under the applied gas pressure to allow oxygen to flow directly through to the hydrogen electrode. This type of event constitutes a breakdown of the oxygen management process and typically leads to a short circuit in a nickel-hydrogen cell.

Separator concepts have been employed that attempt to combine the gas flow behavior of low and high bubble pressure materials to obtain improved oxygen management in nickel-hydrogen cells. Examples of such composite separators include Zircar/asbestos and asbestos/Zircar layers. With the Zircar layer against the nickel electrode, oxygen is channeled to the edges of the electrodes without building up a high internal pressure that could damage the asbestos layer of the hydrogen electrode. Alternatively, the Zircar layer against the hydrogen electrode provides support for the asbestos as it channels the pressurized oxygen gas to the edges of the stack. The supported asbestos is better able to tolerate compression by the oxygen flow without being disrupted and will buffer the hydrogen electrode from any damage by the compression of the asbestos.

7.3.3 Explosive Oxygen Recombination (Popping)

One of the key issues that must be resolved in any successful nickel-hydrogen cell design is the management of oxygen recombination with hydrogen in an

7.3 Oxygen Gas Management

environment where both gases may be present in close proximity at high partial pressures. During normal cell operation, the hydrogen electrode and the gas reservoirs in a nickel-hydrogen cell typically contain 500–1000 psi of hydrogen at the times when the cell is in overcharge and oxygen evolution can occur in the nickel electrodes. Thus, during overcharge the void regions in the nickel electrode will typically contain a high pressure of relatively pure oxygen gas. The oxygen gas will flow out of the nickel electrode, mixing with hydrogen gas, until eventually the oxygen contacts the platinum catalyst in the hydrogen electrode. The platinum catalyst will initiate rapid electrochemical recombination of the oxygen with hydrogen gas.

If there is a pocket containing a mixture of oxygen and hydrogen gas at the surface of the platinum catalyst, it is possible for the catalyst to initiate a chain reaction in the gas that allows rapid and explosive recombination to occur. For such a chain reaction to propagate through a "pocket" of gas there must be a sufficient concentration of both hydrogen and oxygen present, the pocket must have a critical size, and the pocket must not contain frequent solid or liquid surfaces that can quench the propagating chain reaction. When such a chain reaction is initiated in a gas pocket in a nickel-hydrogen cell, the result is referred to as "popping" because of the audible explosive sound made by the event. A number of different types of popping are possible in nickel-hydrogen cells, depending on the location and the type of gas spaces in which the events occur.

When Zircar separator is compressed against a hydrogen electrode, there are many microscopic gas spaces where these two components come into direct contact. While these gas spaces are not large, occasionally they have sufficient size to enable a "micro-popping" event to be initiated in them during cell overcharge. These are extremely small popping events that release small amounts of energy; they often result in the formation of a small "pinhole" in the catalyst layer on the hydrogen electrode. These pinholes are readily seen by simply holding a used hydrogen electrode up to a light, which will make them appear as small lighted spots on the black surface of the hydrogen electrode. They do not degrade electrode performance significantly, as long as their frequency is not excessive.

The popping events most likely to cause failure of nickel-hydrogen cells are referred to as "edge-popping" events. They occur whenever oxygen builds up in the relatively large gas space at the edge of a hydrogen electrode. Because this gas space is large, it can contain significant amounts of oxygen gas, and it can initiate a popping event that has a significant shock wave and dissipates a significant amount of energy. Such edge-popping typically occurs when a significant pool of electrolyte accumulates at the edge of the hydrogen electrode, such as may occur if a cell is excessively filled with electrolyte or if it is operated on its side in a gravitational field. Oxygen will flow from the separator and nickel electrode into the pool of electrolyte, where it will form a growing bubble of high-pressure oxygen gas that is protected from facile mixing with hydrogen by the liquid electrolyte. Eventually the bubble will grow large enough to touch the surface of the platinum catalyst in the hydrogen electrode, which will initiate a large popping event. It is not unusual for edge-popping events of this kind to melt or disrupt the edges of the gas screens, the separators, and the hydrogen electrodes. If such edge-popping is allowed to continue as the nickel-hydrogen cell is charged and discharged, it will eventually result in the accumulation of enough damage to the edge of the electrode stack and separator for a short circuit to occur in the cell.

The oxygen flow dynamics described earlier for low bubble pressure separators assumed a uniform flow of oxygen from the nickel electrode and through the separator to recombine at the hydrogen electrode in the form of either gaseous or dissolved oxygen. In reality, the flow of oxygen may not be uniform. There are often gas channels within the nickel electrodes that open to the surface at the sites of cracks or other microscopic surface openings. During overcharge, oxygen can be channeled to and stream from these openings at rates significantly greater than those expected if there were a uniform flow of oxygen. These streams of oxygen will impinge upon the separator and create a localized oxygen-rich region in the separator. The separator must disperse such oxygen streams before they impinge upon the platinum catalyst in the hydrogen electrode, or disruptive popping events can be initiated by rapid recombination of oxygen and hydrogen in these streams. If oxygen is streaming at sufficiently high rates from channels in the nickel electrode, it is even possible for the explosive recombination of oxygen with hydrogen to propagate back into the channels in the nickel electrodes, typically resulting in damage to the nickel electrodes. These in-stack popping events can in some cases disrupt the surface of the nickel electrode and the separator enough to allow nickel electrode fragments to penetrate the separator and cause a cell short circuit.

A final type of popping event involves the escape of oxygen into the gas reservoirs in the nickel-hydrogen cell. Oxygen does not generally escape in significant quantities into the gas reservoirs, because to get into them it must flow along the narrow gap between the pressure vessel wall and the edge of the electrode stack. In flowing along the electrode stack, the oxygen is stripped from the gas stream by reaction with hydrogen at the edges of each negative electrode in the stack. If this did not occur, it would be possible to build up oxygen in the gas reservoirs to levels that could sustain an explosive chain reaction with the hydrogen gas, which would probably result in explosion of the pressure vessel. Such a situation could potentially arise from runaway overcharge, particularly in combination with a cell design that inappropriately allowed excessive oxygen to escape into the gas reservoirs. The only known incidents that may have occurred from such a cell-level popping event are the explosion of the nickel-hydrogen battery photographically documented and reported[7.1] on the Soviet Ekran-2 satellite in June of 1978, as well as similar explosive events attributed to nickel-hydrogen batteries on the Cosmos-

1691 and Cosmos-1823 satellites. It should be noted that these nickel-hydrogen cells had a design that is significantly different than those used in the non-Soviet nickel-hydrogen cells made during the past thirty years, which have never exhibited any internal explosive event that breached the pressure vessel.

7.3.4 Recombination Wall Wicks

A design feature that has been used in some nickel-hydrogen cell designs is the recombination wall wick, which includes stripes of platinum catalyst sites in the wall wick on the cylindrical portion of the inner pressure vessel wall. The purpose of these catalytic sites on the pressure vessel wall is to enable any oxygen that escapes from the cell stack to undergo recombination with hydrogen directly on the pressure vessel wall. This feature allows a portion of the heat produced during overcharge to be deposited directly on the pressure vessel wall, where it can be more easily dissipated to the environment. However, these recombination wall wicks also significantly affect the management of oxygen in the nickel-hydrogen cell.

The functioning of the recombination wall wick has been examined using the first-principles model outlined in chapter 8 for nickel-hydrogen cells. This model includes the convective and diffusive transport of oxygen and hydrogen needed to carefully examine how these recombination wall wicks function. A finite-element cell model was constructed that included realistic geometries and structures for the nickel electrodes, separators, and hydrogen electrodes in a 100 A h cell design that included four stripes of platinum recombination sites along the stack from the top of the cell to the bottom. This design assumed both positive and negative electrode leads connected to "rabbit-ear" terminals at the top of the cell.

The model was run to determine the dynamics of oxygen recombination during steady-state overcharge at a C/10 rate, both with and without the recombination wall wicks. During overcharge, the concentration of oxygen in the nickel electrodes was quite high (40–50 atm), while the overall gas pressure within the cell was about 65 atm. The concentration of oxygen predicted by the model at the surface of the wall wick was examined as a function of position, from the top of the cell stack to the bottom of the stack, giving the results in Fig. 7.3.

It is clear from this analysis that the recombination sites on the wall wick significantly reduce the amount of oxygen that can move out of the electrode stack and through the gap between the stack and the pressure vessel wall. The model shows that the wall wick recombination sites act as chemical scrubbers, effectively removing most oxygen from the gas that streams from the stack, through the wall gap, and into the gas reservoirs above and below the stacks. At high overcharge rates, the amount of oxygen that is removed from the gas stream (that would otherwise have to recombine at the edges of the hydrogen electrodes) becomes significant, accounting for about 10% of the total overcharge heat for the C/10 overcharge condition shown in Fig. 7.3.

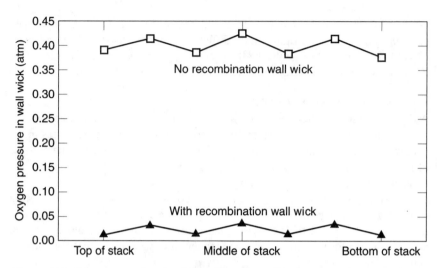

Figure 7.3. The oxygen level predicted at the wall wick surface in a 100 A h nickel-hydrogen cell model (C/10 overcharge, 10°C), with and without recombination stripes in the wall wick. The four low points for each design are at each recombination stripe location, and the three higher points are between the stripe locations.

7.4 Thermal Management

During normal operation of nickel-hydrogen cells, heat is produced during discharge from the electrochemical discharge process, and during recharge heat is either consumed (causing cell cooling) or produced (causing cell heating). Heat is also produced by self-discharge and by oxygen recombination processes during both charge and discharge. Additionally, there are significant local thermal effects arising from the resistance of the cell and electrode leads, as well as the work of compression or expansion of hydrogen gas in the fixed-volume pressure vessel. The dynamics of these interacting thermal processes can make thermal management of nickel-hydrogen cells a challenge. This section discusses some of the key factors involved in the dynamics of nickel-hydrogen cell thermal management.

7.4.1 Latent Heat

Some chemical or electrochemical processes produce heat, but because of kinetic limitations, these processes store a significant portion of their expected heat output in the form of reactive chemicals. This type of heat is latent heat, because it will be released at a later time when these chemicals react, irrespective of how the system is managed. The concept of latent heat is not unique to nickel-hydrogen cells; it is also commonly seen in nickel-cadmium and other types of battery cells.

7.4 Thermal Management

In nickel-hydrogen cells, latent heat is commonly generated by the evolution of oxygen gas during overcharge. The oxygen is clearly unstable in the hydrogen-filled pressure vessel of the nickel-hydrogen cell, so it is generally assumed to rapidly result in heat generation from subsequent reaction with hydrogen. However, in some situations the release of heat from the reaction between oxygen and hydrogen gases can be kinetically inhibited, leading to a lengthy delay between the actual overcharge process and the subsequent release of heat from the overcharge. In other situations it is possible to dramatically accelerate the recombination of accumulated oxygen, leading to a heat-generation rate many times that expected based on the charge or discharge current passing through the cell.

The fundamental cause for the storage of latent heat in a nickel-hydrogen cell is the accumulation of oxygen gas within the cell as the nickel electrodes are brought to a high state of charge. As the nickel electrodes near full charge, they begin to evolve oxygen at an increasing rate. While some of this oxygen will diffuse from the porous nickel electrodes to recombine at the hydrogen electrode, a significant amount of oxygen will remain to displace hydrogen gas and some of the electrolyte filling the pores of the nickel electrodes. When a nickel electrode is in overcharge at a C/10 rate, models of the dynamics of oxygen movement indicate that approximately 80% of the gas within the nickel electrode can be oxygen gas. This reservoir of oxygen gas is capable of responding dynamically to any sudden pressure changes within the cell to either precipitously release the oxygen, or cut back on its rate of release. Any sudden changes in the rate of oxygen release will tend to dramatically affect the instantaneous heat production by the cell, because oxygen is converted to heat relatively quickly in the nickel-hydrogen cell after it escapes from the nickel electrode.

The most dramatic example of the thermal effects caused by latent heat in a nickel-hydrogen cell is seen when a cell that has been on recharge at a normal rate (i.e., C/10) is suddenly switched to a high rate of discharge. When this occurs, calorimetric measurements of the heat generation by the cell show that there is a sudden burst of heat that lasts many minutes before the heat evolution settles down to the rate expected for the discharge processes. This behavior is shown in Fig. 7.4, where a cell that is only about 50% recharged is switched to a moderate discharge rate, giving a thermal transient more than twice as great as the heat normally expected from discharge. In this case more than 30 min of discharge are required before the thermal transient has settled down to the heat production level normally expected for the discharge reaction.

The cause of the thermal transient in Fig. 7.4 is the release of accumulated oxygen gas. As evidenced by the gradual rise in heat production toward the end of the recharge period in Fig. 7.4, oxygen was being actively evolved and was being accumulated in the nickel electrode as the cell pressure increased during recharge. The increasing cell pressure in this situation actually acts to partially inhibit the escape of oxygen from the nickel electrode. Once the cell is switched

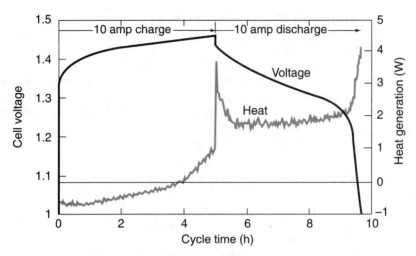

Figure 7.4. Calorimetry measurement showing the large thermal transient resulting from recombination of accumulated oxygen when switching from recharge to discharge for a nickel-hydrogen cell. ©2002 IEEE. Reprinted with permission.

to discharge, however, the decreasing pressure in the cell causes a burst of oxygen to be released from the nickel electrode, which shows up as a burst of heat. This effect becomes even more pronounced when higher discharge rates are used, because the rapidly falling cell pressure more rapidly draws oxygen out of the nickel electrode. For the case shown in Fig. 7.4, more than 30 min are needed before the oxygen in the nickel electrode has fallen to a level low enough that it does not contribute significantly to extra heat production.

Unfortunately, the sudden release of oxygen at the start of discharge, as shown in Fig. 7.4, is not simply a thermal peculiarity of nickel-hydrogen cell operation. The burst of oxygen at the start of discharge has in some instances been associated with an increased incidence of popping events. In addition, cases have been noted where nickel-hydrogen cells have failed because of popping damage within the first 5 min of discharge, even when recharge rates were thought to be safely below the threshold where popping damage would be severe. Clearly it is possible for a sudden release of oxygen gas (in response to a sudden pressure drop) to result in a transient oxygen production rate that is many times the safe level deduced from recharge current alone.

7.4.2 Heat from Pressure Changes

During the charge and discharge of nickel-hydrogen cells, the hydrogen pressure in the fixed-volume pressure vessel can change by up to 1000 psi. Along with this pressure change is an accompanying amount of work that is done on the gas to compress it during recharge, and an equivalent amount of work that is done by the gas on its surroundings as it expands during discharge. This work of compression or expansion is manifested as heat generated or absorbed by the gas, heat that is subsequently transferred to the walls of the gas storage spaces in the cell. During a full charge/discharge cycle, the heat generated by compressing the gas during recharge exactly balances the heat absorbed during discharge as the gas expands. Thus, the pressure-volume heating effects do not increase or decrease the net cyclic heat balance, but simply shift some of the net heat generation during the discharge portion of the cycle to the recharge portion of the cycle.

The pressure-volume work in watts to compress n moles of an ideal gas pressure from P_1 to P_2 in a vessel with constant volume in a time interval Δt is given by Eq. (7.1), where R is the gas constant and T is the absolute temperature.

$$Q = (101.33nRT / \Delta t)\ell n(P_1 / P_2) \qquad (7.1)$$

The amount of heat generated by expansion or compression of the gas is proportional to the rate of pressure change in the cell, as well as the volume of the cell. Because the amount of heat generated is proportional to the rate of pressure change, the thermal effects during a constant current discharge become much greater as the hydrogen pressure in the cell falls toward zero, because the heat capacity of the gas drops as its pressure falls. Because most of the heat from expansion or compression of the gas is deposited on the domes of nickel-hydrogen cells, which are in closest contact with the gas volume, cell dome temperatures can display thermal signatures that are significantly different from those expected from the electrochemical processes. This is the reason that it is not unusual to see a dramatic drop in the dome temperature of a nickel-precharged cell toward the latter stages of a high-rate discharge when the internal pressure is rapidly falling, at a time when the electrochemical heat and total heat generated in the cell is quite high and increasing. This signature is illustrated in Fig. 7.5, where the cell dome temperature actually falls significantly during the last hour of a C/4 discharge, while the temperature on the pressure vessel wall at the middle of the cell stack continues to rise as heat generation increases during discharge. However, even the midcell temperature falls slightly at the end of the discharge when the hydrogen pressure in the cell falls rapidly to nearly zero, and the heat capacity of the hydrogen gas decreases to the point where the cooling from hydrogen expansion begins to influence the cell wall temperature.

The effects of pressure-volume work from expansion of hydrogen during discharge may also be seen in the sudden cooling that commences when discharge is initiated. This effect is seen in Fig. 7.5 as a decrease in cell dome temperature of about 0.1°C at the start of the discharge. This drop in temperature comes at a time when the cell as a whole is producing a large amount of heat, and the midcell wall of the pressure vessel is rising rapidly in temperature in response to the sudden onset of electrochemical discharge heating. While thermal effects from

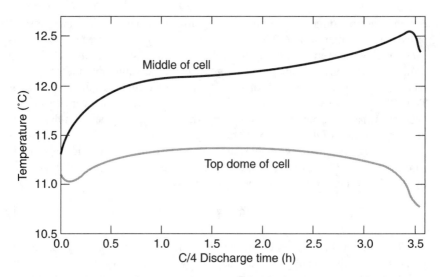

Figure 7.5. Temperature changes on the middle of the cylinder (adjacent to electrode stack) and on the top dome of a nickel-precharged nickel-hydrogen cell during a C/4 discharge at 10°C.

pressure-volume work do not change the overall round-trip thermal balance of a nickel-hydrogen cell, these effects are important to consider in thermal models that are used to predict battery and power system thermal environments in satellites during charge or discharge.

7.4.3 Effects of Thermal Gradients

Thermal gradients across a nickel-hydrogen cell result from either a dynamic interaction between the heat generated in the cell and the thermal dissipation system in which it resides, or in some instances from an externally applied thermal gradient. Thermal gradients across a nickel-hydrogen cell can have a number of effects on the performance of the cell. These include introducing gradients in current density, and overcharge rates, as well as potentially initiating several mechanisms for transfer of water from one region of a cell to other regions. All of these processes will degrade cell performance, causing either short-term loss of capacity or voltage or hastening a wear-out condition that will prematurely end the useful cycle life of the cell.

One of the most obvious effects of a thermal gradient is to produce a potential gradient through the cell as a result of the temperature coefficient of the cell potential. Because this temperature coefficient is negative, the colder electrodes in the cell will have a higher characteristic potential, and therefore will deliver more current and run at a deeper depth of discharge if all other variables are equal. However, the charge efficiency, the stability of the various charged phases in the nickel electrode, and the conductivity of the active materials, electrolyte, and leads are also dependent on temperature. In addition, dynamic processes such as diffusion and convection within the cell are also temperature dependent. The net consequence of these temperature dependences is generally an adjustment of the chemical state within a cell to the thermal gradient, with the result being varying depths of discharge and phase compositions within the electrodes of different temperature. Throughout many years of cycling a cell with a thermal gradient, physical and chemical degradation of the cell is generally accelerated relative to what is normally expected in a more uniform thermal environment.

One area where thermal gradients can significantly accelerate cell degradation is the wear accumulated from overcharge, which is typically the dominant factor contributing to long-term cell wear. When a cell is operated with a significant thermal gradient, the evolution of oxygen (overcharge) will occur in the warmer electrodes long before oxygen evolution begins in the colder electrodes. The average overcharge rate and ampere-hours of overcharge experienced by the warmest electrodes can be a factor of two to three greater than that seen by the colder electrodes, and can result in an accelerated rate of degradation in usable capacity as a cell ages. The oxygen evolved during overcharge in a cell can be concentrated within just several of the electrodes, possibly resulting in damage to the electrodes from popping events, and in extreme cases reinforcing the thermal gradient to cause thermal runaway.

Another issue that becomes a concern in nickel-hydrogen cells that are operated with a thermal gradient involves transfer of water from the warmer part of the cell to the colder part. Such transfer of water can result in electrolyte dry-out and cell failure from increased resistance, and in other cases it can flood regions of the cell and cause severe popping damage in the flooded regions. The obvious method for transfer of water is condensation of water vapor on any sufficiently cold surfaces within a cell. This type of condensation will occur whenever the water vapor pressure over the warmest electrolyte in the cell exceeds the vapor pressure of water at the temperature of the coldest surface, and can be theoretically determined from the vapor pressure curves of the electrolyte compared to pure water. For 31% potassium hydroxide electrolyte, condensation will occur whenever the coldest surface in a cell is about 9°C below the temperature of the warmest electrolyte in the electrodes. For 26% potassium hydroxide, only about a 6°C thermal gradient within a cell is required to initiate water condensation.

Measuring the rate of propagation of ultrasound through the pressure vessel dome while the dome is held at a temperature colder than the rest of the cell has recently been used to measure the rate of condensation of water in actual operating nickel-hydrogen cells.[7.2] The results of these measurements are shown in Fig. 7.6 for a state-of-the-art nickel-hydrogen cell containing 26% potassium hydroxide electrolyte. The measurements for this cell are consistent with the 6°C thermal gradient required for water condensation based on the vapor pressure

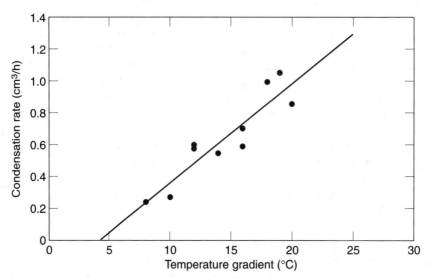

Figure 7.6. Rate of water condensation measured ultrasonically in the cold dome of a nickel-hydrogen cell containing 26% potassium hydroxide electrolyte as a function of the thermal gradient between the stack and the dome.

of 26% electrolyte. These measurements show that the condensation rate is low near the 6°C minimum thermal gradient for condensation, but increases linearly as the magnitude of the thermal gradient increases. Therefore the significance of condensation depends not only on the magnitude of the thermal gradient, but also on the amount of time that the cell is exposed to the thermal gradient.

While many nickel-hydrogen cell designs include wall wicks, which can wick condensed water back into the electrode stack, these wall wicks do not wick the liquid rapidly, and they may not be able to keep up with condensation when sufficiently large thermal gradients are present. The rates at which wall wicks in different cell designs can redistribute condensed water to the cell stack can be measured by using an ultrasonic probe to determine the volume of condensate in the cell.[7.2] The rate at which wall wicks can redistribute water is highly dependent on both the details of the cell design (e.g., wall wick thickness, electrolyte concentration, stack-to-wall contact), as well as the temperature. Figure 7.7 shows the range of wicking rates that have been observed over a range of temperatures for differing cell designs. It is clearly important to understand the dynamics of wall wick operation as well as condensation to understand how a cell design will respond to a specific thermal environment during its operation.

If the temperature of the coldest surface within the cell falls significantly below 0°C, the water that is condensed in that region can freeze, as suggested by the lines in Fig. 7.7 dropping to zero at low temperatures. If freezing of the water

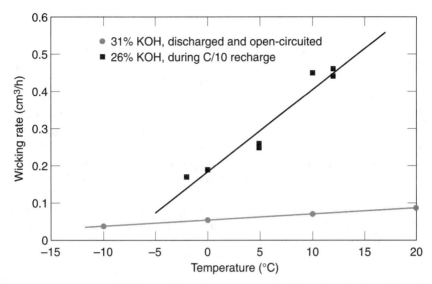

Figure 7.7. Rate at which condensed water is redistributed to the stack by the wall wicks in two different nickel-hydrogen cells, as a function of the cell temperature, from ultrasonic measurements of condensate volume in the cell dome.

occurs, the wall wicks will lose all effectiveness in dissipating the condensate, and accumulation of the ice will continue until the cell dries out or until the temperature is increased enough to melt the ice. If ice that is formed by this mechanism is rapidly melted by an increase in cell temperature, there is a risk of temporarily flooding the electrodes that are closest to the built-up ice. As suggested by the data in Fig. 7.7, freezing of condensate is more likely in cells with lower electrolyte concentration.

Another mechanism that exists for transferring water between different regions of a cell when a thermal gradient is present is transfer by oxygen gas. This mechanism is operative in a nickel-hydrogen cell that contains several stacks of electrodes that are physically isolated from each other. One example would be a dual-stack design where the upper stack runs at a higher temperature than the bottom stack. In this case, oxygen is generated earlier in the upper stack to produce a gradient in oxygen partial pressure between the two stacks. Although the oxygen that can actually escape from the upper stack to recombine in the lower stack may be very small for any one cycle, during many years of operation under such conditions a significant amount of water can be transferred. This type of water transfer will leave the electrolyte in the warmer stack more concentrated, and that in the colder stack less concentrated.

A final issue with thermal gradients can arise when cell temperatures are significantly below 0°C. The freezing point of 26% potassium hydroxide electrolyte is about –40°C, and that of 31% potassium hydroxide is about –60°C. Normally one would not expect issues to arise from freezing of the electrolyte until cell temperatures approach these levels. However, when cell temperatures are below −10°C, the efficiency for charging to the γ-NiOOH phase becomes very high and it is possible to convert a large amount of the active material in the nickel electrodes to this phase after extended periods of overcharge of trickle charge. Because the γ-NiOOH phase contains one atom of potassium for every three atoms of nickel, the formation of large amounts of this phase in the coldest regions of a cell can significantly reduce the concentration of the electrolyte, as shown in Fig. 7.8 for a cell containing 3 g/A h of 26% potassium hydroxide electrolyte. As the concentration of potassium hydroxide falls, its freezing point rises until at some point the electrolyte freezes in the coldest parts of the cell. For example, as indicated in Fig. 7.8, when a nickel electrode is operated at a temperature of −20°C, conversion of about 43% of the active material to the γ-NiOOH phase will lower the electrolyte concentration a sufficient amount to freeze the electrolyte in the nickel electrode. Cells containing 26% potassium hydroxide are more susceptible to this phenomenon than are cells with 31% potassium hydroxide, as a result of the lower amount of KOH in the electrolyte.

If the cell is on trickle charge, oxygen evolution will be frozen out for the electrodes containing the frozen electrolyte, resulting in an increased overcharge rate for the warmer electrodes that remain active. However, the increased overcharge rate will increase the rate of γ-NiOOH formation and further drop the electrolyte concentration, with the result that more nickel electrodes will freeze.

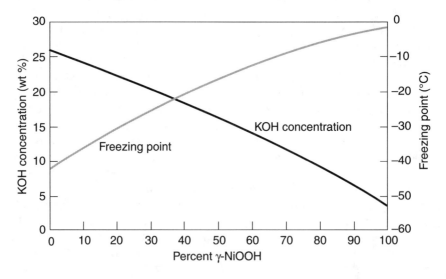

Figure 7.8. Change in potassium hydroxide electrolyte concentration and freezing point as γ-NiOOH forms, initially assuming 26% potassium hydroxide initially and an electrolyte quantity of 3.0 g/A h.

This process will continue until the trickle charge has been concentrated onto a few electrodes and the heat generated from the overcharge is sufficient to keep the electrolyte in these electrodes from freezing even if all the active material is charged to the γ-NiOOH phase. Of course, the volume change that occurs in the active material in these plates is extremely large, resulting in a large amount of electrode swelling, deformation, and active-material extrusion. When this degradation is coupled with the high local rates of overcharge, it actually becomes possible to experience short-circuit cell failures during conditions of trickle charge when cells are sufficiently cold. Minimizing thermal gradients and keeping cells from operating at excessively low temperatures are both actions that can prevent this failure mode from developing.

7.4.4 Thermal Runaway

Thermal runaway in battery cells is the process by which positive feedback between the heat generated by self-discharge causes a cell to catastrophically discharge all its stored energy in a very short period once the temperature exceeds a threshold where the heat generated by the self-discharge cannot be dissipated. This can occur in nickel-hydrogen cells at significantly lower temperatures than for most other types of battery cells, because nickel-hydrogen cells have an intrinsically higher self-discharge rate than most other cells. However, thermal runaway in a nickel-hydrogen cell is also not generally as catastrophic as in other types of cells because discharge causes the cell pressure to drop.

In a nickel-hydrogen cell, thermal runaway can occur at temperatures as low as 30°C or less. The runaway temperature is strongly dependent on the amount of cooling available that must be overcome by the self-discharge heating, as well as how highly charged the cell is initially. The thermal runaway process typically involves rapid oxygen evolution from the decomposing nickel electrodes as the cell heats, as well as accelerated direct self-discharge of the nickel electrodes by hydrogen gas. The rapid evolution of oxygen can cause cell short-circuiting as a result of popping damage; however, cells are often seen to survive thermal runaway without shorting, and with little loss of functionality. The reason for this is that before the cell temperature can reach levels where the charged active materials can explosively decompose[7.3] (near 90°C), the self-discharge process has depleted the stored energy. Thermal runaway does not impact the cell pressure vessel, because the hydrogen pressure drops rapidly as the cell self-discharges, in spite of the increasing temperature. For these reasons, thermal runaway, while undesirable, does not generally create the safety hazard in nickel-hydrogen cells that it does in many other types of battery cells.

7.5 Electrolyte Management

The dynamics of electrolyte management in nickel-hydrogen cells are critical for the proper functioning of the cell, and they are closely linked to the dynamics of gas flow in the cell. Electrolyte concentrations can range from 26% to 38% potassium hydroxide by weight, although most modern cells use either 26% or 31% electrolyte. Electrolyte must be maintained at a proper concentration in the electrodes and the separator to enable the level of ionic conductivity required for cell operation. The concentration is typically maintained by diffusion and ionic migration processes between the electrodes. The nickel-hydrogen cell is electrolyte-starved in the sense that it contains only enough electrolyte to largely fill the pores in the electrodes, separator, and wall wick. Essentially no free electrolyte should normally be present in a properly designed nickel-hydrogen cell. In such a cell, 95–98% of the void space in the electrodes and separator is typically filled with liquid electrolyte.

The correct amounts of electrolyte are maintained in each cell component by the capillary forces within the pores of each component. The component with the smallest pores is the nickel electrode, which has essentially all its internal void volume filled with electrolyte, except when actively evolving oxygen. The catalyst layer on the hydrogen electrode has the next smallest pore sizes, and its pores also largely remain filled with electrolyte during cell operation. However, the hydrogen electrode also has a subdistribution of larger pores that are relatively hydrophobic; these serve as hydrogen gas channels. The separator has relatively large pore sizes, and because it typically is not completely filled with electrolyte, it is thus able to absorb any electrolyte forced from the electrodes during cell operation by gas evolution. Any electrolyte that is displaced from the electrode stack can be wicked back into the stack through the wall wick, which is in contact with the separator edges at the wall of the pressure vessel.

In nickel-hydrogen cells with wall wicks, any electrolyte that escapes from the stack as a result of thermal gradients or other forces will be gradually wicked back into the stack. Nickel-hydrogen cell designs that have no wall wicks are susceptible to degradation as a result of electrolyte or water escaping from the stack, because they contain no mechanism to return liquid to the stack. In such cases, it is possible to return liquid water to the stack by warming the regions of the cell containing the water for several weeks, thus allowing vapor transport to return the water to the stack.

7.6 The Effect of Pressure Fluctuations on Cell Performance

A significant source of stress that contributes to wear in nickel electrodes is the volume change in the active material, a change that accompanies the charge and discharge of this material. In a nickel-hydrogen cell, the fluctuations that occur in the cell pressure during charge and discharge can contribute added volume changes in these relatively compressible active materials, volume changes that can be an important factor in causing long-term electrode degradation.

The effects of such pressure fluctuations on cumulative active-material volume changes have been modeled[7.4] based on the known compressibility of the active materials, and for the different types of pressure fluctuations expected during battery operation in various types of satellite orbital scenarios. The model results

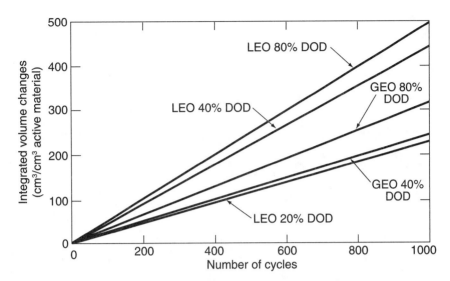

Figure 7.9. Cumulative volume changes experienced by one cubic centimeter of active material when cycled in different orbital scenarios, from both pressure changes and charge/discharge volume changes. More than 50% of the cumulative volume change results from the effect of pressure fluctuations.

suggest that the volume changes caused by compression and expansion of the active materials are at least as great as the volume changes that occur from changes in the state of charge of the active materials.

Figure 7.9 shows the cumulative volume changes experienced by one cubic centimeter of active material from both charge and discharge as well as from compression and expansion in the changing pressure environment of the nickel electrode. More than half of the volume change indicated here arises from compression and expansion of the active-material deposit during the pressure cycle experienced in the cell environment.

7.7 References

[7.1] *Aviation Week and Space Technology*, 9 March 1992, pp. 18–19.

[7.2] V. Ang and J. Nokes, *Proc. of the 2005 Space Power Workshop* (Manhattan Beach, California, April 2005), describes the ultrasonic method. Data are from a personal communication with V. Ang.

[7.3] S. W. Donley, J. H. Matsumoto, and W. C. Hwang, *J. Power Sources* **18**, 169 (1986).

[7.4] A. H. Zimmerman, "Effects of Pressure Fluctuations on Nickel Electrodes," Report SD-TR-87-42, The Aerospace Corporation, 30 April 1987.

8 Nickel-Hydrogen Cell Modeling

Models for nickel-hydrogen cells typically fall into several categories. The first general type, used for predicting the performance of a nickel-hydrogen cell of a given design, is referred to as a performance model. Several methods have been successfully used to predict cell voltage and capacity performance. Empirical methods typically employ either a performance database or some predetermined functional dependence of cell performance on operational parameters. Alternatively, first-principles models predict cell performance from the basic chemical and physical processes within the cell, and depend on semiempirical methods to obtain key information about these basic processes. This chapter describes examples of each of these methods for modeling cell performance, as well as the advantages and limitations of each approach.

The second general type of model is used for predicting cell cycle life. The most successful models of this type have been parametric wear-out models. Life models of this type fit cell lifetime, as established by life-test results, to a parametric wear equation. Wear models of this general type are presented here, in discussions of both single-parameter and multiparameter models, as well as empirical methods to use these wear models for cell performance prediction. In addition, this chapter examines the statistical analysis of cell wear-out rates, which is a key concept in defining cell reliability and failure probability. The second section of this chapter will discuss a wear-based model of nickel-hydrogen cell degradation during long-term storage, and the final section will discuss a wear-based model for predicting cycle life.

A final area of modeling, which has been an elusive goal in most first-principles performance model methods, is the full modeling of the fundamental chemical and physical mechanisms responsible for cell degradation and thus, by extension, the accurate modeling of both cell performance and how that performance changes as a cell degrades during its lifetime. Some attempts have been made to realize this goal, and they are described in this chapter. However, the development of a complete first-principles model—one that allows a 15–20 yr life test of a nickel-hydrogen battery cell to be completed purely by computer simulation—remains a goal that has not yet been attained.

8.1 First-Principles Models of Cell Performance

All first-principles models start with the same fundamental chemical and physical processes that occur within the nickel-hydrogen cell—processes that are accurately described by basic kinetic, thermodynamic, and chemical-balance laws. Different modeling approaches make different assumptions as to what reactions or processes to consider or ignore, what cell geometrical complexities to consider, and what details of the microscopic and macroscopic structure of the cell components to explicitly include in the model.

Two different approaches are discussed here. The first simplifies the chemical processes and physical dimensions of a real cell significantly to obtain continuous temporal and spatial solutions to the coupled equations describing the principal cell processes. The second approach attempts to include all known chemical and physical processes that can act within a structure of finite elements that make up a realistic cell, but settles for approximate finite-difference solutions to the equations describing the processes occurring within and between finite elements. Each of these approaches has advantages and disadvantages that are described here.

8.1.1 Models Based on Continuous Mechanics
The modeling of nickel-hydrogen cells using continuous mathematical solutions to the temporal and spatial equations that describe key processes occurring within the cells has been extensively developed, as described in Gomadam et al.[8.1] Models of this kind are typically based on the macroscopic description of a nickel-hydrogen cell as a one-dimensional stack consisting of a porous nickel electrode, a porous separator, and a porous hydrogen electrode. The single macroscopic dimension that is modeled involves the distance through the thickness of these components, which can be filled with active material and electrolyte to the desired levels to simulate realistic nickel-hydrogen cells.

Thermodynamic and kinetic equations are used to model the electrochemical charge-transfer processes at the nickel and hydrogen electrodes, with appropriate mass and energy balance requirements during the charge-transfer reactions. The electrochemical processes result in concentration gradients for both chemical and ionic species within the electrolyte, with the concentration gradients extending through the thickness of the porous electrodes and the separator. Nickel electrode models also typically include a microscopic dimension that describes the gradients through the thickness of the active-material layers and particles that are present within the pores of the nickel sinter. Models of this kind have been used to describe a number of the key performance characteristics of nickel-hydrogen cells.[8.2-8.7]

The transport processes in this type of nickel-hydrogen cell model include proton migration and diffusion within the solid nickel electrode active material, ionic movement and mass transport of hydroxide ions in the electrolyte phase within all the porous components, mass transport of hydrogen in the gaseous and liquid (dissolved) phases, and thermal diffusion throughout the entire cell. Proton diffusion within the nickel electrode active material is modeled to account for solid-state charge transport and charge redistribution processes within the nickel electrodes.

Oxygen transport in the nickel-hydrogen cell is typically modeled by diffusion of oxygen that is dissolved in the liquid electrolyte phase through the separator. Oxygen evolution at the nickel electrode during overcharge initiates this

8.1 First-Principles Models of Cell Performance

transport process, which is completed by electrochemical recombination of the dissolved oxygen at the hydrogen electrode with the resulting release of heat. Convective transport of oxygen and hydrogen in the gas phase, which is driven by pressure differentials between the cell components, is typically not modeled because the convective flow equations introduce significant added complexity.

The electrochemical processes in the negative and positive electrodes are modeled using Nernstian dependences of the reversible potentials on the concentrations and activities of the electroactive species. The rates of the electrochemical processes are modeled using Butler-Volmer kinetic expressions that describe how the current depends on the overpotential. Arrhenius temperature dependences are assumed for the kinetic parameters describing the rates of kinetic and transport processes.

In addition to the primary charge and discharge reactions in the nickel-hydrogen cell, a model for the self-discharge is paramount in realizing an accurate description of real cell behavior. Continuous-mechanics models typically include the self-discharge resulting from oxygen evolution from the nickel electrode, as well as a chemical self-discharge process involving the direct reaction of hydrogen gas with the charged active material in the nickel electrode. Inclusion of these self-discharge processes, along with the appropriate Arrhenius temperature dependences, can provide an accurate model of nickel-hydrogen cell self-discharge.[8.2,8.6]

Nickel electrodes, particularly when operated in nickel-hydrogen cells, can exhibit a pronounced lower discharge plateau that is 0.3–0.4 V below the normal discharge voltage plateau. This lower plateau has been modeled either as resulting from changes in active-material conductivity with changing state of charge,[8.3] or as being caused by the accumulation of oxygen within the cell.[8.4]

While the latter of these two models describing the second plateau is able to mimic the lower-plateau behavior, the model is not consistent with the thermal behavior seen in real nickel-hydrogen cells during second-plateau discharge. In addition, modeling the lower plateau as being caused by the oxygen-reduction reaction ignores the known gas-phase kinetics of reaction between hydrogen and oxygen, which typically initiates and rapidly propagates by a gas-phase chain-reaction mechanism. This mechanism would rapidly lead to explosive oxygen and hydrogen recombination in a pressure vessel that contained significant quantities of both gases, as must be the case if significant levels of oxygen persist in the cell from the time of evolution during overcharge until the latter stages of cell discharge.

One of the most challenging aspects of modeling nickel electrode behavior in nickel-hydrogen cells using continuous-mechanics models is to adequately model the complexity of the electrochemical processes involving the solid-state phases in the nickel oxyhydroxide active material. A relatively simplistic model that includes only the redox couple involving β-Ni(OH)$_2$ and β-NiOOH can produce reasonable cell behavior when the solid-state lattice is treated as a solid solution. However, to obtain a more realistic cell model, the conversion of β-NiOOH to γ-NiOOH, and the subsequent discharge of γ-NiOOH to α-Ni(OH)$_2$, must be accurately modeled, along with the recrystallization of α-Ni(OH)$_2$ back to β-Ni(OH)$_2$. While these additional processes add considerable complexity to the governing equations in a model based on continuous mechanics, when they are considered,[8.4,8.7] they can substantially improve the agreement between the model results and the performance of real nickel-hydrogen cells.

Additional complexities, such as the formation of deactivated nickel electrode active-material phases, and the separation of the various phases into discrete layers within the active-material structure that fills the pores of the nickel electrode, are not typically included in these types of models, primarily to keep the governing equations tractable. However, these types of processes are in fact responsible for significant changes in nickel electrode performance in response to the previous cycling history of the electrode, creating many of the "memory effects" frequently seen during nickel electrode operation.

8.1.2 Finite-Element-Based Models

In view of the significant advances in understanding the details of how a nickel-hydrogen cell works, it has become possible to develop a high-fidelity model of this type of cell based on its fundamental properties and its components. The previous section described simplified models with this goal. While a simplified model is quite useful, it is not capable of handling multiple phases of charged and discharged material in the nickel electrode, the spatial variations in chemical characteristics endemic in nickel electrodes, or the distributions of gas and electrolyte and their flow in a realistic three-dimensional cell.

The use of simplified models—ones that make significant simplifications to the known chemistry and the complex microscopic and macroscopic geometrical structures within a nickel-hydrogen cell—has been found to give results that are only an approximation to the actual data obtained from real cells. For this reason, The Aerospace Corporation embarked on an effort in the early 1990s to develop a finite-element model of the nickel-hydrogen cell that would include all known chemical and physical processes, would use realistic porous microstructures within the electrodes, and would be capable of being coupled with an accurate representation of the three-dimensional arrangement of components in a real cell or power system.

To make this model computationally tractable, a number of numerical approximations were made in solving the large number of coupled equations describing the cell processes. This model, which was essentially completed in the late 1990s, provided the capability to simulate the performance of a real nickel-hydrogen cell to a level of accuracy equivalent to its performance reproducibility from cycle to cycle. In 1998 this computer model could simulate the operation of a complex nickel-hydrogen cell at a speed of about twice that of real-time cell operation in a standard desktop computer. However, with advances in computer technology, by 2005 this model could run the same simulation at a speed nearly 1000 times that of real-time cell operation. These advances make it now possible to consider using this model to run a life test as a computer simulation, or to run highly detailed models of cells, batteries, and entire power systems. This modeling method will be described in this section.

8.1.2.1 General Model Description
The guiding objective of this effort was to develop a model architecture that could be applied to the full range of chemical, physical, and electrochemical processes possible within the nickel-hydrogen cell. The architecture had to allow the model to be based on the actual size, geometry, and design details of the real cell, in full three-dimensional microscopic and macroscopic detail, where appropriate. The general approach selected for modeling the nickel-hydrogen cell defines the cell in terms of a three-dimensional assembly of rectangular finite elements.

In this approach, six types of finite elements are used to model a nickel-hydrogen cell and its thermal environment:

- gas space elements
- nickel electrode elements
- separator elements
- hydrogen electrode elements
- solid electrical-conduction elements
- solid thermal-conduction elements

Electrode leads, cell terminals, thermal interfaces, wall wicks, recombination catalyst sites, gas screens, and thermal control systems can all be treated as subsets of these six basic building blocks. By choosing the proper number, geometric arrangement, properties, and sizes of these elements, a designer can model any nickel-hydrogen cell in a realistic thermal environment. For each type of element, the computer is fully aware of all the physical and chemical processes that can occur within that specific element type.

In actual practice, it is desirable to keep the number of elements to the minimum that will adequately describe the system of interest, so that computation and data-analysis time will be reasonable. It is assumed here that within each finite element all properties (such as porosity, state of charge, and pressure) are constant. The gradients between adjoining elements provide the forces that drive mass transport by convection, diffusion, and capillary action. The time evolution of the nickel-hydrogen cell during charge, open circuit, or discharge is developed using a finite-difference method that recalculates all cell characteristics after periodic small time intervals. The model dynamically adjusts the simulation time interval as it runs to ensure convergence of all numerical calculations. This general framework for the model readily allows any number of mass transport, chemical, or electrochemical processes to be added simply by introducing additional electrical components to each element, or interfacial transport components between elements.

The model architecture for the nickel-hydrogen cell includes the basic processes of electronic conduction in the electrode supporting structures and nickel electrode active material; ionic conduction in the electrolyte; charge and discharge processes; oxygen evolution and recombination; heat generation; solid-state proton diffusion in the nickel electrode active material; and hydrogen oxidation and reduction at the platinum catalyst electrode. In addition, capillary, convective, and diffusive movement of hydrogen, oxygen, water, liquid electrolyte, and heat are modeled using a three-phase system consisting of solid, liquid, and gaseous materials. The model includes the reversible β-Ni(OH)$_2$ to β-NiOOH electrochemical process, electrochemical conversion of β-NiOOH to γ-NiOOH, discharge of γ-NiOOH to α-Ni(OH)$_2$, and chemical conversion of α-Ni(OH)$_2$ to β-Ni(OH)$_2$. Additionally the model fractures large crystallites of nickel electrode active material in response to volume changes, and then allows β-Ni(OH)$_2$ crystallite size to increase by an Ostwald ripening process. While oxygen is assumed to recombine at the hydrogen electrode, no polarization of the hydrogen electrode from this recombination process is considered. The model also includes both direct and electrocatalytic self-discharge reactions of hydrogen with the charged nickel electrodes.

The model also does not specifically include long-term degradation processes such as plate expansion, active-material porosity changes and extrusion from the sinter structure, or sinter corrosion in the nickel electrode. These processes provide a formidable list of improvements that must eventually be considered in any high-fidelity model of the nickel-hydrogen cell if it is to properly simulate real cell behavior throughout the life of the cell. Thus, it is critical that the general architecture be capable of readily accommodating the factors governing long-term cell degradation as described here into a more advanced model. The following sections describe the key aspects of a first-principles model that have been found important for accurately predicting the performance of nickel-hydrogen cells.

8.1.2.2 Model Architecture
A cell model consists of an assembly of rectangular finite elements, each with dimensions that are specified based on the physical size and shape of the cell and its components. The density of elements within any particular region of a cell component is best dictated by the gradients in the physical parameters associated with that region. Regions having large gradients in properties are best modeled using a higher density of elements, while very few elements are needed if gradients are small or can be acceptably averaged.

An element n may be a gas space element, a nickel electrode element, a separation element, a hydrogen electrode element, a conductive lead element, or a thermal element. Figure 8.1 shows the simplest possible model of a nickel-hydrogen cell. This four-element model consists of a simple stack of rectangular plates having

Ni	Sep	H₂	Gas
1	2	3	4

Figure 8.1. Geometry of elements in a simplified nickel-hydrogen cell.

the desired dimensions; this stack is placed against a gas space, with the gas space being adjacent to the hydrophobic side of the platinum catalyst electrode. In the gas space (element 4), only gas is allowed to be present. Both hydrogen and oxygen gas may exist at partial pressures $P_{H,n}$ and $P_{O,n}$, respectively.

In nickel electrode elements (such as element 1 in Fig. 8.1), the element properties are characterized by state of charge S_n; fractional active-material composition F_β, F_γ, and F_α; partial pressures $P_{H,n}$ and $P_{O,n}$; electrolyte concentration C_n; porosity ε_n; electrolyte volume V_n; loading level L_n; and tortuosity T_n. In all porous components (nickel electrode, separator, and platinum electrode), a realistic pore size distribution $D_{F,n}$ may be defined in terms of capillary pressure as a function of electrolyte fill fraction F. Other parameters, such as active-material density and conductivity, electrolyte viscosity and density, and gas viscosity, are calculated as volume averages within each element based on these properties.

In separation elements (such as element 2 in Fig. 8.1), the properties describing the elements are partial pressures $P_{H,n}$ and $P_{O,n}$, porosity ε_n, tortuosity T_n, electrolyte volume V_n, and electrolyte concentration C_n. These same parameters apply to the hydrogen electrode elements (such as element 3 in Fig. 8.1). It is assumed that in the hydrogen electrode, each time interval during which charge transfer occurs involves the recombination of all oxygen that is present within hydrogen electrode elements. This is done by setting $P_{O,n}$ to zero in these elements, adding the generated heat, and removing hydrogen and adding water according to the stoichiometry of the recombination process. Thus, any oxygen that diffuses or is convectively carried into the hydrogen electrode is assumed to undergo rapid recombination with hydrogen on the platinum catalyst surface.

8.1.2.3 Mass and Thermal Transport
The thermal and mass transport connections between the elements making up the cell are specified in terms of a connection matrix **B**, whose entries b_{ij} indicate which elements share a common interface through which thermal and mass transport can occur. Associated with each interface ij between elements i and j is a mean porosity ε_{ij} [Eq. (8.1)], a mean tortuosity T_{ij} [Eq. (8.2)], and partial pressure differences $\Delta P_{H,ij}$ and $\Delta P_{O,ij}$. In Eqs. (8.1) and (8.2), Δx_i is the mean distance from the center of element i to the interface between element i and element j. For

all interfaces where neither i nor j is a gas volume element, there is an electrolyte concentration difference ΔC_{ij}. For interfaces between two nickel electrode elements, a state-of-charge difference ΔS_{ij} will also generally exist. For the simple example of Fig. 8.1, the matrix **B** is given in Eq. (8.3).

$$\varepsilon_{ij} = \frac{\varepsilon_i \varepsilon_j \left(\Delta x_i + \Delta x_j \right)}{\varepsilon_i \Delta x_j + \varepsilon_j \Delta x_i} \tag{8.1}$$

$$T_{ij} = \frac{T_i T_j \left(\Delta x_i + \Delta x_j \right)}{T_i \Delta x_j + T_j \Delta x_i} \tag{8.2}$$

$$\mathbf{B} = \begin{bmatrix} 1 & 0 & 0 \\ -1 & 1 & 0 \\ 0 & -1 & 1 \end{bmatrix} \tag{8.3}$$

The matrix in Eq. (8.3) is a 3×3 matrix because there are three interfaces between elements in Fig. 8.1. The elements of the **B** matrix can also be made zero or nonzero to either allow or preclude movement of materials between elements, even when two adjacent elements share a common boundary.

Each process that involves the movement of material across an element interface is expressed in terms of a material flux J_{ij}. The interfacial material flux is obtained by diagonalizing the flux matrix **J** in Eq. (8.4). This equation contains the resistance vector **R** describing the resistance to material movement through each interface for each transport process in the system. The vector **X** contains the gradients in state of charge, concentration, or pressure that can cause mass or thermal flow across each interface.

$$\mathbf{J} \bullet \mathbf{R} = \mathbf{X} \tag{8.4}$$

For the diffusion of potassium hydroxide, oxygen, or protons within the nickel electrode, Eq. (8.4) becomes simply a matrix expression of Fick's Law for diffusion over a small but finite distance, as indicated in Eq. (8.5),

$$J_{ij,\ell} = \frac{A_{ij} \Delta C_{ij}}{\Delta x_i + \Delta x_j} D_\ell \varepsilon_{ij}^{T_{ij}} \tag{8.5}$$

where the flux for transport of species ℓ is related to the interfacial area A_{ij}, an interface-specific porosity and tortuosity, the diffusion coefficient D for species ℓ, the concentration gradient, and the distance across which diffusion must occur.

8.1 First-Principles Models of Cell Performance

Similarly, liquid and gaseous convective flux f_{ij} may be expressed in terms of interfacial area, viscosity of the liquid or gaseous medium η, transport distance between elements, porosity and tortuosity, and the pressure difference ΔP_{ij} between elements i and j, as indicated in Eq. (8.6).

$$f_{ij} = \frac{A_{ij} \Delta P_{ij}}{\eta(\Delta x_i + \Delta x_j)} \varepsilon_{ij}^{T_{ij}} \tag{8.6}$$

Electrolyte movement caused by capillary forces between the porous components is handled like convective movement. The capillary pressure difference $\Delta \Gamma_{ij}$ across each interface is obtained as indicated in Eq. (8.7). The capillary pressure Γ_g within each porous element g is defined by Eq. (8.8) as a function of the size of the unfilled pores in the element and the wettability of the pores, and thus is a function of the pore size distribution D and the fraction F of the pore volume that is filled.

$$\Delta \Gamma_{ij} = \Gamma_i - \Gamma_j \tag{8.7}$$

$$\Delta \Gamma_g = Z(F, D) \tag{8.8}$$

For each component in the cell, the function $Z(F, D)$ is computed as a function of electrolyte fill from actual pore size distribution measurements, based on the measured pore sizes as a function of fill volume and the standard equation for capillary pressure. For hydrogen electrodes, 20% of the element volume is assigned a negative capillary pressure to give it the hydrophobic characteristics associated with the microporous Teflon backing on the platinum catalyst electrode. Thus, the pore size distribution is a physical characteristic of each component that has a significant effect on where electrolyte tends to go in the cell, and it is expected to change with electrode aging or manufacturing variations.

Because all elements except those defining lead elements, thermal elements, and the cell gas spaces contain both liquid and gaseous phases, the convective flux in these elements from Eq. (8.6) must be partitioned into liquid and gaseous flux components. This partitioning is based on the relative viscosity of the gas and the liquid, as indicated in Eq. (8.6). Thus, the relative convective movement of electrolyte depends on the relative amount of liquid, and that of the gaseous components depends on both gas pressure and composition. Movement of electrolyte and gas by capillary forces simply involves a displacement of gas by an equivalent volume of electrolyte.

Mass transport may be pictorially represented in terms of transport resistances between the finite elements, as indicated in Fig. 8.2 for the simple cell assembly

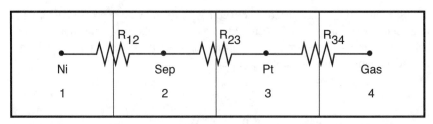

Figure 8.2. Mass transport paths for the simplified nickel-hydrogen cell example of Fig. 8.1.

of Fig. 8.1. In Fig. 8.2, the transport resistance $R_{3,4}$ represents the pure movement of hydrogen and oxygen gas out of the stack and into or through the gas space. Such pure gaseous transport resistances can be fully modeled in more complex assemblies of elements to determine the rate of oxygen movement into and through gas-containing regions, or simplifying assumptions can be made to lump all the gas space into a single reservoir that has a single oxygen and hydrogen pressure throughout. Indeed, this was done in the example of Figs. 8.1 and 8.2, where only gas transport to and from the platinum catalyst electrode from the gas reservoir in element 4 is allowed. Each of the resistances in Fig. 8.2 is calculated for each species undergoing diffusive transport, or for each phase undergoing convective transport, from Eqs. (8.5) and (8.6) respectively. These resistances then provide the elements of the resistance vector in Eq. (8.4).

8.1.2.4 Conduction and Charge Transfer

The movement of electrical charge through the nickel-hydrogen cell is modeled within each element using a specific electrical component for each process that involves movement or electrochemical reaction of any charged species. For the simplified cell that has served as an example in this discussion, Fig. 8.3 indicates the electrical components involved.

In Fig. 8.3, R_1 and R_4 refer to the electronic resistance of the conductive current-collector substrate in the nickel and platinum catalyst electrodes, respectively. The resistance R_2 refers to the ionic resistance from the site in element 1 where charge transfer occurs, into the separator element. The resistance R_3 refers to the ionic resistance between the separator element and the charge-transfer site in the hydrogen electrode element. The impedances Z_N and Z_H refer to impedances to charge transfer and other electrical processes in each nickel or hydrogen electrode element, which are defined by nickel electrode or hydrogen electrode charge-transfer models. The electrical pathways through the cell are thus expressed in terms of k electrical components, each having a resistance q_k. The current through each of these components is determined by diagonalizing the resistance matrix **Q**, which contains resistance elements determined by the electrical process involved and the interconnections in the electrical pathway.

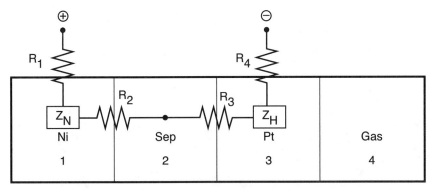

Figure 8.3. Electrical pathways for electronic conduction, ionic conduction, and charge transfer in the simplified nickel-hydrogen cell of Fig. 8.1.

The vector **E** in Eq. (8.9) contains the present voltage values for each electrical component in the model and thus is a combination of reaction overpotentials and reversible potentials when electrochemical processes are involved. The current vector **I** in Eq. (8.9) contains the current passing through each electrical component in the model.

$$\mathbf{I} \bullet \mathbf{Q} = \mathbf{E} \tag{8.9}$$

8.1.2.5 Hydrogen Electrode Model

The impedance of each charge-transfer component in a hydrogen electrode element, Z_H in Fig. 8.3, is assumed to involve the oxidation and reduction of hydrogen gas, and the electrochemical recombination of oxygen gas at the platinum catalyst sites. The rate of the hydrogen reaction is given by Eq. (8.10), and the rate of the oxygen recombination reaction is given by Eq. (8.11).

$$\frac{I_H}{FA} = -k_3[H_2O]\exp\left[\frac{-\beta FE_H}{RT}\right] + k_{-3}[H_2]^{0.5}[OH^-]\exp\left[\frac{\alpha FE_H}{RT}\right] \tag{8.10}$$

$$\frac{I_O}{2FA} = k_{-2}[O_2]^{0.5}[H_2O]\exp\left[\frac{2\beta FE_O}{RT}\right] \tag{8.11}$$

The positive term in Eq. (8.10) refers to hydrogen evolution, or cell charging, while the negative term refers to hydrogen consumption, or cell discharge. E_H is the overpotential ($E - E^{oH}$) for the hydrogen reaction, where E is the electrode voltage. From an electrical point of view, Eq. (8.10) can be represented by two parallel diodes, one carrying the charge current and the other carrying the

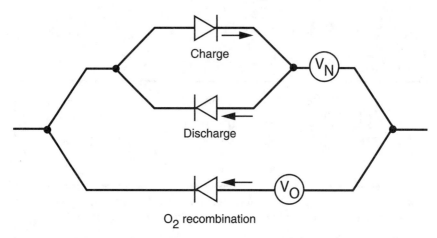

Figure 8.4. Schematic for hydrogen electrode element impedance Z_H in Fig. 8.3.

discharge current, in series with the reversible potential of the hydrogen reaction, as indicated in Fig. 8.4.

Eq. (8.11) indicates the electrochemical oxygen recombination current on the platinum catalyst sites in the hydrogen electrode, where E_O is the overpotential for the oxygen reaction $(E - E^{oO})$. The hydrogen electrode is almost operating near the hydrogen potential, which is more than 1.2 V below the potential for the oxygen recombination process, resulting in a very large oxygen recombination overpotential for nearly all conditions of cell operation. Therefore the rate of oxygen recombination is controlled by the rate of oxygen transport to the hydrogen electrode rather than by the overpotential in the exponential term of Eq. (8.11). The lower diode in Fig. 8.4 represents the pathway for oxygen recombination current.

Because the localized recombination of oxygen gas on the hydrogen electrode at high rates is known to cause popping damage to the electrodes in the cell, a red flag indication is activated during model operation to indicate whenever the localized oxygen recombination rate exceeds C/5 at any hydrogen electrode element. Operation of the model for long periods with this flag activated indicates an increasing likelihood of cell failure as a result of short-circuit formation from accumulated popping damage to the electrodes.

The gas screen used in real nickel-hydrogen cells to ensure facile transport of hydrogen gas to and from the catalyst electrode is modeled by a gas space finite element of the appropriate dimensions and porosity inserted at the back side of each hydrogen electrode element. If electrolyte is allowed to flow into or flood the gas screen on the back side of the catalyst electrode, the model will produce a significant polarization of the hydrogen electrode resulting from hydrogen gas depletion if the discharge rate is sufficiently high.

8.1.2.6 Nickel Electrode Electrochemical Model

The impedance of each nickel electrode element, Z_N in Fig. 8.3, is somewhat more complicated. Oxidation and reduction of the β-Ni(OH)$_2$ to β-NiOOH is assumed to occur at a rate I_β, governed by Eq. (8.12). In this equation, A is the area of the active-material particles in the nickel electrode, θ is the fraction of the active material that is β-NiOOH, κ is the fraction of the active material that is β-Ni(OH)$_2$, and E_{Ni} is the overpotential for the β-Ni(OH)$_2$ to β-NiOOH reaction at the nickel electrode $(E - E^{oNi})$.

$$\frac{I_\beta}{FA} = k_1 \kappa \left[OH^- \right] \exp\left[\frac{\alpha F E_{Ni}}{RT} \right] - k_{-1} \theta \left[H_2O \right] \exp\left[\frac{-\beta F E_{Ni}}{RT} \right] \quad (8.12)$$

Further oxidation of β-NiOOH to γ-NiOOH is assumed to occur at a rate I_γ that is given by Eq. (8.13), where E_γ is the overpotential for the β-NiOOH to γ-NiOOH reaction at the nickel electrode $(E - E^{o\gamma})$. Equation (8.13) assumes that this process is not microscopically reversible at a measurable rate, which is consistent with experimental studies of the discharge behavior of the nickel electrode.

$$\frac{3I_\gamma}{2FA} = k_\gamma \theta \left[OH^- \right] \exp\left[\frac{2FE_\gamma}{3RT} \right] \quad (8.13)$$

The γ-NiOOH is assumed to undergo a reduction reaction, discharging to α-Ni(OH)$_2$, which can be reversibly recharged back to γ-NiOOH. The rate of this process, I_α, is represented by Eq. (8.14), where υ is the fraction of the active material that is α-Ni(OH)$_2$, and E_α is the overpotential for the α-Ni(OH)$_2$ to γ-NiOOH reaction at the nickel electrode $(E - E^{o\alpha})$.

$$\frac{3I_\alpha}{5FA} = k_\alpha \upsilon \left[OH^- \right]^2 \exp\left[\frac{5\alpha F E_\alpha}{3RT} \right] - k_{-\alpha}(1 - \theta - \kappa - \upsilon)\left[H_2O \right]^{5/3} \exp\left[\frac{-5\beta F E_\alpha}{3RT} \right] \quad (8.14)$$

In parallel with the reaction of the active material in Eqs. (8.12)–(8.14), oxidation and reduction of oxygen are also assumed to occur at a rate I_O governed by Eq. (8.15), where E_O is the overpotential at the nickel electrode for the oxygen reaction.

$$\frac{I_O}{2FA} = k_2 \left[OH^- \right]^2 \exp\left[\frac{2\alpha F E_O}{RT} \right] - k_{-2}\left[O_2 \right]^{1/2} \left[H_2O \right] \exp\left[\frac{-2\beta F E_O}{RT} \right] \quad (8.15)$$

While active-material charge and discharge are reasonably reversible in the nickel electrode, the reduction of oxygen is not a facile process. It is therefore expected that an extremely low exchange current constant k_{-2} will be associated with this process (the second term in Eq. [8.15]).

The charge-transfer processes represented by Eqs. (8.12) to (8.15) can occur at any location within the active-material layers within the pores of the sintered substrate, as long as electrons and ions can carry charge to and from the charge-transfer sites at an adequate rate. If the electronic conductivity is too low, there will be a significant voltage drop through the active material that will prevent electrons from reaching charge-transfer sites far from the sinter surfaces. This model assumes that the active material consists of a porous deposit of active-material particles that completely fills the irregularly shaped voids within the sintered substrate, and that the void spaces between the particles are uniformly filled with electrolyte according to the equilibrium distribution of liquid and gas in the nickel electrode at any point in time.

8.1.2.7 Microscopic Model of Porous Nickel Electrode

The charge-transfer current will then be distributed through the thickness of the active-material layers within the pores of the sintered substrate by the active-material conductivity. For this purpose, a transmission-line model of the nickel electrode charge-transfer process is adopted, as in Fig. 8.5, where the active material is divided into n layers. The charge-transfer current in each layer is obtained by diagonalization of an $n \times n$ subelement matrix in each nickel electrode element.

The components in the boxes labeled "Nikl" in Fig. 8.5 are schematically outlined in Fig. 8.6, where the reversible potential of each reaction has been included as a source potential in series with the charge-transfer reaction impedance (which is represented as a pair of parallel diodes for reversible charge-transfer reactions). Because the resistance resulting from the electrochemical processes in

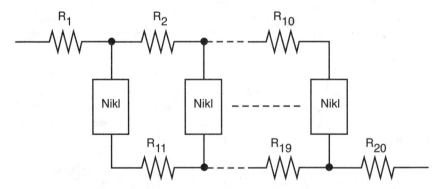

Figure 8.5. Nickel electrode element impedance Z_N in Fig. 8.3, showing how the active-material charge-transfer sites are distributed over n subelements (n = 10 here), with R_1 conducting electrons to the current collector, and R_{2n} conducting ions to the electrolyte interface.

8.1 First-Principles Models of Cell Performance 219

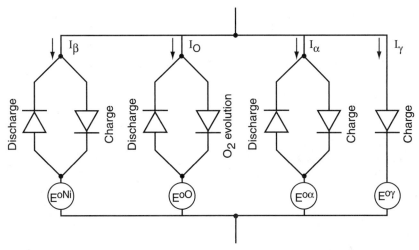

Figure 8.6. Charge-transfer paths used to model the electrochemical processes in the nickel electrode.

Because Fig. 8.6 is nonlinear (exponentially decreasing with increasing overpotential), the time intervals between computations are always kept small enough (no more than a 1 mV change in voltage) to ensure that a linear approximation to the charge-transfer resistance is sufficiently accurate. This linear approximation is dynamically updated at each time in a simulation to properly reflect the exponential I/V relationship for the charge-transfer processes.

The distributed nickel electrode model of Fig. 8.5 shows 10 subelements for simplicity. In the model, the number of subelements can be varied as needed to define the processes in the active material; however, 40 subelements have been found to describe the gradients experienced in the active material of typical nickel electrodes used in nickel-hydrogen cells. The resistances R_1–R_{10} at the top of Fig. 8.5 refer to the electronic resistance of the active material in the nickel electrode, as the electrons flow from the current collector to the charge-transfer sites that are out in the bulk of the active-material layers. The resistances R_{11}–R_{20} in Fig. 8.5 refer to the ionic conductivity within the solid deposit of active material, with ions being carried by solid-state proton diffusion from the charge-transfer site to the interface of the solid active material and the electrolyte.

If the active-material resistance is very low, current will be uniformly distributed through the 5 to 10 µm active-material layer. If the active-material resistance is high, then (as Fig. 8.5 suggests) most of the electrochemical reaction will occur near the current-collector surface. If the ionic resistance is low, there will be little variation in current through the film from electrolyte resistance. However, if the film is quite dense, or if it does not contain sufficient electrolyte, the ionic resistance will force either the charge/discharge reaction or oxygen evolution to occur only near the surface of the nickel electrode.

The thickness range for the active-material layers in sintered nickel electrodes can vary from less than 1 μm up to 40 or more μm within extremely large pores. The effect of variable pore sizes on performance is modeled by using the actual image of a realistic porous nickel electrode as a basis for deciding the proximity of each molecule of active material to the current-collector surfaces. Figure 8.7 contains a typical image obtained from scanning electron microscopy imaging of an actual nickel electrode cross section.

A composite distance for the distance from the active material in each black pixel of the image to the white sinter particles is determined by a two-dimensional ray-tracing analysis of the image. Rays are extended in all directions (in 1 deg increments) from each pixel containing active material to the point where each ray either intersects the edge of the image or impinges upon the conductive sinter in the image.

It is assumed that if a ray strikes the edge of the image, the ray would have to travel an added distance equal to the average pore size in the image before eventually striking a conductive surface. These 360 rays are used as parallel paths

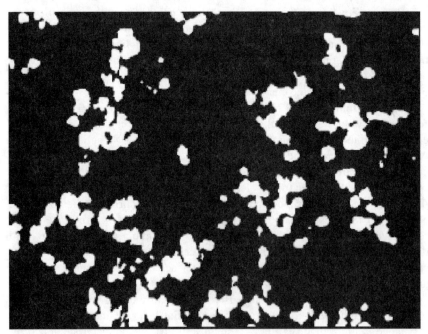

Figure 8.7. Typical distribution of active material (black regions) in sinter (white regions) used to model a porous nickel electrode. The region shown is 40 by 60 μm.

for electronic conduction through the active material, thus defining the composite distance that electrons are required to move by conduction to reach any active material in the image. The subelements in Fig. 8.5 are distributed throughout the range of composite distances that the active material falls within.

Thus, if a nickel electrode image containing large pores between the sinter particles is used to model the nickel electrode, much of the active material will reside far from the sinter, and conductivity gradients will be large, making utilization of the active material relatively low. Conversely, a nickel electrode model with smaller pores will exhibit better utilization and much less residual capacity that cannot be discharged at high rates. Because the model allows a different porosity distribution image for each nickel electrode element, it is possible to separate a nickel electrode into finite-element layers through its thickness or across its area, and to explore the effects of varying sinter porosity characteristics on electrode and cell performance.

8.1.2.8 Nickel Electrode Particle Size Model

The electrochemical activity of the active material in the nickel electrode is dependent on the degree of crystalline order and surface area of the active-material particles. Significant changes in active-material volume can cause fracturing of particles and a decrease in long-range crystalline order. This effect is modeled by keeping track of the average particle size in each subelement that is in each nickel electrode element, and assuming that the average particle size drops in proportion to the volume change during charge or discharge. The volume change involved in converting the β-NiOOH to the γ-NiOOH phase will completely fracture the active material into crystallites about 130 Å in diameter.

The discharged α-Ni(OH)$_2$ phase is assumed to restructure to the β-Ni(OH)$_2$ phase with a time constant τ_α that is inversely dependent on the KOH electrolyte concentration in each nickel electrode element. The time constant τ_α (= 5000 sec in 31% KOH) gives reasonable agreement with nickel electrode data.

The model also includes processes by which the crystallites of discharged active material can ripen to gradually increase their average size and the degree of long-range order. This ripening process is assumed to occur with a time constant τ_r that is dependent on the absolute temperature T in each specific nickel electrode element as indicated in Eq. (8.16). This equation gives τ_r = 3600 sec at a temperature of 25°C. The constants in Eq. (8.16) were determined from the changes in nickel electrode recharge voltage curves after allowing the fully discharged electrode to ripen for varying periods of time.

$$\tau_r = 0.5 + 3599.5 \exp\left[-36.839 + \frac{9167}{T}\right] \qquad (8.16)$$

The upper size limit for a fully ripened active-material crystallite diameter was assumed to be 1300 Å, which was estimated from transmission electron microscopy measurements.

Because the nickel electrode model allows layers of differing phases to form during charge and discharge through the subelements within each nickel electrode element, particle sizes can also vary significantly through the active material. This has been found to be a key factor influencing the fidelity of the voltage predictions made by the model. The formation of layers of differing particle sizes and composition has also been found to play a major role in determining how the behavior of the nickel electrode is influenced by its previous history.

8.1.2.9 Active-Material Conductivity Model

The electronic conductivity of the active material in nickel electrodes varies with the state of charge (as discussed in chapter 4), the temperature, and the applied electronic potential. The conductivity model may be tailored to match the characteristics of nickel electrodes using the logarithmic dependence of resistivity on state of charge shown in Fig. 8.8. This model consists of two regions of different slopes, the state of charge and resistivity level where the two regions intersect (S_L, ρ_L), the resistivity limit at zero state of charge ρ_H, and the resistivity at a state of charge of 1.0, ρ_u. These constants may be adjusted as needed to obtain the expected nickel electrode performance, although if realistic nickel electrode structures

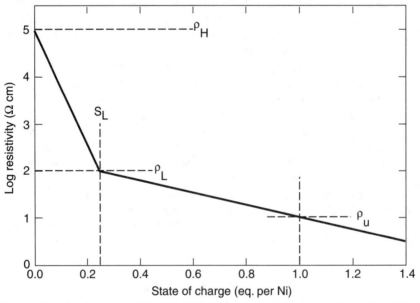

Figure 8.8. Active-material resistivity model used for modeling the nickel electrode.

are used, the best resistivity model has been found to match the measured resistivity of active material isolated from nickel electrodes. Temperature variability was assumed, giving a resistivity increase F_ρ for each 1°C drop in temperature relative to the dependence designated in Fig. 8.8 at 10°C. Typically, F_ρ was about –0.05% per degree Celsius.

The active material in the nickel electrode is a semiconducting material. When it is charged above the transition point (at 0.25 state of charge in Fig. 8.8), it is assumed to be a degenerate semiconductor for which the conductivity does not depend on the voltage drop across a layer of the material. Below the transition point, the resistance across each subelement layer of active material decreases exponentially as the voltage across the layer ΔV increases, as indicated in Eq. (8.17).

$$R = R_0 \exp\left[\frac{-F\Delta V}{RT}\right] \qquad (8.17)$$

The behavior indicated in Eq. (8.17) is responsible for causing the frequently seen second discharge plateau in the nickel electrode, and its inclusion in the nickel electrode model is important to model this behavior. The second plateau results when a layer of active material at the interface with the sinter surfaces becomes depleted and develops a significant voltage drop through a thin interfacial layer. The voltage drop effectively increases the conductivity of the layer, thus turning on additional capability to conduct electronic charge into the interior of the active material for continued discharge. This process continues until the depleted layer thickens to the point where continued discharge cannot be supported.

8.1.2.10 Modeling the Effects of Cobalt Additive
Cobalt additives have been found to influence the behavior of the nickel electrode in a number of ways, as described in chapter 4. Cobalt additives are typically present at levels of 5–10% of the nickel in the active material. The cobalt additive is assumed to influence the behavior of the nickel electrode by (1) decreasing the standard redox potential of the charge and discharge processes, (2) increasing the oxygen evolution potential, and (3) providing trivalent defect sites in the active material that improve electronic conduction as the state of charge becomes depleted. The effects of this additive can significantly influence the charge efficiency of nickel electrodes, the ability to efficiently discharge stored capacity, and the accessibility of each charged and discharged phase in the nickel electrode during typical operation and cycling.

The cobalt additive reduces the redox potential of the nickel electrode that is used in Eqs. (8.12) through (8.14) for the electrochemical charge and discharge processes. One assumes in modeling the cobalt additive that it is present as a solid solution in which trivalent cobalt species occupy nickel sites within the active-material lattice. In modeling the rates of the redox reactions, one assumes that the standard potential drops 4 mV for every 1% increase in the level of cobalt in the nickel hydroxide active material. This rate of change is reasonably accurate up to cobalt levels of about 15%, which is where discrete cobalt-containing phases can begin to form. The decreased redox potential plays a significant role in improving charge efficiency, which is a major reason for the wide use of cobalt additives.

The cobalt additive also slightly increases the potential at which oxygen is evolved from the charged active material in the nickel electrode, which also improves the charge efficiency by lowering the oxygen evolution rate. This effect is modeled by assuming a 0.4 mV increase in the oxygen redox potential of Eq. (8.15) for every 1% increase in the cobalt additive level. While this effect is relatively small, it is behavior that is straightforward to model and that does have a noticeable effect on nickel electrode performance.

The presence of trivalent cobalt defect sites in the active-material lattice provides sites that help enable electronic and proton conductivity as the active material becomes depleted of charge. When the active material is above the transition point where it is a degenerate semiconductor in Fig. 8.8, these cobalt sites have a negligible effect on conductivity. However, when the active material begins to drop to a state of charge low enough to act as a semiconducting material, the cobalt sites help maintain conductivity. This essentially decreases the state of charge where the transition in Fig. 8.8 occurs. In this model of nickel electrode performance, each 1% increase in cobalt additive is assumed to drop the state-of-charge transition point 0.0075 equivalents per nickel atom. This results in a decrease in the residual capacity in the nickel electrode that cannot be discharged at high rates when significant levels of cobalt additive are used.

A key assumption that is made regarding the role of cobalt additives in most nickel electrode models is that the cobalt additive is uniformly dispersed throughout the nickel electrode at both a microscopic and macroscopic level. This is almost never the case in real nickel electrodes, and a realistic model must allow the cobalt additive level to be altered in response to both nickel corrosion and other cobalt segregation processes. Nickel sinter corrosion often occurs during electrode manufacturing processes to result in reduced cobalt near the nickel sinter. Even if the electrode is built with a relatively uniform cobalt composition, nickel corrosion occurs as the electrodes age. Additionally, it is possible to effect significant cobalt depletion near the nickel sinter surfaces as a result of exposure of the electrodes to hydrogen gas when they are fully discharged, particularly if the nickel precharge in the cell has become depleted. Figure 8.9 shows the cobalt distribution profiles that can result from these types of processes.

In the model described here, a profile of varying cobalt level may be specified as a function of the microscopic distance of the active material from the individual sinter particles. This feature has been found to be highly effective in modeling the influence of cobalt segregation or long-term nickel corrosion processes on the

8.1 First-Principles Models of Cell Performance 225

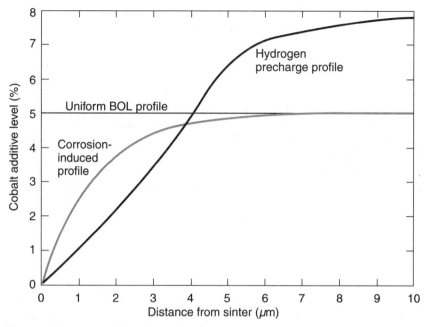

Figure 8.9. Types of distributions of cobalt additive that can result in nickel electrode active material assuming a uniform 5% level at beginning of life, from either nickel sinter corrosion or from cobalt segregation caused by exposure to hydrogen precharge during discharged storage.

performance of nickel-hydrogen cells. Cobalt gradients in the nickel electrode active material can reduce cell capacity by up to 30% and are an important factor contributing to the reduction in charge efficiency typically seen as nickel electrodes degrade during years of cycling.

8.1.2.11 Self-Discharge Model
Self-discharge in nickel-hydrogen cells occurs by three different mechanisms, as described in chapter 4. The first of these processes results from the evolution of oxygen from the charged nickel electrode and the subsequent recombination of the oxygen with hydrogen at the hydrogen electrode. The rate of this process is controlled by the rate of oxygen evolution, which is given by Eq. (8.15). The rate of oxygen evolution is strongly dependent on the voltage of the nickel electrode, and it drops rapidly as the voltage falls during discharge. However, during periods of charged open-circuit stand at states of charge greater than 80%, the evolution of oxygen is generally the major contributor to self-discharge.

In the second process, at lower states of charge, the self-discharge tends to be dominated by the direct reaction of hydrogen gas with the charged active material in the nickel electrode, which is linearly proportional to the state of charge as well as the hydrogen pressure (activity) in the cell.

At very low states of charge, a third self-discharge process can be dominant. This process is the electrocatalytic reaction of hydrogen gas at the nickel metal surfaces in the nickel electrode, which act as an extremely poor hydrogen electrode that is directly shorted to the nickel electrode. The rate of this process is independent of state of charge and is also proportional to the activity of hydrogen gas at the nickel metal surfaces. This electrocatalytic process is responsible for driving the nickel electrode potential to the hydrogen electrode potential when it is fully discharged.

The rate of these self-discharge processes is modeled using the expression in Eq. (8.18). The first term gives the rate of the direct reaction with hydrogen gas, and the second term gives the rate of the electrocatalytic process.

$$I_{sd} = [H_2]\gamma \left\{ k_{dir} M_s \exp\left[\frac{-E_{dir}}{RT}\right] + k_{ec} A_{Ni} \exp\left[\frac{-E_{ec}}{RT}\right] \right\} \quad (8.18)$$

This expression gives the self-discharge rate in milliamperes, where γ is the activity coefficient of hydrogen gas, k_{dir} is the rate constant for the direct reaction of hydrogen at 10°C (2.08 mA/mmol·atm), M_s is the millimoles of charged active material, E_{dir} is the activation energy for the direct process (4.39 kcal/mole), E_{ec} is the activation energy for the catalytic process (21.3 kcal/mole), k_{ec} is the rate constant for the catalytic reaction at 10°C (8.29 × 10^{10} mA/cm^2·atm), A_{Ni} is the area of catalytic nickel metal, and T is the absolute temperature. This expression has been found to accurately model the self-discharge rate of the nickel electrode for most of the conditions encountered in the nickel-hydrogen cell.

8.1.2.12 Modeling Cell Performance
Using a method of finite time differences, the model simulates the processes defined in the previous sections to follow the time evolution of cell characteristics during charge, discharge, or open-circuit periods. After initially specifying the geometry of the elements making up the cell and the initial physical and chemical properties of all the elements that compose the cell, the model iteratively computes the resistances of all the electrical components and the voltages across all components using Eq. (8.9), starting with initial estimates for component currents. During the initial computations or any subsequent computations where significant instantaneous changes in current or voltage occur, the system is allowed to iteratively reach a stable solution before any time changes are allowed to occur. While the first voltage calculated for the cell may not be accurate if an incorrect current distribution is initially chosen, rapid convergence to consistent current and voltage distributions is typically obtained.

8.1 First-Principles Models of Cell Performance 227

After computations have iteratively reached a stable solution, a small time interval Δt is chosen that is consistent with the rate of change in element properties. While this iterative time interval is initially quite small, it is dynamically adjusted by the model to reflect the rate of change in the cell and eventually will increase to a maximum that scales according to rate, and which is 2 sec for a C/2 rate. The process of computing the cell characteristics at time $t + \Delta t$ involves the following computations of updated cell properties.

The capillary pressures associated with the electrolyte in the porous components are computed, and electrolyte is allowed to flow convectively for the period Δt to equilibrate capillary pressures as much as possible in this time interval. Capillary pressures are based on the pore-size distribution in each element, the hydrophobicity of each element, and the fraction of the pore volume filled with electrolyte. Convective capillary flow is modeled using Eq. (8.6).

The electrochemical processes expressed by Eqs. (8.10) through (8.15), as well as Eq. (8.18) for self-discharge, are then allowed to proceed for the time interval Δt, according to the current through each component. The quantities of reagents in each element are adjusted to reflect electrochemical production or consumption, as well as migration of K^+ and OH^- to maintain electroneutrality. At this point, updated electrolyte quantities and concentrations, gas partial pressures, element porosities, phase compositions, particle sizes, and state of charge in the nickel electrode elements are all recalculated based on the coulombs of charge through each element during the time interval Δt.

Because convective forces are the strongest for transporting materials through the cell, the next step in the iterative procedure is to calculate the convective fluxes for both liquid and gas phases across all element interfaces from Eq. (8.4), using Eq. (8.6) to obtain the elements of the convection resistance matrix. For this calculation, the vector consisting of the pressure differences across each interface is used as the driving force **X** in Eq. (8.4). The calculated convective fluxes are then allowed to flow for the time interval Δt, resulting in either a full or partial equilibration of the pressure differences between adjacent elements. If pressures are not able to fully equilibrate in the allowed time interval, it can be assumed that some stress is being exerted on the solid structures and active-material deposits in the direction of the pressure gradient. Such stress is an important part of determining the rate of degradation of the electrodes throughout cycle life.

The weakest force causing material transport in the cell is diffusion (gravity and weaker forces are not considered here). The diffusive fluxes for movement of oxygen in the hydrogen atmosphere, movement of potassium hydroxide in the electrolyte, and movement of lattice protons in the porous active material of the nickel electrode are all calculated from the gradients in concentration for each of these materials between pairs of adjacent elements. The concentration of each of these species is then allowed to relax at the calculated diffusion rate for the time interval Δt. In this way steady-state diffusive gradients of ionic and neutral species within the gaseous, liquid, and appropriate solid phases in the nickel-hydrogen cell can be obtained during operation of the cell.

It is interesting to note that, using this approach for modeling solid-state proton diffusion, one finds that the coefficient for proton diffusion in the nickel electrode active material, which was chosen to be 2×10^{-9} cm^2/sec, is not extremely critical to cell performance. A lower diffusion coefficient will slow the rate of internal charge redistribution in the 5 to 20 μm layers of active material in the nickel electrode, thus affecting the rate of recovery from a depleted condition. A greater diffusion coefficient will enable a much more uniform discharge of the nickel electrode active material, thus reducing residual capacity. A lower proton diffusion coefficient tends to terminate the discharge capability of the nickel electrode more abruptly when the active material that is in contact with the sinter surfaces begins to lose its electronic conductivity as it depletes its charge. The sinter surface area, pore sizes, and the uniformity of pore size are just as important as proton diffusion rate for controlling nickel electrode performance in this model.

The series of computations described here form a single time iteration in the process of simulating the time development of the cell during its operation. After each iteration is completed, another iteration is begun involving recalculation of the currents through each component based on the previous voltages, after which new component voltages are obtained. A new time interval Δt is dynamically chosen on the basis of the changes within the cell, and the entire iterative procedure described here is repeated. This procedure is continued at the prescribed current until the voltage across the cell terminals reaches a preestablished limit during either charge or discharge, or until the required charge or discharge time duration is reached.

8.1.2.13 Validation of the Model
The model described in this section was developed during a period of about 10 yr to include all those processes that have been observed to influence the operating performance of nickel-hydrogen cells. For this reason, it has been extensively validated based on the observed behavior of nickel-hydrogen cells throughout a wide range of conditions. These behavior characteristics include cell voltage, capacity and impedance for charge and discharge rates from C/20 to C, temperatures from 0 to 20°C, and performance changes during several hundred realistic cycles and subsequent cell reconditioning.

The model typically predicts cell performance as accurately as it can be obtained from repeated measurements. For example, as a result of small variability in temperature and starting state of charge, repeated capacity cycles can frequently give up to ±5 mV variability in the voltage of a cell and several percent variability in cell capacity.

The model typically predicts behavior within the cycle-to-cycle spread of observed cell behavior. An example is indicated in Fig. 8.10, where the predicted cell voltage for two different initial states of charge is correlated with

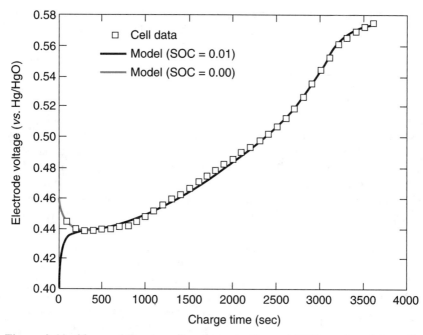

Figure 8.10. Observed C-rate cell recharge voltage at 10°C, and model predictions for two different initial nickel electrode states of charge.

the observed C-rate recharge behavior of a nickel-hydrogen cell at 10°C. For this example, the model predicts the observed cell voltage within ±5 mV when the starting state of charge is essentially 0.0, but exhibits significant voltage deviation early in the recharge period when the cell is initially assumed to be 1% charged. This initial voltage deviation is associated with the high initial resistivity of the active material in the nickel electrode when the cell is fully discharged. This deviation arises because the cell to which the model simulation was compared was fully discharged before the recharge was begun.

In the following section, a number of applications of this model to predicting and describing the details of actual cell operation are presented to illustrate the types of questions and issues that can be resolved and explored using high-quality nickel-hydrogen cell models.

8.1.3 Applications of Models to Nickel-Hydrogen Cell Performance

A high-fidelity model of the nickel-hydrogen cell can communicate information about the factors that control cell performance at a microscopic chemical and physical level that cannot be easily studied in real cells during operation. To illustrate how useful such a model can be, this section discusses a number of areas where modeling has been used to closely examine cell behavior. Although they provide such illustrative examples, these areas are only a small selection of the many detailed processes internal to the cell that can be explored using model-based analysis.

8.1.3.1 State-of-Charge Gradients in Nickel Electrodes

When a nickel electrode is discharged to the point where the voltage of a nickel-hydrogen cell drops to low levels, the cell is customarily referred to as a depleted cell and the nickel electrode is regarded as discharged. However, the nickel electrodes in the cell are not really fully discharged, as can be shown by the recovery of the cell voltage, if the cell is allowed to stand open-circuited for a period of time. After such recovery, the cell may be discharged further to tap more of its residual charge. The processes responsible for the depletion and subsequent recovery of capacity may visually be followed by modeling the behavior of nickel electrode elements in a nickel-hydrogen cell while it is charged, discharged, and allowed to stand open-circuited.

Such a model uses the microstructure for the sinter and active material within the porous nickel electrodes that is indicated in Fig. 8.7. A simplified 100 A h nickel-hydrogen cell model consisting of 59 individual elements is used to examine the behavior of the nickel electrodes during a 16 h charge at C/10, followed by a C/2 discharge to 1.1 V, a 1 h open-circuit stand period, then a repeated discharge to 1.1 V. The initial C/2 discharge decreases the average state of charge of the nickel electrodes in the cell to 16.75%. Following the open-circuit recovery, the C/2 discharge can be supported for nearly another minute, decreasing the average nickel electrode state of charge to 16.5% before the 1.1 V limit is again reached. While the average state of charge is 16.75% after the initial C/2 discharge, Fig. 8.11 indicates the real picture of how this residual charge is distributed in the nickel electrodes.

At the interfaces of the sinter particles with the active material, the state of charge of the active material is about 3%, while the active material most distant from the sinter particles is at a state of charge of about 28%. This distribution of charge is dictated by the distance of the material from the conductive sinter particles, combined with the changing conductivity of the active material as it undergoes discharge. When the depleted structure in Fig. 8.11 is allowed to stand open-circuited, proton diffusion occurs to restore the state of charge in the thin depleted layer of active material immediately adjacent to the sinter surfaces. Subsequent C/2 discharge will allow more capacity to be discharged as the depletion layer is again reformed. The example of Fig. 8.11 shows precisely why the buildup of unusable, or residual, capacity in nickel electrodes is so sensitive to sinter pore size distributions, sinter degradation, and loading level.

8.1.3.2 γ-NiOOH Formation in Nickel Electrodes

Frequently analysis of the behavior of nickel electrodes examines the formation of the γ-NiOOH phase during charge and cycling of cells. A model of the cell

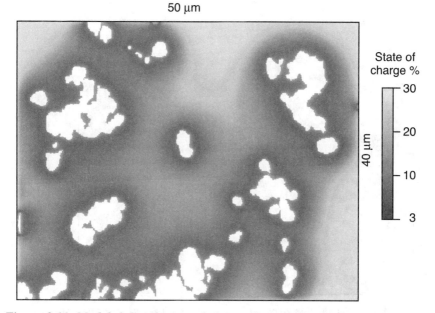

Figure 8.11. Modeled distribution of charge through the active material in a nickel electrode after a C/2 discharge to 1.1 V. The white regions are sinter particles, and the shading of the active material indicates its state of charge.

allows the formation of this phase to be examined in detail as the cell is charged, discharged, and cycled. The same nickel-hydrogen cell model described in the preceding section has been used to examine the formation of γ-NiOOH during a 16 h C/10 recharge, as is indicated in Fig. 8.12. As can be seen in this figure, the γ-NiOOH begins forming at the surfaces of the sinter particles when the cell reaches full charge, and as overcharge continues, it propagates as a thickening layer away from the sinter. While the average level of γ-NiOOH after this recharge was 13%, it varied from nearly 30%, at the surfaces of the sinter particles, to essentially 0% for the active material far from the sinter.

When the layers of γ-NiOOH shown in Fig. 8.12 are discharged, they result in a layered structure of α-Ni(OH)$_2$, which can be either recharged to γ-NiOOH or partially recrystallized to β-Ni(OH)$_2$ if allowed to stand for any significant period of time before being recharged. During repeated cycling it is possible to form relatively complex layers of these charged and discharged phases within the active material.

8.1.3.3 Gradients in the Self-Discharge Rate
The self-discharge rate within the nickel electrodes in the nickel-hydrogen cell is accurately predicted by Eq. (8.18), which indicates that the self-discharge rate

232 Nickel-Hydrogen Cell Modeling

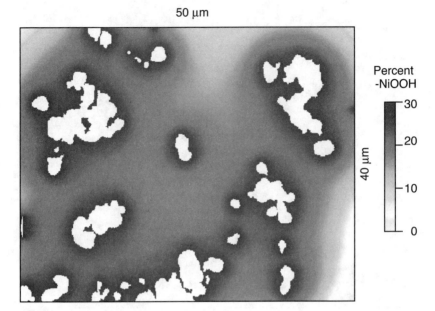

Figure 8.12. Distribution of γ-NiOOH in the active-material microstructure of a nickel electrode after recharge at C/10 for 16 h at 10°C, with 31% potassium hydroxide electrolyte.

is directly proportional to the pressure of hydrogen. However, when an accurate model of the nickel electrode is closely examined in an operating nickel-hydrogen cell, one discovers that the hydrogen gas pressure or activity is not constant throughout the nickel electrodes. When the cell is near full charge, or is being trickle-charged under conditions where there is significant oxygen evolution in the nickel electrodes, a significant fraction of the gas pressure within the nickel electrodes will be made up of oxygen gas.

A model that accurately simulates the convective flow and diffusion of both gaseous and liquid phases in the cell shows that there is a gradient in the ratio of oxygen to hydrogen that varies significantly as one moves from the interior of the nickel electrode to its surface, into the separator, and finally to the surface of the hydrogen electrode. The resulting gradient in hydrogen pressure can lead to significant differences in the rate of self-discharge of the nickel electrodes at their surface compared to within their interior.

The finite-element modeling approach described here has been used to examine the gradients in self-discharge rate and state of charge that can develop within the nickel electrode as a result of hydrogen pressure gradients. These gradients were studied by using a model that consisted of ten finite elements through the thickness of the nickel electrode, thus enabling gradients to be examined through the nickel electrode structure. These ten finite elements each had differing sinter porosity and pore size distributions,* simulating the porosity characteristics typically seen in slurry sinter.[8.8] A uniform active-material loading level of 1.65 g/cm^3 void was assumed in the nickel electrode elements. The gas spaces in the model were configured such that the fully charged cell operated at a pressure of about 600 psi. When the cell model was charged at a C/10 rate for 20 h, which put the nickel electrodes into essentially steady-state overcharge for about the last 8 h of the recharge period, the oxygen evolved within the nickel electrode was found to comprise up to 70% of the gas pressure in the gas space within the pores of the nickel electrode.

However, as expected, the oxygen content in the nickel electrode increased in the interior of the nickel electrode compared to its exterior surface where it contacted the separator, as shown in Fig. 8.13. This variation in oxygen pressure creates a gradient in the hydrogen activity through the thickness of the nickel electrode, and thereby produces a gradient in the self-discharge rate. The self-discharge rate is more than an order of magnitude greater at the surface where

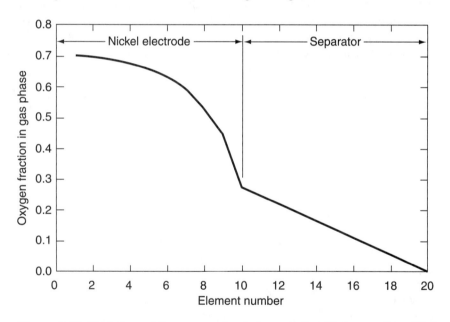

Figure 8.13. Variation of the oxygen level through the thickness of a nickel electrode and its adjoining separator during steady-state overcharge at C/10, at a temperature of 10°C.

---

*Slurry sinter typically has larger pores and higher porosity in its interior, and a surface "skin" of significantly lower porosity and smaller pore sizes.

the nickel electrode contacts the separator than it is in the interior of the nickel electrode. The self-discharge gradient is largely compensated for by a balancing gradient in the charge current, but it can produce a characteristic composition gradient through the thickness of the nickel electrode, as shown in Fig. 8.14.

The self-discharge gradients that can dynamically occur through the nickel electrodes in an operating nickel-hydrogen cell are an important source of variations in charge efficiency during recharge.

8.1.3.4 Computer Modeling to Help Design Improved Cells
A computer model of the nickel-hydrogen cell allows the modeler to examine in detail the origin of voltage losses, diminished charge efficiency, reduced usable capacity, cell resistance changes, or any other key measure of cell performance. The cell design can be adjusted in the computer model to determine how any given design change either improves or degrades cell performance. The details provided by the model allow the cell components that actually limit cell performance to be identified, thus allowing these components to be the target of possible design changes. In this way the design of a nickel-hydrogen cell can be optimized for a particular application. Once the computer cell design has been optimized, it is possible to construct cells that have the computer-specified optimized design, and then to test these cells to validate the model predictions.

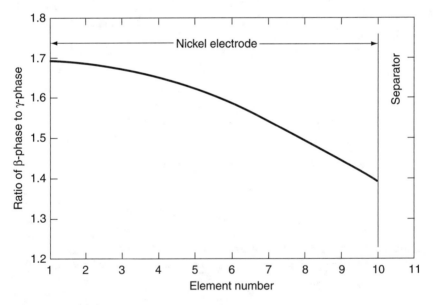

Figure 8.14. Relative amounts of β-NiOOH and γ-NiOOH through the thickness of the nickel electrode during a model simulation after 20 h of recharge at a C/10 rate at 10°C.

8.1 First-Principles Models of Cell Performance

The process of computer optimization of a nickel-hydrogen cell has been performed with the goal of reducing cell impedance and optimizing cycle life for operation at high depths of discharge in low Earth orbit satellite applications. In such applications, a typical charge-discharge cycle can occur every 90 min, with about 30 min of discharge and 60 min for recharge. An optimum cycle life requires that the cell temperature is kept low (0 to −5°C) to minimize the rate of corrosion processes, and that amount and rate of cell overcharge is minimized to reduce long-term nickel electrode damage from active-material volume changes. When standard cell designs are cycled in the relatively rapid low Earth orbit cycle at high depths of discharge and low temperatures, cycle life tends to be prematurely limited by reduced charge efficiency as the cells age. The reduced charge efficiency eventually makes it impossible to adequately charge the cells in the 60 min recharge period without driving the cell voltage so high that the overcharge stresses severely limit cycle life.

This issue with standard cell designs was addressed using a 57-element computer model of a nickel-hydrogen cell to optimize the design. The design parameters that were varied in the model included the electrolyte concentration, the amount of electrolyte in the cell, the thickness of the electrodes and the separator, and the lead conductivity. A number of other design parameters were not varied, thus enabling the cell design to use existing electrode components. These included the nickel electrode, hydrogen electrode, and Zircar separator. These are key cell components for which optimized and highly specific manufacturing procedures and processes are presently in place. Thus, thickness variations of the electrodes or separator were limited to incrementally added or deleted layers of the existing components.

The overall goal of the optimization was to reduce the impedance of the cell during operation sufficiently to allow full recharge during 60% depth-of-discharge cycling in a low Earth orbit profile at −5°C without allowing the cell voltage to rise above 1.52 V. In addition, the changes in active-material volume during a cycle, which are a major source of stress to the nickel electrode during cycling, could not exceed the cumulative volume changes that occurred during a similar 40% depth-of-discharge cycle at +10°C, conditions for which it is known that the cell components are capable of more than 60,000 cycles. Thus, the minimum cycle-life goal of this cell optimization was 60,000 cycles at 60% depth of discharge. It was recognized that the optimization process was likely to increase the specific weight of the cell. The goal in the area of cell weight was to maximize cell performance while keeping the cell weight penalty to less than 10%.

The design optimization used an iterative procedure of changing the thickness of the electrodes and separator in increments of one layer, with the thickness of each layer constrained by the thickness range available for standard cell components. For nickel electrodes, this thickness was 30–36 mils; for Zircar separators, 11–14 mils; and for hydrogen electrodes, 5–6 mils. Iterative optimization also included KOH electrolyte concentration, with the design starting at 28% by weight. The thickness of the leads attached to each electrode was also allowed to vary. Each design point in the optimization procedure involved running a 500-cycle simulation to stabilize cell performance in a 90 min low Earth orbit cycle at 60% depth of discharge, with a –5°C temperature for the base plate on which the cells are mounted.

The result of the model-based optimization procedure was a nickel-hydrogen cell design that is unique in the sense that it combines components in a cell configuration that had never before been built or tested. The computer-developed design used an electrolyte concentration of 26%. This concentration was essentially dictated by a trade between the reduced stress on the nickel electrodes from volume changes at lower concentration, and the reduced diffusional and ohmic electrolyte resistance at higher concentrations. The nickel electrode thickness was optimized at a single 35–36 mil layer, with both sides of the electrode being electrochemically active (i.e., in contact with a separator/hydrogen electrode unit). This electrode-stacking configuration is generally referred to as a dual-anode design,[8,9] in contrast with the more standard back-to-back design that uses two nickel electrode layers placed back to back to give an overall unit that has twice the thickness. The dual-anode nickel electrode design gave major performance improvement by doubling the active superficial electrode area, as well as by cutting the distance for diffusion and conduction through the nickel electrode thickness by a factor of two.

The optimized separator thickness used a single 12-mil-thick layer of Zircar separator against each face of each nickel electrode. The reduced thickness provided half of the electrolyte resistance of double-layer Zircar separator designs, while having a separator on each side of each nickel electrode doubled the active area for ionic current conduction and for transporting oxygen between the electrodes during cell overcharge. This design also enabled the cell stack of electrodes to hold about 3.5 g of electrolyte per ampere-hour of cell capacity, a level high enough to essentially preclude concerns about dry-out or electrolyte freezing.

The optimized negative electrode design employed a single hydrogen electrode for each side of each nickel electrode, thus using twice as many hydrogen electrodes as does a standard back-to-back design. The extra hydrogen electrodes constitute the major weight penalty in the optimized cell design. Fortunately, the hydrogen electrode unit is the lightest cell component; however, the additional hydrogen electrode layers add about 15% to the specific weight of a standard cell. Furthermore, the computer model predicted that the optimized design would provide an improvement of about 5% in cell energy use as a result of the reduced electrochemical and ohmic impedances. Thus, the overall design had a specific energy that was only about 10% lower than achievable for standard designs, but it was predicted to be capable of more than twice the cycle-life throughput of a standard cell design.

8.1 First-Principles Models of Cell Performance 237

Other key points in the design optimization included the electrode leads and the cell terminal configuration. The leads were increased in thickness to the limit of the space available in the core of the 3.5 in. diameter cell to reduce the cell impedance. The computer model also indicated that the thermal gradients across the cell became unacceptably large for 60% depth-of-discharge operation unless the terminals were placed axially at each end of the pressure vessel, and the sleeve mounting-flange used for thermal dissipation from the cell was placed at the center of the pressure vessel. To prevent water condensation in the pressure vessel, which can rapidly degrade cell performance and cycle life, the thermal gradient between the hottest part of the stack and the coldest part of the cell must be kept below 6°C when 26% potassium hydroxide electrolyte is used. The optimized cell design described here maintained this thermal gradient to less than 5°C.

The model for the optimized cell also included a number of important features that are generally regarded as standard in nickel-hydrogen cells. These include wall wicks to redistribute any free electrolyte back into the electrode stack, gas screens to allow ready access of hydrogen gas to the negative electrodes, 40-mil-thick aluminum sleeves to transport cell-produced heat to a −5°C mounting plate, nickel precharge, and low-resistance cell terminals.

The final cell design that emerged from the optimization procedure was predicted by the computer model to have an internal impedance about one-half that of a standard cell design of similar capacity, with the absolute impedance depending on the number of electrode units in the stack (cell capacity). The final design was also predicted to be capable of full recharge back to the starting state of charge after 500 cycles in a 60% depth-of-discharge low Earth orbit cycle at −5°C with the peak recharge voltage limited to 1.515 V at beginning of life. The peak thermal gradient predicted in this cycle was 4.8°C.

The volume change predicted for the active material in the nickel electrodes during cycling (a key stress factor) was highly dependent on the recharge ratio, but at a 1.015 recharge ratio was similar to that expected in a standard 31% potassium hydroxide cell cycled at 40% depth of discharge with a 1.03 recharge ratio at +10°C. The computer model predicted, however, that at beginning of life it should be possible to maintain a state of charge adequate to support a 60% depth-of-discharge cycle at −5°C with a recharge ratio of less than 1.01, thus potentially reducing overcharge stresses and extending cell lifetime considerably compared to the use of a 1.02 recharge ratio.

Table 8.1 provides a summary of the key design parameters that the computer model suggested were critical for a cell design intended to optimize cycle life at high depths of discharge. The performance projections in the table are estimates based on the stress factors that are known to contribute to cell degradation, and they must be validated by testing to ascertain the accuracy of the model predictions. It is important to note that none of these individual design characteristics is unique to this cell design; however, the combination of all of them had never been

Table 8.1. Key Design Parameters and Projected Capability for a Computer-Designed Nickel-Hydrogen Cell

Design parameters
26% KOH electrolyte (~3.5 g/A h)
Single 35-mil–thick nickel electrodes in dual-anode stacking arrangement
Single-layer Zircar separator against each surface of each nickel electrode
A hydrogen electrode against each separator layer
Axial cell terminal configuration
Improved conductivity for electrode leads (limited by space in core)
Thermal flange center-mounted on pressure vessel
Approximately 10% increase in cell specific weight
Increased stack length (because of added negative electrodes)
Twice the life-cycle throughput of standard cells projected

Projected capability
Double the cycle life of standard cells at 60% depth of discharge

previously suggested or attempted. All primary cell components in the design are standard, thus eliminating any need for costly redesign and requalification of the individual cell components.

Without conducting performance testing and life tests to validate the capability of this design, one finds that this cell optimization effort is only a theoretical exercise. Substantiation of the model projections requires that cells be built to the computer-generated design specification, and that these cells be tested to determine how they meet the model projections and the design goals. To obtain such test data, an effort was made to have some nickel-hydrogen cells constructed to the specifications indicated in Table 8.1 by the computer model. This design was communicated to all the U.S. manufacturers of nickel-hydrogen cells in the year 2000 to be considered for production of test cells, and for potential future applications involving up to 60,000 or more cycles at up to 60% depth of discharge.

One of these manufacturers constructed five test cells to the specifications in Table 8.1. Performance characterization of these 60 A h cells gave results for cell resistance, voltage, and capacity behavior that closely matched the model predictions. For example, the computer model predicted a resistance for this 60 A h cell of 0.57 mΩ, while the actual resistance was measured to be 0.60 mΩ. This resistance is about 48% lower than that expected for a standard cell of similar capacity, in excellent agreement with the model prediction of 50% lower resistance.

8.1 First-Principles Models of Cell Performance 239

The five cells were placed in a 60% depth-of-discharge life test with a low Earth orbit profile at –5°C in the year 2001, and at the end of 2008 had completed over 40,000 cycles with continuing good performance. Because these cells are expected to provide the best life capability possible using standard cell components, this life test was designed to evaluate the ultimate cycle-life capability of nickel-hydrogen cells. Toward this end, the cycle profile was designed to minimize the amount of overcharge and the overcharge voltage to which the cells were exposed. Each cell was given whatever recharge ratio it required to keep it cycling between about 15% and 75% state of charge.

As the cells slowly degrade, the recharge ratio and the peak recharge voltage are gradually increased to keep the cells adequately recharged. The increase in the recharge ratio through the early lifetime of the cells is shown in Fig. 8.15. As predicted by the computer model of this cell, at beginning of life the required recharge ratio was well below 1.01. Through the cycling period in Fig. 8.15, the peak cell recharge voltage had to be increased from 1.515 V (precisely what was predicted by the computer model at beginning of life) to 1.519 V. It is noteworthy that cell 1, which has required a slightly higher recharge ratio (see Fig. 8.15), is operating about 1°C warmer than are the other cells. The recharge fractions shown in Fig. 8.15 have remained stable up to the present 40,000 cycle point in the test.

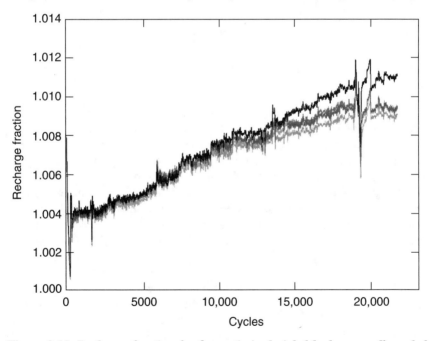

Figure 8.15. Recharge fraction for five optimized nickel-hydrogen cells cycled at 60% depth of discharge under conditions of minimum excess overcharge.

At more than 40,000 cycles into their lifetime, these cells have demonstrated more than 100% of the cycle life typically obtainable from cells of more standard designs at 60% depth of discharge. The goal of this test is 60,000 cycles, of which 67% have now been completed. Most important, however, these cells do not yet show any evidence of significant degradation. Lifetime projections based on the increases in either recharge ratio or the rate of pressure rise (indication of nickel electrode corrosion rate) suggests that the cells could easily provide 80,000 cycles at 60% depth of discharge. While continued testing will determine the ultimate cycle-life capability of these cells, they have already accumulated enough cycles to surpass the cycle life of standard cell designs.

These cells clearly demonstrate the feasibility of using a computer model to optimize a nickel-hydrogen (or other type) battery cell for any particular performance need, whether that need is optimum cycle life, high depth of discharge, minimum weight, or some other critical performance characteristic. High-fidelity battery cell performance models are expected to be more routinely used in the future for cell optimization, particularly as an alternative to the traditional optimization process that involves life-testing a large matrix of cell designs.

8.1.3.5 The Effect of Cobalt Gradients on Cell Performance
Cobalt hydroxide is added to the active material in nickel electrodes because it has been found to significantly improve their performance. When a cell (and its nickel electrode) is new, the cobalt additive is presumed to be uniformly present at either a 5% or a 10% level, depending on the manufacturing procedure. However, several chemical processes have been identified that can cause cobalt levels to be reduced at the interface between the active material and the sintered current collector in porous nickel electrodes.

These processes include active-material reduction by hydrogen gas during cell storage and nickel substrate corrosion during long-term cycling operation of the cell. In either case, the resulting gradient in the cobalt additive (see Fig. 8.9) can influence cell performance through a number of mechanisms. A cell model that is capable of accurately modeling the influence of cobalt additives at the microscopic level in porous electrodes is also capable of informing the modeler precisely what mechanisms are responsible for the performance degradation. This information can subsequently be used to explore possible methods to reverse the cell degradation, or to make nickel electrodes that are more resistant to changes in the distribution of cobalt additives.

The finite-element cell model described earlier in this chapter allows the modeler to specify either a uniform distribution of cobalt additive in the active material, or an arbitrary distribution that can be tailored as desired to match a degradation process of interest. For example, the two distributions shown in Fig. 8.9 have been developed to match the cobalt profiles observed from analysis of nickel electrodes that were exposed to extreme conditions of two particular degradation

8.1 First-Principles Models of Cell Performance

processes. The model applies the prescribed distribution to the active-material microstructure within the porous structure of the nickel electrodes. The modeler may change the cobalt distribution to examine how it affects cell performance, or to simulate gradual electrode degradation or recovery. In this way, the changes in cell performance from a changing cobalt distribution can be separated from other cell characteristics that may change with time.

The effect of a nonuniform cobalt distribution on cell performance can be examined at two different levels. The first level corresponds to the formation of a nonuniform cobalt distribution with essentially no other changes taking place in the cell. This situation enables one to easily separate the effects of the cobalt distribution changes from the effects of other chemical or physical changes in the cell. The second level corresponds to the situation that would be expected to arise from a process such as sinter corrosion. Sinter corrosion not only changes the cobalt content near the corroding sinter surfaces, but it also affects the electrolyte concentration, the active-material loading level, the distance of the active material from the sinter, the sinter electrical-conduction paths, and the cell pressure. Sinter corrosion is, in fact, a key aging process in the nickel-hydrogen cell that can cause nickel electrode expansion and electrolyte redistribution as a cell ages. Modeling cell performance at this level is a key aspect of running a life test in a computer simulation, and it requires adjustment of many of the chemical and physical parameters in the cell model as the cell ages throughout years of operation. This level of modeling will be discussed in a later section of this chapter in the context of carrying out cell life tests through computer simulations.

However, at the first level, it is possible to examine the effect that the development of a nonuniform cobalt additive distribution has on cell performance by comparing a distribution such as that in Fig. 8.9 (hydrogen precharge profile) to a uniform cobalt distribution. If this comparison is done with no changes to any other cell parameters in the model, we can isolate the effects that the changing cobalt distribution has on performance. In an actual nickel-hydrogen cell this situation could arise from chemical changes and cobalt migration that can be induced in the nickel electrodes during cell storage with a hydrogen gas precharge. The key cell performance characteristics that are of practical interest in this situation are the cell voltage and the cell capacity.

If the cobalt profile indicated in Fig. 8.9 resulting from hydrogen precharge is put into a cell model, the cell performance is degraded as indicated in Fig. 8.16. The cell model used consisted of a simplified model similar to that in Fig. 8.3, except that all the cell components were separated into two finite elements. Thus, this model represents a reasonable simulation of a small cell containing a single electrode stack unit. The model was otherwise consistent with a design containing double-layer Zircar separator, with 3.5 g/A h of 31% electrolyte, and a maximum cell operating pressure of about 600 psi. The porous microstructure of the nickel electrode is as shown in Fig. 8.7. As shown in Fig. 8.16, the nonuniform cobalt

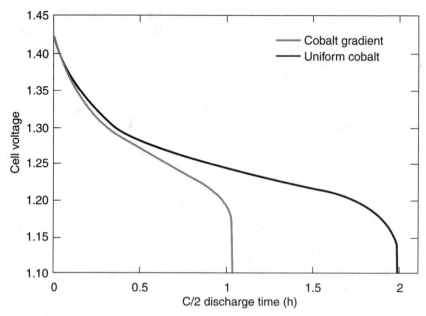

Figure 8.16. Effect of a cobalt gradient such as indicated in Fig. 8.9 resulting from hydrogen precharge on the high rate discharge of a nickel-hydrogen cell.

distribution tends to increase cell voltage during charge and discharge, and it decreases the usable cell discharge capacity by about 50%. These predicted changes in cell performance are consistent with those typically observed for actual nickel-hydrogen cells exposed to the conditions that cause cobalt to redistribute within the nickel electrodes.

However, the availability of a good computer model of the cell makes it possible to determine the root cause for the reduced capacity by examining the effects of the cobalt distribution on the electrochemical processes within the active material. When this is done, we find that gradients in cobalt concentration cause a wide range of gradients in composition and potential through the active material. Typical examples are shown in Fig. 8.17, where the state of charge and the amount of γ-NiOOH are plotted as a function of distance from the sinter surfaces in the porous nickel electrode, following a C/2 discharge to 1.1 V following a C/10 recharge for 16 h. Figure 8.17 illustrates how the cobalt distribution can directly influence the active-material phase composition and state of charge during cell operation.

The underlying causes for the effects of cobalt shown in Figs. 8.16 and 8.17 are the changes in the redox potential of the active material, and the changes in active-material conductivity caused by cobalt additives in the cell model. When used in this way, a cell model can provide an excellent diagnostic tool to pinpoint

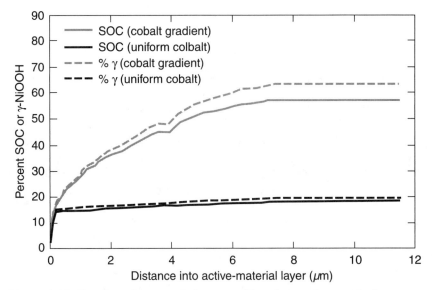

Figure 8.17. Gradients in state of charge and how it correlates with the amount of residual γ-NiOOH through the active-material layer thickness in nickel electrodes with depleted cobalt at the sinter surfaces.

the root causes for changes in cell voltage and capacity. The model predictions can be compared to the observed cell signatures, and the correct root cause is anticipated to be the one that correctly simulates the observed cell behavior.

8.1.3.6 Emergent Behaviors from Cell Models

"Emergent behavior" is a term that is used to identify behavior seen in computer simulations of complex systems that was not programmed into the computer code that runs the simulation or the physical structure of the system model. Such behavior typically "emerges" during a simulation as the result of complex interactions between the basic materials, structures, and processes that are explicitly included in the model. A model that includes all the known processes in the nickel-hydrogen cell, as described earlier in this chapter, has the potential for predicting cell operating behavior and performance characteristics that may be both unanticipated and not explicitly programmed into the simulation code. A number of such examples of emergent behavior have been seen during the use of the finite-element model described earlier in this chapter for modeling nickel-hydrogen cells. Several of these examples will be discussed here to illustrate the range of behavior that can be explained by a sufficiently detailed model of fundamental cell processes.

Calorimetry measurements of the heat generated during nickel-hydrogen cell operation have found that there is typically a large heat spike observed when a cell is switched from recharge to discharge.[8,10] This heat spike can rise to several times the heat dissipation level normally expected during discharge and can last for 10–15 min or more. When the finite-element nickel-hydrogen cell model described in this chapter is run, and the predicted heat generation is monitored as a function of time, a thermal transient is seen at the transition from charge to discharge that is very similar to that seen in the calorimetry measurements on real nickel-hydrogen cells. This thermal behavior was not put into the model, but it emerged from the interacting convective gas transport processes, gaseous composition gradients, and the electrochemical processes involving hydrogen and oxygen within the cell model.

Careful examination was made of the processes taking place in the model at the point when recharge was ended and discharge begun, and it revealed that the thermal spike was initiated by the sudden drop in cell pressure caused by high-rate discharge. This pressure drop results in a burst of oxygen gas being convectively driven from the oxygen-rich pores of the nickel electrode as a result of the pressure differential. The convectively driven oxygen flow generates heat by recombination with hydrogen when it comes in contact with the hydrogen electrode. About 10–20 min of discharge are typically required for the oxygen gas accumulated during overcharge within the pores of the nickel electrode to be fully dissipated, with this time constant depending on discharge rate (rate of pressure drop) and the detailed porous structure of the nickel electrode.

A second example of emergent behavior can be seen in efforts to model the "charge memory" effect that can be seen in nickel-hydrogen cells. For example, cells that are repeatedly cycled to a maximum state of charge that is lower than 100% gradually begin to go into overcharge at the peak state-of-charge point, whatever it might be. For example, when a cell is repeatedly cycled to a peak state of charge of 50%, the cell will eventually start to perform like a cell with only 50% of its normal capacity and go into overcharge at 50% of the state-of-charge level. This is a key reason why nickel-hydrogen cells that begin to walk down in capacity during battery operation cannot be easily brought back into capacity balance with the other cells in the battery.

The charge memory effect described here, while clearly a key factor in cell and battery performance, was not fully understood until it was reproduced in a computer model of a nickel-hydrogen cell that accurately simulated the repeated cycling between the discharged state and a peak state of charge of about 50%. Within 50–100 simulated cycles in this regime, the cell model clearly showed the cell prematurely entering overcharge and oxygen evolution at about 50% state of charge, just as is seen in actual cells. The origin of the charge memory effect was not even understood when the computer code was assembled to describe the processes within the cell. However, this effect clearly emerged from the simulation results, just as it should if the model really includes all fundamental processes that are important to cell operation.

8.1 First-Principles Models of Cell Performance   245

The model also enabled the root cause of the charge memory effect to be clearly identified. The nickel hydroxide active material in the nickel electrode exists in the pores of the sinter as small crystallites. The size of the crystallites, and thus their electrochemical activity, depends on their prior time history of state of charge, phase composition, and temperature during cycling. Repeated cycling to a peak state of charge of only 50% results in the gradual formation of a distinct layered structure within the active material. A layer consisting of the 50% of the active material closest to the conductive sinter particles develops into a high-surface-area layer of very small and active particles that support most of the cycling capacity of the cell. The remaining 50% of the active material comprises a layer more distant from the nickel sinter that gradually ripens into larger crystallites because they are not being repeatedly cycled. The larger crystallites in this more distant layer have a significantly reduced surface area, and a significantly lower electrochemical activity.

Consistent with the behavior seen in real nickel-hydrogen cells, the model exhibited a well-cycled layer that could be charged very efficiently, as well as a more remote layer of active material that could only be recharged with extensive oxygen evolution. The oxygen evolution resulted in an onset of poor charge efficiency and a large amount of heat production when the state of charge in the cell model reached 50%. Thus, a key cell behavior was again found to emerge from the interactions between the charge-transfer and charge-transport processes, the chemical transformations, and the active-material conductivity in the nickel electrode as it was modeled in the nickel-hydrogen cell environment.

A final example of emergent behavior in a cell model explains examples of electrolyte freezing that have been seen in actual nickel-hydrogen cells at temperatures 20–30°C higher than the expected freezing temperature, based on the concentration of electrolyte added to the cell when it was activated. In fact, computer model simulations first identified the root cause of unexpected cell failures at low temperatures as likely being the freezing of the electrolyte. Actual nickel-hydrogen cells, when charged continuously at low temperatures and low rates (~C/50 to C/200) for long periods of time, can suddenly fail as a result of the formation of an internal cell short circuit. The short-circuiting appears to begin as a result of extensive swelling and damage to some of the nickel electrodes in the cell. The temperature at which this failure mode can occur can be more than 30°C higher than the temperature where the electrolyte would normally be expected to freeze.

When the low-temperature, long-term overcharge condition is simulated for charge rates between C/50 and C/200 using a high-fidelity nickel-hydrogen cell model, the model is found to predict cell failure as a result of electrolyte freezing after weeks to months of low-rate overcharge, as has been seen in real cells. The model also predicts a temperature threshold for this failure mode that is highly dependent on electrolyte concentration, the overcharge rate, the level of cobalt additive, the amount of electrolyte in the cell, and the physical cell size (stack length). The model correctly predicts that the electrolyte can freeze at temperatures more than 30°C above the expected freezing point of the electrolyte.

The model also provides a clear picture of why nickel-hydrogen cells can fail by this somewhat surprising failure mode. The fundamental root cause is the facile formation of γ-NiOOH at extremely low temperatures, coupled with the chemical incorporation of 1/3 molecule of potassium hydroxide into the γ-NiOOH lattice structure for each nickel atom in the active material lattice. This causes the electrolyte concentration to drop, and for some cell designs (those with the lowest concentration and quantity of electrolyte), the electrolyte concentration can eventually drop to the level where its freezing point can become elevated 20–30°C or more above that expected based on the "as manufactured" electrolyte concentration.

The model predicts electrolyte freezing as a result of the electrolyte viscosity rising to extremely high values when the freezing temperature is reached at any location within the cell model while it is being run. When this occurs, it essentially cuts off the ionic conductivity and diffusion necessary for cell operation. This freezing process basically turns off all electrochemical activity in that particular element in the cell model.

For a simple cell model that contains a single nickel electrode, this freezing process suggests a high-impedance failure mode if freezing occurs. However, when a more realistic cell model is used that contains a large number of nickel electrodes distributed throughout a typical stack length, a very different failure mode emerges. The thermal gradient across the electrode stack will result in "freezing" of the coldest nickel electrode unit first, thus turning off all its electrochemical activity. This forces all overcharge current to be channeled into the remaining electrode units. The higher overcharge current accelerates the formation of γ-NiOOH and forces the electrolyte freezing point still higher in the remaining electrode units.

This continuing process can cause additional nickel electrode units to freeze, and it can eventually result in the electrochemical activity of all nickel electrodes being "frozen" except for one or two at the warmest end of the electrode stack. These final electrodes cannot freeze because all the overcharge current has been channeled into them, causing enough localized heating to prevent freezing.

When a simulation is run long enough to achieve this condition, which is typically about 1 month of simulated time for the most susceptible cell designs (those with the lowest concentration and quantity of electrolyte), the model continuously displays a red flag indicating that the localized recombination rate of hydrogen and oxygen exceeds a C/5 rate, which is the threshold in the model for indicating accumulation of popping damage to the electrode stack units. Long-term model simulations with this flag activated are predicted to eventually result in cell failure by short-circuiting as a result of the accumulation of popping damage. This is in agreement with the short-circuiting failure mode seen in actual nickel-hydrogen cells.

Model simulations have thus clarified the process, and the consequences, of electrolyte freezing in a nickel-hydrogen cell, in spite of the fact that neither the short-circuit failure mode nor the link between freezing and runaway overcharge were put into the simulation code. This is therefore another example of emergent behavior, for which the interactions between thermal gradients, electrolyte viscosity and conductivity, stack length, oxygen evolution rate, and popping damage thresholds conspire to produce a cell failure mode that would not otherwise be expected. The agreement between the predictions of a relatively complete first-principles model and actual cell behavior in the area of emergent signatures is a good indication that the computer model indeed captures all the key fundamental processes that control nickel-hydrogen cell performance. The examples discussed here to illustrate this agreement are just a few of the many subtleties of cell operation and performance that can be explored through the use of computer models.

8.1.3.7 Life Testing by Computer Simulation
A first-principles model that is capable of accurately simulating nickel-hydrogen cell performance over a range of electrode, electrolyte, and other cell properties offers the possibility of carrying out an actual life test of a cell entirely as a computer simulation. Such a "virtual" life test offers the possibility of predicting cell life in a small fraction of the 10–15 yr or more that may be required for a real-time cycle-life test of an actual nickel-hydrogen cell. An accelerated assessment of cell life done in this manner, if validated, would aid immensely in transitioning new designs and new cell technology into use.

To carry out such a virtual life test, a model must have a number of capabilities. The first such capability is a technique to simulate the long-term changes and degradation in the cell components during life. The finite-element modeling method described earlier in this chapter is capable of determining the performance of a cell after running enough cycles to stabilize all the chemical and physical processes within the cell. If cycling is continued beyond the point of achieving stabilized performance, the model will not predict any further changes that would limit the cell lifetime. A true model of the cell would therefore have to include each of the slow chemical and physical changes within the cell that eventually result in cell failure.

Corrosion of the nickel sinter, swelling of the nickel electrode structure, extrusion of active material from the nickel electrode into the separator, and cumulative popping damage to the electrodes are all examples of slow degradation processes known to occur throughout years of cycling to eventually limit cell lifetime. These chemical and physical changes will require gradual changes in the sizes, porosity, and loading of the nickel electrode elements; modified electrolyte quantity and concentration; changes in the hydrogen pressure and precharge; modified pore size distributions in the electrodes and separators; changes in the active-material distribution; cobalt-additive distribution; and sinter conductivity in the porous nickel electrode. No first-principles model presently exists that can autonomously make all these changes at the correct rates to accurately simulate what really happens as a nickel-hydrogen cell degrades during life-cycling.

However, an alternative method can be developed and used to carry out a life test as a computer simulation. This method requires a new model of the cell to be manually assembled for ever-increasing amounts of cell degradation by using the known or estimated rates for each known fundamental degradation process. For example, it is possible to generate ten models for a nickel-hydrogen cell throughout its lifetime, each model including the changes resulting from 10% increments in the known degradation processes: nickel corrosion, electrode swelling, active-material extrusion, and cumulative overcharge damage. Each of these ten models can be run either in series or in parallel to represent the performance of the cell for potential end-of-life conditions such as capacity loss or inadequate voltage. Each model must be run for enough cycles to indicate the fully stabilized cell performance of the cell at that particular point in its lifetime. In any particular cycle profile, the end of cell life is assumed to occur when the cell voltage at the end of discharge falls below 1.1 V.

A virtual life test has been carried out[8,11] based on an initial computer model of a back-to-back nickel-hydrogen cell design that initially contained 3.5 g/A h of 31% potassium hydroxide electrolyte and a double-layer Zircar separator. The initial model also contained 5% cobalt additive uniformly distributed through the nickel electrode active material, with a realistic nickel sintered plaque structure that was 35 mils thick and 78% porous with a 1.7 g/cm^3 void initial loading level. Additional models can be derived from the initial model with varying amounts of degradation processes, as shown in Table 8.2. The relative rates and amounts of each of these degradation processes was estimated from analysis of components removed from nickel-hydrogen cells that had undergone low Earth orbit profile cycling at 60% depth of discharge at 10°C.

Each model can be run for 500 cycles at 60% depth of discharge with a 1.04 recharge ratio to determine the stable capacity and voltage performance of a cell with a designated amount of internal degradation. The model is assumed to reach end of life when the stabilized end-of-discharge cell voltage at the chosen depth of discharge falls below 1.1 V during the computer simulation. Failure thus occurs at the point where the capacity margin in the cell in excess of the amount required to support the desired depth of discharge reaches zero. Figure 8.18 shows the results of these virtual life tests at 40% and 60% depth of discharge using cycling conditions similar to those for NWSC Crane test packs 5002E and 3603X. The rates of degradation indicated in Table 8.2 were determined from the depth of discharge, the peak overcharge rate, and the calendar life of the cells.

8.1 First-Principles Models of Cell Performance

Table 8.2. Key Degradation Processes in Each Model for Simulating Nickel-Hydrogen Performance Changes Throughout Life

Model ID	60% DOD Cycles	% Corrosion	% Swelling	% Extrusion	% Capacity Margin
Initial	0	0	0	0.00	25.17
1	5319	5	1.92	1.92	44.17
2	10,638	10	4.83	4.43	47.64
3	15,957	15	8.74	7.53	45.57
4	21,277	20	13.64	11.23	30.93
5	26,596	25	19.54	15.51	14.90
6	31.915	30	26.43	20.39	4.33

The results indicated in Fig. 8.18 from these virtual life tests predict a cycle life of 61,920 cycles at 40% depth of discharge (1.04 recharge fraction), and 34,096 cycles at 60% depth of discharge (1.04 recharge fraction). In these 10°C simulations, the peak recharge voltage was limited to 1.595 V, and the peak recharge rate was C/2 at 40% depth of discharge and C-rate at 60% depth of discharge. The initial increase in capacity for the curves in Fig. 8.18 is the result of the nickel hydroxide produced by corrosion, which increases the amount of active material. Roughly 30% of the way through life the capacity begins to drop as the sinter conductivity and charge efficiency degrade and active material is extruded

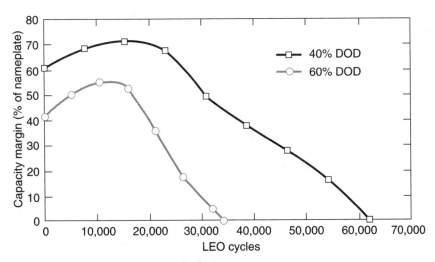

Figure 8.18. Capacity margin predicted by a computer model during virtual life tests of nickel-hydrogen cells at different depths of discharge at 10°C. Reprinted courtesy of NASA.

from the porous sinter. These degradation processes continue at rates determined from the model, until there is insufficient capacity margin in the cell to support the cycling depth of discharge.

The method described here enables a virtual life test to be done at the expenditure of about 8–10 h of computer time (using a 3 GHz PC). This is a much more rapid method for evaluating cycle life than carrying out an actual real-time life test, which runs at a rate of about 5800 cycles per year. However, the model could not be used to do a virtual life test without the data on the rates of degradation processes that have been obtained from destructive physical analyses of life-tested cells. Thus, even the best computer model can never replace the life-testing that is needed to obtain the rates of the key chemical and physical degradation processes that occur within the cells as they cycle. For a computer model to provide reasonable accurate life predictions, it is necessary that the model realistically capture how all the various degradation processes influence cell performance.

While the determination of cell degradation rates does not require that a life test be run to the point where cell failure occurs, life tests do have to be run to completion to measure cycle life to determine how accurately the virtual life test predicts cell cycle life. Real-time life tests have been carried out at 40% and 60% depth of discharge for nickel-hydrogen cell designs similar to those modeled in Fig. 8.18. Figure 8.19 shows a typical result for a 60% depth-of-discharge life test on a pack of 10 cells at 10°C.

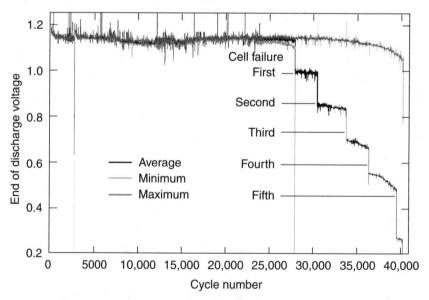

Figure 8.19. Life-test results for a typical 60% depth-of-discharge life test for a pack of 10 nickel-hydrogen cells (Crane test pack 3603X).

Life tests at 60% depth of discharge typically give about 30,000 cycles until the first cell fails. However, as shown in Fig. 8.19, each cell tested does not fail at the same point in the test. The cell failures shown in Fig. 8.19 in the 10-cell pack 3603X fail over the cycle range of about 28,000 to 40,000 cycles, and they have a mean life of 36,000 cycles. A number of cell manufacturing and test variables can contribute to this variability in cycle life. Thus, the mean lifetime of 34,096 cycles predicted by the model at 60% depth of discharge is in relatively good agreement with the 36,000 cycle mean life obtained from the results of test pack 3603X, which was limited to a peak recharge voltage of about 1.56 V. Another test pack, 3604X, gave a mean cell lifetime of about 29,300 cycles; however, this pack allowed the cell voltages to be driven up to about 1.62 V during overcharge, which is predicted by the model to cause significantly more stress than the situation where the maximum recharge voltage is limited to 1.58 V.

Similarly, the 61,920 mean cycle life predicted by the virtual life test at 40% depth of discharge compares favorably to the cycle life seen in Crane test pack 5002E, where the first cell failure in a seven-cell pack was observed at 62,500 cycles.

These comparisons suggest that the virtual life test method is capable of predicting cell lifetime within the accuracy of repeated lifetime measurements on similar cells. Interestingly, the model may also be used to explore the sensitivity of life to variations in cell components or test conditions to better understand the possible reasons for the wide range of cell lifetimes indicated in Fig. 8.19.

8.2 Nickel-Hydrogen Cell Storage Model

Before nickel-hydrogen cells are put into cycling service, they are frequently stored for years after they have been activated with electrolyte. A number of degradation processes are known to occur during such storage. No first-principles model has been developed to predict the effects of storage on cell capacity and eventual cell lifetime. However, an empirical model has been developed that predicts the rates of each known degradation process as a function of cell storage conditions, thus enabling cumulative degradation during storage to be evaluated. This model attributes cell degradation during storage to a collection of wear-out processes that have been identified during destructive physical analyses of cells that have experienced long-term storage at various temperatures. This model predicts the changes within a nickel-hydrogen battery cell during storage before it is placed into a cycling regimen, and it attributes wear to the following processes:

- nickel metal corrosion in the nickel electrodes (above 0.1 V $vs.$ H_2)
- reduction of Co(III) additives to Co(II) in the nickel electrode (below 1.1 V $vs.$ H_2)
- reduction of nickel electrode active material to metal (below 0.1 V $vs.$ H_2)
- corrosion of platinum metal at the negative electrode (above 1.0 V $vs.$ H_2)

252 Nickel-Hydrogen Cell Modeling

This model provides a prediction of the fraction of the cell life that is initially lost as a result of ground storage, which can last for 5 yr or more in some instances. It also provides a prediction of the capacity loss that a cell may experience as a result of its ground storage timeline. The following inputs are required for this model.

- cell diameter
- nameplate cell capacity
- number of nickel electrodes
- percent cobalt additive in the nickel electrodes
- maximum expected operating pressure at beginning of life
- total percent nickel precharge at beginning of life
- loading level for nickel electrodes
- nickel electrode thickness
- nickel electrode plaque porosity
- platinum loading level
- current and temperature timeline during storage

The model computes the rate for each degradation process over a specified cell storage timeline, providing the expected effects of these changes on cell capacity. The rates of the following mechanisms are included in the model.

8.2.1 Corrosion of the Nickel Electrode Sinter

When the potential of the nickel electrode is greater than about 0.13 V $vs.$ H_2, the nickel metal in the sintered plaque is thermodynamically unstable and will slowly oxidize to nickel hydroxide. This process tends to increase the hydrogen precharge or decrease the nickel precharge in the cell. However, because the nickel metal passivates in potassium hydroxide electrolyte, this corrosion process is quite slow and thus allows nickel electrodes to operate for decades without excessive degradation.

Modeling of the nickel corrosion process is based on the diffusion-controlled growth of a passivating oxide layer on the surfaces of the nickel sinter. As the passivation layer thickens, its growth is balanced by physical fracturing of the layer as a result of its volume changes during charge and discharge, and as a result of oxygen evolution during overcharge. The passivation layer is assumed to maintain a minimum thickness of at least 10 Å at all times. All nickel metal that oxidizes by this process resides either in the passivation layer or in the active-material reservoir as nickel hydroxide with no cobalt additive.

The rate of growth R for the nickel passivation layer as a function of temperature is modeled by Eq. (8.19), where L_N is the existing thickness of the passivation layer and T is the temperature in degrees Celsius.

$$R(\text{Å}/\text{h}) = \frac{0.0673}{\sqrt{L_N}} \left[1.5^{(T-5)/15} \right] \tag{8.19}$$

The rate constant in Eq. (8.19) was derived from observations during destructive physical analysis of a cell and from cell pressure changes after 8 yr of low Earth orbit cycling indicating that about 20% nickel corrosion had occurred at 5°C. The temperature dependence gives a 50% increase in corrosion rate for every 15°C increase in temperature, which is about the percentage change in cycle life observed between +10 and −5°C low Earth orbit tests, and it also corresponds reasonably well to the average change in pressure growth at different test temperatures. The total amount of nickel that has corroded during a time increment Δt is then given by Eq. (8.20):
where A is the area of sinter surface in cm² (about 100 times the plaque area), ρ

$$X_{\text{corr}}(\text{moles}) = 10^{-8} A R \Delta t \rho / M \tag{8.20}$$

is the density of nickel hydroxide (4.14 g/cm³), and M is the molecular weight of nickel hydroxide (92.71 g/mole).

The rate of disruption of the nickel passivation layer is assumed to be proportional to the charge throughput C_t, with 100% of nameplate capacity (C_{NP}) throughput disrupting 20% of the passivation layer thickness greater than 10 Å. Eq. (8.21) indicates the adjustment made to the passivation layer thickness for both growth and disruption between an initial time t_1 and a later time t_2.

On the basis of this model, Fig. 8.20 indicates the changes in the thickness of

$$L_N(t_2) = L_N(t_1) + R \Delta t - 0.2 \left(L_N - 10 \right) \frac{C_t}{C_{NP}} \tag{8.21}$$

the passivation layer on nickel sinter during typical low Earth orbit (40% depth of discharge) and geosynchronous orbit (70% depth of discharge) cycles at 0°C starting from a C/100 trickle-charge condition. Figure 8.20 also indicates the total nickel corrosion predicted to accumulate during these cycles.

The growth of a thick nickel passivation layer, as well as the accumulation of disrupted nickel corrosion layer products, will result in losses of both usable capacity and charge efficiency over time. The total corrosion also contributes to capacity loss by producing additive-free material at the sinter surface, by decreasing the sinter area, and by diminishing sinter conductivity. Because the thick passivation layers that can build up during long periods of storage are readily disrupted by cycling while total corrosion products only accumulate over time, there is a significant recoverable capacity loss as well as a permanent capacity loss. The

254 Nickel-Hydrogen Cell Modeling

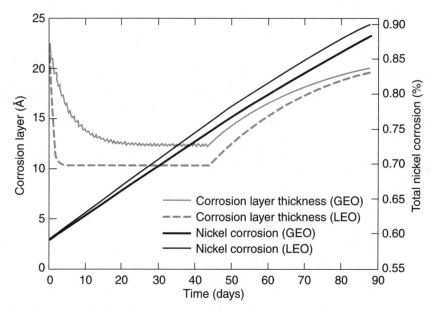

Figure 8.20. Nickel corrosion layer thickness and total accumulated corrosion for 44 days of low Earth orbit and geosynchronous orbit cycling at 0°C, then 44 days of C/100 trickle charge. Reprinted courtesy of NASA.

recoverable capacity loss is modeled by assuming that it is proportional to the thickness of the nickel passivation layer in excess of 10 Å, while the permanent capacity loss is assumed to be proportional to the overall percent of the nickel that has undergone corrosion in the nickel sinter. Thus, the percent capacity loss from nickel corrosion ΔC_{Ni} is given by Eq. (8.22):

$$\Delta C_{Ni}(\%) = K_r L_N + K_p \frac{X_{corr}}{X_{tot}} \qquad (8.22)$$

where X_{tot} is the total moles of nickel in the sinter, K_r is the rate constant for recoverable capacity loss, and K_p the rate constant for permanent capacity loss. We have found that $K_r = 0.005$, corresponding to a 5% capacity loss for a 1000 Å thick passivation layer, works well, although it is rare for passivation layers on nickel metal to exceed 100 Å for very long during electrical cycling. It also appears that $K_p = 50$ is reasonable for permanent capacity loss, based on cell destructive physical analysis studies, corresponding to a 10% capacity loss for 20% corrosion of the sinter. When corrosion levels exceed about 30% in nickel electrodes, the probability for failure by short-circuiting appears to become more important than that caused by capacity loss.

8.2.2 Self-Discharge Model

During ground storage of nickel-hydrogen cells it is critical to have a self-discharge model that can accurately track how the state of charge and the expected voltage of the cell change with time, because each of the degradation modes within the cell can only occur in a specific voltage range. A model can be developed that predicts the self-discharge of the nickel electrode by three mechanisms:

- oxygen evolution, S_{oxy}
- direct reaction of charged active material with hydrogen gas, S_{dir}
- catalytic reaction of hydrogen gas on the metal surfaces of the nickel sinter, S_{ec}

The oxygen evolution reaction is assumed to follow the same Tafel kinetic behavior seen during overcharge, where the rate is exponentially proportional to voltage with a Tafel slope of about 43 mV/decade. Thus, self-discharge from oxygen evolution is dominant at high states of charge when the cell has a high voltage, but it becomes negligible as the cell state of charge decreases. Throughout most of the middle state-of-charge range of the cell, the direct reaction of the charged active material with hydrogen gas tends to dominate the self-discharge. However, at low states of charge, particularly in hydrogen-precharged cells, the catalytic self-discharge process can become the dominant process and is the process that ultimately brings the depleted nickel electrode down to the hydrogen potential.

The rates of each of these processes have been measured in nickel-hydrogen test cells as a function of temperature, hydrogen pressure, and state of charge. The results of these measurements are described by the kinetic expressions in Eqs. (8.23) to (8.25), which give the rates (in mA) of each of the individual self-discharge mechanisms.

$$S_{oxy}(ma) = 1.72 \times 10^{-5} A_e e^{k_{ox}(E - E^\circ_{ox})} \tag{8.23}$$

$$S_{ec}(ma) = 3 \times 10^{-6} P_{H_2} e^{k_R E_a} \tag{8.24}$$

$$S_{dir}(ma) = 8.5 \times 10^{-4} P_{H_2} F_{SOC} X_b e^{k_R E_b} \tag{8.25}$$

where $k_R = \dfrac{1}{R}\left\{\dfrac{1}{283.15} - \dfrac{1}{T + 273.15}\right\}$, R is the gas constant, T is temperature, and P_{H_2} is the hydrogen pressure in atm. In Eq. (8.23) S_e is the total nickel electrode area, $k_{ox} = 57(293/T(\text{Kelvin}))$ is the Tafel constant for oxygen evolution, E is the nickel electrode potential, and the reversible potential E°_{ox} equals $0.401 - (dE_0/dT)(T - 25)$. In Eq. (8.24), $E_a = 21.3$ kcal/mole is the activation energy for the catalytic reaction of hydrogen. In Eq. (8.25), F_{SOC} is the fractional state of charge of the nickel electrode, X_b is the number of mmoles of active material, and E_b = 4.90 kcal/mole is the activation energy for the direct reaction. The overall self-discharge rate is the sum of the rates from Eqs. (8.23) to (8.25).

8.2.3 Nickel Electrode Active-Material Reduction

When the potential of the nickel electrode is held below about 0.1 V *vs.* H$_2$, the metal hydroxides that comprise the active material may undergo reduction to the metals. This condition typically occurs in the nickel-hydrogen cell when the nickel electrode is fully discharged and hydrogen gas is present in the cell (a hydrogen-precharged condition). In this situation the nickel electrode is held at the hydrogen potential by hydrogen molecules that are chemisorbed on the reduced surfaces of the nickel sinter. This situation has been previously discussed in detail[8.12] and is often referred to as "hydrogen sickness" because it can ultimately lead to significant capacity loss.

In this model, capacity loss resulting from active-material reduction at low voltages is based on the gradual growth of a layer of reduced active material from the sinter surfaces during periods when a hydrogen-precharged cell is allowed to rest at low potentials. Recharge of the reduced layer results in the formation of a soluble cobalt hydroxide compound that can dissolve from the region at the sinter surface and redeposit elsewhere in the nickel electrode. The layer where the active material was reduced thus becomes depleted of the cobalt additives that help maintain good electrode performance, with the result being capacity loss. This capacity loss can be recovered with extensive cycling that serves to physically remix the depleted layer with the rest of the active material; however, such cycling puts significant wear on the electrodes. Both the growth of a cobalt-depleted layer and its gradual dissipation by mixing processes are modeled to describe these cobalt redistribution processes.

The rate of growth R_g of a cobalt-depleted layer when the nickel electrode is less than 0.1 V is modeled as being inversely proportional to the square root of the layer thickness L_h and directly proportional to the square root of the hydrogen pressure, as indicated in Eq. (8.26).

This layer grows very slowly from the sinter surface into the active material,

$$R_g (\mu / \text{day}) = 2.48 \times 10^{-4} 2^{0.1(T-20)} \sqrt{\frac{P_{H_2}}{L_h + 0.5}} \qquad (8.26)$$

typically requiring many weeks before noticeable capacity loss occurs. However, it can progress during long periods of time until a depleted layer several microns thick has formed, producing up to a 30% capacity loss.

Capacity loss is modeled by including a direct capacity loss ΔC_H that results from the depleted layer itself. The amount of capacity loss is proportional to the level of cobalt additive in the nickel electrode X_{Co}, with the capacity of 10% cobalt-containing electrodes being much more sensitive to cobalt-depleted layers than 5% cobalt electrodes. The direct capacity loss effect can be gradually recovered by cycling the cell sufficiently; however, the rate of layer-mixing drops rapidly as the cobalt-depleted layer becomes a significant fraction of the total active-material layer thickness. The rate of dissipation of the cobalt-depleted layer is modeled as indicated in Eq. (8.27).

$$R_d(\mu/\text{day}) = \frac{24 J L_h I}{C_{NP}} e^{-10 L_h / L_T} \tag{8.27}$$

where J is the fraction of the layer thickness dissipated by a full 100% depth-of-discharge cycle when the layer is very thin, I is the current in amperes, L_T is the total active-material layer thickness (typically ~5 µm), and the exponential mixing term drops the mixing to zero as the depleted-layer thickness approaches the total active-material layer thickness. Eq. (8.28) indicates the adjustment made to the depletion-layer thickness for both growth and mixing between an initial time t_1 and a later time t_2.

$$L_h(t_2) = L_h(t_1) + \Delta t \left(R_g(t_1) + R_d(t_1) \right) \tag{8.28}$$

There is also an indirect contribution to capacity loss resulting from the dissolution and redistribution of cobalt additives in the nickel electrode. The indirect capacity loss results from the total cobalt lost to reduction processes, including all that has been mixed within the active-material layer. This indirect capacity loss is assumed to be directly proportional to the total amount of cobalt additive that has undergone reduction. The direct capacity loss ΔC_D and the indirect capacity loss ΔC_I are given by Eq. (8.29):

$$\Delta C_D(\%) = 4 P_{Co} \left(1 - e^{-L_h/0.8} \right) \qquad \Delta C_I(\%) = 0.1 L_{Co} \tag{8.29}$$

where P_{Co} is the overall percentage of cobalt additive in the active material and L_{Co} is the total amount of the cobalt that has undergone reduction during the life of the nickel electrode. There is no mechanism for recovery of the capacity lost by the indirect process. The total capacity loss resulting from cobalt reduction and redistribution is given by Eq. (8.30).

$$\Delta C_{Co}(\%) = \Delta C_D + \Delta C_I - \frac{1}{4 P_{Co}} \Delta C_D \Delta C_I \tag{8.30}$$

The capacity losses expressed by Eq. (8.30) can amount to up to 20% of the capacity of a cell with 5% cobalt additive in the nickel electrodes, and twice this amount when 10% cobalt additive is used. Capacity loss that is this great requires the formation of a cobalt-depleted layer that is several microns thick and is very difficult to recover by cycling or any other actions. However, the thinner cobalt depletion layers (those with thickness of 0.2 µm or less) that can result from less than 1–2 months of storage can often be almost completely recovered by cycling or other electrical activity.

8.2.4 Platinum Dissolution

When a nickel-precharged nickel-hydrogen cell is stored in the discharged state, the partially charged nickel electrode maintains a low pressure of oxygen in the cell that holds the platinum negative electrode at an oxidizing potential. The potential of the platinum can be high enough (> 1 V vs. H_2) to allow oxidation of platinum metal to platinum oxide. Platinum oxide is slightly soluble (~6 millimolar) as platinate ions in 31% potassium hydroxide electrolyte, and thus will dissolve from the platinum surface. The platinate species are capable of reacting with charged cobalt and nickel sites in the nickel electrode to form a nickel cobalt platinate complex.[8.12] The formation of this complex, while it can change the level of active nickel precharge within the cell, does not appear to degrade capacity. The fraction of the active material in this complex $F_{Co/Ni}$ is modeled as a function of storage time by Eq. (8.31).

$$F_{Co/Ni} = 1 \times 10^{-4} P_{Co} P_{precharge} \left(1 - e^{-t(yr)/3}\right) \qquad (8.31)$$

8.2.5 Cobalt Reduction with Nickel Precharge

During sufficiently long storage periods (> 5–7 yr), the state of charge within the nickel electrode of a nickel-precharged cell will eventually fall low enough that the nickel electrode potential will fall below the Co(III)/Co(II) redox potential of 1.1 V vs. H_2. When this happens, the Co(III) additive sites in the active material begin to undergo reduction to Co(II) at the sinter surfaces, and a cobalt hydroxide layer will gradually form, allowing the soluble Co(II) species to migrate away from the sinter surfaces. This process occurs very slowly, however, because the hydroxide layer that results has very little electrical conductivity. By this mechanism, capacity loss by cobalt depletion can slowly occur in very old cells with nickel precharge. This mechanism adds an additional term R'_g to Eq. (8.26) that becomes active if the nickel electrode potential falls below 1.1 V vs. H_2, one that is normally negligible but can become important after more than 5–7 yr of storage for nickel-precharged cells. This additional term is given by Eq. (8.32).

$$R'_g (\mu / day) = 7.34 \times 10^{-5} 2^{0.1(T-20)} \sqrt{\frac{1}{L_h + 0.001}} \qquad (8.32)$$

8.2.6 Validation of Storage Degradation Model

The model described here for degradation during storage captures the known chemical processes that occur during the storage of nickel-hydrogen cells. One key issue regarding cell storage is how long cells can be safely stored, and whether storage life can be prolonged by using reduced-temperature storage. Figure 8.21 shows the model prediction for the effect of reduced temperature on capacity loss during storage for a typical nickel-hydrogen cell that contained 15% nickel precharge at activation.

As indicated in Fig. 8.21, both corrosion and capacity loss gradually increase until the active nickel precharge is exhausted. Thereafter, corrosion stops but accelerated capacity loss as a result of slow reduction of CoOOH in the active material begins. By reducing the temperature to typical levels found in freezers, it is possible to more than double the storage life with active nickel precharge from 4 yr out to about 9 yr.

If the storage model described here captures all the key processes that can occur during the storage of nickel-hydrogen cells, then it should be capable of accurately reproducing the degradation that has been observed to result from different modes of ground storage. This approach can be used to validate this model. Three examples of different storage situations are presented for validation in the following discussion, along with the results from the model predictions and the observed changes in cell capacity and precharge.

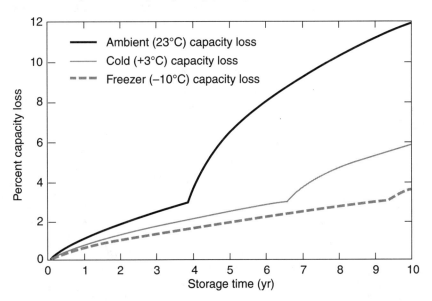

Figure 8.21. Increasing capacity loss during continuous open-circuit nickel-hydrogen cell storage at different temperatures, starting with 15% nickel precharge. The increase in slope occurs when the nickel precharge becomes exhausted. Reprinted courtesy of NASA.

8.2.7 Example 1. Nominal Five-Year Cold Storage of Nickel-Precharged Cells

The first example to test the model considers the five-year cold storage of nickel-precharged cells. Storage of nickel-hydrogen cells that contain about 15% nickel precharge has been observed to typically give from three to more than five years of storage before the active nickel precharge is exhausted. Three to five years has been seen when storage is at ambient temperature, or when the storage is periodically interrupted to check cell capacity. Storage at temperatures of about 3°C has been found (by cell destructive physical analysis) to result in the loss of about 70% of the active nickel precharge in 5 yr of continuous storage in cells containing 5% cobalt in the nickel electrodes. Figure 8.22 indicates the results of model calculations for the change in capacity and precharge in these cells.

The capacity loss predicted by the model must be subtracted from the capacity gain that accompanies the decrease in nickel precharge to accurately predict cell capacity after storage. Initially this cell had a nominal capacity that was 127% of nameplate. After 5 yr of storage the capacity was found to increase to 138% of nameplate. The model indicates a capacity that increased 11% because of the decrease in nickel precharge, which agrees well with the observed capacity. Low amounts of corrosion are not expected to degrade capacity significantly, but act as an initial source of wear that must be considered in determining the ultimate life of the cell.

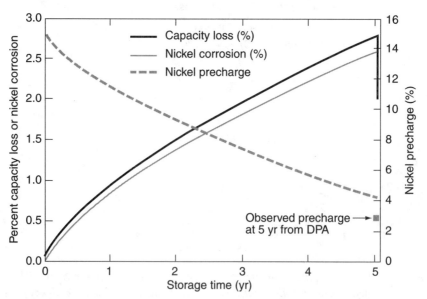

Figure 8.22. Predicted nickel corrosion and changes in precharge during 5 yr of cold storage for a nickel-precharged cell. Reprinted courtesy of NASA.

8.2.8 Example 2. Fourteen-Year Ambient Storage of Nickel-Precharged Cells

The second example involves the storage of several nickel-precharged cells for 14 yr at ambient temperature. These cells contained 5% cobalt additive in the nickel electrodes and had an initial 15% level of nickel precharge. After the 14 yr of storage, the cells were found to contain no hydrogen; however, they became hydrogen-precharged as soon as they were recharged. The cells also exhibited about a 15% capacity loss. This behavior was modeled using the methods developed here, providing the results in Fig. 8.23. The model predicts that after about 6 yr of storage, the active nickel precharge becomes exhausted. At this point, the nickel electrode potential drops such that it becomes controlled by the Co(III) species in the active material and any oxygen gas that is present. The rate of corrosion drops significantly, but additional capacity loss begins to accumulate as the Co(III) slowly is reduced to Co(II) in the nickel electrodes. The 16% capacity loss predicted by our model is in good agreement with the observed 15% capacity loss for these cells.

The example of Fig. 8.23 is particularly noteworthy because it suggests that this cell could have been stored about 6 yr with essentially no beginning-of-life capacity loss. Most of the permanent capacity loss that was observed is expected to have developed after loss of active nickel precharge during years 7–14 of storage. This example suggests that there is indeed a real limitation on the storage life

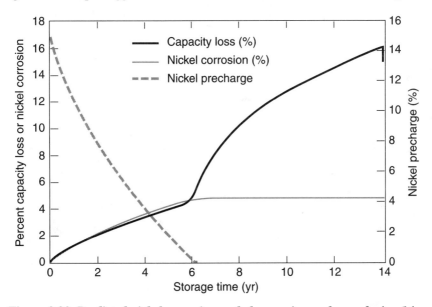

Figure 8.23. Predicted nickel corrosion and changes in precharge during 14 yr of ambient storage for a nickel-precharged cell. Reprinted courtesy of NASA.

of nickel-hydrogen cells, and that this limitation can be extended significantly by lower-temperature storage.

8.2.9 Example 3. Three-Year Ambient Storage of Hydrogen-Precharged Cells

The final example of changes in nickel-hydrogen cell performance during ground storage examines the behavior of a hydrogen-precharged cell. This cell was built with 50 psia hydrogen precharge, then stored at ambient temperature for about three years. Following the storage, the capacity of the cell was found to be about 30% low. This cell contained 10% cobalt in the nickel electrodes. Figure 8.24 indicates the capacity loss and corrosion predicted in this cell by the model during the three years of storage. The nickel corrosion was zero because the hydrogen gas held the nickel electrode below its corrosion potential. The capacity loss was significant and resulted from the formation of a relatively thick layer of reduced active material. This layer was about 2.5 µm thick for the simulation of Fig. 8.24. Destructive physical analysis of nickel electrodes from this cell[8.13] revealed cobalt-depleted layers that were 2–4 µm thick, in good agreement with the model's results. The capacity loss of 30% was also reproduced in the model results.

The results from the three preceding examples indicate that this ground storage model captures, with reasonable accuracy, the principal processes that are known to occur in nickel-hydrogen cells during storage. This model can be used

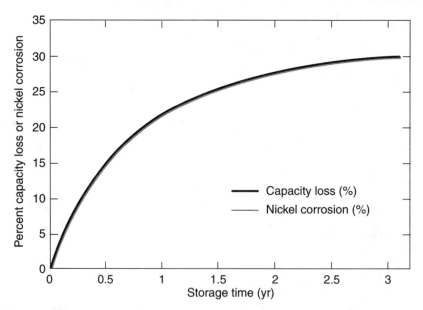

Figure 8.24. Predicted nickel corrosion and capacity loss during 3 yr of ambient storage for a hydrogen-precharged cell at 50 psia. Reprinted courtesy of NASA.

to predict initial wear on a nickel-hydrogen cell and to provide inputs to wear models that predict nickel-hydrogen cell cycle-life capability, such as those described in the next section.

8.3 Wear-Out Models for Predicting Cell Cycle Life

Wear-out models have proven they are able to provide one of the best ways to capture the life-test experience for nickel-hydrogen cells, and to provide a cycle-life prediction method based on the test database. A wear-out model predicts cell lifetime based on the wear rates for one or more factors that contribute to cell wear-out. The wear rates and how they depend on the operating environment of the cell are determined from the available test database. An extensive database allows the wear rates to be determined with a high degree of accuracy, while limited databases constrain the predictive accuracy of wear-out models.

In the following sections a simple wear model is described that is based on a single wear parameter, then a more complex wear-out model is discussed that includes wear rates for multiple wear-producing processes, as well as failure by several mechanisms. In addition, the influence of statistical cell-to-cell variability on lifetime predictions from wear-out models is discussed, enabling wear models to provide the probability of cell failure as a function of cycle lifetime.

8.3.1 Simple Wear-Out Model

A simple wear-out model has been used to predict cell lifetime for low Earth orbit–type cycling based on a single stress factor, cell depth of discharge. This model, which was applied successfully to nickel-hydrogen cells by Thaller and Lim,[8.14] assumes that the rate of cell wear-out is determined primarily by the stresses associated with the repetitive depth of discharge during cycling. This model predicts cycle life based on the expression of Eq. (8.33):

$$L = \frac{1 + M - D}{RD} \quad (8.33)$$

where L is the predicted cycle life, D is the fractional depth of discharge at which the cell is cycled, M is the fractional capacity margin of the cell design above nameplate capacity, and R is the amount of wear per cycle characteristic of the particular cell design that is being modeled. Nickel-hydrogen cells are often built with an actual capacity that is 10–20% higher than nameplate capacity, which will allow them to provide more cycles to failure than a cell having little or no extra capacity. This effect is accommodated in the model by using 0.1–0.2 for the parameter M in Eq. (8.33).

In addition, different values of R can be used for different cell designs to reflect a greater or lesser ability to withstand the stress associated with each cycle. For example, Fig. 8.25 indicates the cycle life predicted by Eq. (8.33) for four different nickel-hydrogen cell designs with $M = 0$ (no capacity margin beyond nameplate), each of which experiences a differing amount of wear per cycle of operation. The standard design, which corresponds to the typical cell design used to power high-reliability satellites, is predicted to have a lifetime of about 43,000 cycles at 40% depth of discharge. This lifetime is consistent with the 40,000 to 50,000 cycles typically obtained from this type of cell design in life tests. This simple wear model does not explicitly include effects of operating temperature on cycle life, but does allow different values of R to be used to reflect differing wear-out rates at different temperatures.

The other designs in Fig. 8.25 indicate the trade in cycle life that can occur if cells are modified to cost significantly less, or if lightweight nickel electrodes are used that wear out more quickly. Similarly, it is possible to optimize the nickel-hydrogen cell design for better cycle life as described earlier in this chapter, typically at the cost of some weight increase. The advanced cell design illustrated in Fig. 8.25, for example, can provide 33,000 cycles at 60% depth of discharge and more than 75,000 cycles at 40% depth of discharge. The cell design that was optimized for maximum cycle life (as described earlier in this chapter) is expected to give significantly better cycle life than even this "advanced" cell design, with a realistic goal of 60,000 cycles at 60% depth of discharge ($R = 0.0000111$).

While the wear-out model of Eq. (8.33) does provide a good qualitative prediction of cycle life, it is not capable of accounting for the effects of different

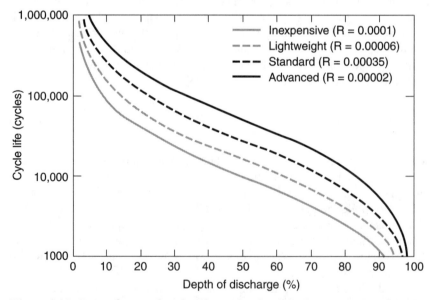

Figure 8.25. Dependence of cycle life on depth of discharge predicted by the wear-out model of Eq. (8.33). Reprinted courtesy of NASA.

operating temperatures, various amounts or rates of overcharge, or even the effect of simple calendar time on the wear rate, and thus on the cell cycle life. Additionally, this model predicts the mean cycle life of a cell, and it would require the incorporation of a statistical lifetime spread around this mean to predict how the probability of cell failure increases as more cycles are completed. To resolve these issues and to make the wear-out modeling method more flexible and quantitative, the multiparameter wear model described in the following section was developed.

8.3.2 Multiparameter Wear-Out Model

A model of nickel-hydrogen cell wear-out rate during low Earth orbit cycling can be developed based on a series of parametric tests carried out by NASA between 1990 and 2002 at the U.S. Navy test facility in Crane, Indiana. These test results are used to determine how depth of discharge, temperature, amount and rate of overcharge, and calendar life influence cell wear-out rate. These test packs and their performance are summarized in Table 8.3.

The cells in packs 3600X–3605X are dual-stack 65 A h cells containing 31% potassium hydroxide, dual-layer Zircar separation, and back-to-back stacks. None of these cells were tested at temperatures lower than 10°C. The last two packs (3831E and 3835E) were included to provide an indication of any cycle-life improvement that could be expected at lower temperatures of operation. The cells in the last two packs are a similar 81 A h design that used a composite separator and a wall wick that included an oxygen recombination catalyst for improved thermal control. Each test pack contained eight cells.

In Table 8.3, the depth-of-discharge numbers are all based on the discharge during each cycle as a percent of the nameplate cell capacity. Recharge fraction, in the third column, is the average observed during the test, defined as ampere hours recharged divided by ampere hours discharged. The temperature is an average across the entire test in degrees Celsius. The rate of pressure rise is expressed in units of psi per 1000 cycles, and the pressure when the first cell failed is given in psi. The last column gives the number of cycles achieved when the first cell in these eight-cell packs failed. Failure is defined here as the time when the end-of-discharge voltage of a cell falls below 1.0 V.

A cell wear-out model[8.15,8.16] was developed from the data in Table 8.3 by assuming that each of the processes that are known to cause degradation of the nickel-hydrogen cell are operating at rates that are proportional to a number of stressing parameters that can be monitored during the test.** These stressing parameters are:

**The multiparameter wear-out model was first described in "Lifetime and Failure Mode Prediction Method for Nickel Hydrogen Batteries," a presentation delivered at the 2003 Space Power Workshop (Redondo Beach, CA, April 2003).

Table 8.3. Performance of Battery Test Packs Used to Develop Wear-Out Model

Pack ID	DOD (%)	Average RR	Average Temperature	Peak Charge V	Pressure Rise Rate	Pressure at Failure	Cycles to Failure
3600X	35	1.0483	12	1.5227	14.37	1240	64,000*
3601X	35	1.0303	12.6	1.5642	12.93	1320	79,400*
3602X	35	1.0386	11.6	1.5684	12.86	1220	63,000*
3603X	60	1.0398	14	1.5614	28.16	1500	27,938
3604X	60	1.0403	13	1.6276	19.87	1215	25,330
3605X	60	1.041	13.8	1.6272	23.16	1278	22,307
3831E	35	1.04	12	1.58	8	750	38,300
3835E	35	1.04	−2	1.635	4	800	58,500

*Estimated from voltage signatures

- Temperature (*T*): All reactions have some temperature dependence.
- Coulombic overcharge in excess of that needed to balance self-discharge (*A*): This causes active-material phase and volume changes, oxygen pressure, corrosion, and heat.
- Depth of discharge (*D*): This causes active-material volume changes and heating.
- Rate of overcharge (*R*): This causes oxygen pressure stress and corrosion.
- Calendar life (*L*): Ni corrosion occurs at a background level throughout life.

For the fourth parameter, the rate of overcharge (*R*), the excess peak recharge voltage above 1.52 V (V_p) was used as an indication of overcharge rate. The rationale for this parameter is the known exponential relationship between the rate of oxygen evolution from nickel electrodes and the voltage of the electrodes. The amount of coulombic overcharge used in this analysis corresponds to the measured percent overcharge minus the percent of nameplate capacity needed to balance self-discharge, as determined from Eq. (8.34) as a function of temperature. In Eq. (8.34), D_{SD} is the percent of nameplate capacity lost during each 90 min low Earth orbit cycle from self-discharge.

$$D_{SD} = 8.074 \times 10^{-5} T^3 + 2.1905 \times 10^{-4} T^2 + 2.1696 \times 10^{-2} T + 0.43246 \tag{8.34}$$

The temperature was averaged over the entire life of the cell in these life tests, because it typically did not vary by more than about 3°C.

Each of the five parameters listed here was assumed to contribute to the instantaneous wear-out rate by either of two mechanisms. The first involves failure by the formation of a short circuit in the cell as a result of the wear imposed by cycling. For this mechanism, cell failure is assumed to occur at a given wear level irrespective of the remaining cell capacity or voltage performance. This wear rate is W_s, and it is expressed by Eq. (8.35) as a fraction of cell life per kilocycle (kc). The second mechanism involves failure from gradual loss of usable capacity until the remaining capacity is insufficient to support the required loads. This wear rate is W_c, and it is expressed by Eq. (8.36).

$$W_s = K_t e^{-E_a/T_k} \left[K_{Os} + K_{As}A + K_{Ds}D + K_{Rs}V_p + K_{Ls}L \right] \tag{8.35}$$

$$W_C = K_t e^{-E_a/T_k} \left[K_{Oc} + K_{Ac}A + K_{Dc}D + K_{Rc}V_p + K_{Lc}L \right] \tag{8.36}$$

The units for *W* are life-fraction/kc, and T_k is temperature in degrees Kelvin. The constants in Eqs. (8.35) and (8.36) were determined by fitting to the cycle-life and parametric data in Table 8.3 using multiple-regression analysis. For mechanism 1, failure by short-circuit formation, the wear was assumed to be independent of discharge capacity margin (unused capacity remaining at the end of discharge) and results in cell failure when the cumulative wear has reached 100%. In this model the wear rate is simply the reciprocal of the observed cycle life. For mechanism 2, failure by capacity loss, the wear was assumed to result in cell failure when the cumulative wear has reached the percent capacity remaining at the end of discharge. For example, when cycling at 60% depth of discharge, failure is assumed to occur by capacity loss when the cumulative wear reaches 40%, because at this point insufficient capacity remains to support discharge. For mechanism 2 the wear rate is given by Eq. (8.37).

$$W_c = (1.0 - 0.01 \times \text{DOD})/(Kc \text{ to failure}) \tag{8.37}$$

The data and results from the regression analyses for both of these mechanisms are summarized in Tables 8.4 through 8.7. The cumulative cycle life for each test in Tables 8.4 and 8.6 was determined by summing the instantaneous wear-out W_s or W_c for each cycle across the duration of each life test. Tables 8.5 and 8.7 summarize the regression constants used in this wear-out model.

As indicated in the last two columns of Tables 8.4 and 8.6, this wear-out model provides an excellent correlation between the observed and predicted life of the first cell to fail (out of eight cells—which corresponds to an 87.5% reliability point) in these tests, whether by short-circuit or capacity-loss mechanisms. Clearly, this regression method is capable of accurately fitting the data from these test packs. At levels less than about 30% depth of discharge (because of low stress) and greater than about 65% depth of discharge (because of low capacity margin), the capacity-loss mechanism governs the cycle-life prediction. Short-circuit failure mechanisms are most likely at intermediate states of charge, and they govern the cycle-life prediction.

The first six tests summarized in Table 8.3 were run at 10–12°C. Very little test data were found at the lower temperatures more characteristic of where nickel-hydrogen batteries are often operated. To obtain the temperature dependence indicated in Tables 8.5 and 8.7, data obtained at –2 and +12°C for test packs 3831E and 3835E were used. Because these cells were of a different design than the cells in packs 3600X–3605X, the data were not combined with the 3600X–3605X data. The ratio of cycle life at these two temperatures was used to obtain the constants K_t and E_a in Tables 8.5 and 8.7, which correspond to a 52.5% increase in cycle life as the temperature was reduced from +12 to –2°C. Thus, reduced temperature in itself can significantly improve cycle life.

Examination of the parameters developed in this wear-out model provides some interesting correlations. For instance, the constant K_o is a measure of any initial wear-out that may exist for cells before they begin cycling—an example

8.3 Wear-Out Models for Predicting Cell Cycle Life

Table 8.4. Regression Analysis Results for Short-Circuit Wear-Out Model of Eq. (8.35)

Pack ID	Overcharge	Life (yrs)	Average Temperature	Excess charge V	Wear/Kc	Observed Kc	Predicted Kc
3600X	0.6905	11.52	12	0.0027	0.015625	64	64.78
3601X	0.0605	11.53	12.6	0.0442	0.012594	79.4	82.27
3602X	0.351	11.53	11.6	0.0484	0.015873	63	60.61
3603X	1.388	5.91	14	0.0414	0.032051	31.2	31.04
3604X	1.418	5.45	13	0.1076	0.039401	25.38	24.61
3605X	1.46	4.94	13.8	0.1072	23.5	23.5	24.3

Table 8.5. Regression Constants for Eq. (8.35) (Short-Circuit Model)

Regression Constant	Value
K_t (Life-fraction/Kc limit at high temp)	3620.6
E_a (Activation energy, K^{-1})	2335.4
K_{os} (Initial wear at beginning of life)	0.0 assumed here
K_{As} (Life-fraction change/% of nameplate overcharge)	0.013199
K_{Ds} (Life fraction change/% DOD)	0.0001152
K_{Rs} (Life fraction change/excess recharge voltage)	0.121279
K_{Ls} (Life fraction change/yr of calendar life)	0.0104198

270 Nickel-Hydrogen Cell Modeling

Table 8.6. Regression Analysis Results for Capacity Loss Wear-Out Model of Eq. (8.36)

Pack ID	Overcharge	Life (yrs)	Average Temperature	Excess charge V	Wear/Kc	Observed Kc	Predicted Kc
3600X	0.6905	11.52	12	0.0027	0.010156	64	63.68
3601X	0.0605	11.53	12.6	0.0442	0.008186	79.4	79.48
3602X	0.351	11.53	11.6	0.0484	0.010317	63	63.26
3603X	1.388	5.91	14	0.0414	0.012821	31.2	31.28
3604X	1.418	5.45	13	0.1076	0.01576	25.38	24.57
3605X	1.46	4.94	13.8	0.1072	0.017021	23.5	24.19

Table 8.7. Regression Constants for Eq. (8.36) (Capacity-Loss Model)

Regression Constant	Value
K_t (Life-fraction/Kc limit at high temp)	3620.6
E_a (Activation energy, K^{-1})	2335.4
K_{os} (Initial wear at beginning of life)	0.0 assumed here
K_{As} (Life-fraction change/% of nameplate overcharge)	0.0064997
K_{Ds} (Life fraction change/% DOD)	−0.000155
K_{Rs} (Life fraction change/excess recharge voltage)	0.049787
K_{Ls} (Life fraction change/yr of calendar life)	0.011013

8.3 Wear-Out Models for Predicting Cell Cycle Life

would be following lengthy ground storage. For the analyses performed in this section, K_O was assumed to be zero, but it can be determined from the storage model described earlier in this chapter.

The constant K_A indicates the amount of wear-out per kilocycle that occurs for each percent of overcharge applied to the cells every cycle (overcharge is expressed here as a percentage of nameplate cell capacity). This term dominates the others in this model in terms of contributing to cell degradation and wear-out, suggesting that significantly longer cycle life could have been obtained in the tests summarized in Table 8.3 by further reducing overcharge. For example, reduction of the recharge fraction to about 1.02 for test pack 3600X (10°C and 35% depth of discharge) could have extended life to substantially more than 100,000 cycles, while a recharge fraction as high as 1.15 would have shortened life for this pack to only about 20,000 cycles. It is also noteworthy that this model predicts that the likelihood of having cells fail by short circuit increases significantly as the amount of overcharge increases. This model therefore predicts that different conditions of cell operation can have a dramatic effect on cycle life.

Surprisingly, the constant K_D, which represents the added degradation caused by every increase in percent depth of discharge, is nearly zero; in fact, it is slightly negative for the capacity-loss mechanism—a condition that may reflect gradual capacity increases often seen in nickel-hydrogen cells as they cycle, and one that is thought to result from the buildup of nickel corrosion products in the nickel electrodes over time. This result suggests that for the conditions of these tests, simply increasing the depth of discharge by itself does little to hasten cell failure by either mechanism. Rather it is the increased amount of overcharge, higher charge voltage, and added heating (which all typically accompany increased depth of discharge) that actually reduce cycle life for higher depth-of-discharge operation. This further suggests that if cells are run according to conditions where increased overcharge and temperatures do not control cycle life, the intrinsic effects of elevated depth of discharge alone may become more clearly observable.

The constant K_R indicates the fraction of cell life lost per volt of recharge voltage in excess of 1.52 V. This term accounts for accelerated degradation from high-rate oxygen evolution, which is known to damage nickel electrodes. This constant is significant for both the short-circuit and capacity-loss mechanisms, as expected, clearly indicating that it is advantageous for optimum cycle life to recharge at as low a cell voltage as is possible. However, while important to cycle life, the peak voltage is of somewhat lesser importance than is the amount of overcharge, unless very high peak charge voltage is allowed. For example, with the conditions of test pack 3600X, variation of peak recharge voltage between 1.52 and 1.65, while keeping the amount of overcharge constant, gives about a 60% reduction in cycle life. While this is significant, it is much less than the 600% cycle-life reduction that results when the recharge fraction is increased to 1.15 for this pack. It is also predicted by the model that higher overcharge voltage will increase the likelihood of having a short-circuit cell failure instead of experiencing the more gradual capacity-loss failure mode.

The final constant in this model, K_L, indicates the fraction of cell life lost for each year of active calendar life to which the cells are exposed. This term does not begin to have an appreciable effect on cell life until it exceeds about 10 yr, and beyond about 20 yr, it begins to be a very important and eventually dominant factor influencing life. Of course for a nickel-hydrogen cell to last more than 20 yr, stresses from other factors must have been kept quite low.

8.3.3 The Statistics of Cell Wear-Out

To determine the reliability or failure probability of a battery consisting of a number of series-connected cells, the statistical distribution of failure probabilities for the individual cells must be known as a function of cycle life. Equation (8.38) relates the reliability of an N-cell battery R_{batt} to the cell reliability R_{cell} and the failure probability of the individual cells in the battery F_{cell}, assuming that the battery fails when the first cell drops below 1 V. While this is not strictly true (because most batteries are built with at least 1 cell voltage margin so they can still operate with one cell shorted), at the point where the first cell fails the battery performance is sufficiently degraded that battery replacement should be expedited. If the battery can continue to function adequately after the first cell fails, additional terms can be included in Eq. (8.38) to reflect the improvement in reliability.

$$R_{batt} = R_{cell}^N = (1 - F_{cell})^N \qquad (8.38)$$

Cell reliability as a function of cycle life can be determined from the test results summarized in Tables 8.4 and 8.6 if one assumes a normal statistical distribution of cells in each battery. The cycles to failure reported in Tables 8.4 and 8.6 are for the first cell out of eight to fail, and they thus correspond to the cycles to failure at 87.5% cell reliability. The distribution of all cell failures in these test packs can be used to provide the needed failure statistics. Fortunately, tests 3603X, 3604X, and 3605X were continued to the point where all cells failed. Table 8.8 gives the cycles to failure for each of the cells in these packs, along with the mean cycles and the standard deviation for each pack.

From Table 8.8, it appears that the standard deviation becomes significantly greater as the mean increases. For this reason, it was decided that the standard deviation should be expressed as a fixed percentage of the mean, and the percentage used was simply the average of that observed for the three test packs in Table 8.8, which was 11.63% of the mean. Using this statistical approach, one could determine the fraction of cell life that corresponded to each different level of reliability. The results are provided in Table 8.9 for four commonly used cell reliability benchmarks, and the corresponding 22-cell battery reliability. Note that in Table 8.9 a cumulative cell wear of 100% corresponds to a 50% probability of cell failure, or the mean cell failure point.

The results in Table 8.9 show that it is necessary to have quite high cell reliability if one demands 99+% battery reliability. Similarly, with the statistics used here, only about 60% of the available mean cell life can be consumed if high battery reliability needs to be maintained. The relationship between cumulative wear and the probability of cell failure in Table 8.9 allows the probability of cell failure

Table 8.8. Cycles to Failure for Each Cell in Packs 3603X, 3604X, and 3605X

	Pack 3603X Failures	Pack 3604X Failures	Pack 3605X Failures
Cell	Cycles at Failure	Cycles at Failure	Cycles at Failure
1	42,300	32,720	28,780
2	41,700	25,460	22,460
3	35,350	28,900	28,800
4	27,938	31,200	28,100
5	33,780	30,470	27,860
6	30,472	25,350	30,600
7	37,800	27,970	22,260
8	38,800	27,140	28,230
Mean	36,018	28,651	27,136
Std. deviation	5129.23	2679.31	3065.05
SD (% of mean)	14	9.35	11.3

Table 8.9. Relationships Between Cumulative Cell Wear, Cell Reliability, and 22-Cell Battery Reliability

Cell Reliability	Battery Reliability	Cumulative Wear
50.00%	2.38×10^{-7}%	100.00%
87.50%	5.30%	86.60%
99.00%	80.16%	72.95%
99.90%	97.82%	64.06%
99.99%	99.78%	56.74%

to be tracked as a function of cycle life as wear accumulates based on the results of a wear-out model.

8.3.4 Validation of Multiparameter Cell Wear-Out Model

As shown in Tables 8.4 and 8.6, this multiparameter wear model does a very good job of predicting cycle life for the test packs used to develop the model. However, the real value of a wear-out model is in its ability to predict the life of other nickel-hydrogen cells operated under a variety of conditions, and potentially including a range of cell design variations. This section provides this correlation with several test packs that cover a range of low Earth orbit cycling conditions.

Figure 8.26 indicates the predictions of this model for NWSC test pack 5002E, which was run at 40% depth of discharge until cells were observed to fail starting at about 62,500 cycles. These cells had a capacity of 50 A h, and they had 31% potassium hydroxide electrolyte, 5% cobalt additive in the nickel electrodes, back-to-back electrode stacking, and dual-layer Zircar separator. However, these cells contained only a single stack, rather than the tandem stack used in the larger cells. This was nominally a 10°C test, and the cells actually operated at an average temperature of about 12°C. The average recharge fraction during the test was

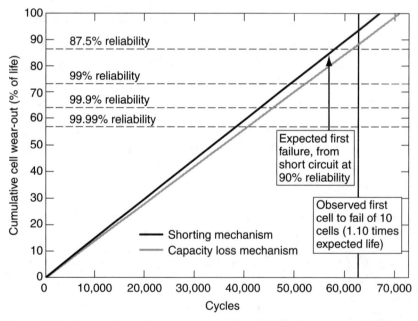

Figure 8.26. Comparison of test results for the NWSC Crane Pack 5002E with model predictions. These cells are 50 A h capacity, and they were operated at 40% depth of discharge with a 1.04 nominal recharge fraction. Reprinted courtesy of NASA.

1.0393 and the average peak recharge voltage was 1.5467. The model predicts that for these operating conditions the first cell should have failed after about 57,000 cycles. Thus, there is only about a 10% difference between the model prediction and the observed test result, in spite of the slightly different cell design represented in pack 5002E.

The model predicts a slightly greater likelihood of failure from short-circuiting rather than simply from a loss of usable capacity. The actual observed failure mode appears to be the formation of parasitic high-impedance shorts, based on the drop-off of cell pressure observed near end of life, and thus is consistent with the model prediction.

A second example where the model predictions can be correlated with test data is provided by the results from NWSC test pack 3314E, which is operated for 76,300 cycles, until the first cell failed. These cells are similar in design to those in pack 5002E, but they have been operated near 1.03 recharge fraction rather than the higher 1.04 recharge fraction used for pack 5002E. Similarly, the cells have been operated near 12°C at a 40% depth of discharge. The average recharge fraction through the test has been 1.0307, and the average peak recharge voltage has been 1.5591. The model predicts the first cell in this pack will fail after about 70,000 cycles, most likely by a short-circuit failure mode. Again the model prediction is within 10% of the observed cycle life.

Validation of this model would not be complete without the inclusion of at least one life prediction for a nickel-hydrogen design that is significantly different from the examples described thus far. A dual-anode cell design has been on test for more than seven years (see Fig. 8.15) under conditions chosen to minimize the wear-out rate and provide the maximum achievable life for a nickel-hydrogen cell. This test operates five 60 A h cells between 0 and −5°C at 60% depth of discharge. The test recharges each cell only to the recharge fraction that is required to maintain the end-of-discharge voltage and thus minimizes the stresses of overcharge and oxygen evolution rate. At present these cells have completed more than 40,000 cycles, with a test goal of at least 60,000 cycles. The cells are presently operating with a 1.010 recharge fraction, as indicated in Fig. 8.15, which shows the recharge fraction history of these cells during the test. Because the charge system maintains the minimum recharge fraction needed by each cell, the gradual increase in recharge fraction during the test gives the best indication of the rate at which the cells are degrading. Cell failure is not expected until the recharge fraction reaches at least 1.03.

When the multiparameter wear model is used to predict the lifetime of these cells, the prediction indicated in Fig. 8.27 is obtained. The model predicts about 80,000 cycles until the first of the five cells fails. Based on the slope of increasing recharge fraction in Fig. 8.15, the minimum recharge fraction required to maintain the cell state of charge at 80,000 cycles will be about 1.036, which is not an extremely high recharge fraction level.

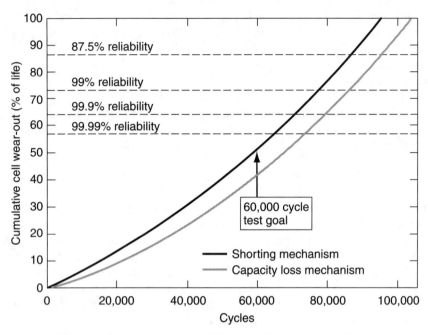

Figure 8.27. Predicted wear rate and life for optimized nickel-hydrogen cells operated at 60% depth of discharge with the recharge ratio trend shown in Fig. 8.15.

Clearly, the goal of 60,000 or more cycles at 60% depth of discharge appears possible for these cells.

It is also possible to combine this multiparameter wear model with ground storage models to determine the effect of initial wear during ground storage on the ultimate cycle-life capability of the cells or battery. This is possible because the model includes the parameter K_o, which corresponds to an initial level of wear at the start of the cycle life. The initial wear may be determined using some other model—for example, a storage model such as that described earlier in this chapter. The effect of the initial wear is to simply offset the wear curves upward by the fraction of the cell life consumed by the initial wear. With proper storage conditions and duration, this initial wear is generally limited to less than 15% of the total wear that a cell can tolerate during its lifetime.

8.3.5 Extension of Wear-Out Model to Geosynchronous Cycling

The model described here has proven quite accurate for predicting the lifetime of cells in a wide range of low Earth orbit life tests, providing a typical accuracy within about ±10% of the observed lifetime. However, it is not clear that this

8.3 Wear-Out Models for Predicting Cell Cycle Life 277

model can be extended to predict the lifetime of a nickel-hydrogen battery operating in a geosynchronous regime, or whether a modification of the model is required for this different cycling profile.[8.17] Battery operation in geosynchronous satellites typically involves 88 eclipses per year divided into two seasons, with one cycle per day during the eclipse season. The rest of the time is typically spent on trickle charge to maintain all the cells at a high state of charge.

The wear factors in the low Earth orbit model described here are also expected to be operative in geosynchronous scenarios to cause the gradual degradation and eventual failure of the nickel-hydrogen cells. However, in the geosynchronous environment these wear factors can have a very different significance in controlling battery life than they do during low Earth orbit operation. Depth of discharge typically is significantly higher in a geosynchronous profile, thus leaving less margin to accommodate cell degradation. However, the actual stress resulting from cyclic discharge and recharge is expected to be much less because geosynchronous cycling involves only 88 cycles per year. In addition, because charge rates are typically low in geosynchronous orbit, the wear factor associated with the overcharge rate is expected to contribute little to accelerating cell wear-out. However, because geosynchronous satellite applications can last longer than 15 yr, calendar life degradation is expected to be more significant than in low Earth orbit. Operating temperature is expected to be an important variable affecting wear rates, just as in low Earth orbit applications.

The greatest uncertainty in the geosynchronous application of a wear model such as that described in the previous section is in how overcharge contributes to cell wear. Analysis of the low Earth orbit databases for nickel-hydrogen cells has separated the wear resulting from overcharge into two contributions: one from the ampere-hour amount of overcharge in excess of that needed to just compensate for self-discharge, and one from the overcharge rate based on the cell voltage relative to 1.52 V. In low Earth orbits, the amount of excess overcharge is typically relatively low, but the overcharge usually occurs at rates high enough to bring peak cell voltages up to 1.52 V or more. In geosynchronous orbits, however, the total amount of excess overcharge can be quite high, but much of that overcharge is at trickle-charge rates that maintain overcharge cell voltages well below 1.52 V. This is particularly true during geosynchronous solstice seasons, where the batteries undergo no cycling and are maintained on continuous trickle charge for long periods of time. Thus, for geosynchronous solstice simulations when the batteries are on trickle charge, the low Earth orbit wear model described here will typically predict a negative stress contribution from the overcharge rate, in conjunction with a positive stress contribution from the total excess ampere hours of overcharge. The only requirement for physically reasonable overcharge stress in the model is that the total stress from these two contributions must remain above zero at all times.

Because the wear model does allow negative stress from low overcharge (trickle-charge) rates to partially compensate for the stress resulting from the excess ampere hours of overcharge, it is possible to have a very low wear rate in geosynchronous while still maintaining a high state of charge using trickle-charge rates that may be two to three times the self-discharge rate. This approach to modeling wear during geosynchronous operation provides a quantitative method for capturing the known stress factors, particularly in comparison with the assumption—often made in accelerated testing—that trickle charge is always a zero-wear condition. However, it must be established by comparison with life-test and orbital performance data whether this wear model accurately predicts battery lifetime in the geosynchronous cycling profile. If any significant inaccuracies are found, the model must be modified for geosynchronous profiles, and it would probably have to be modified for any other cycling profile that differed significantly from low Earth orbit profiles.

The geosynchronous performance database for nickel-hydrogen batteries consists primarily of accelerated ground tests and real-time operation in satellites. The only known wear-related failures are in accelerated ground tests, as described in the following sections. Nickel-hydrogen batteries continue to operate without failure in a number of space missions, as will also be discussed; therefore the orbital experience provides only lower limits on battery life for correlation with the model predictions.

Most geosynchronous ground tests of nickel-hydrogen cells are run under accelerated conditions where most or all of the two annual five-month solstice periods is eliminated, and the batteries are exposed to repeated 44-day eclipse seasons. Numerous examples of such tests have been run in excess of 40 eclipse seasons without experiencing cell failure. In a few instances, battery cell failure has been seen in these life tests. The first life test that will be compared with the model predictions is one of these, and it involved a nine-cell pack of 4.5 in. diameter nickel-hydrogen cells. At the $-5°C$ temperature at which these cells were charged, they had a capacity that exceeded the nameplate capacity by about 10%. The cells were cycled at 91.6% depth of discharge based on nameplate capacity, or 82% depth of discharge based on the $-5°C$ capacity. The cells were recharged after each discharge at about a C/12 rate to a charge return ratio of 1.25, and then put on a C/100 trickle charge until the next eclipse discharge. A 15-day solstice period was used between consecutive 44-day eclipse seasons, with about a 35% depth-of-discharge cycle occurring each day during the solstice period. In addition, this test experienced an anomaly after eclipse season 20 that involved the application of a continuous C/20 overcharge for about 90 days. The cells were put back on test after this anomaly and continued to perform acceptably, although there was a noticeable drop in performance. After 26 seasons on test, one of the cells fell below 1 V, thus indicating failure as a result of insufficient usable capacity.

The timeline from this nine-cell test can be simulated using the wear model described in the previous section. Figure 8.28 shows the accumulated wear predicted by the model for these test conditions. Of interest is the predicted ~50%

8.3 Wear-Out Models for Predicting Cell Cycle Life

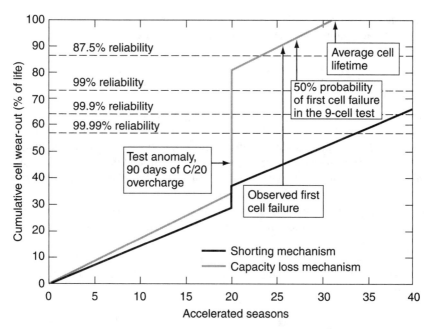

Figure 8.28. The predicted accumulation of wear based on the wear model described here during the test timeline for a nine-cell accelerated geosynchronous ground test of 4.5 in. diameter nickel-hydrogen cells at 91.6% depth of discharge.

wear contributed by the test anomaly. While this may seem like an extreme contribution from an anomaly of this kind, when it is added to the normal wear accumulated during the test, the first cell failure is predicted to occur during eclipse season 27. This is quite good agreement (well within 10%) with the first observed cell failure during eclipse season 26. Not only does the model accurately predict life in this instance, but it also correctly predicts that failure is almost certain to occur as the result of capacity loss rather than the formation of internal cell short circuits. Thus, the wear model appears to be applicable without any significant modification to the conditions of accelerated geosynchronous ground tests typically used for nickel-hydrogen cells.

Accelerated geosynchronous ground tests typically do not involve the test anomaly indicated in Fig. 8.28 and also do not show cell failure as early in life as was seen in the test of Fig. 8.28. An example of such a life test involved six 4.5 in. diameter nickel-hydrogen cells that were run for 30 accelerated seasons with no cell failures. The peak depth of discharge was 91.2% based on the nameplate cell capacity, and 80% based on the actual –5°C capacity, which is the temperature at which the cells were charged. The full recharge rate was about C/15, and the recharge fraction each cycle was 1.30. After recharge to this recharge fraction, a C/100 trickle charge rate was maintained until the next eclipse discharge. This test was operated for 30 eclipse seasons before being terminated with all cells still operating (the lowest cell was at 1.05 V for the longest eclipse).

The results of this test may be correlated with the model predictions, providing the results in Fig. 8.29. The model predicted that after 30 seasons of operation, there would be less than a 0.02% probability of experiencing a cell failure as a result of capacity loss, and failure from short-circuiting would be several orders of magnitude less probable. The model also predicted that an average cell lifetime would be 51 seasons in this test, and that about 48 seasons would be needed to accumulate enough wear to achieve more than a 50% probability of having one cell fail. Because the test was not run to failure, it is not possible to quantify the comparison between the model prediction and the test results. However, it is clear that the model prediction is fully consistent with the test results.

Accelerated life tests have also been performed on numerous 3.5 in. diameter nickel-hydrogen cells that typically are in the 50–90 A h capacity range. One example of such a test on a 23-cell pack run at 82.3% depth of discharge based

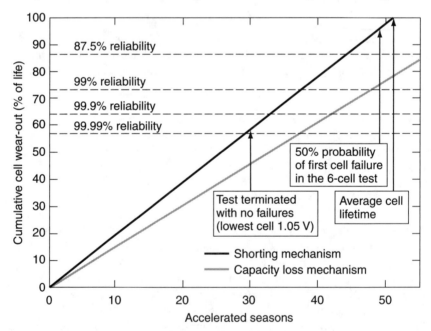

Figure 8.29. Correlation of model predictions of accumulated wear with test results for a six-cell test of 4.5 in. diameter nickel-hydrogen cells at 91.2% depth of discharge (based on nameplate).

on nameplate capacity (70% based on actual +5°C capacity) is indicated in Fig. 8.30. This pack involved repeated 44-day eclipse seasons, with two-week solstice seasons during which the cells were on continuous trickle-charge. Recharge was at about a C/15 rate to a recharge fraction of 1.20, after which the C/100 trickle-charge rate was applied. These cells operated for 39 accelerated seasons before the test was stopped with all cells still operating (the lowest was at 1.05 V during the longest eclipse).

The model can be used to predict the rate at which cell wear accumulated during this test, providing the results in Fig. 8.30. Interestingly, the model predicts that failure is most likely under these conditions from short-circuiting, but was still unlikely after the 39 seasons that were completed. The model predicts a 50% probability of the first cell failing after 44–45 seasons, and an average cell lifetime of 56 seasons. Because short-circuiting is not commonly seen as a failure mode in geosynchronous tests, there may be some question regarding the accuracy of the short-circuit failure-mode model in the geosynchronous environment. However, the test of Fig. 8.30 was not run long enough to provide any quantitative check on whether the predicted short circuits would have actually materialized. The results of this test are consistent with the model prediction. In terms of capacity-loss

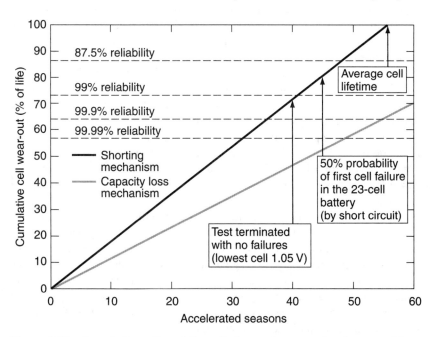

Figure 8.30. Correlation of model predictions of accumulated wear with test results for a 23-cell test of 3.5 in. diameter nickel-hydrogen cells at 82.3% depth of discharge (based on nameplate).

degradation alone, this test would have had to run well beyond 60 seasons before a capacity-loss failure would be anticipated.

While there are numerous other life tests that can be compared to the model, there are very few that have been run long enough to provide actual lifetime to failure. It is therefore of interest to examine the orbital experience with nickel-hydrogen cells. The longest geosynchronous orbital nickel-hydrogen battery performance data for the ManTech-type cells commonly used today covers about 15 yr of operation (as of 2007) at depths of discharge in the 55–60% range. These cells are continuing to perform adequately, showing no evidence of significant degradation or impending cell failure. Additional nickel-hydrogen batteries have been operating in a number of orbiting satellites for periods that range up to 15 yr, again with no reports of any cell failures resulting from normal wear-related processes in properly designed cells. Therefore, such orbital data can only impose a lower limit on the lifetimes of the cells for each set of operating conditions. The lower limit will continue to increase as additional orbital data are accumulated in the coming years.

The life model may be correlated with this orbital database in a number of different instances to evaluate whether the model predictions are consistent with the presently available data. When this correlation is done, it is found that the model is in agreement with the observed orbital experience, which involves no cell failures from wear-out after up to about 15 yr of operation. This correlation is summarized in Fig. 8.31, which shows each orbital condition as a point at the operating depth of discharge and present lifetime of the batteries. The curves in Fig. 8.31 show the model predictions of expected lifetime; the gray data points accompany the gray line that corresponds to the C/200 trickle-charge rate, and the black data points accompany the black line that corresponds to the C/100 trickle-charge rate. The comparison in Fig. 8.31 illustrates that many more years of battery operation are required before the performance data provides a good test of the model predictions. It should be noted that the model predicts that many more years of life are expected with the lower trickle-charge rates, in spite of the lower capacity maintained by lower trickle-charge rate.

In addition to the differences in life expected with different trickle-charge rates shown in Fig. 8.31, the model also correctly predicts a surprising temperature dependence of cycle life that has been seen in a number of life tests. Life tests run at lower temperatures have often given lower cycle life than have life tests run at higher temperatures. This typically occurs when an unnecessarily high trickle-charge rate or recharge ratio is employed in the lower-temperature cycling, which dramatically increases the wear rate from overcharge, because the self-discharge rate is reduced significantly at lower temperatures.

Figure 8.32 shows the predictions of the wear model for geosynchronous lifetime for commonly employed trickle-charge rates at both low and high temperatures. This figure shows that increased temperature produces a dramatic improvement

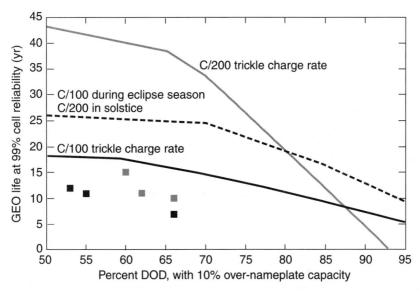

Figure 8.31. Correlation of typical orbital life experience from nickel-hydrogen batteries with lifetime predictions from the wear-out model for several different charge control approaches. These predictions assume 10% initial cell capacity over the nameplate capacity.

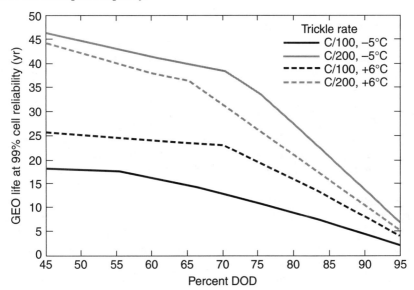

Figure 8.32. Nickel-hydrogen cell lifetime predicted by wear model for geosynchronous operation in different temperatures using different trickle-charge rates.

in cycle life when a C/100 trickle-charge rate is used, essentially because the higher trickle is needed at the higher temperature to offset the increased self-discharge rate. At the lower temperature the C/100 trickle charge rate is excessive, and accelerates wear-out. When a C/200 trickle-charge rate is employed, an increase in temperature is predicted to reduce cycle life. This interaction between temperature, self-discharge, and trickle-charge rate can make the cycle life appear to have a temperature dependence that is the opposite of that expected for some operating conditions.

On the basis of the correlations described in this section, the wear model developed from the low Earth orbit test database predicts lifetimes that are consistent with both the geosynchronous orbital data and the accelerated geosynchronous ground test data that are available. Therefore, it appears appropriate to use the model in its existing form until actual data are obtained indicating that it needs to be modified, which will require a number of years of further orbital experience with nickel-hydrogen batteries. However, it must also be recognized that while this model is consistent with the available geosynchronous data and seems to accurately capture wear in accelerated geosynchronous tests, it has not been quantitatively validated for real-time geosynchronous nickel-hydrogen battery life by a source of independent data, as it has for low Earth orbit operation.

8.4 Summary and Status of Nickel-Hydrogen Cell Modeling

The discussions presented here of the different modeling methods that have been successfully applied to nickel-hydrogen battery cells show that both first-principles and empirical models can be used to predict the performance and the lifetime of the cells. Much of the success of both of these modeling approaches is based on the accumulation of more than 20 yrs of life-test data on nickel-hydrogen cells, which has allowed the models to become quite well validated based on actual cell performance.

The first-principles models have emerged as highly accurate tools for predicting performance, but only when they include all the processes occurring in the cells, along with realistic macroscopic and microscopic structures for the cell components and the cell geometry. Not only can first-principles models of this kind accurately predict performance, but they can also enable the modeler to readily focus on the root causes of unusual performance or cell failure. When sufficient level of detail is included in a first-principles model, it can become capable of predicting a wide range of emergent cell behavior not encoded into the model, which has generally been found to be consistent with actual cell behavior. First-principles models can be used to develop better cell designs, or to optimize cell designs for best performance in key areas.

Today's first-principles nickel-hydrogen cell models are not yet capable of simulating a complete life test within a computer; however, with the inclusion of some empirical degradation rates for the key degradation mechanisms in nickel-hydrogen cells throughout cycle life, they appear capable of predicting cell lifetime within about ±10–15%, which is approximately the statistical variability in cell lifetime that has been seen in life tests. First-principles models allow the root causes for this statistical variability to be traced to sensitivity to operating environment or variability in internal cell components.

Empirical models can provide highly accurate performance predictions as long as there is a good database that applies to the operational conditions of interest. The empirical modeling methods have been most successful in modeling lifetime, whereas wear models have provided an empirical framework for capturing the results of life tests for use in a general predictive framework. Empirical models can easily include statistical distributions of cell behavior, and they thus can predict the probability of cell failure as a function of time or cycle life. The best wear-out models available today can predict cell lifetime with an accuracy of about ±10%. Empirically based wear models offer a powerful tool to help design power systems for high-reliability systems such as satellites. Progress in developing better empirical models is focusing on applying a wider range of existing test data to further refine the models, applying the models to specialized conditions (such as storage), and including additional degradation or wear mechanisms.

8.5 References

[8.1] P. M. Gomadam, J. W. Weidner, R. A. Dougal, and R. E. White, "Mathematical Modeling of Lithium-Ion and Nickel Battery Systems," *Journal of Power Sources* **110**, 267–284 (2002).

[8.2] Z. Mao and R. E. White, "Mathematical Model of the Self Discharge of a Nickel-Hydrogen Battery," *J. Electrochem. Soc.* **138**, 3354 (1991).

[8.3] Z. Mao, P. De Vidts, R. E. White, and J. Newman, "Theoretical Analysis of the Discharge Performance of a Nickel-Hydrogen Cell," *J. Electrochem. Soc.* **141**, 54 (1994).

[8.4] P. De Vidts, J. Delgado, and R. E. White, "A Multiphase Mathematical Model of a Nickel/Hydrogen Cell," *J. Electrochem. Soc.* **143**, 3223 (1996).

[8.5] P. De Vidts, J. Delgado, B. Wu, D. See, K. Kosanovich, and R. E. White, "A Nonisothermal Nickel-Hydrogen Cell Model," *J. Electrochem. Soc.* **145**, 3874 (1998).

[8.6] B. Wu and R. E. White, "Self-Discharge Model of a Nickel-Hydrogen Cell," *J. Electrochem. Soc.* **147**, 902 (2000).

[8.7] B. Wu and R. E. White, "Modeling of a Nickel-Hydrogen Cell: Phase Reactions in the Nickel Active Material," *J. Electrochem. Soc.* **148**, A595 (2001).

[8.8] A. H. Zimmerman, G. A. To, and M. V. Quinzio, "Scanning Porosimetry for Characterization of Porous Electrode Structures," *Proc. of the 17th Annual Battery Conf. on Appl. and Adv.*, IEEE 02TH8576, ISBN 0-7803-7132-1 (Long Beach, CA, January 2002), pp. 293–298.

[8.9] R. F. Gahn, "Performance of a Dual Anode Nickel Hydrogen Cell," *Proc. of the 1991 Space Electrochemical Research and Technology Conf.*, NASA Conf. Pub. 3125 (Cleveland, OH, January 1991), pp. 195–208.

8.10 M. V. Quinzio and A. H. Zimmerman, "Dynamic Calorimetry for Thermal Characterization of Battery Cells," *Proc. of the 17th Annual Battery Conf. on Appl. and Adv.*, IEEE 02TH8576, ISBN 0-7803-7132-1 (Long Beach, CA, January 2002), pp. 281–286.

8.11 A. H. Zimmerman, "Virtual Life Testing of Battery Cells," *Proc. of the 2005 NASA Battery Workshop* (Huntsville, AL, November 2005).

8.12 A. H. Zimmerman, "Mechanism for Capacity Fading in the Nickel Hydrogen Cell and Its Effect on Cycle Life," *Proc. of the 1992 NASA Battery Workshop*, NASA Conf. Pub. 3192 (Huntsville, AL, November 1993), pp. 153–176.

8.13 A. H. Zimmerman and R. Seaver, "Cobalt Segregation in Nickel Electrodes during Nickel Hydrogen Cell Storage," *J. Electrochem. Soc.* **137**, 2662 (1990).

8.14 L. H. Thaller and H. S. Lim, "A Prediction Model of the Depth-Of-Discharge Effect on the Cycle Life of a Storage Cell," *Proceedings of the Twenty Second IECEC, Vol. 2* (Philadelphia, PA, 10–14 August 1987), pp. 751–757.

8.15 A. H. Zimmerman and M. V. Quinzio, "Model for Predicting the Effects of Long-Term Storage and Cycling on the Life of NiH_2 Cells," *Proc. of the 2003 NASA Battery Workshop*, NASA/CP-2005-214190 (Huntsville, AL, 20 November 2003).

8.16 A. H. Zimmerman, "Lifetime and Failure Mode Prediction Method for Nickel Hydrogen Batteries," *2003 Space Power Workshop* (Redondo Beach, CA, April 2003).

8.17 A. H. Zimmerman and V. J. Ang, "Life Modeling for Nickel Hydrogen Batteries in Geosynchronous Satellite Operation," *2005 International Energy Conversion and Engineering Conf.*, AIAA Conf. CD 1090-347500 (San Francisco, CA, August 2005).

Part III

Application and Practice

9 Charge Management for Nickel-Hydrogen Cells and Batteries

The techniques of charge management and how they are implemented for operating nickel-hydrogen batteries are a critical aspect of obtaining a long and reliable lifetime from these batteries. Appropriate charge management is the key to balancing the accelerated degradation caused by excessive battery overcharge against the degraded performance that can result from undercharge, namely increased cell-to-cell capacity imbalance and reduced state of charge. A good charge-management system will maintain required battery performance and at the same time prevent unnecessary stresses that could significantly reduce battery life. A number of charge-management methods have been successfully used with nickel-hydrogen batteries, both for normal operational charge management and also as part of special procedures for maintaining and monitoring state of health. These methods, which are described in detail in the following sections, include:

- temperature-compensated voltage (V/T) charge control
- recharge ratio charge control
- pressure-based charge control
- adaptive or self-optimizing charge control
- ratchet charging
- reconditioning
- storage maintenance charge control

Each of these charge-management methods offers key benefits as well as some notable weaknesses; these will be described after a brief discussion of the theory and operational variables involved in charge management.

9.1 Theory of Charge Management

The key requirement of a good charge-management system is to keep all the cells in the battery recharged to a matched capacity when the battery is fully charged, thus preventing any individual low-capacity cell from limiting the performance of the entire battery by becoming depleted during discharge. For nickel-hydrogen batteries, the charge-control system must ensure that each cell gets enough overcharge to keep it charged, while avoiding excessive overcharge that can overheat the cells or prematurely wear them out. Some overcharge is necessary to replace the relatively significant self-discharge losses in nickel-hydrogen cells, which can vary between cells in a battery as the result of cell-to-cell differences and varying thermal environments across the battery.

The charge-management system may receive its key control inputs either from individual cells (cell-level control) or from the overall battery (battery-level control). Most nickel-hydrogen battery charge-control systems are based on battery-level control and thus control the current that passes through all the series-connected cells within the battery. The key parameters that may be available to the charge-control system, be it hardware-based, software-based, or a combination of these, may include:

- battery voltage
- battery current
- battery temperature (typically provided as an appropriate average based on the temperatures of at least two or three cells within the battery)
- battery pressure (typically provided as an appropriate average based on the pressures of at least two or three strain-gauged cells in the battery; the pressures may or may not be corrected for temperature changes)
- current time sums that reflect the ampere-hours discharged and the ampere-hours recharged (which enable the recharge ratio or the excess ampere-hours recharged during any cycle to be determined)
- individual cell voltages (may be available for some or all of the cells in the battery in some power systems)
- half-battery voltage (may be available in some power systems; this is the voltage difference between one half of the cells in the battery and the other half of the cells; indicates when one or more of the cells in one portion of the battery are significantly out of balance relative to those in the rest of the battery)
- individual cell currents (may be available in systems where some capability exists to control the current passing through each individual cell)

These parameters, as available in each specific charge-control system, are used by the system to properly meter the charge back into the battery cells during recharge.

The ideal charge-control method has the task of bringing each cell in the battery to its required state of full charge, and then maintaining the fully charged state until it is necessary to discharge the battery. Ideally this charge-management function should be performed without any unnecessary overcharge, because overcharge in excess of that required to replace self-discharge losses in the cells will significantly accelerate the wear rate of those cells. While most practical charge-control systems cannot achieve this ideal, each charge-control method typically has a range of selectable charge levels available (V/T levels, recharge ratio levels, pressure levels, etc.) that can be used to select an appropriate level of overcharge that will keep all the cells in the battery adequately recharged. If battery-level charge control is used, all the cells in the battery must receive at least the amount of overcharge needed to keep the cell having the highest internal loss rate (self-discharge) adequately charged. This is a key reason why it is critical to match the initial performance of cells closely within a nickel-hydrogen battery, as well as to provide a thermal control system that minimizes cell-to-cell thermal variations within the battery.

9.2 Temperature-Compensated Voltage Charge Control

V/T charge control typically involves recharge of the battery at a constant high current until a temperature-compensated battery voltage level is reached, whereupon the control system begins to throttle the current back to lower levels. The approaches used most commonly involve either holding the battery voltage at the V/T level and allowing the current to taper as required to maintain the voltage, or simply cutting the current back to a lower rate that maintains the fully charged state. The battery voltage used for this purpose is compensated for variations in battery temperature. Because an increasing temperature will cause the cell voltage to decrease at a given state of charge, overcharge heating can cause thermal runaway if the charge voltage level is not properly temperature-compensated. The slope of this temperature compensation may vary for different power systems, but it is typically in the range of -0.25 mV/°C per cell.

V/T charge control for nickel-hydrogen batteries was in many cases inherited from its earlier usage in nickel-cadmium-cell–based power systems, where it was widely used. While V/T charge control is probably not the system of choice to use in designing specifically for nickel-hydrogen batteries, it can be effectively used for these batteries.

The action taken by the charge-control system when the battery voltage hits the V/T level can vary significantly, depending on the application. In low Earth orbit satellite applications, where recharge time is only about 60 min, peak recharge currents can be quite high (up to C/2 or more) and the system is typically designed to hold the battery at its V/T level and force the battery recharge current to taper down as necessary to maintain battery voltage at the V/T level. In this type of system, the amount of overcharge can be adjusted by moving the V/T level up or down to different V/T settings.

In geosynchronous or medium Earth orbit satellite applications, where there can be 10–22 h of time available for recharge, the recharge rates are much lower. While in some cases, the current is allowed to taper down much as described above, in these systems it is more common to use the attainment of the V/T level as an indication that the battery has reached full charge, and the system responds by switching to a trickle-charge rate. Trickle-charge, which is typically a C/100 to C/300 rate, is intended to maintain the fully charged state in all the cells for the remaining portion of the recharge period. In some power systems, combination approaches have been used, which include stepwise reductions in recharge current at specific V/T settings or combinations of V/T levels with fixed amounts of ampere-hours of recharge following attainment of the V/T level.

There are several potential weaknesses that can create problems during V/T charging of nickel-hydrogen batteries. The first can occur when low peak charge currents (less than C/10) are used for recharging, or if relatively large spacing is used between V/T levels. While battery voltage generally rises as the battery reaches full charge and enters overcharge, the voltage rise signifying the onset of overcharge becomes smaller as the recharge current decreases and as the battery temperature rises. Therefore, when low recharge rates are used, it is possible to have one V/T level that is too low to keep the battery fully recharged while the next higher V/T level is too high to ever be reached during recharge. If this situation occurs, the battery will never trigger the transition to trickle-charge and can go into thermal runaway. For this reason, many charge-control systems have a high-temperature set point that will automatically terminate the high charge rate and either force a transition to trickle-charge or safe the system by turning off all recharge current. One solution to this problem is to design V/T levels that are spaced closely enough for an appropriate level to always be found.

An additional issue with V/T charging of nickel-hydrogen batteries involves how the charge-control system responds to state-of-charge imbalance between the cells in the battery. If cells in the battery have significantly differing self-discharge rates, differing charge efficiencies, or differing temperatures, the cells with the higher internal losses can drop in state of charge and the cells with lower internal losses can increase in state of charge. The lower state-of-charge cells will cease to contribute a clear voltage upturn to signify full charge, and the higher state-of-charge cells will be driven to ever-higher states of charge and overcharge. Over long periods of cycling, particularly at elevated temperatures (those greater than 8°C), this can result in significant capacity imbalance between the higher- and lower-capacity cells in the battery. Ultimately, this condition will result in reduced battery performance and increased wear on the cells, significantly reducing battery life expectancy and reliability.

When V/T charging is used, selection of the appropriate V/T level is critical. The appropriate level is the lowest level that keeps all cells adequately recharged but does not apply unneeded overcharge. The most common method for selecting a reasonable initial V/T level is to base the selection on attaining a recharge ratio or pressure high enough to ensure that all the cells charged. The V/T level may subsequently be optimized by gradually decreasing it until a level is found that is just high enough to keep all cells sufficiently recharged. This optimization process may be based on measurements of cell voltages or cell pressures, or on capacity measurements made during battery reconditioning. It should be recognized that the full charge level desired in a nickel-hydrogen battery for optimum life is typically less than 100% state of charge, often corresponding to a point in the 85–95% state-of-charge range.

9.3 Recharge Ratio Charge Control

Recharge ratio charge control is based on the principle of ampere-hour integration during discharge and recharge. The attainment of full charge is signified by the return of a recharge ratio (capacity recharged/capacity discharged) that is a specified level in excess of 1.00. For example, a recharge ratio of 1.10 means that 10% more capacity was returned into the battery during recharge than was discharged.

A variation on this method returns a fixed excess number of ampere-hours over what was discharged rather than a fixed ampere-hour ratio. When the full charge trigger point is reached, the charge current is typically switched to a trickle-charge rate to hold the battery at full charge. The principle behind this charge-control method is that the charge returned in excess of a 1.00 recharge ratio is sufficient to compensate for internal losses from self-discharge and charge inefficiency.

The recharge ratio charge-control method requires that the power system has a charge integrator at its disposal. This integration function is often performed using a software algorithm, although hardware-based ampere-hour integrators have also been used successfully. A wide range of recharge ratio levels is typically available. The recharge ratio charge-control method has sometimes been combined with V/T charging, particularly in low Earth orbit satellite operation where the high peak recharge rates that must be used to ensure energy balance must be throttled back appropriately as the battery voltage rises into the overcharge region. In these types of charge-control systems, the V/T function is normally a hardware control that tapers the current back as the battery attains the V/T level, while a software-based recharge ratio–based control can then be used to ultimately force a switch to trickle-charge when the required recharge ratio is attained.

Recharge ratio charge control has arguably proven to be the most robust in satellite applications. Appropriate recharge ratio levels have been defined accurately from ground test data for most nickel-hydrogen cell designs. A safe (for long life) battery level recharge ratio can generally be selected by determining the self-discharge losses expected for the thermal environment experienced by a typical cell over the charge/discharge cycle, and then multiplying these losses by a factor that accommodates the range of losses over all the cells in the battery. This factor will ensure that even the cells having the highest internal losses will be adequately recharged.

Typical safe recharge ratios are shown in Table 9.1 for operational conditions typical of nickel-hydrogen battery operation. The safe recharge ratios are sufficiently high that all the cells in the battery should be maintained near 100% state of charge by the recharge. Lower recharge ratios may be used to optimize life at the expense of not keeping all the cells charged fully.

Table 9.1. Safe Recharge Ratios for Keeping Nickel-Hydrogen Batteries Fully Recharged

Temperature (°C)	40% Depth of Discharge	60% Depth of Discharge
10	104.0	103.3
–5	103.0	102.5

As with V/T charging, recharge ratio charge control may be optimized to find the minimum recharge ratio that holds the battery at an adequate state of charge for its particular operational environment. This optimum recharge point will ensure that the battery will provide a long lifetime, and it is determined by gradually decreasing the recharge ratio until the threshold is found that just maintains capacity as determined by pressure, capacity, and voltage measurements. If all the cells in the battery are well matched and temperature is uniform across the battery, the optimized level should be only slightly higher than the recharge ratio that just compensates for self-discharge and charge inefficiency losses. Because recharge ratios are typically under software control, it is often possible to make much finer adjustments to the charge return than is possible for the fixed V/T level adjustments available in most hardware.

The most significant issue that must be considered when using recharge ratio charge control involves the accuracy of the ampere-hour integration. Inaccuracies in integration normally are dominated by the current-measurement accuracy rather than the accuracy of other sources, such as time-interval accuracy. Even small offsets, nonlinearities, or differential scaling factors in the current monitor can have a large effect on the recharge ratio calculation, particularly if these errors are accumulated over long periods.

A good approach for ensuring the required accuracy is to use high-resolution current monitors that are well calibrated, and to use error-correcting or resetting algorithms that prevent inaccuracies from accumulating over long time intervals. If such care is not used, the inaccuracy in recharge ratio can easily be as high as 0.10. Inaccuracy of this magnitude may prevent recharge ratio measurements from being effective for charge control. However, even smaller inaccuracies may increase the difficulty associated with selecting an appropriate recharge ratio, making selection of recharge ratio an empirical process and increasing the risk of inadvertently under- or overcharging the battery before finding the desired recharge ratio setting.

9.4 Pressure-Based Charge Control

The best method for determining the state of charge of nickel-hydrogen batteries is monitoring the hydrogen pressure within the cells. The hydrogen serves as a chemical integrator that tracks the amount of charge put into or taken out of the cells, thus providing a highly linear measure* of cell and battery state of charge. Nickel-hydrogen cells are customarily equipped with strain gauges mounted on the cell domes. A strain gauge detects the small deformations in the pressure vessel dome that are proportional to the internal pressure, providing a signal that is amplified and converted into calibrated pressure readings. A battery of cells will typically contain 2–4 cells with strain gauges, whose readings are typically averaged to obtain an average battery state-of-charge measurement.

The charge-control system may either directly or indirectly use the measurements from the cell pressure monitors. Direct pressure control involves the use of a number of pressure or P/T (temperature-compensated pressure) curves that encompass the range of pressures that a fully charged battery may experience during its lifetime. The appropriate pressure curve is selected (based initially on cell pressure calibration data), and when the pressure set point is reached, the full charge rate is switched down to a trickle-charge rate. For applications where rapid recharge is required, such as low Earth orbit satellites, several pressure set points may be defined, each corresponding to a state-of-charge limit where the current is further decreased, with the final step-down of the current reducing it to the trickle-charge rate.

The indirect use of pressure for charge control has also been commonly employed in satellites. In this situation, ground-based satellite power system operators use pressure measurements to determine the appropriate recharge ratio or V/T level for power system operation. The desired charge-control limit, whether a recharge ratio or a V/T limit, is set in the power system by ground command to the satellite. If, later in battery life, the pressure within the cells rises or falls significantly in response to operational or environmental changes, the charge-control level can be manually updated to a new level that will maintain the correct battery state of charge.

However, pressure-based charge control is not without problems. Clearly, when monitoring pressure in 2–4 cells in a battery with 22 or more cells, one finds a potential problem in that cells that are not monitored may be significantly divergent from the others in state of charge and pressure. This is an undesirable possibility that can be made less likely by matching the behavior of the cells at their beginning of life, and by maintaining a uniform thermal environment for all the cells that is neither too hot nor too cold. If one or more nonmonitored cells do diverge from the others in state of charge and pressure, the divergent situation must be detected from the voltage signatures or the thermal signatures of the cells or battery.

---

*Some corrections for nonlinearities of the pressure vs. state-of-charge function are required at high pressures as a result of deviation of hydrogen gas from ideal gas behavior. Corrections are also necessary for the variations of pressure with cell temperature.

Another area of concern when pressure is used for charge control, or even if cell pressures are simply being monitored for information, is pressure drift. This can result either from a real long-term change in cell pressure or from a drift in the strain-gauge characteristics or the strain-gauge electronic amplifiers (one that does not correspond to real pressure growth). Apparent drift from both types of sources has been seen, and compensation methods must be available to periodically reevaluate and reset the pressure limits that correspond to full charge.

Real pressure drift in nickel-hydrogen cells results from slow corrosion processes in the nickel electrodes over many years of operation, and it causes a gradual rise in internal cell pressure over time. Figure 9.1 shows the typical long-term rise in cell pressure that occurs from these corrosion processes over the operational lifetime of a nickel-hydrogen cell. The upward pressure drift shown here, although real, does not correspond to an increase in the usable state of charge of the cell, and therefore the pressure limit to which the cell or battery is charged must also be periodically adjusted upward to maintain a full state of charge.

Two methods are customarily used to periodically adjust pressure limits for apparent drift, whether real or not. The first involves bringing the battery to a fully charged state as established by an independent means and noting the pressure corresponding to the fully charged state. The most straightforward method for doing this is by recharge until a temperature rise indicating overcharge is seen. The second method involves fully discharging the battery (reconditioning), and noting how the pressure changes for the fully discharged battery over time. The pressure

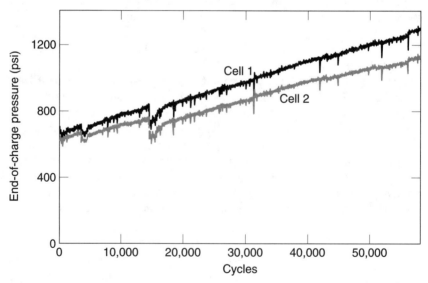

Figure 9.1. Upward pressure drift for two cells in NWSC Crane test pack 5002E (40% depth of discharge, 10°C).

rise from this fully discharged condition is a good measure of the total amount of capacity stored in the battery. Either method typically indicates a slow upward drift in pressure over time, which can be used to periodically adjust the pressure set point used to detect full charge.

Other factors can influence how well pressure indicates usable capacity in nickel-hydrogen cells. When a cell enters overcharge, oxygen is produced in the nickel electrodes. This oxygen, while eventually recombining with hydrogen at the hydrogen electrode, will temporarily produce a small upward pressure indication in the cell that is not directly related to usable cell capacity.

Additionally, as a cell ages, a significant portion of its stored capacity will eventually become unusable at normal discharge voltages during high-rate discharge. This unusable capacity can be discharged at the lower rates used for reconditioning, often at a discharge voltage plateau significantly lower than normal. Hydrogen pressure measures all stored capacity, both that usable during a high-rate discharge and that which is not usable. The effect that loss of usable capacity has on using battery pressure measurements must be ascertained from periodic discharge to measure capacity behavior as a function of pressure. Such capacity measurement discharges should be done at rates as close as possible to those required during normal discharge.

9.5 Adaptive or Self-Optimizing Charge Control

Adaptive charge control is a method that was developed in recognition that any unnecessary overcharge of nickel-hydrogen cells will increase their wear rates and decrease lifetime.[9.1] The basic premise of adaptive charge control is to use a system that is self-optimizing in the sense that it gradually "learns" how to return only the recharge required of each cell in a battery to bring that cell back to its full charge point and fully compensate for its self-discharge losses. If this can be done for each cell within the battery, the charge control will be truly optimized in terms of providing the capability to maximize battery lifetime. To realize this capability in a nickel-hydrogen battery it must be possible to independently control the charge returned to each cell in the battery. This has not been a standard method for charge control in nickel-hydrogen batteries, and it does require some modifications to the standard charge-control systems and the typical nickel-hydrogen battery.

Two methods for implementing adaptive charge control have been tested successfully. In each method a measure of the performance capability of each cell is monitored and trended as the cell is cycled. The amount of recharge (controlled by either recharge ratio or excess ampere-hours returned) is incremented either upward or downward for each cell at the end of each cycle depending on whether that cell has risen above or fallen below a specified performance threshold (e.g., the voltage level corresponding to 20% state of charge). In this way each cell is independently regulated at this performance threshold and the charge-control system learns over time what the optimum recharge needs are for each cell. This is why the method is termed "adaptive" or "self-optimizing" charge control. For a nickel-hydrogen battery test example, some cells in batteries are often seen to run several degrees warmer than the coolest cell as the result of test environment variability. As indicated in Fig. 9.2, an adaptive charge-control system rapidly learns that the warmer cells require a slightly higher recharge ratio during each cycle, and the recharge ratio of all the cells is automatically adjusted appropriately for their average temperatures.

The two methods that have been tested for implementing the individual cell charge control required to independently vary the amount of recharge for each cell have utilized either individual cell V/T charge control, or individual cell excess ampere-hour charge control. In the first method, each cell in the battery is given an appropriate V/T level and a target recharge ratio. The entire battery recharge current is tapered based on the voltage of the cell closest to its V/T level, and when any cell reaches its recharge ratio, its recharge current is individually switched to zero and recharge continues until all cells reach their recharge ratio limit. If the end of discharge voltage falls below a threshold, the recharge ratio target is increased, and if a cell does not reach its recharge ratio limit before the end of the recharge period, its V/T level is increased to allow the recharge ratio set point to be attained for future cycles. These feedback algorithms enable the charge control system to provide the minimum amount of recharge and the lowest

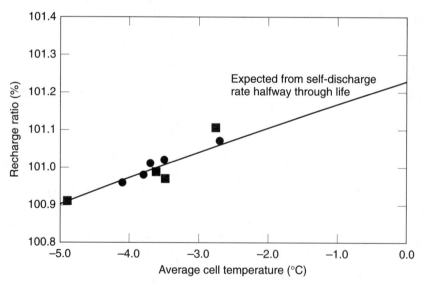

Figure 9.2. Dependence of recharge ratio required to keep nickel-hydrogen cells recharged in a test pack on the individual cell temperatures, after 30,000 cycles at 60% depth of discharge.

9.5 Adaptive or Self-Optimizing Charge Control

recharge voltage needed to maintain each cell in the battery at its minimum specified performance level.

Figure 9.3 shows the results from a test on a nickel-hydrogen battery containing 60 A h cells that is managed by an adaptive charge-control system such as that described above. This test has completed about 36,000 cycles at a 60% depth of discharge each cycle. These cells, whose temperature varies between –5 and 0°C during the cycling, need a recharge ratio that is only about 1.01 to maintain their performance when cycling between about 20% and 80% state of charge. As the cells gradually degrade over their cycle life, the adaptive logic in the charge-control system learns that they need an increasing recharge ratio, and it increases the ratio appropriately. With continued cycling, these cells are expected to demonstrate the ultimate cycle life possible from nickel-hydrogen cells at 60% depth of discharge when operated with optimum charge management.

The second method that has been tested to implement adaptive charge control has been specifically tailored for use in a standard 22-cell nickel-hydrogen battery, with minimal modification to the battery. In this case the standard 22-cell battery was modified by placing a 1 Ω resistor in series with a small relay across each cell, thus enabling a form of individual cell charge control with less than a 0.1% increase in battery weight. The voltage of each cell was monitored to provide the necessary indication of the relative performance of each cell in the battery. The adaptive charge-control algorithm simply turns on the cell resistive

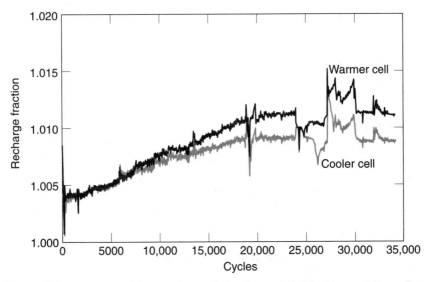

Figure 9.3. Average recharge ratio needed to keep nickel-hydrogen cells cycling between about 20% and 80% state of charge in a 60% depth-of-discharge life test operating with a –5°C end-of-charge temperature.

shunts during recharge for the period of time needed to give an excess ampere-hour differential between individual cells during recharge. Based on the relative performance of each cell, the adaptive system learns precisely how many excess ampere-hours each cell needs to maintain its performance at the specified level. Additionally, the maximum V/T of each cell during recharge is adjusted upward as required to ensure that the needed excess ampere-hours of recharge can be supplied to each cell.

A life test of a spare flight battery from an actual satellite program is being done to evaluate the feasibility of this form of adaptive charge control. At the start of this test, the battery had experienced nearly 6 yr of cold storage, but still gave good performance. Because the flight battery had been designed for a maximum depth of discharge of 25–30%, its thermal dissipation capability had to be significantly enhanced for a test that was planned to operate at a maximum 60% depth of discharge. This was done by retrofitting the battery with a radiator on the top to augment the base plate cooling in the original design. In this way the thermal gradients across the cells could be maintained at less than 5°C for depths of discharge up to 60%, which was necessary to prevent premature failure from electrolyte condensation in the cells.

The battery continues to run in a 15-cycle-per-day profile after completing more than 20,000 cycles, with the depth of discharge varying between 25% and 60% during the daily profile. The highest depth of discharge occurs once per day at the end of a 1.5C discharge pulse. The adaptive charge-control algorithm has effectively maintained all 22 cells in the battery at a well-balanced state of charge by continually adjusting the recharge ampere-hours provided to each cell during the cycling. After about 10,000 cycles, the 22 cells were operating with about a 2.1 A h excess charge applied to the battery, corresponding to an average recharge ratio of 1.035.

After about 10,000 cycles the battery was reconditioned to determine how well the capacity balance between the individual cells was being maintained, providing the results in Fig. 9.4. All cells except cell 11 had the same capacity within a 1.9 A h range and were operating at a maximum state of charge more than 91% during the cycling just before the reconditioning. Cell 11, which was stored with hydrogen precharge prior to start of the test, was the limiting cell in the battery and had a 91% state of charge. The results in Fig. 9.4 clearly show that the adaptive charge-control approach can keep cells well matched in a battery while minimizing overcharge, and that this charge-control approach can be implemented with minimal increase in battery weight.

9.6 Ratchet Charging

Sometimes a nickel-hydrogen battery charge-control system must function in a thermal environment that does not provide sufficient cooling to obtain efficient charging or a high-rate discharge without heating the battery excessively. The

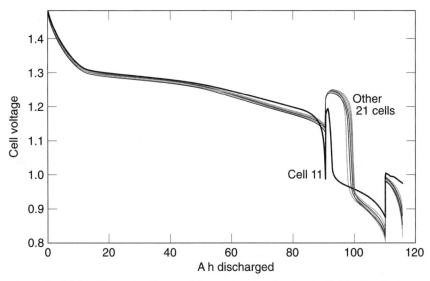

Figure 9.4. Matching of cell capacities maintained using individual cell charge control for a 22-cell battery after 10,000 cycles at a maximum 60% depth of discharge.

most common example of this scenario is battery charging at the launch site as a satellite and its launch vehicle are readied for launch into orbit. The thermal environment in this situation typically involves only the flow of cool air into the battery compartment, with the air temperature rarely being below 55–60°F (13–15°C). At these temperatures and cooling rates it is not possible to fully recharge nickel-hydrogen batteries because they begin to warm up from self-discharge and overcharge well before approaching full charge, forcing recharge to be terminated to avoid thermal runaway.

A charge-control method has been developed that is capable of bringing a battery to a high state of charge in a less-than-optimum thermal environment. This method, "ratchet charging," involves battery recharge at the normal full recharge rate for a limited time or until a prescribed battery temperature rise is detected (typically less than 2–4°F). At this point battery recharge is stopped and the battery is allowed to equilibrate and cool back to the temperature of the air-conditioned compartment. The better the cooling is, the more rapidly the battery cools and the lower are the capacity losses from self-discharge during the cooldown process. After the battery cells have had a chance to equilibrate and cooldown is completed, the full recharge rate is turned back on for another recharge interval until the prescribed temperature rise is again seen. This additional recharge will "ratchet" the battery state of charge up an additional amount. The process of ratcheting the battery state of charge upward can be continued, with cycles being repeatedly applied until the battery stops increasing its state of charge significantly from cycle to cycle.

Figure 9.5 shows the state of charge and thermal profile for a battery during a ratchet recharge sequence compared to a single-step recharge using a C/10 rate. The battery in this situation had very limited cooling, as evidenced by the 10°C temperature overshoot following the single-step recharge up to nearly 90% state of charge, and then the slow cooldown to about 12°C during the next 14 h. When periodic ratchet charge pulses are used, a state of charge well in excess of 85% can be achieved while keeping the temperature less than about 10°C. Figure 9.5 clearly shows that ratchet charging can be used to produce a significantly higher state of charge with less of a battery temperature rise.

Ratchet charging has been successfully used for charging nickel-hydrogen batteries in a number of situations during which only limited battery cooling is available. As discussed here, these situations have included launch-base charging, as well as predeployment battery charging in orbit, on-orbit operation during anomalous thermal situations, and providing extra battery state of charge during periods when higher-than-normal capacity is needed.

9.7 Reconditioning

The practice of reconditioning a nickel-hydrogen battery can be an important part of its charge-management protocol. Reconditioning the battery involves the complete discharge of the cells within the battery. This is typically done at a relatively

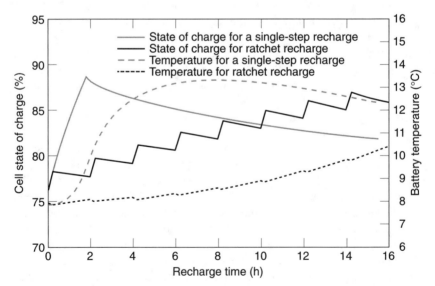

Figure 9.5. State-of-charge and temperature profiles during a 16 h period following a C/10 recharge in a single step, compared to intermittent ratchet charging.

low discharge rate, although stepwise reconditioning involving an initial discharge at a higher rate to remove the majority of the capacity more rapidly is common.

Reconditioning discharge of a battery is generally done by connecting an appropriately sized resistive load (size often being dictated by thermal considerations) across the battery and allowing the entire battery to discharge until it reaches a predetermined cutoff voltage level. Typical low cutoff voltages for reconditioning a 22-cell nickel-hydrogen battery have ranged from 1.6 V to more than 20 V. Generally, lower discharge rates and lower battery voltage levels correspond to a more complete battery reconditioning because these conditions allow more capacity to be discharged. When a nickel-hydrogen battery is reconditioned, it is not uncommon for a significant number of the lower-capacity cells in the battery to be driven into reversal by the higher-capacity cells.

Reconditioning nickel-hydrogen batteries can provide a number of performance improvements that can help significantly with the management and long-term maintenance of the battery. First, and arguably one of the most important functions of reconditioning, is that the complete discharge allows all the cells in the battery to be reset to a well-balanced capacity level at zero state of charge. This process of rebalancing the capacities of the cells in the battery, which requires that they contain hydrogen precharge,** is possible because when a hydrogen-precharged cell becomes depleted it generates hydrogen gas from the depleted nickel electrode during continued discharge at the same rate at which the hydrogen gas recombines at the hydrogen electrode. This chemical cycle, which involves no net energy storage and occurs just below a zero cell voltage, allows the higher-capacity cells to come down to the same fully discharged state as the lowest-capacity cells. Subsequent recharge of the battery will bring all the cells up to a well-balanced state of full charge. This rebalancing of cell capacities can also help with thermal management of batteries, because the higher-capacity cells will not go into overcharge prior to the other cells after they are rebalanced, thus significantly decreasing heat generation as a battery approaches full charge.

The second effect of reconditioning on nickel-hydrogen cells is the conversion of all the active material in the nickel electrodes to a uniform phase composition corresponding to β-nickel hydroxide. This conversion will eliminate other phases that can build up during extended cycling into layers within the active material. These layers can significantly degrade the charge efficiency and the usable discharge capacity of the nickel electrode. The uniform phase that results from

---

**Capacity rebalancing by complete discharge requires that the nickel precharge normally present at beginning of life in modern nickel-hydrogen cells has been replaced by hydrogen precharge, a process that naturally occurs after several years or more of operation. Prior to this precharge change, reconditioning discharge produces no net rebalancing of cells, because the lowest-capacity cells can achieve a negative state of charge during reversal by producing oxygen during continued discharge rather than consuming hydrogen. The oxygen then must be recombined during recharge before the cell can begin normal production of hydrogen, thus restoring the original imbalance.

reconditioning can exhibit significantly higher recharge and discharge voltages, and significantly less capacity that discharges at a lower, second-plateau voltage level.

The improvement in discharge voltage from reconditioning has proven to be an extremely useful charge-management tool for power systems in which there is little margin in the system to tolerate lower battery discharge voltage. This voltage improvement can last for months in a low Earth orbit satellite, and has often been invaluable for maintaining improved performance through the 45-day eclipse seasons that occur for geosynchronous satellite applications. Along with the improvement in the discharge voltage comes a corresponding decrease in battery heat generation, an effect that can significantly improve thermal management of the batteries in some power systems.

A final charge-management feature offered by reconditioning is the information that it provides regarding the actual dischargeable capacity in the cells and the battery, along with the growth in the hydrogen pressure seen after complete discharge. The pressure growth measurement can be used to adjust the pressure set points for charge-control purposes in power systems that control charging based on cell pressure. The pressure growth measurements from reconditioning can also be used to periodically update the correlation between pressure and state of charge that is often used for managing and planning nickel-hydrogen battery usage for both normal and contingency operations in satellites.

9.8 Management in Storage

Appropriate charge management during the storage of nickel-hydrogen batteries prior to their operational deployment is critical for minimizing battery wear and stress that can reduce performance or operational life. The normal storage state for modern nickel-hydrogen battery cells is the nickel-precharged condition in which they are manufactured. If all the cells in a battery are verified to contain nickel precharge,*** the battery should be stored discharged in the open-circuit state. In this storage mode, the precharge in the nickel electrode will maintain normal voltage conditions that maintain the performance of the nickel electrode. The nickel precharge also maintains the hydrogen electrode near the nickel electrode potential (typically about 0.2–0.3 V below the nickel electrode). At this potential, limited oxidation of the platinum catalyst in the hydrogen electrode can slowly occur during storage. However, sufficient excess platinum is present that this has not been seen to cause any performance or life problems. Reduced temperature during storage is highly recommended to decrease the rate of sinter corrosion and nickel precharge loss during extended discharged storage. Temperatures as low as

---

***The procedures used to verify precharge are described in chapter 13.

−20°C have been effectively used to maintain active nickel precharge for as long as 9 yr of storage.

If any cell in a battery is determined to contain hydrogen precharge, the entire battery should be handled and maintained as a hydrogen-precharged battery, and all the cells should be kept at voltages above 1.2 V during storage. While application of a continuous trickle-charge to the entire battery may be used to keep the cells at voltage, the preferred method is to apply periodic top-off charge to the battery. This method involves partial recharge of the battery to an initial partial state of charge (typically between 30% and 70% of full charge), and then subsequent top charge to a pressure limit, an ampere-hour limit, or a temperature rise limit frequently enough that all cells are kept above 1.2 V (typically once or twice per week).

During the open-circuit period between the top-charges, cell voltages should be checked to make sure they do not fall below about 1.27 V (below this level, the open-circuit voltage can drop rapidly, increasing the risk that a cell could fall below 1.2 V between monitoring measurements). If any cell falls below this level while other cells are higher, cell capacity imbalance has probably built up over time and it is necessary to rebalance the cells by fully discharging the battery and then recharging. Following cell rebalancing, top-charge storage can be continued as previously described.

During satellite integration and test activities, batteries are frequently allowed to stand open-circuited, or in a partially or fully charged condition for significant periods of time. This practice is acceptable from a charge-management point of view; however, from a battery point of view it leads to deactivation of the nickel electrodes over a period of time. This deactivation can cause a temporary loss of usable battery capacity, which can lead to a permanent capacity loss if not managed properly. Proper management involves reactivating the cells in the battery by performing a full charge/discharge cycle that includes at least some overcharge. Following this reactivation process, the battery may be placed back into storage; however, nickel precharge in the cells must be reverified each time the discharged open-circuit storage mode is utilized.

9.9 Other Charge-Management Tools and Methods

9.9.1. Adiabatic Charging

Nickel-hydrogen batteries operated in a warm environment are often very difficult to bring to a high state of charge because of a high self-discharge rate and poor charge efficiency at elevated temperature. Adiabatic charging can be used to temporarily bring warm batteries to a state of charge significantly higher than is otherwise achievable.[9.2] Adiabatic charging involves charging the battery at a low temperature where the charge efficiency is relatively high, then allowing its temperature to increase. A significant period of time can lapse before the self-discharge of the battery reduces its state of charge back to the level that normally would have been attained by recharge at the higher temperature.

9.9.2. Electrolyte Redistribution

The exposure of a nickel-hydrogen battery to thermal environments that cause a significant thermal gradient (one greater than 5–7°C) to form across its cells can result in the condensation of water on the coldest surfaces within the cell. If this condensation forms a pool in the cell (i.e., in the dome) that cannot be readily drawn back into the electrode stack by the wall wicks or gravitational action, the situation can result in dry-out of the electrode stack and reduced cell performance. If this occurs, it is sometimes possible to redistribute the liquid back into the electrode stack by warming the colder regions of the cells using heaters or other thermal control methods.[9.3] If the warmer condition is maintained for a long period of time (up to two weeks, for example) it is possible to reverse this condition. When there are wall wicks on the interior surfaces of the cell, performance improvements can also be obtained in some cases by reducing the thermal gradient to the point where the wicking rate of the wall wicks is capable of restoring the electrolyte to the electrode stack in the cell.

9.9.3. Capacity Recovery Cycling

The improper storage of nickel-hydrogen cells or batteries can result in loss of capacity. Recovery of the lost capacity is generally very desirable, and can often be accomplished by performing specific types of charge/discharge cycling. The goal of this type of cycling is generally to reverse the chemical and physical changes that have occurred to cause the capacity loss. These changes typically involve the formation of thin chemically modified active-material layers between the active material in the nickel electrode and the current-collector surfaces that are in direct contact with the active material. For this reason, cycling techniques that are capable of disrupting such layers have been most successful as capacity-recovery methods. Layers within active material can be disrupted by significant volume changes during cycling, which produce some remixing of these modified layers with the bulk material during each cycle. Thin layers can be disrupted to obtain recovery relatively easily, while thicker layers may be impossible to fully recover from.

Two types of cycling approaches have been most successful, and they use two differing methods to disrupt layers that can cause capacity loss. The first method involves the application of significant amounts of overcharge, and it depends on the evolution of oxygen to disrupt layers. This method is epitomized by long-term (48–72 h) charge at rates of C/10 or lower, and may involve several such cycles. The second method depends on the repeated formation and discharge of γ-NiOOH during cycling, which involves large active-material volume changes that can disrupt films that cause capacity loss. Because γ-NiOOH forms most efficiently at low temperatures, this method involves repeated low-temperature cycles, typically in the –10 to 0°C temperature range. It is, of course, possible to combine the effects of these two methods by doing repeated low-temperature cycling that involves significant overcharge. However, extended overcharge at low temperatures is more likely to cause damage from popping, electrolyte redistribution, or electrolyte freezing than is cycling at higher temperatures.

9.9.4. Capacity Verification Testing

During operational use of nickel-hydrogen batteries, it is often useful to verify the performance capability of the batteries at a significantly higher depth of discharge than that to which they are cycled during normal operation. For example, normal operation may only require a maximum depth of discharge of 25%, but it is important to know that the batteries could support operation out to 40% depth of discharge in an emergency. Such verification of operational margin can be obtained by periodically performing a capacity verification test.

The capacity verification test involves using satellite loads to discharge the batteries to a target depth of discharge that ensures the desired margin (40% in the example described here). If this verification test is regularly performed, the performance trending data that it provides (from voltage, temperature, and pressure data) will minimize the potential battery-depletion risk associated with the occasional higher depth of discharge. The trending of the test data will enable the detection and correction of impending performance issues as they approach, not when they become severe enough to start affecting battery performance at the normal operating depth of discharge.

9.10 References

[9.1] A. H. Zimmerman and M. V. Quinzio, "Model for Predicting the Effects of Long-Term Storage and Cycling on the Life of NiH2 Cells," NASA/CP-2005-214190, *Proc. of the 2003 NASA Battery Workshop* (Huntsville, AL, 20 November 2003).

[9.2] C. Lurie, S. Foroozan, J. Brewer, and L. Jackson, "Adiabatic Charging of Nickel-Hydrogen Batteries," NASA/N95-26785 09-44, *Proc. of the 1994 NASA Battery Workshop* (Huntsville, AL, November 1994), pp. 581–598.

[9.3] M. Earl, T. Burke, and A. Dunnett, "Method for Rejuvenating Nickel-Hydrogen Battery Cells," *Proc. of the 1992 International Energy Conversion and Engineering Conf.* (1992).

10 Thermal Management and Reliability

This chapter discusses two general areas that are critical to the proper operation of a power system that uses nickel-hydrogen batteries, but that are most meaningfully discussed in the context of the entire power system rather than at the individual cell level. The first area, thermal management, includes dissipation and efficiency as well as thermal control methods. The second area, reliability, includes design methods for ensuring high reliability and tools for evaluating system reliability.

10.1 Thermal Management

10.1.1 Thermal Dissipation

The factors that contribute to the heat a nickel-hydrogen battery produces are
- heat dissipated within the cells
- heat dissipated by the strain gauges on cells and any strain-gauge electronics within the battery
- dissipation from current flow in the cell-to-cell connections, connectors, and other power wiring in the battery

These heat sources must all be considered when the thermal control system for the battery is designed or modeled.

The heat generated by the strain gauges and their supporting electronics can be calculated using the strain-gauge electronics design. The heat produced by current flow in the battery wiring and connectors can be accurately predicted by coupling their measured resistances with knowledge of the maximum current or the actual expected battery current profile. However, the most complex part of the battery heat-dissipation picture arises from the heat produced within the cells as they operate.

The instantaneous heat dissipation Q from within the battery cells can be predicted from the cell pressure, cell voltage, and a single thermodynamic parameter that corresponds to the enthalpy for the cell reaction. The enthalpy is often referred to as the thermoneutral voltage E_{th} for the cell, which can be used to predict cell heat generation as shown in Eq. (10.1), where I is current, V is the cell voltage, η is the charge efficiency, n is the number of electrons transferred per mole of hydrogen gas, F is Faraday's constant, R is the gas constant, T is the absolute temperature, P is the cell pressure, and t is time.

$$Q = I\eta(V - E_{th}) + IV(1-\eta) + \frac{nF}{RT}\ell n \frac{dP}{dt} \qquad (10.1)$$

The first term on the right-hand side of Eq. (10.1) corresponds to the heat released (enthalpy change) for the energy storage reaction. The second corresponds to the heat from charge inefficiency, which can be estimated from the rate of hydrogen pressure change in the cell. The final term arises from the work involved in compressing or expanding hydrogen gas within the constant volume pressure vessel.

To first order, the thermoneutral voltage of a nickel-hydrogen cell is independent of temperature, state of charge, and current, and it is typically in the range 1.48 to 1.52 V. However, as shown in Fig. 10.1, the thermoneutral voltage can exhibit subtle variations with state of charge and temperature when measured via high-precision calorimetric techniques during cell discharge. The thermoneutral potential averaged over the full range of cell state of charge is about 1.50 V from these measurements.

Equation (10.1) assumes that no heat is stored within the cell in the form of free oxygen gas, an assumption that is reasonably accurate except when entering overcharge or starting discharge following a period of overcharge. During periods of overcharge and the initial stages of discharge there can be significant errors in the instantaneous heat generation as determined from Eq. (10.1). These errors result because Eq. (10.1) ignores the accumulation of pressurized oxygen gas within the pores of the nickel electrodes and the subsequent release and recombination of the accumulated oxygen when discharge begins. These errors can be corrected

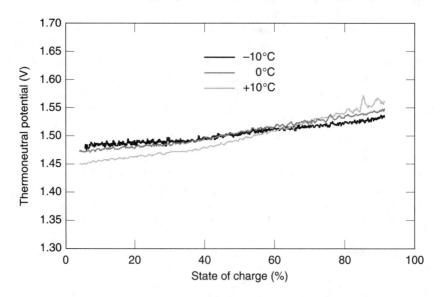

Figure 10.1. Thermoneutral potential measured by calorimetry for a nickel-hydrogen cell as a function of state of charge at various temperatures.

if the rates of oxygen evolution and oxygen recombination can be independently determined from pressure changes or from an accurate cell model. This correction involves replacing the second term in Eq. (10.1), which describes the charge or discharge inefficiency, with the sum of an oxygen generation term having an oxygen generation rate I_g and an oxygen recombination term having a recombination rate I_r. The oxygen generation term causes cooling of the cell and the oxygen recombination term generates heat. These terms are shown in Eq. (10.2), where ΔS is the entropy change for the oxygen reaction.

$$Q = I\eta(V - E_{th}) - I_g(T\Delta S) + I_r(V - T\Delta S) + \frac{nF}{RT}\ell n\frac{dP}{dt} \qquad (10.2)$$

When heat dissipation in nickel-hydrogen cells is modeled for battery thermal control system design or verification purposes, it is important to recognize that the heat generated by the terms in Eqs. (10.1) and (10.2) is not always uniformly distributed throughout the cell. The heat from all but the last term does tend to be uniformly distributed on a macroscopic scale within the electrode stack, with the exception of some of the oxygen recombination heat in cells that have recombination wall wicks on the inner walls of the pressure vessel. In this type of cell, a significant portion of the oxygen recombination heating (the third term in Eq. [10.2]) is deposited directly on the pressure vessel wall where these recombination sites are placed. The final term in Eqs. (10.1) and (10.2), which corresponds to the heat from the work of compression or expansion of hydrogen gas, thus involves the direct heating or cooling of the hydrogen gas. Therefore the thermal effects from this work term tend to be localized to the gas reservoir regions in the cell, and primarily affect temperatures measured at the domes of the pressure vessel.

Examples of the effects of the various processes contributing to cell heat generation are readily seen in the thermal signatures obtained from calorimetry measurements. Figure 10.2 illustrates the thermal signatures from the first three terms in Eq. (10.2) during a single charge/discharge cycle. The first term typically results in moderate cooling during the early portion of the recharge period when charge efficiency is high. The effects from the second and third terms in Eq. (10.2) cannot be separated during overcharge because as the oxygen is being generated it also begins to recombine after a time lag, as shown by the rise in heating during the latter stages of recharge in Fig. 10.2. The time lag arises because some time is required for oxygen transport from the generation sites in the nickel electrodes to the recombination sites in the hydrogen electrodes.

Figure 10.2 clearly shows the thermal signature that arises as a heat spike during the early stages of discharge from the third term in Eq. (10.2) as oxygen recombines at a high rate. The high oxygen recombination rate early in the discharge arises because the drop in internal cell pressure from discharge enables oxygen entrained within the porous nickel electrodes to rapidly escape and recombine on the hydrogen electrodes.

The final term in Eqs. (10.1) and (10.2), which corresponds to the heat from the work of compression or expansion of hydrogen gas in the constant volume of the pressure vessel (*PV* work), will produce heating as the hydrogen is compressed during recharge and cooling as it expands during discharge. The net heat contribution over a complete charge/discharge cycle is thus zero. The effects of this term are most visible when the temperatures of the pressure vessel domes are monitored, because this is where most of the gas resides. These effects are also greatest when the cell pressure is low, as shown at the end of discharge in Fig. 10.2. Figure 10.3 shows this last signature more clearly, where a large cooling effect is detected on the cell pressure vessel at the end of discharge, in spite of the fact that the cell stack is producing a large amount of heat from cell discharge at that time.

10.1.2 Thermal Control Issues and Requirements

The thermal control system for a nickel-hydrogen battery must be designed to maintain an appropriate operating temperature for all the cells, as well as acceptable thermal gradients across each cell and across the battery. Before discussing the design approaches commonly used for thermal control of nickel-hydrogen batteries, it is necessary to identify the preferred operating temperature ranges,

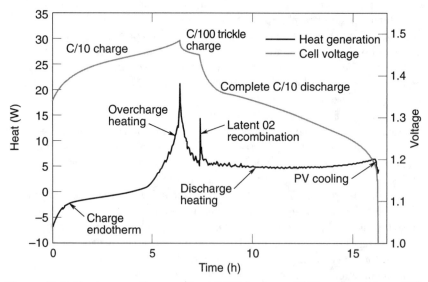

Figure 10.2. Heat-generation profile at 10°C during a C/10 recharge from 50% state of charge, followed by a 1-hour trickle charge period and a complete C/10 discharge.

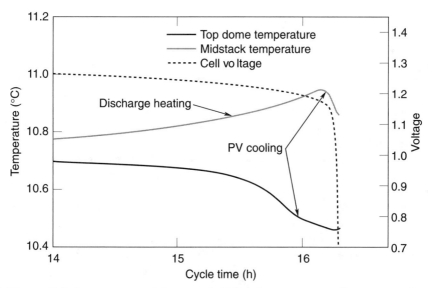

Figure 10.3. Pressure vessel dome and midstack temperature signatures at the end of a C/10 discharge for a nickel-precharged cell, in which the hydrogen gas is depleted at the end of discharge.

as well as the issues that constrain the acceptable thermal gradients within and between the cells in the battery.

Nickel-hydrogen batteries must be designed to operate at relatively low temperatures to provide good capacity and charge efficiency. Operating temperatures above 10°C should generally be avoided, because above this temperature good capacity and state-of-charge balance in a battery can be difficult to maintain. The capacity performance of cells in batteries continues to improve as temperature is reduced; however, care must be taken when operating below –5 to –10°C to avoid conditions that could cause electrolyte freezing.[10.1] Operation below –10°C is not recommended, because electrolyte freezing issues, oxygen management and popping issues, and elevated cell impedance can all potentially create cell performance problems or cell failure. The ideal operating temperature range for nickel-hydrogen batteries is between –5 and +5°C.

The thermal control system for a battery must also maintain an acceptably small thermal gradient between the warmest cell in the battery and the coldest one. A good thermal control system should keep the cell-to-cell maximum thermal gradient below 5°C. If the gradient becomes greater than this, the difference in charge efficiency between the warm cells and the cold cells makes it difficult to properly charge the battery without unduly stressing the cold cells in an attempt to keep the warmer cells fully charged. The result is either excessive overcharge stress on some cells in the battery, or acceptance of a low battery capacity that is controlled by the warmer cells.

Another danger from excessive thermal gradients between cells arises if the coldest cells fall below −10°C, particularly if they are not instrumented for temperature measurements and their low temperature is unanticipated, which makes their anomalously low temperature invisible to the charge-management system and the battery operators. In this situation, processes can occur within the coldest cells to make them unexpectedly and prematurely fail by internal short-circuiting as the result of electrolyte freezing.[10.2]

The final key thermal control requirement of a nickel-hydrogen battery is to properly control the thermal gradient within each cell. This gradient is typically measured between the hottest location within the electrode stack and the coldest surface within the cell. The need to control this gradient in cells stems from the condensation of electrolyte that can occur on cold surfaces inside the cell if this gradient exceeds a critical threshold. The critical thermal gradient threshold that can initiate water condensation is a function of the concentration of the alkaline potassium hydroxide electrolyte concentration within the cell, and is slightly dependent on the temperature of the surface onto which the water condenses, as shown in Fig. 10.4.

The critical gradient that can initiate water condensation in cells with 26% electrolyte concentration is about 6°C, and in cells with 31% electrolyte it is about

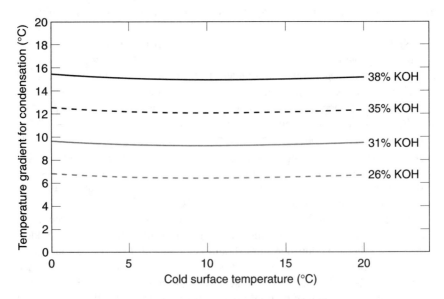

Figure 10.4. Temperature gradients needed to cause water condensation in nickel-hydrogen cells, based on vapor-pressure data from Ref. 10.2.

9°C. It is very important to keep gradients below these levels because water condensation, if it occurs over any significant time period within a cell, can lead to cell failure as a result of stack dry-out and increased cell impedance.[10.3] This is a particularly important concern if the cold surfaces in the cell are significantly below 0°C, because the condensed water can then freeze. While cells typically have wall wicks to return condensed liquid back to the stack, the rate of return by the wall wick is not rapid, and it drops to zero if the condensate freezes. Additionally, significant dilution of the electrolyte within a cell can occur under some conditions as the result of potassium hydroxide molecules being temporarily chemisorbed into the nickel-electrode active materials. The diluted electrolyte is then more prone to exhibit water condensation on cold interior cell surfaces at significantly lower thermal gradients than would otherwise be expected.

10.1.3 Battery Thermal Control Methods
The most commonly used method to remove heat from the cells in a nickel-hydrogen battery involves mounting each pressure vessel in a thermally conductive sleeve that is clamped around and thermally bonded to the pressure vessel. The sleeves are typically made from aluminum, but carbon composite sleeves have also been used to reduce the sleeve weight. The sleeve must be electrically insulated from the cell wall, generally by thin layers of plastic insulating material such as Cotherm or Kapton. If this electrical isolation is not maintained, an electrically conductive path will exist between the battery structure and the pressure vessel that can result in pressure vessel galvanic corrosion. The sleeve is generally long enough to make full contact over the length of the cylindrical section of the pressure vessel. A sleeve thickness must be chosen that is thick enough to ensure adequate thermal conductivity to maintain an acceptable thermal gradient between the pressure vessel wall and the point where the sleeve mounts to its thermal control surface. A higher heat-generation rate from the cells during operation (i.e., higher current or higher depth of discharge) requires a thicker sleeve to maintain acceptable thermal gradients.

Cell mounting sleeves can be designed to mount to a thermal control surface either at the midpoint of the pressure vessel (center mount) or at the end of the sleeve (end mount) using a mounting flange that is an integral part of the sleeve. The center mount sleeve provides shorter thermal paths and better thermal control, but does not allow cells to be as efficiently packed into as small a mounting footprint as can be achieved with the end mount configuration. For this reason, the end mount configuration is more widely used.

Nickel-hydrogen cell mounting sleeves are typically bolted onto a thermally conductive base plate that is cooled either by radiative coupling to deep space, or by heat pipes that are connected to an appropriate radiator. Although not as common, lightweight battery designs have been constructed in which heat pipes mount directly to the cell sleeves, thus eliminating the need for a base plate and its accompanying weight penalty. Additionally, cells have been designed that use carbon composite sleeves with built-in heat pipes that can be directly attached to a radiator for cooling. When either heat pipes or passive systems are used to keep the battery cool, it is common to mount appropriately sized heaters on each cell, which are thermostatically controlled to come on if the battery temperature falls below a selectable set point. Such systems are typically designed so that the active or passive thermal paths can handle the peak thermal dissipation that is predicted with an acceptable temperature rise. At times when the battery dissipation is low, the battery temperature is maintained at the heater set point by on-off cycling of the heaters. In some cases, variable-conductance heat pipes have been used to decrease the heater electrical loads during periods when the battery dissipation is low.

Other methods have been adopted to cool nickel-hydrogen batteries, including the use of louvers to control the amount of deep-space cooling seen by the battery radiators. Additionally, in louver-based thermal control systems, battery cooling has been designed based on passively balanced conductive coupling to a radiator. Such systems will work well if the conduction paths and all interfaces, surfaces, and thermal environments remain at or near their design point throughout life. However, it is frequently found that surfaces (from contamination) and environments will change over many years in space, resulting in significant thermal drift in thermally unregulated systems that operate based on passive thermal balance.

The reliability of the components used in the battery thermal control system is critical to achieving good battery performance over life. The passive components, such as sleeves, base plates, and radiators generally do not have reliability problems. The thermal reliability issues are more frequently associated with the thermal interfaces between the components. The thermal interface design must be carefully qualified by exposure to thermal cycling, shock, and vibration extremes that are well outside the normal battery operating extremes. The workmanship of the thermal interfaces must be validated for each battery by exposure to acceptance-level environments that encompass the worst-case expected battery operational environments. Verification that all thermal interfaces remain intact is best done by thermal vacuum testing, and by demonstrating repeatable battery capacity measurements over the extremes of temperature through which the battery is required to operate. Components such as heat pipes, relays, thermostats, and heaters must have proven reliability, and tested performance, and they should always be designed into the battery system with sufficient redundancy so that the failure of any one component (e.g., heat pipe, heater circuit, relay, etc.) will not result in battery failure or degradation that would compromise the particular mission that the power system supports.

10.2 Other Battery Components

10.2.1 Cell Bypass Devices

Cell bypass relays and cell isolation switches have been used in batteries to ensure that the failure of a single cell (open-circuit or short-circuit) cannot cause the entire battery to fail. This is required in power systems that have only a single battery, but is often not required in multibattery power systems that are sufficiently oversized to tolerate the loss of one battery without compromising the mission. Cell bypass relays are placed in parallel with a cell, but are open unless an anomalous voltage occurs across the cell. If the cell goes to a high-impedance condition (failed open), the high positive voltage (on recharge) or extremely negative voltage (on discharge) can be used to forward bias diodes that can thermally activate closure of the bypass relay. When the power system uses a relatively low charge rate, bypassing cells in the charge current direction is not necessary because the current and heating can be simply handled by the forward biased diodes. This significantly reduces any risk associated with bypassing a fully charged cell. When a bypass relay is activated during battery discharge, the cell must be at a voltage and state of charge low enough to ensure that the cell does not produce a large surge current. Thermal activation by a diode can turn on a bypass relay when the cell voltage falls below -0.7 V, a voltage that assures the cell is fully discharged.

An alternative to bypass relays uses an isolation switch that removes a cell from a series battery circuit, leaving the battery circuit intact with $n - 1$ cells in series. An isolation switch can take a cell out of the battery even when the battery is fully charged without short-circuiting the cell. Passing a firing current through a fusible element that, when fired, releases a spring-driven contact assembly is the mechanism that activates the isolation switch. Isolation switches can only be fired once, resulting in permanent removal of the cell from the battery, and cannot be reset.

Cell bypass relays or isolation switches can be designed for activation by command from the ground. This capability can be useful if a cell fails because of a short circuit, but still has a high enough internal impedance to put a significant load in series with the other cells in the battery (yet does not automatically activate the bypass relay or isolation switch). Manual activation will eliminate the load from the failed cell as well as the anomalous heating that it produces in this situation. If manual-commanding capability for bypass or isolation devices is designed into the battery, appropriate safeguards must be put into place to prevent a single command or signal from inadvertently bypassing or isolating a good cell, or bypassing a cell containing significant stored charge. The reliability of the bypass relays, isolation switches, and their control system must be included in the analysis of the battery and power system reliability.

10.2.2 Cell Pressure Vessels

The pressure vessel in which a nickel-hydrogen cell is contained is a critical component of cell reliability. The overriding concern with a pressure vessel is that it could develop a leak, causing the cell to fail by losing hydrogen gas. Cell pressure vessels are made from Inconel (which provides leak-before-burst behavior) and are designed to have a vessel wall thick enough that it should not develop a leak at normal cell pressures unless there is a critical flaw in the pressure vessel material or one of its weldments (girth weld, terminal weld, or fill tube weld). Most cell designs use a pressure vessel having a margin of safety more than 2.5 to 1 between the normal maximum expected operating pressure and the burst pressure. However, screening pressure vessels and cells for all flaws that could eventually cause a leak, as well as screening for outright leaks, is a key part of ensuring high reliability for nickel-hydrogen cells.

Nickel-hydrogen cell pressure vessels, which are hydroformed from Inconel sheet stock, must be 100% inspected for surface flaws, contamination, or cracks that could propagate into leaks as the vessel is pressure-cycled during cell operation. The pressure vessels are thoroughly cleaned of all greases, lubricants, and surface scale. They are then inspected using special dye-penetrant methods qualified to detect any flaw equal to or greater than the critical size that could propagate into a leak during cell cycling.

The critical flaw size that must be detected is a function of the vessel wall thickness, and it is determined from an appropriate stress-mechanics analysis. Any pressure vessel found to have a critical-sized flaw must be rejected. Additionally, statistical analysis of the number of flaws found in any single lot of pressure vessels should be done to evaluate the probability that a critical flaw could go undetected. As more flaws are found in any single vessel lot, the probability of not detecting a critical flaw rises. If this probability becomes unacceptably high, the entire lot of pressure vessels should be rejected.

Pressure vessel welds create several areas where inspection and verification of reliability is a challenge. Voids within welds cannot be reliably detected by surface dye penetrant inspection, but are detected best using X-ray screening methods. Gas bubbles, or any of a variety of contaminants in the weld area, can cause voids. For electrolyte fill tubes, contamination can come from residual potassium hydroxide electrolyte in the fill tube weld area. During welding, the potassium hydroxide vaporizes to leave behind a variety of voids. These voids can cause a cell gas leak and subsequent cell failure, either at the beginning of cell life or after many years of cell operation. Such voids can be detected by X-ray inspection, and if they are found to be unacceptably large, the welds can be reworked or the cells can be rejected. The best way to eliminate voids in the fill tube welds is to ensure that all potassium hydroxide contamination is cleaned from the weld area prior to welding.

The girth welds and terminal welds are also subject to flaws caused by contaminants in the weld areas. The most common contaminants that have been observed to cause cell leaks are particulates of zirconium oxide (from separator edges) or sulfur-based lubricants on the surfaces (from hydroforming processes). Because these types of contaminants are typically not easily seen by visual or X-ray inspection, it becomes very important to make sure that good cleaning procedures are in place. The surface regions of all parts to be welded should be carefully cleaned to remove any particulates before welding. If this is not done, weld flaws and cell leaks will eventually occur.

Another problem can arise if the sulfur-based lubricants used in the hydroforming process are not properly cleaned from the Inconel prior to welding. If such lubricants are in the heat-affected zone of the weld, the sulfur can react with the Inconel metal to produce a sulfide inclusion or flaw that can work its way through the vessel wall to cause a leak when the vessel is pressure-cycled during operation. Therefore, such lubricants must be cleaned from the Inconel surfaces before welding. Recommended cleaning procedures typically involve ultrasonic cleaning of the Inconel parts in a cleaning solution at appropriate agitation intensity.

After nickel-hydrogen cells have been fully constructed, but prior to activation with electrolyte, they are customarily checked for leaks using a helium-leak–detection procedure. This is a highly sensitive method for detecting any existing leak, and any cell found to have a detectable leak rate (typically greater than 10^{-7} standard cm^3/sec) should be rejected. However, this method will not detect flaws that could cause a leak later in life after operational pressure cycling, and it does not inspect the weld sealing the fill tube following cell activation, because the open fill tube is used for the helium leak test. Leak inspection methods involving hydrogen gas sniffing may be used after completed cells have been charged to operational pressures; however, such methods are generally relatively insensitive. The most sensitive leak-detection method in a finished cell is probably the observation of a steadily decreasing capacity from repeated standard capacity tests.

10.2.3 Terminal Seals

Several types of terminal seals are found in nickel-hydrogen cells. The Zeigler seal, which is a molded nylon seal that is compressed around the terminal post within an Inconel boss assembly, has been used for many years. Another terminal seal in common usage is a seal with a Teflon compression fitting. A third terminal seal that utilizes a ceramic-to-metal brazed feed-through seal is also used in some nickel-hydrogen cell designs.

10.2.3.1 The Zeigler Seal

The Zeigler seal has provided a robust terminal seal in nickel-hydrogen cells from the early COMSAT designs through the state-of-the-art ManTech designs used today. The key to ensuring a good Zeigler seal is to provide adequate compressive force where the molded nylon seal material contacts the metal surfaces of the terminal and the boss assembly that holds the seal and the terminal. The compression of the nylon must be sufficient to maintain seal integrity at the lowest temperature that the cell may experience. Because the metal parts of the terminal seal preferentially contract as the temperature is reduced, lower temperatures will reduce the compressive forces in the seal area. Additionally, the compressive forces will undergo gradual relaxation over time as a cell ages. For these reasons, it is important to qualify nickel-hydrogen cells involving thermal cycling at temperatures at least 10°C less than the minimum expected operating temperature for the cell.

The geometry and finish of the metal seal surfaces on the terminal and the boss over which the nylon Zeigler seal is molded are important to ensure a high-reliability seal. Any deep grooves or scratches on these surfaces can potentially provide a leak path if the nylon cannot be sufficiently compressed into these features. Additionally, threaded terminal posts must have smooth and well-machined threads into which the nylon can be molded and compressed.

The Zeigler seal, while mechanically strong, cannot withstand temperatures high enough to begin softening the nylon seal material. Temperatures greater than 200°C can make this material soften or melt, and will result in the release of hydrogen gas from a nickel-hydrogen cell. Exposure to temperatures above the glass transition of nylon (about 50°C) can accelerate the relaxation of the compressive forces within the seal. For these reasons having to do with terminal seal reliability, nickel-hydrogen cells should not be exposed to temperatures greater than 45°C during any test procedures.

10.2.3.2 The Teflon Seal

Another type of terminal seal that has been extensively used in nickel-hydrogen cells, starting with the early Air Force cell design, is the Teflon seal. This seal utilizes a threaded terminal nut to compress a Teflon washer against the sealing surfaces of the terminal post and the boss. The Teflon washer undergoes cold flow to conform to and compress against the metal seal surfaces. This type of seal may be released and retightened after the cell has been built.

At the lowest extremes of temperature, the Teflon terminal seal can also suffer from an inadequate rate of Teflon cold flow to compensate for the differential thermal contraction of the metal parts in the seal. Thus Teflon seals can leak when exposed to low temperatures, and then reseal when the temperature is increased. Again, it is important to qualify the seals by testing them in cells at temperatures well below (at least 10°C below) the lowest expected cell operational temperatures. Because Teflon is chemically stable to relatively high temperatures, few problems have been reported as the result of exposure of this type of seal to high temperatures of cell operation.

Teflon seals are extremely susceptible to any damage to the metal or Teflon surfaces that must mate to provide the seal. Additionally, any debris or foreign particles in the seal region will generally result in a leaking seal. It is critical to maintain clean and defect-free surfaces for the Teflon seal to provide the high reliability needed for nickel-hydrogen cells.

10.2.3.3 The Ceramic Seal

The final type of terminal seal that has been and continues to be used in nickel-hydrogen cells is a brazed ceramic-to-metal seal. A ceramic insert is brazed to the boss and the terminal post, thus providing gas-tight electronic insulation between the terminal post and the pressure vessel. This type of seal is reliable if properly fabricated. It requires that the brazing be properly done so that all mating surfaces are fully wetted and filled with the braze material. It suffers from the susceptibility of the ceramic insulator to cracking if it is overstressed or exposed to excessive shock or impact. With proper handling and inspection of cells and seals, ceramic seal cracks should not have a major impact on cell reliability.

10.3 Cell Reliability

The long-term reliability of nickel-hydrogen cells is normally controlled by the rate of wear-out of the nickel electrodes in the cells. If other components such as seals, pressure vessels, hydrogen electrodes, and separators are properly screened and inspected, and if stack assembly is properly controlled, these other components should not have a large influence on long-term cell reliability. Early life performance, acceptance testing, and cell matching criteria are generally capable of detecting and eliminating most of the problems that can arise from defective components.

The nickel electrodes are the only component in the individual pressure vessel cell that is subject to significant degradation and wear processes, and that can thus significantly affect cell reliability during normal operation. These processes include the gradual corrosion of the metal sinter in the nickel electrodes, electrode swelling, and extrusion of active material into the separator. These degradation processes are generally lumped into an overall wear-out rate that can vary with cell design or operating environment, which can be integrated to reflect how the total wear from all degradation processes accumulates over the lifetime of any particular cell design.

While the nickel electrodes are responsible for the primary wear processes, other components and design features, such as the number of separator layers, the amount and concentration of electrolyte in the cell stack, and the thermal control system, can significantly impact the susceptibility of a cell design to the changes caused by the nickel-electrode wear. Similarly, different nickel-electrode designs (i.e., those with varying sinter porosity, loading level, or thickness) can cause significant differences in how readily nickel electrode performance is degraded by the normal wear processes.

The overall wear rate of individual pressure vessel nickel-hydrogen cells has been found from statistical analysis of multiple life tests to depend on (in order of decreasing significance): the amount and rate of overcharge, the operating temperature, the depth of discharge, and pure operational calendar life.[10.4] The reliability of a battery is governed by the cell reliability as a function of life, the statistical variation in the reliability of the cells throughout the entire battery, and the redundancy built into the battery to accommodate one or more degraded cells or cell failures. The reliability of an entire system containing multiple batteries is controlled by the level of redundancy built into the system to tolerate degradation or loss of a battery. This section examines how each of these aspects of battery system reliability can affect the likelihood of failure from accumulated cell wear over its lifetime.

The reliability of an individual nickel-hydrogen cell is controlled by the rate at which wear accumulates during cell operation over its lifetime. The rate at which wear accumulates depends on how the cells are operated. Figure 10.5 shows several examples of wear accumulation for differing conditions of operation. In Fig. 10.5, 100% wear corresponds to the level of wear damage that results in the failure of 50% of the cells in the battery. The results in this figure illustrate that, while the nickel-hydrogen cells exhibit long life, quite a wide variation in lifetime is anticipated depending on the depth of discharge and the stresses of accumulated overcharge over years of life.

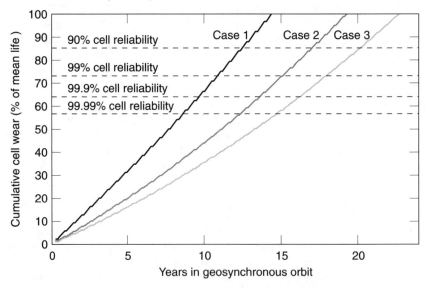

Figure 10.5. Wear accumulation for nickel-hydrogen cells in different geosynchronous satellite operating conditions. Case 1 is 65% peak depth of discharge with a C/100 trickle-charge rate; case 2 is 72% peak depth of discharge with a C/200 trickle charge rate; case 3 is 65% peak depth of discharge with a C/200 trickle-charge rate.

10.3 Cell Reliability

The examples of Fig. 10.5 were evaluated for a ManTech cell design that contains two layers of Zircar separator and 31% electrolyte concentration. Other cell designs that possess either lesser or greater robustness to wear may tolerate significantly different amounts of wear before reaching the 50% failure probability threshold. For example, a cell in a single pressure vessel nickel-hydrogen battery (22 cells within a single pressure vessel) appears to reach the mean probability of failure after accumulating only 77.6% of the wear damage that can be tolerated by the ManTech cell design shown in Fig. 10.5.[*] Analysis of life-test data for single-layer Zircar separator cells with 26% electrolyte concentration indicates that this design has tolerance to wear similar to that shown for the design in Fig. 10.5. Thus, key cell-design changes can make cells either more or less tolerant to the wear damage that accumulates over life. The rate of wear and the critical thresholds where each cell design is expected to fail are best determined from statistical analysis of life tests for each cell design.

An important aspect of cell reliability is the variability observed in the wear threshold over an ensemble of cells tested under identical electrical and environmental conditions. Some cells in this ensemble will fail before reaching the 100% wear threshold shown in the example of Fig. 10.5, while others will continue performing acceptably well past the 100% wear level. This ensemble variability can be statistically extracted from life-test results on groups of battery cells. For nickel-hydrogen cells the ensemble variability in life has been found to be essentially invariant over operational conditions and environments, giving a lifetime standard deviation across the ensemble that is 11.63% of the average lifetime. If this cell-to-cell lifetime variability is included in the cell reliability analysis, the cell reliability can be predicted as a function of cell age, as is shown in Fig. 10.6 for the wear accumulation conditions in Fig. 10.5.

In Fig. 10.6, the sharpness of the reliability drop-off as a cell reaches its end of life is controlled by the ensemble standard deviation in cycle life. Typically, cells of a given design that are built more reproducibly, and thus have less cell-to-cell variability in cycle life, will produce higher-reliability batteries. Conversely, cells that have more variability will give lower-reliability batteries, assuming all other variables are equal.

Battery reliability may be determined from knowledge of how cell reliability changes with age, as is shown in the curves of Fig. 10.6, as long as the number of cells allowed to fail before the battery fails to support its required voltage or power performance is known. Batteries may be designed to fail when the first cell fails in power systems that contain a sufficient number of batteries that the loss of one battery can be tolerated by the power system. Significantly improved

---

[*]The relative wear rate of single pressure vessel batteries is based on statistical analysis of the SPV batteries in packs 3003L and 3004L from the NWSC Crane test database, compared to similar statistical analysis from Ref. 10.5 for ManTech cells.

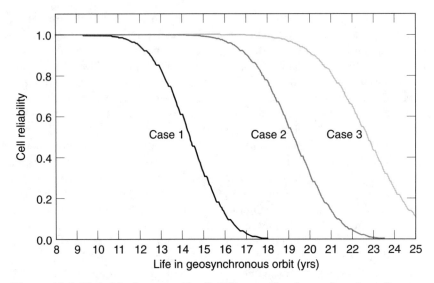

Figure 10.6. Nickel-hydrogen cell reliability predicted as a function of operating life in a geosynchronous satellite for the conditions shown in Fig. 10.5.

system reliability is obtained if the battery is designed to accommodate the failure of one or even two cells, and still provide adequate power to support the system. Cell-level redundancy in a battery is frequently used to improve overall battery system reliability.

Alternative design rules for power systems with a single battery (or too few batteries to allow normal operation with one failed) typically allow two or more cells to fail before the battery voltage falls to levels too low to support the needs of the power system. Examples of how these different cell-redundancy scenarios affect battery level reliability are shown in Fig. 10.7. These examples clearly demonstrate the battery reliability improvement that can be realized by including enough cells in the battery so that several may be lost before the battery ceases to provide adequate voltage or power.

10.4 Cell Design Guidelines

Software cell design packages have been used to capture some of the reliability guidelines and cell wear-out principles discussed in the previous section. One of these design packages[10.5] allows an appropriate cell design to be easily chosen for the required power system life and performance. Once an appropriate cell type is chosen, the cell capacity, depth of discharge, battery configuration, and operating temperature can be chosen to be consistent with the required cell cycle and calendar lifetime. This software allows a simple wear model,[10.6] such as that shown in Fig. 10.8 as determined from life tests on a range of cell designs, to be used with

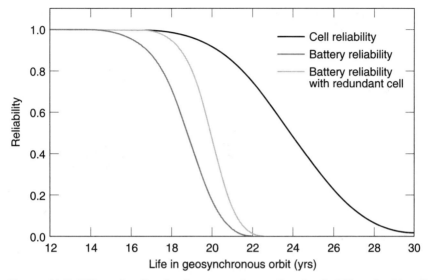

Figure 10.7. Effect of cell redundancy on the predicted reliability of a 22-cell nickel-hydrogen battery as a function of operating life in a geosynchronous satellite (case 2 conditions in Fig. 10.5).

any desired power system architecture to enable selection of a cell of the appropriate type and size to meet all power system requirements.

Cell designs that correspond to the lower-reliability curves in Fig. 10.8 may be inexpensive commercial designs that have a lower level of quality control associated with their production, or they may be lightweight designs, common pressure vessel designs, or single pressure vessel designs. Lightweight cell designs typically reduce weight by using lower-porosity nickel sinter or higher loading levels in the nickel electrodes, both changes that make nickel electrode performance decrease more rapidly as wear accumulates. Lightweight cell designs may also reduce separator thickness and thus decrease the quantity of electrolyte in the cell, a change that also hastens cell failure in response to wear-related changes over life.

Common pressure vessel cell designs are subject to several added degradation and failure modes not present with individual pressure vessel designs:

- migration of the oxygen generated in one of the two common pressure vessel stacks to recombine in the other cell stack
- ionic conduction of parasitic currents from one stack to the other if an electrolyte bridge occurs between the two stacks
- condensation of water from a thermal gradient that transfers liquid from one stack to the other

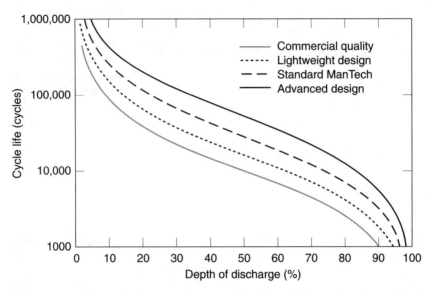

Figure 10.8. Life prediction from wear model of Ref. 10.7 as a function of depth of discharge for four nickel-hydrogen cell designs. Reprinted courtesy of NASA.

These processes can cause imbalance in the amount and concentration of electrolyte as well as major state-of-charge imbalance between the two stacks in the common pressure vessel cell. These types of problems are typically accentuated by greater overcharge or thermal gradients, and they make such cells more sensitive to charge-control system or thermal control system deficiencies than are individual pressure vessel cells.

Single pressure vessel nickel-hydrogen batteries are in principle subject to the same issues associated with gas and electrolyte transfer between the cell stacks as in common pressure vessel cells. However, single pressure vessel batteries typically include design features that help minimize water or oxygen transfer between any of the cells in the single pressure vessel battery, and help to maintain thermal balance between cells in the single pressure vessel battery. Nevertheless in general the reliability levels for those systems that have cells in shared pressure vessels can at best only approach the reliability level for an individual pressure vessel cell of similar stack design.

It should also be recognized when selecting a cell design and capacity for a particular application that each cell manufacturer designs its cells with some added capacity margin beyond that required. This is to ensure that all the cells comfortably meet the capacity and power requirements over the entire operating temperature range. Such margin must accommodate the cell-to-cell variability within each production lot, as well as the variability in average capacity that can occur from lot to lot, and the variability that can occur during cell storage and handling. Greater variability within and between cell lots typically causes a manufacturer to make cells having greater capacity margin, thus guaranteeing that cell rejections seldom occur because of insufficient capacity. It is important when designing a power system to recognize that the margin between the rated capacity and the actual capacity of the cells is there to ensure high cell reliability and should not be included as additional power free for use in the normal power system operation.

Each nickel-hydrogen cell manufacturer also has its own design guidelines for cell sizing and the details of how the cells are assembled. These guidelines typically depend on the requirements specified for the cell and the battery in which the cell will be used. These guidelines are usually based on experience with cell and cell lot reproducibility, life-test data, and cell-level performance data. Adhering to such experience-based guidelines is important for ensuring a reliable nickel-hydrogen cell design.

10.5 Cell Safety

Nickel-hydrogen has historically been one of the safest types of battery cells used to power satellites. The primary reason for this safety record is the fact that nearly all conditions of cell abuse result in a decrease in cell pressure and loss of stored energy before the cell can go into a runaway condition. This section discusses the response, in terms of battery and system safety, of nickel-hydrogen cells to the common conditions of abuse or environmental extremes to which they may be subjected.

Exposure of a nickel-hydrogen cell to temperature extremes generally does not result in an unsafe condition. Low temperature extremes (those below the electrolyte freezing point) typically stop all electrochemical activity in the cell. High temperature extremes can, in principle, result in thermal runaway of nickel-hydrogen cells. However, in practice, as nickel-hydrogen cell temperature rises, the self-discharge loss rate accelerates at temperatures so far below the point where active materials experience runaway thermal decomposition (~90°C for charged nickel electrodes[10.7]) that very little stored energy remains to support a violent thermal decomposition when the cell temperature does attain the critical decomposition threshold. No instance has been documented where the Inconel pressure vessel of a nickel-hydrogen cell has been breached by rapid thermal decomposition that was initiated by high temperature exposure or cell short-circuiting.

However, not surprisingly, some safety hazards are associated with exposure of nickel-hydrogen cells to high temperatures. The seal materials in nickel-hydrogen cells are nylon, Teflon, or brazed ceramic, depending on the seal design. The ceramic seal is stable to high temperature exposure from a safety point of view.

The nylon seal, if heated above the melting point of nylon (about 230°C), can release the hydrogen gas from the pressure vessel into the surrounding air, thereby creating the potential for an explosive mixture or hydrogen fire if an ignition source is present. Such heating can result from either a catastrophic cell short circuit, or from external exposure of the cell to high temperature. Cell short circuits typically have not created a significant hazard because by the time the terminal temperature is hot enough to melt the seal, very little hydrogen gas remains to leak from the pressure vessel.

External heating, if rapid enough, can melt nylon seals and cause the escape of significant amounts of hydrogen gas; several instances of this have been observed. The external heating situations that have occurred have been the result of runaway heating caused by the failure of a thermal control chamber, or by direct exposure of a cell to flame. Teflon seals have a much higher softening point (about 450°C), and thus are much less likely to release hydrogen gas during extreme conditions of heat exposure. Teflon seals typically will leak hydrogen at relatively low rates when thermally abused. No instances have been reported where Teflon seals have catastrophically failed and suddenly released large amounts of hydrogen gas from a nickel-hydrogen cell.

A somewhat related safety condition could result from internal thermal runaway, or the ignition of a hydrogen/oxygen flame internal to the cell pressure vessel. This condition may in principle be caused by runaway overcharge, which can generate large amounts of oxygen within the high-pressure hydrogen gas in the cell. As these gases recombine (often in explosive internal popping events) the temperature of a cell can be driven up to a point where thermal runaway is possible. Such popping events typically result in a hard short circuit in the cell stack from physical damage to the separator, causing the stored cell capacity to run down during a high-temperature cell excursion.

These types of situations have occurred for nickel-hydrogen cells, but have never been observed to breach the Inconel pressure vessels used in modern nickel-hydrogen cells. However, in such cases cell disassembly has frequently revealed melted stack components as well as evidence of hydrogen/oxygen flames that have burned either explosively or nonexplosively within the cells. The observed safety of the cells during these types of events probably is largely a result of the high strength and the leak-before-burst characteristics of the Inconel pressure vessel. It should be noted that early nickel-hydrogen cell designs used in Soviet Union satellite applications, which utilized stainless steel pressure vessels, had a history of pressure vessel explosions in response to runaway overcharge. However, catastrophic pressure-vessel containment failure in response to extreme overcharge has not been observed when using Inconel pressure vessels in any cases with the COMSAT, USAF, or ManTech cell designs.

10.6 References

[10.1] A. H. Zimmerman, "New Low-Temperature Failure Mode for Nickel Hydrogen Cells," *Proc. of the 2006 Space Power Workshop, The Aerospace Corporation* (Manhattan Beach, CA, April 2006).

[10.2] M. Earl, T. Burke, and A. Dunnett, "Method for Rejuvenating Nickel-Hydrogen Battery Cells," *Proc. of the 1992 International Energy Conversion and Engineering Conf.* (San Diego, CA, August 1992).

[10.3] P. Bro and H. Y. Kang, "The Low-Temperature Activity of Water in Concentrated KOH Solutions," *J. Electrochem. Soc.* **118**, 1430 (1971).

[10.4] A. H. Zimmerman and M. V. Quinzio, "Model for Predicting the Effects of Long-Term Storage and Cycling on the Life of NiH_2 Cells," NASA/CP-2005-214190, *Proc. 2003 NASA Battery Workshop* (Huntsville, AL, 20 November 2003).

[10.5] A. H. Zimmerman and L. H. Thaller, "Expert System for Nickel Hydrogen Battery Cell Diagnostics," NASA/CP-1999-209144, *Proc. 1998 NASA Battery Workshop* (Huntsville, AL, November 1998), pp. 297–316.

[10.6] L. H. Thaller and H. S. Lim, "A Prediction Model of the Depth-of-Discharge Effect on the Cycle Life of a Storage Cell," *Proc. of the 22nd International Energy Conversion and Engineering Conf., Vol. 2* (Philadelphia, PA, 10–14 Aug 1987), pp. 751–757.

[10.7] S. W. Donley, J. H. Matsumoto, and W. C. Hwang, "Self Discharge Characteristics of Spacecraft Nickel Cadmium Cells at Elevated Temperatures," NASA/N87-11072 02-33, *Proc. 1985 NASA Goddard Space Flight Center Battery Workshop* (Greenbelt, MD, November 1985), pp. 195–214.

11 Cell and Battery Test Experience

Nickel-hydrogen cells and battery packs have been life-tested under a wide range of real-time and accelerated operating profiles since they were first developed in the early 1970s. Over the years, these tests have provided a significant database allowing the performance and life of many of the major nickel-hydrogen cell types to be defined reasonably well, including COMSAT cells, ManTech cells, dual-anode cells, common pressure vessel cells, and single pressure vessel cells. Cell capacities have ranged from 6 to 350 A h.

Design variations for these cell types have included single or double layers of separator; potassium hydroxide electrolyte concentrations from 26% to 38%; cell diameters of 3.5, 4.5, and 5.5 in.; dual-stack configurations; recombination wall wicks; axial and rabbit-ear terminals; and various low-resistance electrode lead and terminal configurations. Cells have been life-tested at depths of discharge between 10% and 90%; temperatures from –10 to 20°C; and a wide range of recharge rates, recharge voltage limits, and recharge ratios. The life tests have included cycle throughput that has ranged from about 1 cycle per day (in accelerated geosynchronous satellite profiles) up to 32 cycles per day (in accelerated low Earth orbit satellite profiles).

All these design variations and different operating conditions affect cell and battery cycle life to some degree. For this reason it is important to understand how the test data for each cell type indicate both the susceptibility of that design to the wear processes, as well as how rapidly any specific operating environment causes wear to accumulate within the cell. In this chapter, the range of performance changes that have been seen in tests of nickel-hydrogen cells is reviewed, and the physical and chemical changes within the cells that are responsible for these changes are discussed. This test database can be used to empirically model the wear rates and operational lifetimes of nickel-hydrogen cells,[11.1] as was discussed in chapter 8. It can also help validate first-principles models of nickel-hydrogen cell performance.

Much of the work done to develop the life-test database for nickel-hydrogen cells that is publicly available for analysis has been conducted at the Naval Weapon Support Center (NWSC) in Crane, Indiana,[11.2,11.3] and has been sponsored by NASA and the U.S. Air Force. In addition, other U.S. government agencies have sponsored these tests over many years in support of their satellite and spacecraft development programs. Life tests have also been conducted at facilities of cell manufacturers (Hughes/Boeing, EaglePicher, Gates Energy Products, Yardney), as well as at The Aerospace Corporation, Phillips Labs, major NASA centers, and facilities of major aerospace contractors (Lockheed/Martin-Marietta, Boeing, TRW/Northrop-Grumman, and Ford Aerospace/Loral). While not all of these data are publicly available, data from many of the government-sponsored life tests provided representative data sets that were available for review and detailed discussion here.

The review and discussion of the life-test data for nickel-hydrogen cells that is provided in this chapter is organized into sections that focus on each of the major variables that influence cell lifetime, or the major wear-related signatures seen during cell operational life.

11.1 Cell Discharge Voltage Signatures

The key performance characteristic of a battery cell that must be maintained during its discharge is a voltage level adequate to support the loads connected to the power system in which the battery operates. A number of factors influence the discharge voltage behavior and cause it to degrade or improve over the lifetime of the battery. These factors can include capacity loss, increases in impedance, reconditioning effects, and performance degradation from operation at high or low temperature extremes. This section discusses the signatures associated with major changes in discharge voltage for nickel-hydrogen cells as they cycle and degrade over their operational lifetime.

11.1.1 Capacity Loss Signatures

A common cause for degraded performance in a nickel-hydrogen battery is loss of capacity in one or more of its cells. The lowest-capacity cell in the battery typically limits battery performance; thus minimizing capacity loss in all cells is important for maintaining good performance. Capacity loss over life typically occurs for two reasons related to nickel electrode wear. First, the nickel electrode degradation results in an increasing amount of the stored capacity becoming impossible to discharge at usable voltage levels. It is not unusual when this occurs to see a significant amount of capacity discharged at lower, second-plateau levels. The second consequence of nickel electrode wear is a decrease in charge efficiency, which results from both chemical corrosion processes that occur over time as well as physical swelling and damage that accumulate in the nickel electrodes during long-term use.

While there is no real method to recover from the first form of capacity loss, it is possible to temporarily recover capacity lost to reduced charge efficiency by increasing the amount of overcharge. However, that is only a temporary solution, because greater overcharge increases the wear rate and will drive cells faster toward their eventual failure, whether from capacity loss, electrolyte dry-out, or short-circuiting.

The performance signature seen when a nickel-hydrogen battery experiences capacity loss is very similar to that seen from most forms of degradation, which is rapid loss of voltage toward the end of discharge. Because battery discharge early in life typically does not occur to a great enough depth of discharge to drive any cells into capacity depletion, capacity loss is frequently not detected until it becomes fairly significant later in life. Periodic reconditioning of a battery[*] will

---

[*]Reconditioning involves a complete battery discharge at a relatively low rate, and thus enables the capacity of the cells in the battery to be measured and trended.

enable the degradation in the capacity of the cells to be measured and its trending to be noted as the cells age. An example of the typical trend in cell reconditioning behavior over life is shown in Fig. 11.1, where the cell was periodically reconditioned during a long-term life test. This figure shows a typical capacity reduction signature as the cell ages. What is not shown in Fig. 11.1 is the growth of a lower-voltage discharge plateau as some of the lost capacity transitions from the upper usable voltage level to a level below 1 V that has very limited usability.

Other changes in cell performance typically accompany the capacity loss shown in Fig. 11.1. These changes, discussed later in this chapter, can include an increase in the upper-plateau voltage during charge or discharge, as well as a significant increase in internal cell hydrogen pressure. The increases in upper-plateau voltage during discharge and recharge, as well as the hydrogen pressure increase, arise from the chemical changes that accompany sinter corrosion in the nickel electrode.

11.1.2 Increased Cell Impedance and Electrolyte Dry-Out

The internal resistance of a nickel-hydrogen cell can increase over life if there is not enough electrolyte in the cell to keep the separator adequately wetted as the nickel electrodes undergo swelling and corrosion during years of operation. Nickel-

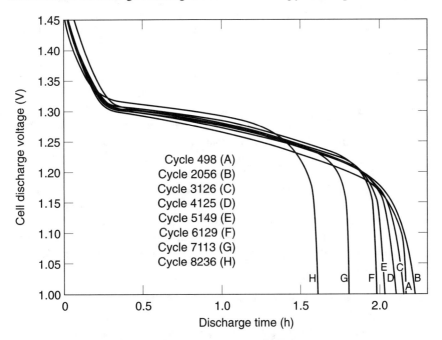

Figure 11.1. Typical variation in capacity through the life of a nickel-hydrogen cell when operated in a life test at 80% depth of discharge.

hydrogen cell designs that have two layers of Zircar separator typically do not display this signature because the two layers of separator provide a large enough reservoir of electrolyte to compensate for dry-out as nickel electrodes expand or corrosion occurs as cells age. However, even a cell with two layers of Zircar separator can exhibit a dry-out signature if significant electrolyte migration or water condensation occurs within the cell because of thermal gradients. Other separator designs, specifically single-layer Zircar or compressible separators, such as nonwoven nylon or asbestos, are the most likely designs to display dry-out–related performance signatures and degraded cell lifetime as the result of the dry-out.

Electrolyte dry-out can become a problem and lead to significant increase in cell impedance if the percentage of the pore volume in the separator that is filled with electrolyte becomes reduced to 25–30% or less by nickel electrode degradation or electrolyte migration processes. At this level, the fibers of separator material begin to lose the continuous electrolyte conduction paths across the separator, and thus lose the capability to effectively dissipate the concentration gradients that form between the positive and negative electrodes during high-rate charge or discharge. This condition will lead to increased resistance, and thus a depressed voltage plateau during high-rate discharge, and an increased voltage during high-rate recharge. This performance signature is shown in Fig. 11.2 for nickel-hydrogen cells both before and after extended operation in a life test over ~10 yr of operation at 60% depth of discharge.

As the amount of electrolyte within the separator is further decreased, the cell will eventually lose the capability to support high-rate operation for an extended duration because of reduced electrolyte transport capability and poor conduction through the separator, as shown in the failed state in Fig. 11.2. This condition is referred to as "diffusion-limited" cell performance, and is associated with a critical threshold of cell current that can be supported by the diffusion of electrolyte. At currents above this threshold, cell voltage will collapse to a low value and the cell will fail because the electrolyte concentration drops to zero at one side of the separator. If a failure occurs because of diffusion-limited current, a rest period that allows the electrolyte concentration gradients to equilibrate can enable the recovery of significant additional capacity. At currents below this diffusion-limited current threshold, the cell can continue to operate; however, the performance will be degraded significantly because of higher-than-normal cell impedance.

11.1.3 Deconditioning and Reconditioning Signatures

An important discharge voltage signature commonly seen during the testing and operation of nickel-hydrogen cells is often referred to as "deconditioning." This is a type of "memory effect" whereby nickel electrodes appear to remember the depth of discharge to which they are repeatedly cycled. The capacity that has been

11.1 Cell Discharge Voltage Signatures

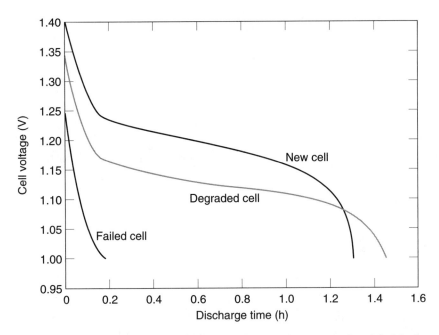

Figure 11.2. Degradation in the C-rate capacity performance of a nickel-hydrogen cell with a single layer of Zircar separator as it degrades and fails during 10 yr of operation in a long-term life test.

regularly cycled between the maximum depth of discharge and the full charge state discharges at a normal level, and all the capacity that has not been regularly cycled because it is beyond the maximum depth of discharge will discharge at a lower voltage level (typically about 40 mV lower than normal). Figure 11.3 shows the normal cell discharge voltage behavior, as well as the discharge voltage behavior that develops after about 2000 cycles to a 50% depth of discharge. The problems that can result from deconditioning include inadequate battery voltage if the power system is not designed with sufficient margin to tolerate the 40–50 mV cell voltage depression that can develop at the maximum depth of discharge, or the added heat generation associated with the depressed discharge voltage.

The deconditioning of the discharge voltage is caused by the gradual conversion of the active material that is not cycled in the nickel electrode from the β-phase to the γ-phase. The γ-phase material stores additional capacity, but discharges the capacity at a voltage plateau about 40 mV lower than that from the β-phase. Thus, the γ-phase is thermodynamically about 40 mV more stable than the β-phase, and is ultimately the more stable state for active material. During overcharge the β-phase is slowly converted to the γ-phase, and therefore during each recharge cycle a small amount of γ-phase forms. Because the β-phase has a

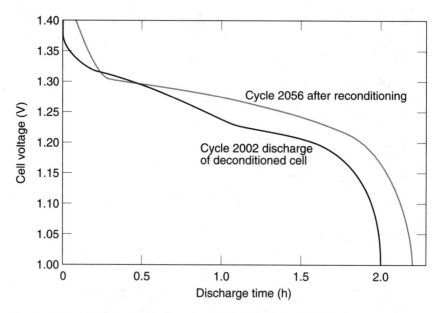

Figure 11.3. Capacity discharge of a deconditioned nickel-hydrogen cell after 2002 cycles to 50% depth of discharge, then again after a full discharge to recondition the cell. Approximately the last half of the capacity is delivered at a depressed discharge voltage before the reconditioning.

higher discharge potential, it is preferentially discharged each cycle, and thus over hundreds of cycles the nickel-electrode active material segregates into two layers: a β-phase layer that is repetitively cycled, and a γ-phase layer that is not cycled.

If a deconditioned cell is completely discharged, a process referred to as "reconditioning," the γ-phase layer is completely discharged and will recrystallize back to the β-phase. Recharge following the reconditioning will completely reestablish the normal discharge voltage response shown in Fig. 11.3, thus completely reversing the voltage changes (or memory effect) and the thermal effects associated with deconditioning. This is one reason for regularly reconditioning nickel-hydrogen batteries during periods when cycling is not being performed. Life tests of nickel-hydrogen batteries should always be performed using the same frequency and method for reconditioning that is planned during operation of the actual satellite power system.

11.1.4 Second-Plateau Signatures
The discharge of a nickel-hydrogen cell after significant cycle life typically shows a normal discharge plateau in the 1.15–1.3 V range. As the cell gradually degrades, its capacity that is available above 1.0 V decreases; however, some of the lost capacity is seen in a secondary discharge plateau between 0.8 and 1.0 V.

11.1 Cell Discharge Voltage Signatures

This capacity is typically associated with a voltage too low to be of significant use during operation, and additionally is accompanied by the production of large amounts of heat. The effect of the secondary discharge voltage level that can develop is shown in Fig. 11.4 for cell 7 in Crane nickel-hydrogen cell life-test pack 5004L after years of operation. While the secondary voltage plateau is not at usable voltages, it can be maintained for many years of operation and does contribute to the overall battery voltage.

The cause for the discharge plateau below 1 V is the formation of a semiconducting depletion layer in the layer of nickel-electrode active material that is in direct contact with the conductive sinter matrix.[11.4] This depletion layer forms when the diffusion rate of protons through the active material is not high enough to maintain an adequate density of charged sites in this layer. This happens as the state of charge drops during discharge; however, electrode degradation can make this transition occur earlier and earlier during the discharge. When the transition does occur, the nickel-electrode voltage will collapse. After the voltage has fallen 0.3 to 0.4 V, thus establishing this potential gradient across the depleted layer, the voltage gradient will sharply increase the electronic conductivity of the semiconducting material, thus allowing continued discharge of charged active material that is in contact with the depleted layer. This discharge will continue to increase the thickness of the depleted layer until the electrode runs out of capacity, or until the layer becomes so thick that its electronic conductivity becomes inadequate to support continued discharge.

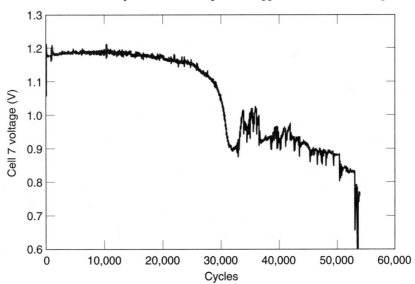

Figure 11.4. Normal and second-plateau operation for one of the cells in 10-cell test pack 5004L, which operated at 10°C with a 60% depth of discharge and a 1.05 recharge ratio.

While it is normal in a new nickel electrode to see a portion (5–10%) of the capacity discharged in the second plateau, in a nickel-precharged nickel-hydrogen cell this plateau is not normally seen, because the cell runs out of hydrogen before the second-plateau discharge occurs. However, as shown in the 60% depth of discharge test of Fig. 11.4, after extensive cycling significant capacity can be lost from the normal discharge plateau. This change cannot be reversed, because it results from the normal aging processes in the nickel electrode: sinter corrosion, electrode swelling, active-material migration, etc. Reconditioning the cell can change the magnitude of the second plateau in situations where it is magnified by layers of differing phase composition in the nickel-electrode active material. However, reconditioning will not reverse the normal wear processes that make the magnitude of the lower-plateau capacity increase with cell age at the expense of the usable capacity.

11.1.5 Electrolyte Freezing Signatures

If nickel-hydrogen cells are operated at temperatures well below the freezing point of water, several processes can occur to cause voltage signatures associated with freezing of water or electrolyte within the cell. The first such process can occur when a high-rate discharge is initiated following a long period of trickle-charge at temperatures less than about -8 to $-10°C$. During the trickle-charge period all the components within the cell become equilibrated to the low temperature. When the high-rate discharge starts, large amounts of pure water are produced from hydrogen gas in the hydrogen electrode. This water can freeze on the cold surfaces of the hydrogen electrode, particularly under conditions of dry-out or when the electrolyte diffusion rate is inadequate to prevent the freezing. The transport processes that must function in the hydrogen electrode to bring hydrogen gas to the catalytic sites and transport water away from them can become completely blocked by the layers of ice that can form in the hydrogen electrode structure under these conditions.

This freezing process produces a characteristic voltage signature during the discharge, which is shown in Fig. 11.5. The voltage will rapidly collapse during the early portion of the discharge as the hydrogen concentration and/or hydroxide ion concentration becomes depleted at the catalytic sites. Under these conditions, the impedance of the cell can temporarily become quite high, and it is possible for the cell voltage to fall to a very low level or even become negative for a short period of time. However, the voltage will typically recover after only a few minutes at the low level because the large amount of heat generated in the high-impedance region of the hydrogen electrode, combined with normal cell discharge heating, will melt the ice blockage and significantly increase electrolyte and gas transport rates. Typical voltage recovery is shown in Fig. 11.5 following a period of 5–10 min of discharge.

The electrolyte-freezing signature shown in Fig. 11.5 is generally regarded as fully reversible, and does not directly lead to any irreversible damage or

11.1 Cell Discharge Voltage Signatures

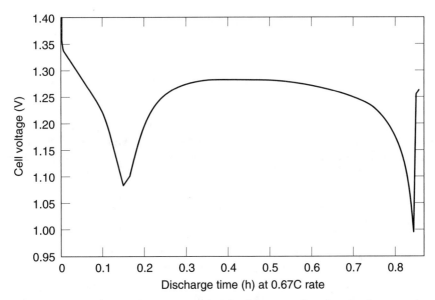

Figure 11.5. Voltage signature associated with water freezing in the negative electrodes. This discharge began at a temperature just below –20°C.

degradation to the nickel-hydrogen cell. Another freezing process in nickel-hydrogen cells can, however, cause cell damage and failure. This condition results from generalized electrolyte freezing within the electrode stack in the cell. This process typically occurs during long periods of cold trickle-charge, and begins in localized regions of the electrode stack and propagates through the stack until arrested by the heat produced from trickle-charge. The freezing effectively terminates all electrical activity in the frozen regions of the stack, and causes the trickle-charge current to be concentrated on just a few of the electrodes and to produce large rates of localized overcharge and oxygen evolution in these locations. The high rate of localized overcharge can rapidly cause cell failure from stack damage and the formation of short circuits. Either a sudden or gradual drop in cell voltage during long periods of trickle-charge is the typical signature for this type of failure.

A final electrolyte freezing condition results from thermal gradients within a cell that can result in condensation of frozen water on the cold pressure vessel domes. The ice cannot be dissipated by the wall wicks on the inner pressure vessel surfaces, and thus can build up over time to form a large block of ice within the cell. This can induce high impedance from dry-out of the electrode stack. The heating from the high impedance during discharge, or sudden environmental heating, can cause the ice to rapidly melt and flood a portion of the electrode stack. This can be a real problem in space, where there is no gravitational force to keep a pool of melt water in the bottom dome from flooding the stack. When the flooded regions of the stack are subsequently overcharged, popping damage can occur and cause stack damage and cell short-circuit failures.

11.2 Cell Recharge Behavior

11.2.1 Recharge Rate

The recharge rate is a key variable used to keep nickel-hydrogen cells charged during cycling operation, and to minimize stresses that could reduce their cycle-life capability. Recharge is typically done at a full charge rate until the voltage or state of charge of the cells in a battery reaches a threshold that triggers subsequent reductions in the charge rate to control the wear and heating caused by overcharge. The full charge rate must be sufficient to return the maximum capacity discharged within the time available for recharge.

For a 24 h geosynchronous orbit during which the battery is discharged to a 60% to 80% maximum depth of discharge, the full charge rate can be as low as C/20 (0.05C) while still providing sufficient recharge capacity to return what was discharged plus 30% extra to ensure that full charge can be maintained at the end of battery life when charge efficiency is lowest. In contrast, for a 90 min low Earth orbit there may be only 60 min total time for recharge. If the maximum depth of discharge is 40%, the full recharge rate must be about 0.6C to safely return the battery to full charge without undue stress from overcharge.

The key in both extremes is to reduce the full recharge rate when the cells begin to approach full charge and begin to produce heat from overcharge. In the geosynchronous case this is often done by simply dropping the charge rate to a trickle-charge condition to maintain full charge once the voltage or pressure of the cells is sufficiently high. In the low Earth orbit case, the full charge rate can be reduced in steps at fixed charge return points, or tapered back at a predefined battery voltage level. In either case the wear on the nickel-hydrogen cells is predominantly caused by the amount and rate of overcharge toward the end of the recharge period, rather than by the full-rate recharge early in the recharge period.

Based on this discussion of the effect of recharge rate on nickel-hydrogen cell wear, it is no surprise that the method for controlling recharge can have a significant effect on cycle life. This is particularly true for the low Earth orbit type of cycling, during which recharge rates are often quite high. Life tests that allow cells to rise to higher voltages during recharge typically display significantly reduced cycle life compared to life tests that reduce the recharge rate more aggressively to keep peak recharge voltages lower.

This is illustrated by the comparison of cell performance from two such life tests as shown in Fig. 11.6. In a typical cell from one of these life tests (pack 3603X), the recharge rate is linearly decreased after 95% charge return to give a net 1.04 recharge ratio. As the cells degrade over their life, the peak charge voltage

11.2 Cell Recharge Behavior

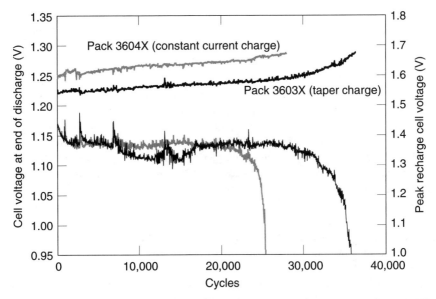

Figure 11.6. Comparison of two life tests, with one going to more than 1.6 V throughout its life and the other being lower through most of its life. The lower peak charge voltages give longer life.

rises, until it is about 1.67 V for the cells at the end of their life. In the other test (pack 3604X), the recharge rate is maintained at a constant level through the entire recharge period, while maintaining a similar 1.04 recharge ratio. The tapering of the current produced a peak recharge voltage about 80–90 mV lower in the first test at any given point in the cell life, a difference that resulted in a significantly lower wear rate and about a 40% improvement in cycle life. This difference in wear rate is attributed to the exponentially increasing rate of oxygen evolution with increasing voltage in overcharge. The reduction in cell lifetime is thus a consequence of the higher overcharge rate. Even more aggressive tapering of the recharge current can be used to further decrease the wear on nickel-hydrogen cells and optimize life by keeping the overcharge voltage at the lowest possible level.

11.2.2 Cell Recharge Voltage

While it is not possible to separate the recharge voltage of battery cells from their recharge rate, some clear differences in recharge voltage have been observed to occur over the cycle life of nickel-hydrogen cells. The recharge voltage as well as the overcharge voltage both have relatively strong inverse temperature dependence, and thus the voltage level to which a cell may be safely charged increases as the temperature is reduced. This inverse temperature compensation is typically included in power systems that use cell or battery voltage for controlling the recharge, and they are generally referred to as V/T (voltage-temperature compensated) charge-control systems. In ground-based life tests, which are often run in a nearly constant temperature environment, automated temperature compensation of the charge voltage level is typically not required.

Based on the overcharge voltage trends shown in Fig. 11.6, it is not surprising that the recharge voltage of nickel-hydrogen cells commonly increases as the cells age. This recharge voltage increase occurs over the entire state-of-charge range of cell operation, although the overcharge voltage can exhibit a different rate of increase over life than is seen at lower voltage prior to entering overcharge. The voltage increase during high-rate recharge primarily results from increases in cell impedance as the cell components age, and is thus much greater during high-rate recharge than during lower-rate recharge. The impedance increase can give two different types of signatures.

The first signature corresponds to a general increase in cell impedance that increases the recharge voltage and decreases the discharge voltage in rough proportion to the current. This signature is shown in Fig. 11.7 for a low Earth orbit life test at 60% depth of discharge that involves high rates of discharge and recharge. In this figure, a significant increase in cell impedance causes a pronounced drop in discharge voltage as well as a significant increase in recharge voltage, particularly during the initial high-rate portion of the recharge period. This

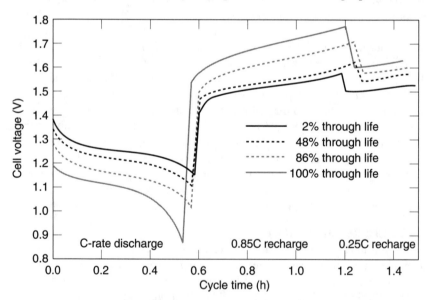

Figure 11.7. Changes in high-rate discharge and recharge voltage for a cell during a life test at 10°C and 60% depth of discharge showing general impedance increase.

signature is typically associated with increasing cell impedance from dry-out of the separator, and in some cases can be caused by water condensation from thermal gradients experienced by the cell during cycling. The behavior seen in Fig. 11.7 is most often seen for cells with asbestos separator or a single layer of Zircar separator, or when large thermal gradients are present across a cell.

A second signature often seen for changes in cell recharge voltages as a cell ages during long-term cycling is shown in Fig. 11.8. This signature involves very little change in the discharge voltage plateau, but an increase in the recharge voltage. The increase in the recharge voltage is most pronounced at the highest recharge voltages, where the oxygen evolution rate is highest. Thus, this figure shows a more pronounced age-related increase in the overpotential for the oxygen evolution reaction than for the recharge reactions that occur prior to overcharge. This signature is not associated with dry-out of the separator, because there is no significant drop in the discharge voltage plateau, but it is typically associated with degradation of the sinter structure and active material in the nickel electrodes within the cell.

If a capacity measurement cycle is periodically performed during a life test, it is possible to evaluate the shifts in recharge voltage that are not primarily the result of increasing impedance. Such a capacity cycle involves discharge of all

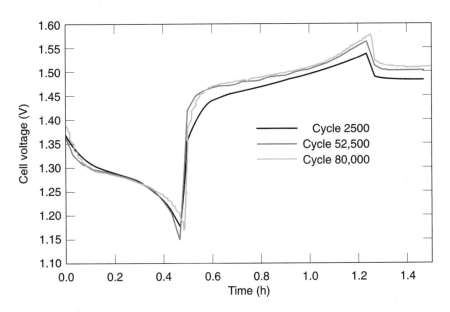

Figure 11.8. Changes in high-rate discharge and recharge voltage for a cell during a life test at 10°C and 40% depth of discharge showing primarily an increase in recharge voltage.

capacity at the high rate used in the life test, then recharge at a low rate (e.g., C/10 for 12 h) until the cells are fully charged. Figure 11.9 shows the results from this type of capacity cycle performed periodically through a life test, and indicates how the recharge and overcharge voltage of a cell typically change for a low-rate (C/10) recharge as the cell degrades over its operational lifetime. The recharge voltage plateau does not change much until the cell is nearly at the end of its life, but the overcharge voltage increases significantly as the cell ages. However, at the end of life the overcharge voltage drops, probably because the cell failed from increased internal losses that significantly decreased the charge efficiency and state of charge. Most of the changes in voltage prior to the rapid shift at end of life are caused by sinter corrosion in the nickel electrode, which weakens the sinter structure, decreases its conductivity, and gradually dilutes the cobalt additive in the active material.

11.2.3 Recharge Ratio, Overcharge, and Charge Efficiency

The interplay between the amount and rate of excess recharge capacity applied, the recharge voltage *vs.* the overcharge voltage, and the internal self-discharge processes control the recharge of nickel-hydrogen batteries. If a nickel-hydrogen cell is brought to a full state of charge, significantly more overcharge must be applied than if the cell must only be recharged to 90% state of charge. Because the internal loss rate and charge efficiency are also highly sensitive to temperature,

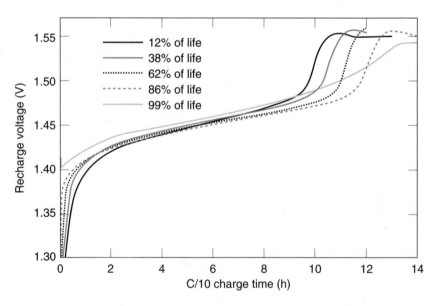

Figure 11.9. Changes in nickel-hydrogen cell recharge voltage behavior at low rate as the cell progresses through a life test to failure. Temperature during recharge is 0°C.

11.2 Cell Recharge Behavior

much less overcharge is required to bring a cell to a high state of charge at low temperatures than is required at higher temperatures.

When a nickel-hydrogen cell is tested, it is often tested at a standard temperature in the −10 to +10°C range, and at a prescribed recharge ratio for the selected charge-control method. In low Earth orbit life tests, recharge ratios have varied from about 1.01 to 1.05. Most tests have selected a fixed recharge ratio high enough to be ensured of supplying more than adequate recharge to keep the cells charged over their entire lifetime. However, a few tests have allowed the recharge ratio to freely vary such that it is just sufficient to keep the cells charged to a required capacity level that may be less than 100% charged. Such recharge may be controlled by directly controlling the recharge ratio or pressure limit, or by varying the peak charge voltage limit.

Figure 11.10 shows a life test done by raising the peak recharge voltage only as required to keep the four cells in the test above about 1.1 V during discharge. This test (Crane pack 3001C) was performed at 40% depth of discharge, and ran with a recharge ratio of 1.018 early in life that gradually increased to about 1.022 toward the end of the test. This represents the minimum recharge ratio that was needed to keep these cells adequately recharged, and thus should help maximize the life of the cells. The test was terminated before cell failure (at 32,000 cycles), and thus the cycle-life capability under these conditions was not determined.

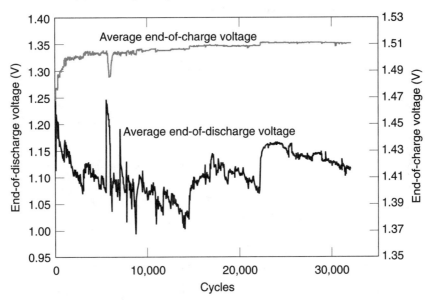

Figure 11.10. Performance in life-test 3001C at 40% depth of discharge and 10°C. Just enough overcharge was applied to compensate for losses (1.018 to 1.022 recharge ratio) by adjusting the peak charge voltage.

11.2.4 Trickle-Charge

Trickle-charge involves applying a low charge rate to a battery to maintain its cells at a full state of charge. To do this, the trickle-charge must compensate for the internal self-discharge losses in the cell within the battery having the greatest losses. To ensure that trickle-charge is effective in keeping all the cells charged, the rate must typically be about twice the average cell self-discharge rate at beginning of life at the maximum expected operating temperature, thus ensuring that even the cells having the highest losses will remain charged.

Figure 11.11 shows the average self-discharge rate of a typical nickel-hydrogen cell as a function of temperature at approximately a 95% state of charge, compared to a C/100 trickle-charge rate. At 20°C even a C/100 trickle-charge rate will allow cells to drop well below 90–95% state of charge. At 10°C, the maximum temperature at which nickel-hydrogen cells normally operate, a C/100 trickle-charge rate is adequate to maintain all cells essentially fully charged. Lower trickle-charge rates (e.g., C/200) would also be effective, but could allow some cells to drop to lower states of charge. At 0°C, a common operating temperature for nickel-hydrogen cells, a C/200 trickle-charge rate provides good margin for keeping all cells charged. At –10°C a trickle-charge rate of C/300 is appropriate. If a trickle-charge rate that is unnecessarily high for the operating temperature is used, it will keep the cells recharged, but at the cost of making them wear out faster than if a lower trickle-charge rate was used.

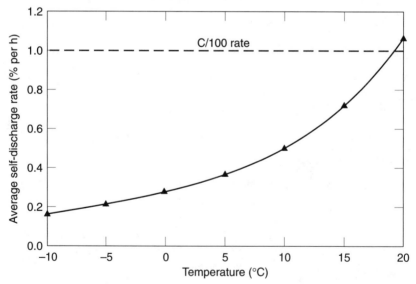

Figure 11.11. Typical self-discharge rate of a nickel-hydrogen cell at about 95% state of charge as a function of temperature, compared to a C/100 trickle-charge rate.

11.3 Cell Pressure Behavior

The pressure of hydrogen within a nickel-hydrogen cell provides a good indication of the state of charge of the cell, as long as the pressure that corresponds to full charge and full discharge are known, and as long as appropriate temperature corrections to the pressure are made. Figure 11.12 shows the typical variation in hydrogen pressure over the course of a single discharge-and-recharge cycle, as measured by the calibrated strain gauges that are often affixed to the dome of a nickel-hydrogen cell pressure vessel. The curve shown in Fig. 11.12 is highly repeatable from cycle to cycle, and provides a quantitative measure of the state of charge based on the gas pressure in the cell.

The pressure variation during charge and discharge of a nickel-hydrogen cell has frequently been used for charge-control purposes. Recharge can be done until the pressure reaches predefined thresholds that will trigger appropriate reductions in charge current or a switch to trickle-charge. This has proven to be a highly effective charge-control approach as long as all the cells in the battery remain well matched in state of charge. If some cells wander up or down in state of charge relative to the cells that are instrumented for pressure, this charge-control method can have problems.

The effective use of pressure for charge control or just for information on the state of the battery over many years of operation depends on maintaining a

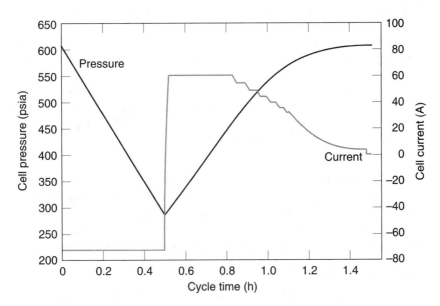

Figure 11.12. Variation of hydrogen pressure in a nickel-hydrogen cell during a 60% depth-of-discharge low Earth orbit cycle. This cell is cycling between about 15 and 75% of its maximum state of charge.

good understanding of how the pressures corresponding to the fully charged and the fully discharged states change over time. The pressure corresponding to full charge is generally determined by charging the battery until there is a noticeable temperature rise from overcharge. The pressure that corresponds to this point is then assigned to the fully charged condition. If this measurement is performed periodically over the years, it will indicate how the pressure level associated with full charge changes over the life of the battery.

Similarly, it is possible to discharge the battery using the normal discharge loads or using a lower reconditioning load to determine the pressure that corresponds to the point at which the battery voltage drops below the usable voltage threshold for the system. This point defines the pressure that corresponds to zero state of charge. If a discharge rate lower than normal system operating discharge rates is used for this measurement, a correction must be made for the added capacity that is available at a low discharge rate. The periodic determination of the zero state-of-charge pressure allows the capacity margin available in the batteries to be accurately tracked over the battery lifetime.

The pressure curves in Fig. 11.12 can be used for purposes other than keeping track of state of charge. The slope of the pressure rise during recharge provides an indication of charge efficiency. The charge efficiency is high and nearly constant early in the recharge, producing a linear increase in pressure, and then drops off as the cell begins to go into overcharge and generate oxygen at the end of recharge. During discharge the pressure also tends to drop linearly in proportion to the capacity discharged. The only major corrections to using the slope of the pressure curve during discharge to define the slope corresponding to 100% efficiency are, first, for the temperature rise that occurs during discharge and, second, for the self-discharge that continues to occur at low rate during the discharge. Pressure growth behavior over many years of operation can also provide an indication of the type of degradation and eventual failure mode for the cell, examples of which are discussed in the following sections.

11.3.1 Pressure Growth
As suggested in the previous discussion, processes that slowly occur during years of nickel-hydrogen cell operation cause the pressure to gradually drift. These processes include the slow corrosion of the nickel sinter in the nickel electrode, which results in an upward pressure drift, and the decreasing charge efficiency of the nickel electrode as it ages, which produces a partially compensating downward drift. The net effect of these two processes is nearly always an upward pressure drift, although a few noteworthy instances have been observed where these two effects nearly offset each other to make a cell exhibit almost zero pressure drift over time.[11.5]

Figure 11.13 shows the typical pressure growth curve seen in nickel-hydrogen cells as they cycle over their full lifetime, eventually failing at end of life as the result of not sustaining adequate usable capacity. In this situation pressure tends

11.3 Cell Pressure Behavior   349

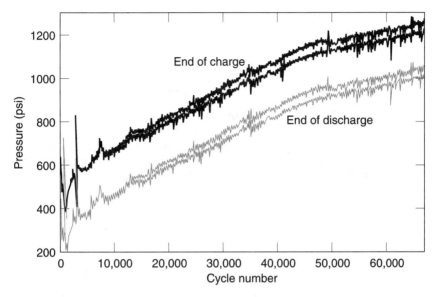

Figure 11.13. Pressure growth behavior over life for two cells in pack 3600X, which operated at 35% depth of discharge and 10°C with a 1.05 recharge ratio.

to rise in a nearly linear fashion over time, occasionally exhibiting a decreased slope late in life because nickel-electrode degradation eventually tends to reduce charge efficiency and cause stored capacity to gradually drop off.

11.3.2 Capacity Walk-Down Pressure Signatures

The pressure growth curve shown in Fig. 11.13 tends to be seen when cells are tested at low temperature (e.g., 0°C) and the recharge voltage is kept relatively low (less than 1.55 V). If the temperature is higher (e.g., 10°C) during cell operation, another interesting pressure signature is often seen, as is shown in Fig. 11.14. This signature is similar to that of Fig. 11.13, except that there is a significant walk-down in pressure during roughly the first 5000 cycles (the first year) of operation. This early-life pressure decrease arises from a walk-down in the cell capacity relative to the capacity at which the cells began cycling. Prior to the start of the test the cells were recharged at reduced temperature to a full state of charge, and then began cycling at a higher temperature (10–12°C). At the higher temperature, the 1.04 recharge ratio used in the cycling was not sufficient to keep the cells as highly charged as when they started cycling. Thus, the cells gradually walked down in capacity to the point where the excess recharge just balanced the losses from self-discharge and overcharge. The behavior shown in Fig. 11.14 should be regarded as a normal response for nickel-hydrogen cells when cycled at an elevated temperature.

350   Cell and Battery Test Experience

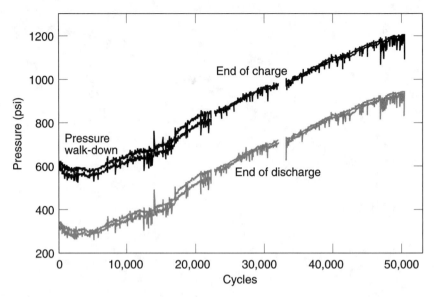

Figure 11.14. Pressure behavior of two cells in Crane pack 3316E (60% depth of discharge, 1.04 recharge ratio, 10°C), showing early-life pressure walk-down response.

11.3.3 Soft-Short Pressure Signatures

The pressure growth behavior in Fig. 11.14 is typically seen when the eventual cell failure results from an increase in the unusable capacity in the cell to the point where the usable capacity (defined in these tests as capacity delivered above 1 V) is insufficient to meet the discharge requirement. This failure mode is usually associated with keeping the overcharge voltage and the amount of overcharge relatively low. A very different failure mode can occur when cells are cycled with much higher overcharge voltages (higher overcharge rate) or much higher amounts of overcharge. This alternate failure mode, which also results in a drop in voltage below 1 V when the cell fails as well as reduced usable capacity, can be most easily recognized by its characteristic pressure growth curve, which is shown for two cells in Fig. 11.15.

These two cells from Crane test pack 3605X were driven to peak charge voltages in excess of 1.65 V near the end of their life. The high overcharge rate at these voltages, coupled to the significant amount of overcharge from the 1.04 recharge ratio, resulted in the formation of soft shorts in these cells after about 25,000 cycles. The soft shorts caused increasing internal capacity losses that eventually could not be offset by the overcharge being applied to the cell pack every cycle. The resulting pressure growth behavior shows a pronounced drop in

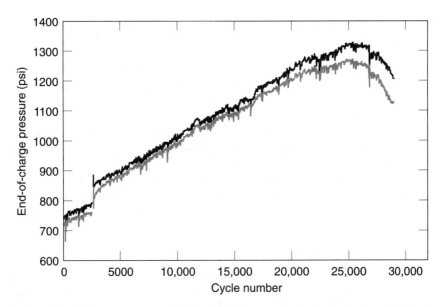

Figure 11.15. Pressure growth behavior for two cells in Crane test pack 3605X, which operated at 60% depth of discharge, 1.04 recharge ratio, and 10°C nominal temperature.

pressure over the last approximately 3000 cycles (~7 months of cycling) in Fig. 11.15, which eventually led to insufficient usable capacity in the cells to support the discharge and keep the voltage above 1 V.

11.4 Charge-Control Methods and Cell Lifetime

The importance of the method and details for the charge-control approach used for recharging nickel-hydrogen cells has been demonstrated by a number of life tests that have been performed over the years. These life tests have provided a convincing database indicating that the key factors that control the cycle life of nickel-hydrogen cells are the amount and rate of the overcharge applied to the cells. These tests have demonstrated that nickel-hydrogen cells are robust in terms of tolerating significant amounts and rates of overcharge. However, overcharge that is not necessary to overcome the effects of self-discharge has been shown to be the principal reason why nickel-hydrogen cells have fallen short of their potential performance lifetimes in numerous life tests.

Some of the longest-running cycle-life performance for nickel-hydrogen cells has been obtained for low–depth-of-discharge applications. These include satellites such as the Hubble Space Telescope (95,000 cycles), as well as life tests that have been run at 15% depth of discharge. The cycling of these cells typically involved recharge ratios of 1.03–1.04. When considering higher–depth-of-discharge demonstration tests, one typically assumed that the recharge ratio that was effective at 10–15% depth of discharge would be just as effective at 40–60% depth of discharge. For this reason most of the nickel-hydrogen cell-life tests that have been started to demonstrate their life capability at higher depth of discharge also used a 1.03 to 1.04 recharge ratio. These high recharge ratios provided an absolute amount of overcharge to the cells per cycle that was 3 to 4 times that experienced in low–depth-of-discharge operation. While cells tested under these stressful conditions gave good cycle life, it was typically far short of the optimistic expectations for this technology at its inception.

With the 1.03 to 1.04 recharge ratio paradigm for cycling double-layer Zircar nickel-hydrogen cells containing 31% potassium hydroxide electrolyte (Man-Tech cells), the cells typically provided 20,000 to 30,000 cycles at 60% depth of discharge, and 50,000 to 60,000 cycles at 40% depth of discharge, as is indicated in Fig. 11.16 for Crane test packs 3604X and 5002E. The Air Force design (single-layer Zircar with 26% electrolyte) was also extensively tested with this same overcharge paradigm, and has provided reasonably similar cycle-life times. The conclusion was drawn that this observed cycle life defined the ultimate capability of nickel-hydrogen cells, and that the wear was controlled by the stresses associated with the discharge. A simplified wear model in which the cycle life

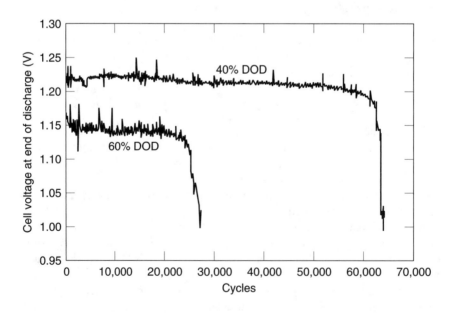

Figure 11.16. Average cycle-life performance for Crane test packs 3604X (65 A h) and 5002E (50 A h), which were run at 1.04 recharge ratio.

was logarithmically proportional to depth of discharge seemed to fit the test data well for either type of cell design. This type of wear model has been widely used over the years to predict the expected lifetime from nickel-hydrogen cells and batteries.[11.6,11.7] The predictions from this type of model have tended to limit the depth of discharge to 35–40% in low Earth orbit power system designs, and have similarly allowed depths of discharge of up to 90% in geosynchronous power system designs.

More recent meta-analysis of a large number of nickel-hydrogen life tests[11.1] has used regression methods to allow the test data to determine the dependence of wear rate (and thus cycle life) on depth of discharge, temperature, amount of overcharge, rate of overcharge, and calendar life. Contrary to earlier analyses that assumed that depth of discharge was key for determining wear rate, this regression-analysis approach consistently found that overcharge dominated the wear rate for all tests, followed by the effect of temperature, with calendar life and depth of discharge being the least significant. It should be noted that the regression analysis analytically removed the known dependence on depth of discharge that results from higher depth of discharge leaving less margin to accommodate capacity loss before failure occurs, a dependence that was found to account for nearly all the influence of depth of discharge on wear rate.

The conclusion from this regression analysis of the nickel-hydrogen life-test database was that the cycle life was not necessarily limited by the depth of discharge, but could be extended significantly by closely controlling unnecessary overcharge, keeping overcharge voltages low, and avoiding extended operation at warm temperatures. Charge-control methods in a number of operational satellites have been modified to take advantage of the longer projected lifetime obtained when overcharge is reduced.

In addition, several alternative charge-control methods have been proposed to provide the minimum necessary overcharge to the nickel-hydrogen cells in a battery while at the same time keeping all cells balanced in their capacity. One of these methods uses an adaptive charge-control technique where the recharge ratio for each cell is varied every cycle to return only the charge needed to keep cell performance at a minimum acceptable level.[11.8,11.9] Other charge-control methods involving pressure, voltage, and recharge ratio techniques for controlling overcharge have also been adapted for the purpose of increasing long-term cell and battery reliability by minimizing overcharge.

11.5 Temperature and Cell Lifetime

Early tests of nickel-hydrogen cells cycled the cells at nominal temperatures of 10°C, and were found to give very good cycle life, although the projected lifetime was not as great as had been hoped for when the technology was first introduced. However, because nickel-hydrogen batteries are generally operated at temperatures much lower than 10°C, the expectation was that at lower temperatures, such as –5 or 0°C, the degradation processes would be slower and the cells would last significantly longer. Surprisingly, this was not found to be the case. Figure 11.17 compares two life tests, one with cell lifetime at 10°C and the other at –5°C, while keeping all other variables constant. The typical result was that the cycle life at the lower temperature was similar to or worse than that at the higher temperature.

While the root cause for the unexpectedly low cycle life at reduced temperature was not initially obvious, it was eventually noted that two factors were probably responsible for the trend. The underlying cause was the use of a constant recharge ratio for testing at all temperatures. While this recharge ratio was more than enough to compensate for the internal cell losses at 10–12°C, it represented an ever-increasing amount of unnecessary overcharge as the cell temperature was reduced. The increased overcharge at the low temperatures caused the second problem, the formation of large amounts of γ-NiOOH. The large volume change associated with the repeated formation of this phase caused accelerated swelling and fracturing of the sinter structure in the nickel electrodes, and thus caused accelerated cell failure. This was, in fact, one of the earliest indications that overcharge can be one of the overriding factors in controlling cell lifetime.

Because of how temperature, charge rate, and the amount of overcharge interact, it is not possible to determine the effect of temperature on cycle life

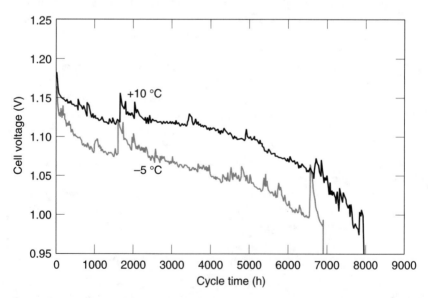

Figure 11.17. Cycle life at different temperatures for low Earth orbit cycling at 60% depth of discharge, determined using a recharge ratio of 1.05 at both temperatures. The data are from NWSC packs 3761E and 3765E.

by directly comparing the results from two life tests for which temperature was the only variable. This comparison can lead one to the erroneous conclusion that reduced temperature operation either has little beneficial effect or significantly degrades cycle life as indicated in Fig. 11.17. To properly compare the cycle life at two different temperatures, one must adjust the cycle life observed at the low temperatures for the added degradation that does not result from the temperature, but from the higher excess overcharge at the lower temperatures. ("Excess overcharge" refers to overcharge in excess of that required to just compensate for self-discharge during the cycle; the self-discharge rate is as shown in Fig. 11.11.) This adjustment has been performed for life tests at 35% and 60% depth of discharge for cells from two different cell manufacturers, giving the results shown in Fig. 11.18.

At 35% depth of discharge both types of cells exhibit about the same improvement in expected cycle life for operation at –5°C compared to 10°C. This is the anticipated improvement in cycle life at lower temperatures, and using a full multiple-regression analysis provides an Arrhenius activation energy of 4.65 kcal/mole[11.1] for the temperature dependence of nickel-hydrogen cell cycle life. Figure 11.18 also shows that the adjusted cycle life decreases with increasing depth of discharge. It is noteworthy that for both types of cells the –5°C cycle life drops more rapidly with increasing depth of discharge than the 10°C cycle life. This effect

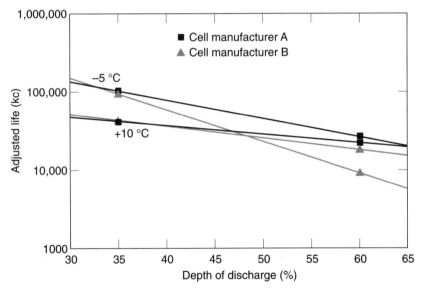

Figure 11.18. Cycle life adjusted for differences in excess overcharge for life tests run at different temperatures and depths of discharge. These data points were determined using the results from NWSC packs 3731E, 5735E, 3761E, 3765E, 5631W, 5635W, 5661W, and 5665W.

is likely to arise from the stress of the high-volume γ-NiOOH, which is more extensively formed and cycled during the 60% depth-of-discharge operation.

A second noteworthy observation from Fig. 11.18 is the more rapid decrease in cycle life with increasing depth of discharge for the cells from manufacturer B. Apparently, the cells made by manufacturer B are more sensitive to the wear produced by γ-NiOOH during cycling. Destructive physical analysis of a cell made by manufacturer B after it had failed in a 60% depth-of-discharge life test after only a few thousand cycles has confirmed that the nickel electrodes were heavily damaged. The damage consisted of large amounts of swelling and fracturing of the sinter structure over approximately 50% of the surface area of all the nickel electrodes within the cell. This degree of electrode damage suggests that large volume changes occurred in the active materials in the nickel electrodes during every cycle, and that the sinter structure did not have the strength to tolerate these volume changes.

11.6 Depth of Discharge and Cell Lifetime

The depth of discharge to which a battery is cycled has an obvious influence on its cycle life. If the life is assumed to be limited by loss of capacity, and it is also assumed that loss of capacity accumulates over the lifetime of the cells in the battery, then operation at higher depth of discharge leaves less available capacity margin before failure is expected. For example, operation at 90% depth of discharge would leave only 10% capacity margin while operation at 80% depth of discharge would leave 20% margin in the cells. For a constant capacity-loss rate over life, the battery operating at 80% depth of discharge would therefore be expected to have twice the lifetime of the battery operated at 90% depth of discharge.

However, this obvious dependence on depth of discharge is known if one knows the maximum required depth of discharge, and can be analytically removed from the picture if one determines the wear rate from the life-test data, rather than simply the absolute lifetime. With this way of looking at the data, the two hypothetical tests discussed here would exhibit identical wear rates if the 80% depth of discharge had exactly twice the cycle life as the 90% depth-of-discharge test. When the life-test database for nickel-hydrogen cells is analyzed in terms of the intrinsic wear rates for different failure modes such as capacity loss or short-circuiting, it is typically found that depth of discharge has only a small influence on the wear rate. The increase in wear that is often cited as being caused by higher depth of discharge actually arises from the decreased capacity margin, the higher charge rate, the increased temperature, and the greater overcharge stresses that typically accompany higher–depth-of-discharge operation.

11.7 Electrolyte Concentration and Cell Lifetime

The concentration of the potassium hydroxide electrolyte used in nickel-hydrogen cells has been found to have a strong influence on the cycle life that can be

11.7 Electrolyte Concentration and Cell Lifetime

obtained from the cells. This effect arises because the potassium hydroxide in the electrolyte does not act only as a simple electrolyte to provide the necessary ionic conductivity between the anode and cathode of the cell. The potassium hydroxide also undergoes an electrochemical reaction with the charged β-NiOOH during nickel-electrode overcharge to produce the γ-NiOOH phase, a process that involves the chemical incorporation of one potassium hydroxide molecule for every three nickel sites in the active material. This process can significantly dilute the electrolyte concentration within the cell if a large percentage of the active material is converted into the γ-NiOOH phase, which is a real concern at low temperatures where charge efficiency is quite high.

The chemical incorporation of potassium hydroxide into the γ-NiOOH phase also means that this phase becomes more stable and formed more and more easily as the potassium hydroxide concentration in the cell increases. Because the γ-NiOOH phase can discharge effectively, this does not appear at first glance to be a significant liability, and it can in fact significantly increase nickel-electrode capacity when higher electrolyte concentrations are used. The problem arises from the fact that the electrochemical formation of the γ-NiOOH phase involves a large change in active-material volume. If this volume change occurs repeatedly as the result of regular cycling of the γ-NiOOH phase, it will lead to rapid swelling and fracturing of the sintered electrode structure, and extrusion of the active material out of the surface pores and into the separator. The result is that nickel-electrode degradation and short-circuit paths through the separator are much more likely to cause cell failure when a higher electrolyte concentration is used.

Analysis of life tests for cell designs using different electrolyte concentrations has allowed the variation in capacity stability during long-term cycling shown in Fig. 11.19 to be developed based on the accelerated-throughput type of cycling. Cells having higher electrolyte concentration display a higher beginning-of-life capacity; however, the higher capacity is much less stable during cycling than when a lower electrolyte concentration is used. Because nickel-hydrogen cells are typically used in power systems designed for long cycle life, the designs in most common usage employ either 31% or 26% potassium hydroxide electrolyte concentrations by weight. The relative slopes of the curves in Fig. 11.19 tend to vary somewhat as the temperature, recharge ratio, and recharge currents change, but the capacity-loss rate is always greater with the higher concentrations. The lower stresses from cycling the γ-NiOOH phase when 26% electrolyte is used make this concentration most appropriate for use with single-layer Zircar separators, for which premature failure by short-circuiting can become an issue with higher concentrations.

Another issue associated with different electrolyte concentrations in nickel-hydrogen cells is the changing susceptibility to electrolyte freezing for different concentrations. When 26% potassium hydroxide electrolyte is combined with

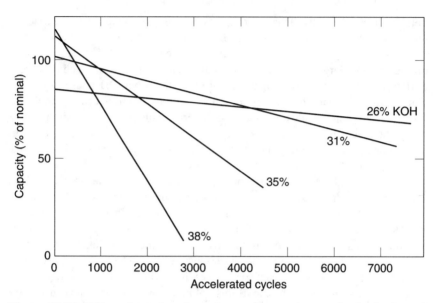

Figure 11.19. Effect of varying electrolyte concentration on the beginning-of-life capacity, and on the capacity stability of nickel-hydrogen cells during long-term cycling.

complete γ-NiOOH phase formation, it is theoretically possible to realize electrolyte freezing at temperatures as high as −5°C. This theoretical threshold drops to below −10°C for 31% electrolyte. However, just as important, it is much more difficult to form large amounts of γ-NiOOH phase in 26% electrolyte than it is in 31% or higher concentrations.

A number of life tests have been run that directly compare the cycle life of nickel-hydrogen cells with 26% electrolyte and 31% electrolyte under real-time cycling conditions using low Earth orbit profiles at 35% and 60% depth of discharge.[11.3] These tests have combined cells that were identical in design, but were activated with differing electrolyte concentrations, into the same test pack. When the tests have been run to the end of cell life, they have generally been consistent with the conclusion that 26% electrolyte gives significantly better cycle life than 31% electrolyte. An example is shown in Fig. 11.20 for Crane pack 3604G, which ran at 60% depth of discharge and 10°C.

One issue in these tests is that the cells with 26% electrolyte tend to have lower initial capacity, and thus in a combined test pack are operated at a somewhat higher depth of discharge relative to the actual cell capacity. However, in such tests that directly compare the two electrolyte concentrations, the cells with 26% electrolyte tend to have up to twice the cycle life of cells with 31% electrolyte, as is indicated in Fig. 11.20.

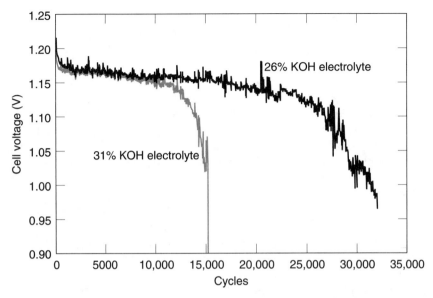

Figure 11.20. Typical effect of potassium hydroxide electrolyte concentration on the cycle life of nickel-hydrogen cells. Data from NWSC pack 3604G, which was run at 60% depth of discharge and 10°C.

11.8 Accelerated *vs.* Real-Time Testing

A number of accelerated test methods have been applied to the problem of evaluating the expected lifetime of nickel-hydrogen cells without having to cycle them for many years. These test methods have included acceleration by increasing different variables: temperature, depth of discharge, cycle frequency, amount of overcharge. None have been successful in developing an acceleration technique that provides consistent acceleration factors that can be unambiguously applied to predict the real-time operational life of the cells.

However, several of the accelerated test methods have been successfully used for screening tests meant to rapidly evaluate the relative robustness of one cell design compared to another (as in Fig. 11.19), or cells from one production lot compared to those from another lot. This approach adopts a standard cycling protocol that is highly stressful or highly accelerated, and it evaluates the relative life of each cell or design in the standard test protocol. More-robust cells are expected to provide longer lifetime in such tests.

Accelerated screening tests for evaluating the life capability of nickel-hydrogen cells have included 80% depth-of-discharge cycling at either 8 or 16 cycles per day until the cell fails to maintain 1.0 V at the end of discharge, as well as the use of a more normal depth of discharge (i.e., 40% or less) with 32 cycles per day. These types of tests typically try to maximize the ampere-hour throughput passed through the cell per unit of test time. Each of these types of screening methods is effective in causing cells to fail much more rapidly than in real time; however, each also has significant uncertainty as to whether the resulting failure mode is the one that will really limit the cell life in its actual application. Thus, such tests provide good relative estimates of cell robustness and can provide a rapid indication of whether cells are similar. However, these types of accelerated throughput tests cannot replace more realistic cycling profiles for quantitatively predicting the life of nickel-hydrogen cells in real satellite applications.

Attempts to use elevated temperature operation as an acceleration factor to determine life capability more rapidly than could be done in real time have been particularly unproductive for nickel-hydrogen cells. These cells are not used in applications at temperatures greater than +10°C. The typical Arrhenius dependence of chemical and physical degradation processes would suggest that life-testing cells at +20°C could provide controlled acceleration of the normal degradation processes.

Unfortunately, in such testing the desired acceleration did not occur. Life tests at +20 to 25°C first required large amounts of overcharge to keep the cells charged. Not only did the overcharge accelerate normal wear processes, but the overcharge condition at elevated temperatures also initiated a new electrochemical dehydration process in the nickel-electrode active material that was not seen during normal cell operation. This irreversible dehydration process led to premature cell failure from capacity loss in these high-temperature life tests, while no evidence of the dehydration products could be found in cells operated at +10°C or lower. Figure 11.21 shows the cycle-life performance typically seen in these high-temperature life tests.[11.10] The data in this figure indicate that the life at 60% depth of discharge is only about 10% of that seen in testing at 10°C (compare with Fig. 11.16).

11.8.1 Low Earth Orbit Profile Tests

Low Earth orbit tests of nickel-hydrogen cells are generally run using an orbital time profile that involves 30–35 min of discharge followed by approximately 60 min for recharge. This allows 15–16 orbits to be run per day. Low Earth orbit satellites typically have design lifetimes ranging from 3 to 10 yr, and therefore low Earth orbit tests must give good cell performance for more than 50,000 cycles to demonstrate high reliability for the longest of this mission type. Low Earth orbit tests have been run for nickel-hydrogen cells at several test facilities, and have explored performance over temperatures from −5 to +10°C, depths of discharge from 15% to 60%, and recharge ratios from 1.01 to 1.10. In addition a range of charge-control techniques have been used in these life tests, including constant rate charging, taper charging with a V/T limit, pressure limit control, and

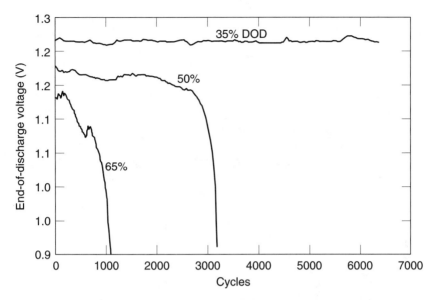

Figure 11.21. Results from accelerated low Earth orbit life tests on nickel-hydrogen cells at +20°C average cell temperature, from Ref. 11.10.

straight recharge ratio or excess ampere-hour charge control. The effectiveness of the charge-control method generally depends on how well it limits the peak overcharge voltage and amount of overcharge to which the cells in a battery are exposed, while still allowing adequate recharge to keep all the cells charged.

Low Earth orbit life tests have been run at the Naval Weapons Support Center test facility at Crane, Indiana (NWSC-Crane), for a range of differing cell designs that have been made by EaglePicher, Hughes/Boeing, Gates Energy Products, and Yardney. These tests have been largely sponsored by the U.S. Air Force and NASA. A relatively standardized low Earth orbit test profile for many of these cells has involved cycling at 35–40% depth of discharge at 10°C, with a recharge ratio of 1.03 to 1.04. Some of the results for different individual pressure vessel cell designs life-tested in this standard low Earth orbit test profile are described here.

11.8.1.1 Separator Design Variations

The separators that have been used in low Earth orbit nickel-hydrogen cell designs are dual-layer Zircar, single-layer Zircar, Zircar-asbestos composite, and asbestos. Of these types, the single-layer Zircar and asbestos separators have generally not been considered sufficiently robust for the stresses of low Earth orbit cycling unless the depth of discharge is relatively low (less than 15%). Life tests have been performed that compare the cycle life of the dual-layer Zircar separator with that of the Zircar-asbestos composite separator. These life tests have been performed at 35% and 60% depth of discharge; however, the 35% depth of discharge tests are examined here because the relatively high amount of overcharge in the 60% depth-of-discharge tests essentially made these tests of cell tolerance to high-rate overcharge.

Zircar-asbestos composite separator was evaluated starting in the early 1990s by NASA as a potential advanced component that could provide improved performance and life in batteries for the International Space Station. A number of life tests of cells containing this separator were begun, using cells made by two different manufacturers. The average cycle life for these nickel-hydrogen cell designs is shown in Fig. 11.22 at 35% depth of discharge and 10°C. The results in this figure are reasonably consistent across all life tests with these separator variations, and show that the dual-layer Zircar separator consistently gave significantly better lifetime than did the Zircar-asbestos composite separator. These test results caused the nickel-hydrogen cell industry to move away from the Zircar-asbestos composite separator by the mid-1990s.

11.8.1.2 Tolerance of Cells to Overcharge and High Depths of Discharge
A number of low Earth orbit life tests have been run on nickel-hydrogen cells at 60% depth of discharge. These tests, which have typically been performed using

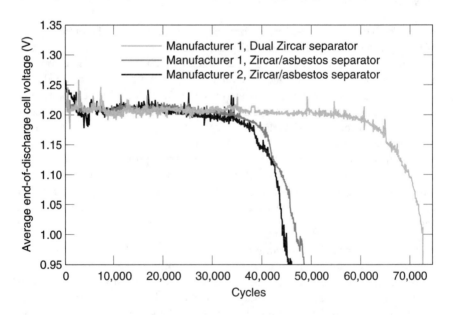

Figure 11.22. Low Earth orbit life-test results for cell designs having different separator types; the tests were run at 35% depth of discharge and 10°C.

a 1.04–1.05 recharge ratio, have given a wide range of cycle life for different cell designs and manufacturers. Because the recharge ratios are relatively high in all these tests, the variability in cycle life has essentially served as a measure of the general tolerance of each design to high-rate overcharge. For this reason, a comparison is provided here of the relative low Earth orbit cycle lifetime for a number of cell designs at 60% depth of discharge and 10°C. This comparison is shown in Fig. 11.23. The upturn in performance just past 11,000 cycles for the cells made by manufacturer 3 is caused by an increase in recharge ratio from 1.04 to 1.06, which only briefly improved performance.

The variability shown in Fig. 11.23 indicates that these cell designs have very different wear rates for the effects of high-rate overcharge. Cells made by manufacturer 1 are more than twice as tolerant to high-rate overcharge than cells made by manufacturer 3, with manufacturer 2 falling about halfway between. Also indicated in Fig. 11.23 for comparison is a cell made by manufacturer 1 having a composite asbestos-Zircar separator, which has significantly less tolerance for overcharge than any of the dual-layer Zircar cells.

Interestingly, the cells made by manufacturers 2 and 3 both contained dry sinter nickel electrodes, while the longer-lived cell made by manufacturer 1 contained slurry sinter nickel electrodes. This result suggests that a large part of

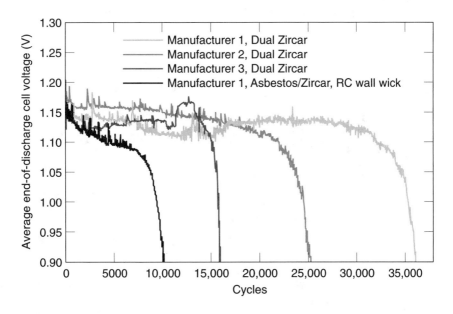

Figure 11.23. Life of various cell designs at 60% depth of discharge (10°C and ~1.04 recharge ratio), which essentially serves as a test of the overcharge tolerance of each cell design.

nickel-hydrogen cell tolerance to high-rate overcharge has to do with the strength of the nickel-electrode sinter structure, because the slurry sinter is structurally stronger than the dry sinter.

It should be pointed out that all the dual-Zircar cell designs shown in Fig. 11.23 have demonstrated good cycle life (greater than 50,000 cycles) with lesser amounts of overcharge and at depths of discharge of 40% or less.

11.8.1.3 Dry Sinter and Slurry Sinter
The results described in the previous section indicated that the stronger sinter structure associated with well-made slurry sinter could give improved tolerance to high-rate overcharge, and thus potentially provide longer cycle life. However, at lower depths of discharge, where the absolute amount of overcharge is typically lower, this may not be the overriding factor that it appears to be under conditions of high stress from overcharge. Figure 11.24 shows a comparison of the average cell performance and life for test packs containing cells with either dry sinter or slurry sinter nickel electrodes. Based on these life test data, it appears that under conditions of lesser overcharge stress (40% depth of discharge), both types of sinter give comparable life performance.

Two key differences are seen in Fig. 11.24 when comparing these two types of cells. The first is the higher voltage response for the cells with the dry sinter nickel electrodes. The higher voltage is likely to result from a greater surface area

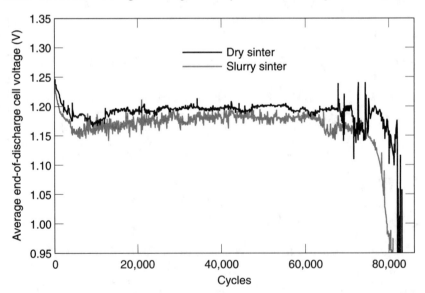

Figure 11.24. Comparison of life-test cells with dry sinter and slurry sinter nickel electrodes, cycled at 40% depth of discharge and 10°C. Cells with dry sinter started exhibiting intermittent short circuits around cycle 70,000.

for the dry sinter, as well as from differences in the loading of the active material into the sinter. The second difference is in how the cells failed. The cell with dry sinter began to exhibit intermittent short circuits around cycle 70,000, but with management lasted more than another 10,000 cycles until eventual failure from short-circuiting. The cell with slurry sinter did not exhibit such short-circuiting prior to failure. This difference in the failure mode is likely to be the result of the greater structural strength associated with the slurry sinter. However, clearly both types of sinter are capable of very long life if abusive stresses from overcharge are avoided.

11.8.1.4 Recirculating and Back-to-Back Cell Stacks
The recirculating stack design is not frequently used in today's state-of-the art nickel-hydrogen cells because the back-to-back cell-stack design can be made with a lower weight. However, life tests have been performed to evaluate which of these stacking arrangements is capable of providing the best cycle life. The most direct comparison evaluates 3.5 in. diameter, dual-layer Zircar-separated cells containing 31% electrolyte, and slurry sinter nickel electrodes. These cells, in NWSC test packs 5002E and 3314E, were tested using a 40% depth-of-discharge low Earth orbit cycle, at a temperature of 10°C. The comparison of the cells in these two tests is shown in Fig. 11.25. Both test packs were run using a recharge ratio of 1.04, with some upward adjustments at the end of the testing in an attempt to keep the cells cycling longer.

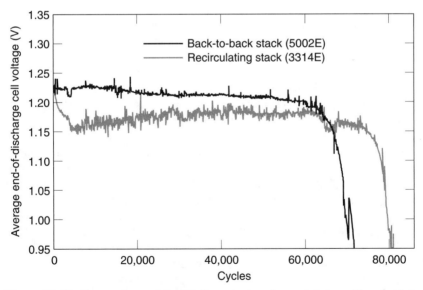

Figure 11.25. Comparison of the life from test packs containing either back-to-back or recirculating electrode stacking arrangements; the test was at 10°C and 40% depth of discharge.

The results displayed in Fig. 11.25 indicate the recirculating stack cells had a longer life (by roughly 10,000 cycles) than did the back-to-back cells. Performing a statistical analysis of the cells in these packs, one finds that the recirculating-stack cells in pack 3314E had a 14.6% longer cycle life than the similar cells in pack 5002E with the back-to-back design. While this is not a large difference in life, it does indicate that the recirculation of electrolyte through the recirculating stack design does not introduce electrolyte concentration gradients that could be detrimental to performance, and therefore that the wall wicks in these cells are capable of effectively equilibrating the water recirculation in this design. From these tests, it is not possible to determine whether the improved cycle life for the recirculating stack design was caused by improved oxygen management or by the electrolyte concentration gradient that naturally occurs in this cell design during continuous cycling.

11.8.1.5 Cell Diameter: 3.5 in. vs. 4.5 in.
Individual pressure vessel nickel-hydrogen cells are made with two different diameters, 3.5 in. and 4.5 in.** The 4.5 in. diameter is used for cells having capacity greater than 50 A h for single-stack designs, and greater than 100 A h for dual-stack designs. The scale-up of the 3.5 in. diameter cell to the larger diameter was engineered such that the gradients and stresses associated with transport, conduction, and thermal management would not be greater in the larger cell size. If this scale-up is done properly, the cycle life of the 4.5 in. diameter cells should be equal to or better than that of its 3.5 in. precursor. Low Earth orbit life tests have been performed that directly compare the cycle life of these two different cell sizes, although in most such instances design variables other than cell diameter were also altered, making a direct comparison difficult.

The most direct comparison between 3.5 and 4.5 in. diameter cells is for NWSC packs 5002G and 5402G. The 3.5 in. diameter cells (pack 5002G) have a 50 A h capacity, while the 4.5 in. diameter cells (pack 5402G) have a 90 A h capacity. Both these cell types use a recirculating stack design, with dual-layer Zircar separator, dry sinter, and 31% electrolyte. The packs were tested in a standard low Earth orbit test profile at 40% depth of discharge, and at a temperature of 10°C. The results from these two life tests are compared in Fig. 11.26. These tests were terminated after about 42,000 cycles, and at that point all the 3.5 in. diameter cells in pack 5002G had failed, while none of the 4.5 in. diameter cells in pack 5402G had begun to drop off. These two life tests suggest that the 4.5 in. diameter cell is capable of longer cycle life than is the 3.5 in. diameter cell.

Other tests have been run that also compare the two different cell diameters in combination with other variables that were also changed. If the pronounced improvement in cycle life implied in Fig. 11.26 consistently arises for 4.5 in. cells

---

**A larger 5.5 in. diameter cell has also been built, but has not been extensively tested in low Earth orbit cycling profiles.

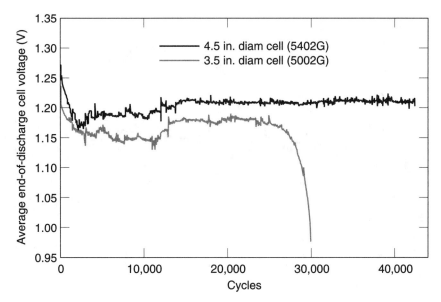

Figure 11.26. Low Earth orbit cycle life for nickel-hydrogen cells having a 3.5 in. diameter (50 A h) and a 4.5 in. diameter (90 A h) at 40% depth of discharge and 10°C.

compared to 3.5 in. cells, it should also be seen in some of these other tests. One may look for such consistency by comparing the cycle life of the cells in pack 5002H (3.5 in. diameter) to that of the cells in pack 5402H (4.5 in. diameter). The cells in these two packs were similar in design to those shown in Fig. 11.26, except that they both had a different manufacturer and the smaller-diameter cells had a back-to-back stack rather than a recirculating stack (which is expected to cause only a 14.6% improvement in life for the larger-diameter cell).

The results from these two life tests are shown in Fig. 11.27. Unfortunately, these two life tests were discontinued with most of the cells still working well after about 42,000 cycles. For pack 5402H (4.5 in. diameter), all ten cells in the pack were functioning well and showed little evidence of degradation when the test was terminated, as indicated by the average behavior shown in Fig. 11.27. For pack 5002H (3.5 in. diameter), four of the ten cells in the pack failed between 35,000 and 39,000 cycles, and most of the other six cells were beginning to show evidence of the voltage degradation that signals end of life. While one cannot quantify the cycle life of the 4.5 in. diameter cells relative to that of the 3.5 in. cells from this test, it is clear that the larger-diameter cells had much more than the 14.6% improvement in cycle life anticipated from the effect of the recirculating stack.

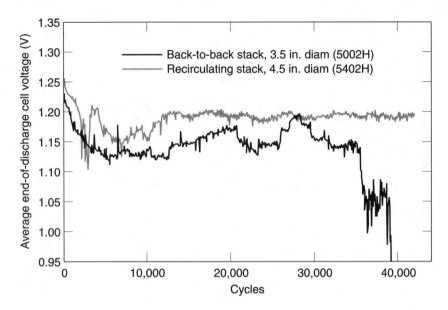

Figure 11.27. Low Earth orbit cycle-life performance comparison for nickel-hydrogen cells having a 3.5 in. diameter (50 A h back-to-back) and a 4.5 in. diameter (90 A h recirculating) at 40% depth of discharge and 10°C.

Thus, the results of this test are consistent with that suggested by Fig. 11.26. The larger, 4.5 in. diameter cell size appears to provide significantly improved cycle-life capability, suggesting that the scale-up of cell size was successful in producing a cell design that in no way compromised the cycle life of the smaller-diameter cell design.

11.8.1.6 Longest-Running Nickel-Hydrogen Cell Life Test

A number of life tests of nickel-hydrogen cells have been operated in excess of the typical ten-year lifetime needed for the longest low Earth orbit missions. The one that has the record for the longest-running nickel-hydrogen test is pack 5000H at NWSC. This test has operated since 1988, and has run approximately 115,000 cycles for this 10-cell pack over the past 19 yr. This test is operating using a standard low Earth orbit cycling profile at 40% depth of discharge and –5°C. The cells are 3.5 in. diameter cells containing 31% electrolyte and dual-layer Zircar separator. The stacking arrangement of the internal electrodes is the back-to-back design, and the nickel electrodes use dry sinter. These 10 cells appear to exhibit failures over the range of about 100,000 to 115,000 cycles. The life performance of a typical cell in this pack is shown in Fig. 11.28. This cell gave a cycle life of 113,300 cycles before falling below 1.0 V.

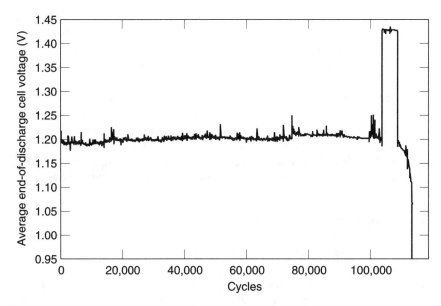

Figure 11.28. Performance of pack 5000H at 40% depth of discharge and –5°C. This is the longest-running nickel-hydrogen cell-life test.

A key to the extremely long life of these cells is the recharge ratio, which was about 1.02–1.03, except for a period between 70,000 and 80,000 cycles when it increased to 1.07. This period of greater overcharge is likely to have added extra wear to these cells. Another period of accelerated wear occurred over about a 5000-cycle interval (103,000 to 108,000 cycles) during which the cells experienced recharge, but essentially no discharge. The end-of-discharge voltage during this period was about 1.42 V, as shown in Fig. 11.28. However, this test demonstrates the extremely long cycle life and ampere-hour throughput possible from nickel-hydrogen battery cells if operated at low temperatures with minimal overcharge stress.

11.8.1.7 Variability in Nickel-Hydrogen Cell Lifetime
Typically, life tests are run on series-connected cells in battery packs, with from 5 to 10 cells in the pack. The cells in a given test pack all experience similar electrical throughput and a similar thermal profile during operation. Thus the variability in lifetime over the cells in a pack provides an indication of how reproducibly cells and their components are manufactured and assembled. The cell-to-cell lifetime variability is a key factor in determining the reliability of a battery. This variability has been analyzed for a number of test packs that were run at NWSC-Crane until essentially all the cells in each pack had failed. A typical test that shows the variability in cycle life between similar cells in a test pack is indicated

in Fig. 11.29 for a 60% depth of discharge in 8-cell pack 3604X. The standard deviation bars shown in Fig. 11.29 were derived from the entire database of nickel-hydrogen cell life tests. This figure shows that the variability seen in a typical life test is consistent with the statistical standard deviation seen across the entire nickel-hydrogen cell life-test database.

While the absolute cell lifetime is very different for different conditions of depth of discharge, temperature, or overcharge, the standard deviation in lifetime for cells made using typical aerospace manufacturing controls has been statistically determined to be 11.63% of the mean cycle life.[11.1] While this lifetime variability has been determined for ManTech cells with two layers of Zircar separator, with similar manufacturing controls the variability in life for other types of cells is expected to be similar.

11.8.2 Geosynchronous Profile Tests

Most aerospace contractors have carried out geosynchronous life tests on nickel-hydrogen cells. The normal geosynchronous operational profile involves a 45-cycle eclipse season every 6 months with a 4.5-month noneclipse season (solstice season) between the eclipse seasons. During the eclipse season the cycles vary in duration and each cycle occurs over a 24 h period. Geosynchronous tests are generally

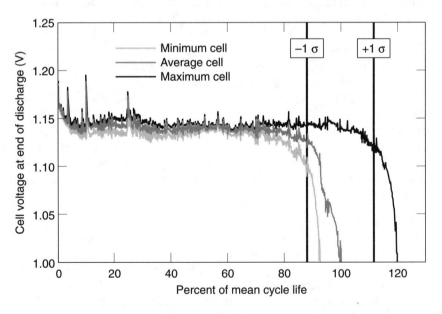

Figure 11.29. Typical variability in lifetime for nickel-hydrogen cells, from the 8 cells in NWSC Crane pack 3604X. A standard deviation of 11.69% of the mean lifetime was derived from the entire nickel-hydrogen cell life-test database.

run using an accelerated time profile that eliminates most of the 9-month solstice time during each year of operation. This acceleration method assumes that most of the wear results from the cycles, and that the wear from the solstice calendar life and overcharge is negligible. This assumption is only reasonable if the trickle-charge rate used during the real solstice periods is low enough that it does not provide significant overcharge in excess of the self-discharge rate.

Very few geosynchronous life tests have been run to the point where any of the cells have been observed to fail, simply because even with an accelerated test profile, the cells degrade very slowly. Accelerated geosynchronous life tests are typically run for 40–50 accelerated eclipse seasons, which involves 5–6.25 yr of calendar life at eight seasons per year, and then they have been halted with the cells still functioning acceptably. The assumption in these life tests is that significant cycle-life margin has been demonstrated over that required during a 15 yr geosynchronous mission (30 eclipse seasons). The potential fallacy in this assumption is that the calendar-life and overcharge-related wear that occurs during the solstice seasons that have been eliminated from the tests is not significant. If the added wear that has been neglected by running an accelerated test rather than a real-time test is sufficiently great, it may be sufficient to offset the margin demonstrated by running the life test for more than 30 seasons, and the battery reliability may be less than what is required at the 15 yr point in the mission.

Because the time required to run a real-time geosynchronous life test is usually prohibitively long, one must typically depend on life projection or wear models to determine reliability under real-life operating conditions. Such a model has been developed, and had been found to be well validated against the available life-test data for both low Earth orbit and geosynchronous test profiles, both real-time and accelerated,[11.1, 11.11] and is described in detail in chapter 8. This model typically predicts that nickel-hydrogen cells with dual-layer Zircar separator and 31% electrolyte, as well as single-layer Zircar separators with 26% electrolyte, are capable of a geosynchronous lifetime of more than 15 yr if unnecessary overcharge is avoided during solstice periods.

Other types of orbital profiles, such as medium Earth or highly elliptical orbits that typically have a 12 h orbital period with two eclipse seasons per year, have been used in a few life tests for nickel-hydrogen cells. These life tests generally are run in an accelerated manner eliminating most or all of the solstice period, much as for geosynchronous tests. Therefore these life tests are potentially subject to the same issues having to do with the effect of the wear during the solstice periods, which may be evaluated with an appropriate wear model.

11.9 Effects of Cell Storage on Cycle Life

The effects of cell storage on cycle life have been evaluated by three different methods. The first has involved life-testing cells both with and without long-term ground storage. The second method has involved destructive physical analysis of cells that have been stored for long periods to measure the physical changes that occur in the cells during storage that could affect cycle life. The final method has involved models that predict the chemical and physical changes during storage based on the rates from destructive physical analysis studies, and that translate these changes into a projected loss of cycle life. The projections from these models may then be validated by comparison with the available life-test data. The modeling approach to evaluating the effects of storage on cell cycle life has been discussed in detail in chapter 8, and will not be discussed here.

The major chemical changes seen to occur in nickel-hydrogen cells are loss of the nickel precharge, oxidation of platinum metal in the anode (hydrogen electrode), and oxidation of nickel metal in the cathode (nickel electrode). The loss of nickel precharge results from the combination of the platinum and nickel metal oxidation, as well as some isolation of nickel precharge such that it is no longer electrochemically active. The overall rate at which nickel precharge is lost is sensitive to temperature, as is shown in Fig. 11.30. The data points in Fig. 11.30 were measured from destructive physical analysis studies on cells that had been stored from 5 to 9 yr at various temperatures. These results show that the 12–15% nickel precharge initially present in a cell can typically last for 4 yr of storage at room temperature and can remain active for well beyond 10 yr if most of the storage

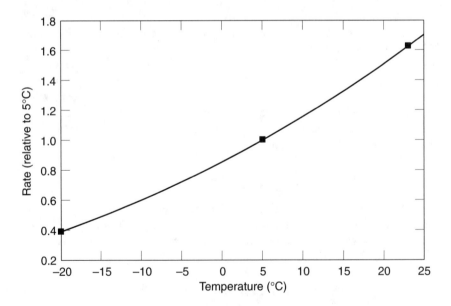

Figure 11.30. Rate at which nickel precharge is lost in nickel-hydrogen cells during discharged storage as a function of storage temperature.

11.9 Effects of Cell Storage on Cycle Life

time is in a freezer at –20°C. The maintenance of nickel precharge is important for maintaining optimum cycle life of nickel-hydrogen cells, because it enables the cells to be stored safely at a minimal state of charge while preventing damage to the nickel electrodes from exposure to a reducing chemical environment during storage.

Another chemical change that accompanies the extended storage of nickel-precharged nickel-hydrogen cells is oxidation of platinum metal in the anode (hydrogen electrode). The platinum slowly oxidizes over years of storage. It builds up in the alkaline electrolyte to a millimolar concentration as platinate complex ions, and is chemisorbed into the nickel electrode to a level commensurate with the amount of γ-NiOOH to form an oxidized Ni-Co-Pt compound that is thermodynamically stable as long as the nickel precharge is not discharged. Eventually the nickel precharge will be lost as the cell is operated, and the Ni-Co-Pt compound is discharged, making it lose its associated platinum. The platinum is subsequently plated back onto the hydrogen electrode during extended periods of cell recharge.

The final chemical change in the nickel-hydrogen cell during storage is the slow corrosion of nickel metal in the sintered structure of the nickel electrodes. This corrosion process, which is driven by the oxidative potential of the nickel precharge, is very slow and requires decades to significantly degrade the performance of the nickel-hydrogen cell. However, because this same process is a key contributor to normal wear-out of nickel-hydrogen cells, some loss of lifetime is inevitable as a result of storage. Destructive physical analysis of failed nickel-hydrogen cells from long-term life tests has indicated that the cells tend to fail (when approximately a C/2 discharge rate is used) after about 30% of the nickel metal in the sinter of the nickel electrode has undergone corrosion. On this basis, observation of 3% corrosion in the sinter after a period of storage would suggest that the cells would have about a 10% lower cycle life than cells that were not stored.

Several life tests have been carried out on cells that were divided into two groups. One group was life-tested with no storage, and the other group was stored five years and then put into the identical life test. Comparison of the performance and eventual lifetime of these cell groups should allow a direct measurement of the effects of storage on cycle life, and enables an important validation method for storage models. Figures 11.31 and 11.32 show life tests that examine a 150 A h cell design operating at 25% depth of discharge that was made by two different cell manufacturers to the same general design specification, but used the nickel and hydrogen electrodes, and other cell assembly and activation procedures specific to each manufacturer. This design used a single layer of Zircar separator, back-to-back stack design, and 31% electrolyte concentration.

The first test (Fig. 11.31) suggests that there is approximately a 20% loss of cycle life as the result of 5 yr of storage at room temperature when one looks at the lowest-performing cell of the five in each test. However, when one examines the

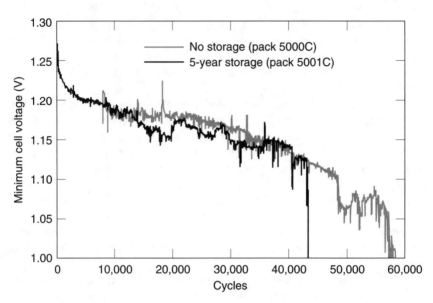

Figure 11.31. Life tests at 25% depth of discharge with and without 5 yr of room temperature discharged storage for cell type 1.

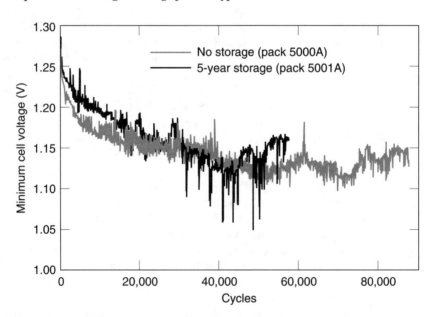

Figure 11.32. Life tests at 25% depth of discharge with and without 5 yr of room temperature discharged storage for cell type 2.

test data with no storage more closely, one finds that the drop in voltage between 55,000 and 60,000 cycles was caused by only two of the five cells in the test, and these two degraded cells recovered performance at around 70,000 cycles and operated at a stable end of discharge near 1.15 out to the end of the test at about 87,000 cycles. The test data with 5 yr of storage showed a drop below 1 V for one of the five cells near 43,000 cycles. While this cell continued to operate for the 57,000 cycles the test was run, it stayed below 1.0 V at the end of discharge and did not recover. However, the other four cells continued to give stable and good performance out to the end of the test at 57,000 cycles. Thus, it is somewhat uncertain how much of the observed degradation was because of out-of-family cells, as opposed to true storage effects on cycle life.

The cell from the other manufacturer (Fig. 11.32) illustrates a frequent problem with these long-term life tests. Here, the tests were terminated with all the cells still functioning well, and thus do not give a clear picture of how the 5 yr of storage affected cell lifetime. All one can glean from these tests is that the lifetime with no storage at 25% depth of discharge was greater than 87,000 cycles, and the lifetime after 5 yr of storage was greater than 57,000 cycles. These tests ran for well over 15 yr, but eventually were terminated before they gave the desired results because of a lack of funding to maintain spacecraft nickel-hydrogen battery technology.

11.10 Effects of Specialized Cell Design Features on Lifetime

The common pressure vessel nickel-hydrogen cell design was pioneered by EaglePicher Industries. In it, the two stacks within a dual-stack individual pressure vessel cell are connected in series rather than in parallel, thus providing a cell with twice the voltage but half the capacity of the traditional individual pressure vessel cell. Only a limited number of life tests on the common pressure vessel cell design have been performed. It would be expected that the common pressure vessel design could, at best, only approach the life of its individual pressure vessel cousin, because of its additional degradation modes from gas and electrolyte transport between the two stacks in the pressure vessel.

Two life tests of common pressure vessel cells are indicated in Fig. 11.33 at 40% depth of discharge and at temperatures of +10 and –5°C. These tests both used a recharge ratio of 1.04. The test run at –5°C (pack 3006L) was stopped after completing about 56,000 cycles with all six of the cells in the pack still functioning, although as indicated in Fig. 11.33, the voltage was starting to drop off. The pack operated at +10°C (pack 3005L) exhibited initial cell failures at about 45,000 cycles. The performance of this pack can be compared directly to that of individual pressure vessel cell pack 5002E (also tested at NWSC Crane), which was operated under identical conditions of temperature, recharge ratio, and depth of discharge. The first individual pressure vessel cells in pack 5002E failed after

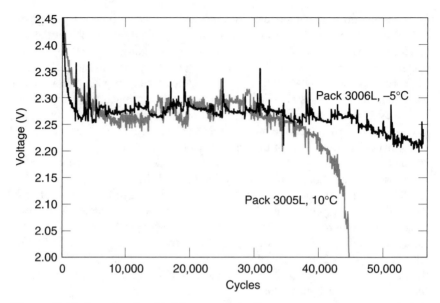

Figure 11.33. Low Earth orbit life-test results for common pressure vessel nickel-hydrogen cells at 40% depth of discharge.

about 63,000 cycles, indicating that the common pressure vessel cell is capable of only 71% of the life of state-of-the-art individual pressure vessel cells.

Extrapolation of the voltage trends in Fig. 11.33 for pack 3006L suggests that failure would probably have occurred between 60,000 and 64,000 cycles. This indicates about a 35% longer cycle life at the lower −5°C temperature. This result is reasonably consistent with the approximately 50% cycle-life improvement expected at the lower temperature (based on the individual pressure vessel database), offset back to about 35% improvement by the 1.04 recharge ratio, which is considerably higher than needed at the lower temperature.

Another type of nickel-hydrogen battery, which was pioneered by Johnson Controls and then eventually transitioned to EaglePicher, is the single pressure vessel design. This design features 22 series-connected cells contained within a single 10 in. diameter pressure vessel. The cells each contain their own reservoir of electrolyte within each stack container, but they share the hydrogen gas in common. Each cell container is equipped with special gas vents that minimize the transport of oxygen gas or water from one stack to another.

The single pressure vessel battery, in both 50 and 60 A h capacities, has been used successfully in the satellites of the Iridium constellation. These batteries continue to operate after approximately 10 yr of use in these satellites. The test data that indicate how long the single pressure vessel battery will last compared

11.10 Effects of Specialized Cell Design Features on Lifetime

to batteries made from individual pressure vessel cells are not extensive. The best comparative data come from two tests conducted at Crane, tests 3003L and 3004L. The performance of the single pressure vessel batteries in these two tests is shown in Fig. 11.34.

The comparison of the results from these two tests to the individual pressure vessel database is complicated by the fact that the distribution of performance degradation between the individual cells in the single pressure vessel batteries is not known. If the average cell voltage when the first cell in an 8–10 cell individual pressure vessel pack fails is typically below about 1.1 V at these depths of discharge, one can assume that when the single pressure vessel voltage drops below 22 V, two of the cells in the single pressure vessel battery have probably dropped below the 1 V individual cell failure level. This assumes that the distribution of cells in the single pressure vessel is similar to the 11.63% spread within individual lots of individual pressure vessel cells. When this assumption is made, the data from the tests in Fig. 11.34 can be analyzed to obtain the life of the single pressure vessel relative to the life of individual pressure vessel cells run under identical test conditions. This analysis indicates that the lifetime of the single pressure vessel is 77.6% of the lifetime that is possible from a 22-cell battery of individual pressure vessel cells.

Another nickel-hydrogen cell design that has shown promise of providing significantly longer lifetime than the standard ManTech cells is the dual-anode

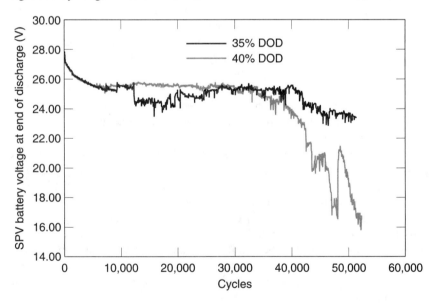

Figure 11.34. Performance of single pressure vessel batteries 3003L and 3004L at 35% and 40% depth of discharge respectively. The recharge ratios were 1.04 and the operating temperature was 10°C.

design. This design replaces the back-to-back nickel electrode unit in the standard stack unit with a single nickel electrode, thus equally using both faces of each nickel electrode for current flow. This produces significantly lower impedance, more efficient transport, and presumably much lower wear rates in the nickel electrodes. Because nickel electrode wear almost always is the cause of nickel-hydrogen cell failure, this design seems capable of giving much-improved lifetime. With even a single layer of Zircar separator against each face of each nickel electrode, the cell contains as much electrolyte as the normal double-layer Zircar-separated ManTech cell. With two layers of Zircar separator, the dual-anode design contains much more electrolyte than do standard designs.

The principal drawback of the dual-anode design is that it is at least 10% heavier than the standard-design cells, largely from the added weight of the additional hydrogen electrodes required by the stack design. Additionally, if this design is to last longer than the 10–15 yr it takes to wear out standard cells, it will take more than 15 yr to gather the data needed to prove the reliability of the design. This factor, along with the continual pressures to reduce weight in satellite batteries, has prevented the dual-anode nickel-hydrogen cell design from being adopted in any satellite power systems.

However, dual-anode cells have been built and tested. An early lot of 60 A h cells operated in a life test at 60% depth of discharge for nearly 11,000 cycles before being terminated because of anomalous wear resulting from accidental exposure of the cells to lengthy periods of C/2 overcharge and several temperature excursions to 35°C. A follow-on lot of 65 A h dual-anode cells has been on life test at 60% depth of discharge for 7 yr, and has accumulated more than 40,000 cycles of stable operation. These cells are operating at –5°C with a recharge ratio of about 1.011, and are cycling between about 15% and 75% state of charge. Model projections indicate that these cells could operate at 60% depth of discharge for 90,000 to 100,000 cycles before failure under their present operating conditions.

Because of the ever-present pressure to reduce battery weight in satellite power systems, nickel-hydrogen cells have been built with a single layer of Zircar separator. The single-layer–separated cells with 26% electrolyte have given lifetimes about as long as have dual-layer–separator cells with 31% electrolyte. Because the 31% electrolyte gives better cell capacity (see Fig. 11.19), cells have also been built with 31% electrolyte and a single layer of Zircar separator. This design may be significantly more likely to fail from leakage through the separator because particulate extrusion from the nickel electrode into the thinner separator is much more pronounced with the higher electrolyte concentration.

Extensive life testing of the single-layer Zircar separator with 31% electrolyte has not been performed to determine how its life compares with that of the dual-layer separator cells. The best tests run to date are probably those shown in Fig. 11.31 at 25% depth of discharge for cells made by two different manufacturers. The test shown in Fig. 11.31 indicates that the weakest cells gave nearly 60,000 cycles at 25% depth of discharge, but the other cells were still operating well at 87,000 cycles. At 25% depth of discharge, the 60,000 cycle lifetime is only about half that expected for a dual-layer–separator cell. This test suggests that the single-layer separator in combination with 31% electrolyte may result in more variability in cell lifetime than the 11.63% found for dual-layer–separator cells, and that the weakest cells may have up to 50% of the lifetime. The results in Fig. 11.32 for cells made by another manufacturer show no failures after 87,000 cycles for five cells, which suggests that the lifetime for these cells is at least ~75% of that expected for cells with dual-layer separator. One conclusion to be drawn from these test results is that the single layer of separator can make cell lifetime more sensitive to the details of the cell design and assembly procedures.

A final design feature that has been included in some nickel-hydrogen cells is a set of platinum catalyst stripes on the walls of the pressure vessel, which can serve as a catalytic scrubber to remove any oxygen from the hydrogen gas stream that comes out of the side of the stack during recharge. This feature is intended to recombine the oxygen on the wall of the pressure vessel where the heat that is generated can be more easily removed from the cell than if the recombination occurred on the hydrogen electrodes. This wall recombination capability has been demonstrated to work effectively using infrared imaging of the pressure vessel wall during overcharge.[11,12]

It is expected that if a cell is operated with significant overcharge, the recombination wall wick will significantly decrease the wear on the cells from oxygen recombination and heating. However, under conditions of minimal overcharge, such as the ones used when cycle life is to be optimized, the recombination wall wick is not likely to have a large impact on cell life. The recombination wall wick has the potential for decreasing cell reliability as a result of flaking of the platinum from the wall, if the platinum is not well bonded to the wall wick on the inner surface of the pressure vessel. While the flaking is a potential problem, properly manufactured cells have not shown any loss in reliability from the recombination wall wicks.

11.11 Specialized Cell Life Tests

A number of life tests have been run on nickel-hydrogen cells to evaluate special conditions that may arise in unusual operating modes during satellite usage. These have included tests that evaluate cell performance for extremely high-rate pulse discharge, tests that evaluate the effect of large thermal gradients across a battery, tests that determine the effects of various forms of capacity loss during storage on cycle life, and tests that examine the effects of extremely low- or high-temperature operation. This section briefly discusses several such tests that involve pulse discharge and battery thermal gradients.

11.11.1 Pulse Discharge Tests

A life test has been run on a pack of nickel-hydrogen cells that operated at 40% depth of discharge, but in which half of the cells had a several-second discharge pulse of 2C periodically applied to them. Sometimes the discharge pulse would be applied during normal discharge, and sometimes during recharge. Its effects on the state of charge of the cells were negligible over the entire charge/discharge cycle, and all the cells maintained normal performance for more than three years of low Earth orbit cycling.

After 17,000 cycles, one of the cells that experienced the pulse discharges failed from an internal short circuit. Destructive physical analysis of this cell revealed a hard short that formed at the outer edge of a nickel electrode near the top of the stack, immediately across from the location where the current-collecting tab was spot-welded onto the electrode. This is precisely the location on the electrode edge where the highest current densities are expected during the high-current pulses. It is not known exactly how the pulses could have caused this short, but this test suggests the possibility that high-current pulsing can degrade nickel-hydrogen cell cycle life in some situations.

11.11.2 Thermal Gradient Tests

Another life test has been run to examine the impact of thermal gradients across the cells in a battery on the overall performance of the battery. While it is generally agreed that thermal gradients have a negative influence on battery performance, sometimes these gradients occur as the result of unexpected anomalies. For this reason, a test was performed on a four-cell battery of 100 A h cells, in which two of the cells were kept 10°C warmer than the other two by being mounted on a separate cold plate. The temperatures and recharge ratios of the battery were parametrically varied over a range of typical operational levels. After 1000 cycles at 25% depth of discharge for each operating condition to stabilize the performance of the cells, the pack was discharged until the lowest cell fell below 1 V to determine the pack capacity. The results from this test are shown in Table 11.1.

In all cases in this test the battery capacity was limited by the capacity of the higher-temperature cells. The results in Table 11.1 clearly show the detrimental effect that a thermal gradient can have on the capacity of a nickel-hydrogen battery, particularly if the warmer cells are allowed to rise above +5°C. If all the cells are kept cold, battery capacity can be maintained relatively high; however, one must take care not to let the colder cells drop below –10°C or there could be risk of electrolyte freezing in some situations. The results in Table 11.1 also show that stable long-term operation at temperatures well above 10°C can be difficult to maintain, and that stable battery capacity can be very low at these elevated temperatures even if significant overcharge is applied to the battery.

Table 11.1. Stabilized Battery Capacity with a Thermal Gradient for Various Operational Modes

Temperatures (°C)		Recharge Ratios (A h)		
Cold cells	Warm cells	1.08	1.06	1.04
−5	+5	92.38	91.85	89.54
0	+10	78.29	66.03	39.95
+5	+15	54.00	34.83	Not stable

11.12 References

[11.1] A. H. Zimmerman and M. V. Quinzio, "Model for Predicting the Effects of Long-Term Storage and Cycling on the Life of NiH$_2$ Cells," *Proc. of the 2003 NASA Battery Workshop* (Huntsville, AL, 20 November 2003).

[11.2] B. A. Moore, H. M. Brown, and C. A. Hill, "Air Force Nickel-Hydrogen Testing at NAVSURFWARCENDIV Crane," *Proc. of the 32nd Inter-Society Energy Conversion and Eng. Conf.*, Vol. 1 (Honolulu, HI, July 1997), pp. 186–191.

[11.3] B. A. Moore, H. M. Brown, and T. B. Miller, "International Space Station Nickel-Hydrogen Battery Cell Testing at NAVSURFWARCENDIV Crane," *Proc. of the 32nd Inter-Society Energy Conversion and Eng. Conf.*, Vol. 1 (Honolulu, HI, July 1997), pp. 174–179.

[11.4] A. H. Zimmerman, "Discharge Model of the Nickel Electrode: Cause for Lower Plateau Discharge," ISBN 1-56347-091-8, *Proc. of the 1994 IECEC, Amer. Inst. of Aeronautics and Astronautics* (1994) pp. 63–68.

[11.5] L. H. Thaller and A. H. Zimmerman, *Nickel-Hydrogen Life Cycle Testing* (The Aerospace Press, El Segundo, CA, 2003), p. 118.

[11.6] L. H. Thaller and H. S. Lim, "A Prediction Model of the Depth of Discharge Effect on the Cycle Life of a Storage Cell," *Proceedings of the Twenty Second IECEC, Vol. 2* (Philadelphia, PA, August 1987), pp. 751–757.

[11.7] D. P. Hafen, "Nickel-Hydrogen Reliability Update," *Proc. of the 1998 NASA Battery Workshop*, (Huntsville, AL, November 1998), pp. 503–520.

[11.8] A. H. Zimmerman, J. Matsumoto, A. Prater, D. Smith, and N. Weber, "Characterization and Initial Life-Test Data for Computer-Designed Nickel-Hydrogen Cells," *Proc. of the 1997 NASA Battery Workshop* (Huntsville, AL, November 1997), pp. 471–484.

[11.9] A. H. Zimmerman, "Life-Testing of Nickel-Hydrogen Batteries at 60% Depth of Discharge for LEO Satellite Applications," *2005 Space Power Workshop* (Manhattan Beach, CA, April 2005).

[11.10] J. Brewer, personal communication.

[11.11] A. H. Zimmerman and V. J. Ang, "Life Modeling for Nickel-Hydrogen Batteries in Geosynchronous Satellite Operation," AIAA Conf. CD 1090-347500, *Proc. of the 2005 International Energy Conversion and Engineering Conf.* (San Francisco, CA, 2005).

[11,12] J. Shue, D. Sullivan, L. Lee, G. Rao, and J. B. Ramirez, "Thermal Imaging of Aerospace Battery Cells," *Proc. of the 2004 NASA Battery Workshop* (Huntsville, AL, November 2003).

12 Degradation and Failure Modes

A number of conditions and processes are known that can lead to degradation of performance and, eventually, to failure of nickel-hydrogen cells and batteries. This chapter discusses five general categories of these degradation and failure modes:

- manufacturing-related problems
- cell or battery storage-related problems
- problems related to specific environmental exposures
- issues specific to batteries
- cycling and long-term life-related issues

In the following sections, specific problems that are possible and have in fact been observed within each of these general categories are discussed, along with techniques for mitigating each one.

12.1 Manufacturing-Related Problems

While the incidence of nickel-hydrogen cell problems related to manufacturing processes and procedures can vary considerably depending on the specific practices used by each manufacturer, the issues that are discussed here represent real problems that have been seen in producing nickel-hydrogen cells. Each of these potential problem areas should be considered in any manufacturing program that produces high-reliability nickel-hydrogen cells.

12.1.1 Strain Gauge Drift

Strain gauges are often attached to nickel-hydrogen cells to indicate cell pressure, and thus the state of charge of the battery containing the cells. The pressure of hydrogen within the cells is a relatively direct indication of cell, and thus battery, state of charge. Such cell pressure indications can be used for information only, or they can be used as an active part of the battery charge control system. The strain gauges are typically bonded to the dome of the cell pressure vessel, and they detect the subtle flexing of the pressure vessel as its internal pressure goes up or down. Each cell manufacturer generally uses a proprietary process to bond the strain gauge assembly to the cell pressure vessel. If this bonding process is not properly verified and controlled, the strain gauges can drift rapidly with time or usage, or even become detached from the pressure vessel.

Even a properly bonded and cured strain gauge, however, can exhibit drift in its readings during long periods of cell operation. Such drift can arise either from gradual material relaxation in the strain gauge assembly itself or, as is more frequently the case, from slow relaxation of the material that bonds the strain gauge assembly to the pressure vessel. Measurements have shown that repeated thermal cycling, as well as frequent exposure to changes in humidity, can cause such drift. While the strain gauge drift is generally quite slow, throughout years of cell operation it can become significant.

The drift in strain gauge pressure readings can only be unambiguously differentiated from real cell pressure growth, or from corrosion or state of charge shifts, by opening the pressure vessel, and thus independently measuring and comparing the actual pressure to that indicated by the strain gauge. Such measurements have been taken for some nickel-hydrogen cells after long-term life testing, and it is not unusual to find a 100–200 psi drift in the strain gauge pressure reading after 10 or more years of cell operation.

A classic example of how slow strain gauge drift may influence conclusions regarding cell performance is illustrated by the pressure behavior of NWSC Crane test pack 3214E during life testing. This pack, which was cycled at 40% depth of discharge at a temperature of 10°C, displayed pressure readings that gradually appeared to increase more than 400 psi during about 10 yr of cycling in a 90 min low Earth orbit test profile, as is indicated in Fig. 12.1 for two cells in the test pack. This large pressure increase suggested that this particular cell was experiencing significant nickel corrosion in the positive electrodes. However, after about 10 yr of cycling, one of the cells equipped with a strain gauge in this test pack was disassembled. During cell disassembly the actual pressure in the fully discharged cell was measured and compared to that indicated by the strain gauge. This comparison showed that the actual pressure within the fully discharged cell was about 118 psi, which was much lower than the more than 400 psi of pressure growth suggested by Fig. 12.1 and indicated in the discharged cell by the strain gauge. This kind of apparent drift in strain gauge readings may not be unusual during extremely long periods of cell operation.

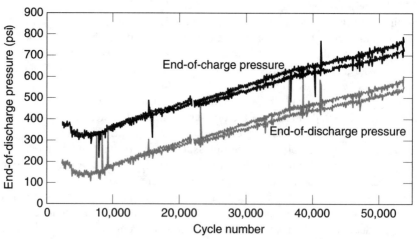

Figure 12.1. Pressure trends observed in Crane Pack 3214E during cycling at 40% depth of discharge.

Strain gauge drift should probably be regarded as a normal aspect of nickel-hydrogen cell operation, and it is one reason why absolute pressure readings cannot be treated as a precise and stable measure of cell state of charge throughout the entire cell lifetime. Periodic calibration of the strain gauge pressure indications against actual cell state of charge, performed either by fully discharging the cell to zero state of charge or by fully recharging into a condition of overcharge, allows the absolute strain gauge readings to be correlated with the actual operational state of charge range of the cell. However, the strain gauge readings do provide an excellent relative indication of changes in cell state of charge throughout the entire cell lifetime.

12.1.2 Cell Contaminants

A number of contaminating materials have been observed to deleteriously influence nickel-hydrogen cell behavior. The most common of these contaminants and their general effect on cell performance are discussed here.

Organic contaminants tend to fall into two general classes. The first consists of materials that are relatively stable in the cell environment, such as pump oils and plastics. This class of chemically inert materials includes oils that affect the wettability of the cell components, and thus the amount of electrolyte that the cell retains during the cell activation process. The performance of the hydrogen electrode can also be affected by oils, which decrease the wettability of the platinum black catalyst on the hydrophilic side of the hydrogen electrode. Similarly, contaminating surfactants (such as Triton X-100) can degrade performance by increasing the wettability of the porous Teflon layer on the hydrophobic side of the hydrogen electrode. In addition, stable plastics that may be present in the porous nickel electrode, either by design or by accident, can block electrochemical access to the active materials.

The second type of organic contaminant is a broad class of materials that can be chemically or electrochemically oxidized or reduced during cell operation. These materials typically end up in the form of carbonate in the electrolyte (if they are oxidized at the nickel electrode), or methane in the gas phase if they are prone to reduction and hydrogenation at the hydrogen electrode. The carbon reservoir that is created by such organic contaminants can shift between these two forms (carbonate and methane) by fully discharging the cell (which burns methane at the hydrogen electrode), or by overcharging (which can slowly hydrogenate carbonates to methane at the hydrogen electrode).

This type of organic contaminant tends not to have a noticeable effect on cell performance unless quite significant quantities are present. For example, up to several percent of the electrolyte in the form of carbonate can be tolerated before cell performance is seen to degrade. Methane levels greater than about 10 psi can result in the observation of a depressed discharge voltage toward the end of discharge as the methane is oxidized rather than hydrogen at the catalyst in the negative electrode of the cell. A key source of carbon contamination is airborne carbon dioxide, which acts as an acidic species when present with moisture that can react with nickel hydroxide active material to produce inactive nickel carbonates.

Another contaminant of significant concern in nickel-hydrogen cells is iron. This element, if present as oxides at levels greater than about 100 ppm in the nickel electrode active material, can significantly decrease charge efficiency by preferentially catalyzing the evolution of oxygen at the nickel electrode. This will increase the self-discharge rate, as well as decrease the cell capacity level that can be achieved during recharge. Iron is also an undesirable contaminant in the hydrogen electrode. At levels greater than about 50 ppm, it can significantly increase the overpotential for the hydrogen reaction on the platinum catalyst in the hydrogen electrode.

Silicate contaminants can originate from many sources that can be generally categorized as silicate-based minerals. Asbestos separator, which was used in early nickel-hydrogen cell designs, provided a diverse mix of silicate minerals that were prone to slow dissolution in the alkaline electrolyte. This process leads to the buildup of silicate ions in the electrolyte, which can replace much of the hydroxide species that are normally present. Other sources of silicate minerals that are of concern in all cell designs include airborne clay or mineral dust, or particulates from unfiltered aqueous processing solutions.

All these sources of silicate-containing minerals will gradually react with the hydroxide in the cell electrolyte to form soluble potassium silicate. At levels greater than about 1% by weight in the electrolyte, the silicate can significantly depress the normal cell discharge voltage, particularly at temperatures less than 0°C. At levels greater than 2–3%, silicate contaminants can increase cell resistance during discharge to levels that can make the cell unusable. This effect is likely to result from immobilization of the electrolyte that is caused by the formation of silicate glass domains within the nickel electrode during discharge.

Sulfate contamination in the nickel-hydrogen cell has been observed[12.1] to cause the appearance of a large second plateau in the discharge of the nickel electrode. This effect appears to be associated with a significant reduction of the amount of cobalt additive in the active material of the nickel electrode. Sulfate contaminants seem to preferentially complex with cobalt in the active material, allowing this additive to preferentially dissolve in the electrolyte and migrate out of the nickel electrode to be deposited elsewhere in the cell. Sulfate can originate from mineral contaminants to which the cell or its components are exposed, or form from acidic airborne pollutants, such as SO_2 or H_2S, that can react with the nickel electrode active materials during electrode storage.

A final contaminant class of significant concern in nickel-hydrogen cells is inert gas. One would think, at first, that inert gases such as helium, argon, and nitrogen should have little effect on cell performance, because they are chemically and electrochemically inert in the cell environment. However, any appreciable pressure of inert gas in a nickel-hydrogen cell will significantly reduce the high-rate cell capacity. For example, 12 psi of nitrogen resulting from air ingress during cell activation has been observed to cause an 8–10% reduction in high-rate cell capacity. The reason for the capacity loss is accumulation of the inert gas in the gas screens of the negative electrode during high-rate discharge, thus blocking the facile convective flow of hydrogen into the gas screens necessary for continuous high-rate discharge. Simple diffusion of hydrogen through the blocking inert gas is generally inadequate to deliver hydrogen rapidly enough to the catalyst to support high-rate discharge. The solution to this problem is to adopt cell activation procedures that preclude the possibility of significant amounts of inert gas in the cell.

12.1.3 Conductive Particles
The presence of conductive particles in nickel-hydrogen cells can be a problem because they can cause a short circuit between the positive and the negative electrodes. Such short circuits can vary significantly in resistance, and can come and go as particles move within the cell, making their detection difficult to be ensured with the standard charge-retention tests employed during cell or battery acceptance testing. The best approach to this potential issue is to eliminate the likely sources of conductive or any other particulate contamination arising from the cell- or component-manufacturing processes.

The most common sources of conductive particulates are nickel electrodes (particularly the edges), platinum electrodes, and the various cell-welding operations. While the electrodes and other components can be cleaned and inspected to ensure they are free of particulates, most electrode-generated particulates come from the edges of electrodes as they are handled, stacked, and compressed onto the core that holds the electrode stack. Careful adherence to optimized stacking procedures is the best way to control this potential problem area. Particles that arise from weld splatter can also be eliminated by adhering to weld schedules and procedures, and by making certain that there are no significant pressure differentials across weldments during the welding operations.

12.1.4 Electrode Stacking
The assembly of a large number of nickel electrodes, separators, and hydrogen electrodes onto a supporting core is a key step in making modern nickel-hydrogen cells. However, it is also a procedure that can significantly affect the reliability of the cells. As these components are stacked onto the core, their edges must be protected from being bent or chipped, because such types of damage can cause either immediate or latent cell shorts.

Once the components are stacked onto the core, they must be compressed to immobilize them in a stack where all the component surfaces are firmly held in intimate contact. Such compression can cause further movement of component edges over the core assembly, again with the possibility of bending or fracturing separator or electrode edges. This stack preloading operation is particularly critical when longer stacks involving a large number of electrodes are being assembled. In this situation, incremental stack compression at several stages of partial assembly can help significantly in ensuring a reliable stacking operation.

12.1.5 Mechanical Stack Stability

Maintaining the mechanical stability of the electrode stack in a nickel-hydrogen cell is critical for ensuring reliable cell performance. The mechanical stability of the stack is provided in the axial, or z, direction of the stack, by the core on which the stack is compressed between end plates, as illustrated in Fig. 12.2. Stability in the direction perpendicular to the core axis (x-y axes) is provided by the weld ring onto which the core is mounted, as well as by the rigidity of the stack assembly itself. This assembly must be stable within the pressure vessel with respect to the vibration and shock environments that the cell may experience.

The proper compression of the stack is one of the most important mechanical requirements of the cell internal to the pressure vessel. This compression must be great enough to immobilize all the stack components into a single rigid unit, but it must also be low enough to allow for some electrode expansion during the

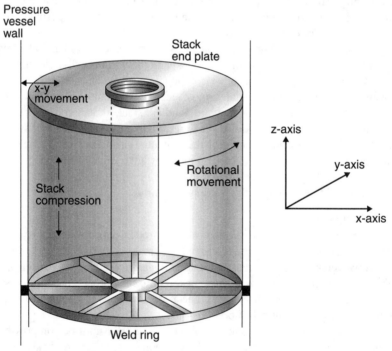

Figure 12.2. Typical mechanical support structures for the stack within nickel-hydrogen cells, and the types of stack or component movement that can occur from vibration or shock.

lifetime of the cell. If the compression is too low, plates can warp and lose the required interfacial contact area, or plates can vibrate and rotate (see Fig. 12.2) within the stack relative to each other when the cell is exposed to vibration or shock. Such interstack component movement can cause electrode leads to break, or short circuits to form as the result of significant damage coupled with lead and stack component movement.

The stability of the stack perpendicular to its z-axis is provided by the attachment of the stack to a fixed weld ring at one or both ends. At the weld ring attachment points, the stack cannot move relative to the pressure vessel walls. However, at the unsupported end of the stack, particularly if it is a longer stack, the stack assembly can vibrate to produce x-y movement toward the pressure vessel walls, as shown in Fig. 12.2.

If the amplitude of this x-y movement becomes large enough, the edges of the stack can hit the inside of the pressure vessel wall, potentially causing damage to both the wick material coating the inner wall and the edges of the stack assembly. Such damage is undesirable because it can compromise the performance of the wall wick and the stack, and it is an indication of a mechanically unstable cell design for particular conditions of vibration or shock.

12.1.6 Pressure Vessel Reliability

A critical component for a reliable nickel-hydrogen cell is a pressure vessel that can reliably contain high-pressure hydrogen gas for the decades of operation demanded of nickel-hydrogen cells. It is clear that pressure vessels and their welds must be carefully inspected and tested to ensure high reliability. The pressure vessels used in modern nickel-hydrogen cells are made by welding together dome/cylinder components that are hydroformed from Inconel sheet stock. The hydroformed dome and cylinder parts are typically helium-leak tested and inspected for flaws before cell assembly is begun. Special dye-penetrant procedures are typically used to inspect for flaws that could propagate into leaks with repeated pressure cycling. The critical flaw size that is allowed in the inspection procedure is a function of the thickness of the pressure vessel material and the expected maximum cell pressure; thinner-walled vessels have a smaller allowed critical flaw size.

The welds where the cylindrical parts of the cell are joined (girth welds), where the terminals pass through the pressure vessel domes, and where the electrolyte fill tube is attached to the dome or sealed off, must also be inspected for potential flaws and leaks. Dye-penetrant and visual (with magnification) inspection of these welds can detect weld flaws that are exposed to the external surface. Helium-leak testing of the fully welded cell can detect weld flaws that result in leaks. Voids or contaminants that are interior to the welds, and that could cause latent leaks later in life, can only be detected by X-ray inspection using a spatial resolution that is appropriate to detect the critical flaw size or critical level of weld porosity. Such inspection should be performed on all cells in a lot, or on a cell sample size large enough to ensure a statistically acceptable risk of undetected critical weld flaws.

The welds where electrolyte fill tubes are sealed are special among the welds in nickel-hydrogen cells. They cannot be checked for leaks using helium-leak tests (because the cell is sealed and activated and can contain no inert gas). These fill-tube welds can be prone to weld flaws as a result of residual electrolyte contamination on the metal surfaces that are welded. While the fill tube is a low-stress region in terms of weld flaws propagating through layers of unflawed Inconel metal, electrolyte contaminants can leave large weld voids that have interconnecting crevices. The combination of dye-penetrant inspection (to detect crevice-to-surface paths) and high-resolution X rays (to detect internal weld voids and crevices) can ensure reliable fill-tube welds.

In general, contaminants should be controlled in the Inconel material used for nickel-hydrogen cell pressure vessels. Chemical analysis of Inconel samples can detect contaminants from hydroforming or other sources that can reduce pressure vessel reliability. Examples include aluminum inclusions or sulfide lubricants that may exist in or on the Inconel. While these contaminants may not, by themselves, be large enough to cause a leak path to develop, their presence can result in significant degradation of the material properties of the Inconel, particularly in the heat-affected zones around the cell weldments. Without the inherent strength of the Inconel alloy, such defects can result in propagation of leaks through the pressure vessel wall in high-stress regions, such as areas around girth welds.

The terminal seals in modern nickel-hydrogen cells are compression seals, whether of the Ziegler type, which compresses a molded nylon material within the terminal assembly, or the Teflon type, which torques Teflon sealing rings within the terminal assembly. All of these seals require clean and particulate-free sealing surfaces, although this requirement is more critical for the Teflon seals because the sealing surfaces cannot easily mold themselves around the contaminants. It is also important that the seal have adequate compressive margin to accommodate the differential thermal expansion or contraction of the seal and terminal materials throughout the temperature range that the cell may experience. It is typically at the lowest temperatures that the compressive forces within the terminal seals are at their lowest. Helium-leak testing of the seals in each cell, particularly when performed at temperatures well below those expected during cell operation, can ensure good terminal-seal hermeticity.

12.1.7 Negative Electrode Sintering

The platinum black catalyst in the hydrogen electrode is typically sintered with a binder onto a metal screen along with porous Teflon backing. If this sintering is not properly carried out or the screen surfaces are not properly primed, the layer containing platinum black will not have the required physical integrity. Although its electrical performance will initially be adequate, it will degrade with extensive operation and can begin to shed conductive platinum-containing particles. The best indication of a sintering issue of this kind can be observed by testing the adherence of the platinum black layer to the negative plate. The platinum black layer can easily be scraped from the underlying screen of an improperly sintered negative plate, while significant force is required to scrape this layer free if it is properly sintered. Alternatively, simply shaking a negative electrode vigorously over a sheet of white paper will yield black fragments if the sintering has not been properly done.

In addition to inadequate sintering, another possible cause of a weak negative electrode may be an improperly prepared screen surface, to which the platinum black layer is to be bonded during sintering. If this surface is not properly primed, or if it accidentally becomes coated with a smooth Teflon layer prior to sintering, the platinum black layer will adhere poorly to the screen.

12.1.8 Weak Nickel Sinter

The sintered nickel plaque used as the starting point in making nickel electrodes must have sufficient strength to contain the active-material loading and withstand its expansion, contraction, and gas evolution during repeated cycling in the cell. If the sinter is weak (i.e., the contact points of the individual nickel particles making up the sinter are not adequately melted together), then the electrodes can swell prematurely and will typically have poorer-than-expected utilization (low capacity). This condition can be detected at the sintering stage by taking bend-strength measurements on samples of sintered plaque, a method that is commonly used by plaque manufacturers to screen for this problem. A secondary screening method that is commonly employed at the finished nickel electrode level is referred to as the "stress test." This test typically subjects a finished nickel electrode to several hundred high-rate cycles that include significant overcharge in a flooded test cell. Electrodes having weak sinter will swell and blister excessively in response to this test.

12.1.9 Nonuniform Sinter

The sintered nickel plaque used to make nickel electrodes is manufactured to an overall porosity requirement that is typically in the range of 76–80% porosity. This porosity is an average through the thickness of the plaque. A number of situations during sinter manufacturing can result in relatively large variability of sinter porosity through the thickness of the plaque. A common form of nonuniform sinter porosity stems from large voids formed in the interior of the plaque—voids that are not visually apparent at its external surfaces. They may be caused by poorly mixed binder or entrapped air bubbles during slurry sinter production. During the production of dry sinter, voids can result from nonuniform packing of the metal particles around grid wires prior to sintering, and differential settling of the particle bed during sintering. Other situations responsible for nonuniform sinter are nickel powder clumping, which can occur in the production of either slurry or dry powder plaque. The presence of foreign particles, whether they are inappropriately sized nickel particles or contaminating particles in the sinter, can also be a problem during either slurry or dry sinter production.

The primary consequence of nonuniform sinter typically is reduced utilization of the active material loaded into the pores of the sinter. This occurs because regions of higher porosity (particularly large voids) have less conductive metal surface area to make contact with the active material. There is also evidence that nonuniform sinter will tend to develop blisters at the locations of interior voids as electrodes age during long-term cycling. The formation of blisters is presumably related to regions of high local stress and reduced local strength that result where interior sinter voids occur. While nonuniform sinter can provide nickel electrodes having acceptable cycle life for some applications, more uniform sinter has been correlated with improved utilization of active material and increased capacity, as well as longer cycle life.

12.1.10 Nonuniform Nickel Electrode Loading

Active material is typically loaded into the pores of sintered nickel plaque at a density of 1.6–1.7 g per cubic centimeter of void volume, which produces nickel electrodes capable of long cycle life in nickel-hydrogen cells. This loading level, which is typically measured by weight gain, is an average through the thickness of the sinter and throughout the 50 or more square centimeters of surface area comprising a finished nickel electrode.

Ideally the active material should be uniformly loaded into all the pores at the same density. However, in practice, electrochemical impregnation processes nearly always result in some variability in loading density, both through the thickness of the nickel sinter, and across the surface, as a result of inhomogeneity within the impregnation vat. Within limits, such loading variability is normal and will not significantly degrade nickel electrode performance or lifetime. However, extreme conditions of nonuniform loading can result in low electrode capacity, blistering during cycling or stress testing, and reduced electrode and cell lifetime.

Nonuniform loading of nickel electrodes is generally not readily detected during the manufacturing flow. At best, a reduction in electrode capacity, or an increase in swelling, may be seen. Association of these issues with nonuniform loading requires a relatively detailed analysis of nickel electrodes through their cross-sectional thickness, or over wide areas of plaques removed from the impregnation vat. For these reasons, loading uniformity (or the lack thereof) is typically associated with a specific loading process, and concerted effort is made to maintain that particular process so that it can be repeated in a controlled way from one lot of electrodes to the next. The repeatability of the loading rate, utilization, capacity, and stress-test measurements on electrode samples are used to ascertain that the process has, in fact, been successfully repeated for each batch of nickel electrodes.

Loading nonuniformity can manifest itself in a number of ways. Nonuniformity through the thickness of the plaque can result in a densely loaded surface and a lightly loaded interior for some off-nominal conditions of flow, pH, concentration, temperature, and current density in the impregnation vat. Other off-nominal conditions of vat operation can result in heavy interior loading and lighter surface loading. Similarly, vat conditions can also lead to variability of loading at a more macroscopic level, causing it to vary significantly from top to bottom or side to side, depending on the location of each plaque area in the impregnation vat.

A secondary consequence of variability in loading level that can sometimes be observed is variability in the concentration of cobalt additive in the active material. Cobalt additive variability stems from differences in the solubility of cobalt complexes in the impregnation solution relative to complexes of nickel. These differences are sensitive to temperature, solution concentration, and pH conditions at the local plaque surfaces in the impregnation vat.

The uniformity issues discussed here are a major reason why it is important to use an impregnation process that is qualified based on a history of repeatedly making well-performing nickel electrodes. Along with this type of process qualification, process controls must also be in place to rapidly detect significant changes in the performance or composition of the electrodes that could signify changes in loading uniformity, or other issues important to producing good nickel electrodes.

12.1.11 Nickel Electrode Formation

After sintered nickel plaque is properly impregnated with active material at the required density, the active material must be formed. The process of stabilizing performance by creating a physically stable active-material deposit in the pores is termed "formation." As deposited, the active material consists of dense layers that coat the metal surfaces within the pores of the sinter. As the electrodes are cycled, these dense layers begin to break up and fall from the metal surfaces of the pores. These physical changes can cause dramatic changes in electrode performance as the dense layers gradually turn into a powdered deposit that uniformly fills the pores of the sintered plaque.

If nickel electrodes are put into nickel-hydrogen cells without being fully formed, the cell may have unstable performance. The cell may also develop increased or variable stack compression, and may be easily damaged by overcharge if some electrodes are more fully formed than are others. Of course, repeated cycling of the electrodes in the nickel-hydrogen cell will eventually result in complete formation.

Each electrode manufacturer has a formation procedure that should be qualified based on verification that it appropriately forms the newly impregnated plaque. Verification should typically involve scanning electron microscopy imaging of the active-material deposits obtained from electrode cross sections after formation. These analyses should demonstrate that the dense "as deposited" active-material layers have been adequately fragmented to result in a relatively uniform filling of the pores with active material.

12.1.12 Electrolyte Quantity

After nickel-hydrogen cells are fully assembled, the final step before sealing them is "activation." Activation involves filling the cell with the correct amount of electrolyte. The correct amount is the quantity that nearly fills (i.e., occupies 95–98% of) the void spaces within all the porous components. Too much electrolyte will leave free liquid to pool in the bottom or sides of the cell (depending on the cell's orientation), and too little electrolyte will leave the separator inappropriately dry (and prone to increased resistance).

Each cell manufacturer has an activation process that should leave each cell with the correct amount of electrolyte. The amount of electrolyte retained by each cell is tracked based on its weight pick-up during the activation process. All cells within each cell lot should have a relatively reproducible amount of electrolyte pick-up, and should be statistically in family with the historical average for that type of cell. The weight of electrolyte pick-up can be strongly influenced by the design specifics of the cell: separator thickness, nickel electrode thickness, electrolyte concentration, cell diameter, etc. The electrolyte pick-up is frequently expressed in terms of grams of potassium hydroxide electrolyte per ampere-hour of cell capacity. Relatively wet cell designs, which tend to be more robust in highly stressful cycling environments (low Earth orbit cycling profiles), tend to have electrolyte fill levels of about 3.5 g/A h. More lightweight cell designs for geosynchronous operation may have electrolyte pick-up levels as low as 2.9 g/A h.

Electrolyte fill levels should not only be verified relative to historical experience for a given cell design, but they should also be verified as being appropriate for the void volume known to exist within the porous cell components. This type of verification can employ a simple model of stack void volume (including wall wick), one that accounts for all stack volume that should be filled with electrolyte. Any design that has less than ~90% or more than ~103% of its expected stack void volume filled with electrolyte should be regarded as suspect. This electrolyte accounting exercise enables lot-to-lot variations in separator thickness, electrode thickness, or nickel electrode porosity to be properly considered in judging the correctness of any given cell electrolyte fill.

12.2 Storage-Related Issues

Once a nickel-hydrogen cell or battery has successfully passed its acceptance test requirements, it is often placed into storage until needed for its eventual application. On some occasions, this storage period can extend for many years. The maintenance of proper battery cell performance during storage requires proper preparation for storage and proper maintenance of storage conditions thereafter. The following sections summarize a number of potential issues related to cell and battery storage. A more detailed description of recommended storage procedures, as well as pitfalls, for various types of cells is provided in chapter 11.

12.2.1 Nickel Electrode Deactivation
The active material in the nickel electrode normally consists of a partially oxidized nickel hydroxide during the storage of discharged cells. If a cell is stored in the dry condition for long periods of time (prior to being filled with electrolyte), it is possible for a dehydration process to occur in the partially oxidized nickel hydroxide. This process results in the conversion of some of the normal active material [$Ni(OH)_2$ + NiOOH] into an electrochemically inactive nickel oxide hydroxide (Ni_2O_3H). Because this material is highly stable and electrochemically inert, a fraction of the cell capacity is lost that corresponds to the amount of Ni_2O_3H formed. This process can be prevented by the simple procedure of making sure that the dried cells (or stored nickel electrodes) are kept dry, but are not continuously exposed to an active desiccant that could drive the dehydration process. Without exposure to the extremely low humidity that an active desiccant can produce, the active materials in a dry, stored cell maintain the traces of humidity that are required to kinetically stabilize the hydroxide structure relative to the more thermodynamically stable oxide hydroxide structure.

12.2.2 Nickel Corrosion
A nickel-precharged nickel-hydrogen cell (standard design for modern cells) maintains a partially charged nickel electrode during storage in the fully discharged state. Under these storage conditions, the potential of the nickel electrode is highly oxidizing relative to the potential where the nickel metal in the sinter can undergo oxidation. Fortunately, the nickel metal passivates in alkaline electrolyte, and the corrosion process is very slow indeed. The lack of cycling during storage makes the corrosion even slower than during normal cell operation. However, the corrosion process does occur steadily (if slowly) during storage, and after many years of storage a significant amount of nickel corrosion can accumulate.

Because corrosion of the nickel sinter is one of the key processes limiting the ultimate life of nickel-hydrogen cells, long-term storage will reduce cell life capability. A quantitative model relating the expected corrosion rate to cycle life losses has been developed.[12.2] However, the test experience on which this model is based suggests that 5 yr of storage at ambient temperature will result in the loss of about 20% of the cycle life capability of a cell. With reduced temperature storage it is likely that the 5 yr could be extended to about 7 yr, with a similar 20% reduction in cycle life.

12.2.3 Platinum Plating
During the long-term storage of nickel-precharged nickel-hydrogen cells in the fully discharged state, the negative electrode that contains platinum catalyst is held at a voltage that is within about 0.2 V of the partially charged nickel

electrode. This highly oxidizing potential is maintained at the negative electrode by the oxygen gas that is given off by the charged active material in the nickel electrode. At this oxidizing potential the platinum catalyst surface can oxidize in the alkaline electrolyte, and the resulting platinum oxide can dissolve to maintain a low level of platinate ions in the electrolyte. The platinate ions can then migrate throughout the cell, and specifically to the nickel electrode, where they can slowly react with the partially charged active material to form a Ni-Co-Pt oxyhydroxide compound within the active-material lattice. This Ni-Co-Pt compound is stable as long as it remains in the oxidized state (i.e., the nickel precharge in the cell remains intact). However, it becomes unstable if the nickel precharge is lost and the nickel electrode is allowed to completely discharge. If the nickel electrode undergoes complete discharge, the Ni-Co-Pt compound will slowly disproportionate to release its platinum back into the electrolyte in the form of platinate ions.

These reactions that involve platinum enable the accumulation of a reservoir of potentially soluble platinum to build up in the nickel electrodes of a nickel-hydrogen cell during long-term storage. Because the active nickel precharge is also slowly depleted during long-term storage, a stored nickel-hydrogen cell can eventually reach a point where it has lost essentially all of its active nickel precharge and has accumulated a significant amount of platinum in the nickel electrode. If the cell is recharged and then fully discharged in this condition, platinate ions will be released by the nickel electrode. The platinate ions will slowly plate back onto the hydrogen electrode when current is flowing at normal cell operating voltages. However, if the nickel electrode is discharged to less than 1.0 V $vs.$ H_2 (which is possible if the cell has developed a hydrogen precharge), then platinum metal can slowly plate onto the nickel electrode. The presence of platinum metal on the nickel electrode leads to a decrease in charge efficiency and an increase in the self-discharge rate.

Fortunately, platinum metal passivates once the nickel electrode is recharged to its normal operating potential, and thus the platinum does not have a catastrophic effect on cell performance. However, it can reduce the capacity that can be obtained from the cell. Because the oxides that passivate the platinum are slightly soluble in the alkaline electrolyte, simply holding the nickel electrode at normal operating potential with a low recharge current can gradually dissolve the platinum metal (as oxides) from the nickel electrode and plate it back onto the negative electrode. Typically, holding a cell in trickle charge for one week will ensure that all platinum metal has been removed from the nickel electrode and will return the cell to its normal performance. Because the nickel electrode can contain a significant reservoir of Ni-Co-Pt compound within its active material, it is possible for some of the platinum to be released to reinitiate the platinum plating cycle each time the nickel-hydrogen cell is fully discharged and held below 1 V for more than several hours.

The platinum plating cycle described here can only occur in a nickel-hydrogen cell during a specific window in its storage and operational lifetime. The cell must have been stored long enough to accumulate oxidized platinum within the nickel electrode active material, and it must have lost essentially all its active nickel precharge. This typically requires about five years of ambient storage, assuming a 12–15% initial level of nickel precharge. Thereafter, the cell is susceptible to the platinum plating cycle whenever it is discharged and held below 1 V for a significant period of time. Each time the cell is held below 1 V, more platinum metal can be plated onto the nickel electrode until the reservoir of oxidized platinum in the nickel electrode active material eventually becomes fully depleted. After this occurs, all of the platinum will have been deposited back onto the negative electrode, and the platinum plating cycle will not recur.

12.2.4 Hydrogen Reduction Processes

At its normal operating potential in the nickel-hydrogen cell, the nickel electrode is always at an oxidizing potential. However, if the cell contains excess hydrogen gas when it is fully discharged (hydrogen precharge), the hydrogen can reduce the nickel electrode potential to that of the hydrogen electrode (thus corresponding to a cell potential of zero). This hydrogen-precharged condition is the eventual state that all cells attain as a result of aging processes such as nickel corrosion and nickel electrode deactivation processes.

The problem with hydrogen precharge is that the hydroxide active materials in the nickel electrode are not stable at the hydrogen potential, and they can slowly be reduced by the hydrogen to form finely divided nickel and cobalt metal particles (cobalt is an additive in the nickel oxyhydroxide). These finely divided metals will undergo oxidation back to the hydroxides when the cell is recharged. However, the cobalt hydroxide is significantly more soluble in the electrolyte than is the nickel hydroxide, and thus it can dissolve and migrate within and out of the active material. The eventual consequence of these processes is loss of the cobalt dopant from the active-material layers that are in direct contact with the metallic sinter surfaces, because it is in these layers where the hydrogen reacts with and reduces the active materials.

The migration of the cobalt additive produces a significant degradation of nickel electrode performance. The migration process produces a gradient in the concentration of cobalt additive, thus producing a gradient in chemical potential within the nickel electrode active material. This gradient will significantly elevate the voltage of the nickel electrode, because undoped nickel hydroxide has a higher redox potential than the cobalt-doped active material. A higher recharge voltage means that more oxygen will be evolved sooner from the nickel electrode during recharge, thus decreasing charge efficiency significantly. In addition, the resistance of the active material increases significantly near the end of discharge when cobalt-depleted layers exist, leaving a significant amount of charge in the nickel electrodes that is not usable at normal discharge rates and voltage levels. Up to 30% loss of cell capacity has been observed to occur as a result of the cobalt migration processes after several years of discharged storage for hydrogen-precharged nickel-hydrogen cells.

The cobalt segregation processes described here result from the formation of chemically modified layers within the active material of the nickel electrodes. These layers can be quite difficult to dissipate, and it is not generally possible to reestablish the cobalt-doped nickel oxyhydroxide solid-solution structure that was initially present in the active materials. However, if the layers in which the cobalt has been depleted are quite thin, it is possible to largely dissipate them and recover lost cell capacity by physically mixing the layers.

This kind of physical mixing can be caused by the volume changes in the active material that result from high-rate cycling, particularly at lower temperatures where significant amounts of the higher-volume γ-NiOOH phase can more readily form. Physical mixing by cycling can be effective for recovering lost capacity if the cobalt-depleted layers are relatively thin (i.e., < 0.1 μm), but it is generally ineffective when thicker cobalt-depleted layers have been allowed to develop. Thicker cobalt-depleted layers are typically associated with a capacity loss of more than 10%, while thin layers typically cause less than 8–10% capacity loss.

12.2.5 Loss of Nickel Precharge

Modern nickel-hydrogen cells are manufactured with nickel precharge. This means that when the cell is fully discharged, anywhere from 8% to 20% residual capacity can remain in the nickel electrode, while the hydrogen gas within the cell is completely consumed. The nickel precharge enables the fully discharged cell to be stored at low voltage without experiencing degradation from reduction of the nickel electrode active material by hydrogen gas. If nickel precharge is limited to about 12–15%, it does not diminish the high-rate cell capacity, because nickel electrodes normally contain 12–15% residual capacity that is not usable during high-rate discharge.

During long-term cell storage, a number of processes occur to slowly consume the active nickel precharge. After a storage period of 4–7 yr, depending on the storage environment and the initial amount of nickel precharge, most nickel-hydrogen cells will lose their active nickel precharge and revert to the hydrogen-precharged condition. This transition generally requires that the cell be subsequently stored at a partially charged state (i.e., with cell voltage maintained above 1.2 V) to prevent rapid degradation.

The principal processes that contribute to loss of active nickel precharge are corrosion of the nickel sinter (which produces hydrogen gas), platinum oxidation in the negative electrode, and physical degradation of the nickel electrode (such as swelling, active-material extrusion, flaking, or blistering). Of these processes, the first two are typically the most important in a low-stress storage environment.

12.2.6 Second Plateau Capacity Loss

One of the most common degradation signatures seen in nickel-hydrogen cells after lengthy storage is a loss of capacity to a lower discharge plateau. This lower discharge plateau, which occurs in the range of 0.8–1.0 V, results from the formation of a barrier layer consisting of semiconducting active material that forms in the nickel electrode as it begins to experience depletion at the end of discharge. This barrier layer, which forms at the junction between the sinter and the active material, requires a voltage drop of 0.3–0.4 V to form across its thickness before it can develop sufficient electronic conductivity to enable continued discharge of charged material more distant from the sinter. Any process that occurs in the nickel-hydrogen cell to make the barrier layer form at a higher average cell state of charge will result in loss of capacity from the normal higher discharge plateau, to the lower plateau.

The primary mechanism contributing to an increased second plateau in nickel-hydrogen cells during storage is loss of nickel precharge. As long as the second plateau increases in duration without decreasing the higher plateau capacity, this is considered normal degradation resulting from slow nickel and platinum corrosion processes. However, physical damage to the nickel electrode active-material structure, or exposure to hydrogen gas at low cell voltages, can cause a dramatic decrease of higher-plateau cell capacity with a corresponding increase in lower-plateau capacity. Typically, capacity losses of this kind are not readily recoverable once they occur.

12.3 Issues Related to the Cell Operating Environment

The operating environment of a nickel-hydrogen cell can have a significant effect on its performance, lifetime, degradation modes, and performance signatures. The following discussion addresses some of the key areas of concern related to the operating environment of the cell.

12.3.1 Popping Damage

An instance of "popping" is a microexplosion within a nickel-hydrogen cell. This phenomenon results whenever the platinum catalyst ignites a bubble or pocket of oxygen gas, causing it to rapidly recombine with the high-pressure hydrogen gas in the cell. Popping occurs as a consequence of oxygen evolution during overcharge of the cell. It is more likely when overcharge rates are high, when cell pressure decreases sharply (forcing oxygen out of the nickel electrodes), when cells are operated on their sides, or at low temperatures. The damage that results from popping is greatest when the oxygen bubbles are larger, or when the internal cell pressure is higher (more oxygen is in the bubbles).

Popping is often associated with free electrolyte in the electrode stack or along its edges. Free electrolyte obstructs the continuous convective transport of oxygen gas from the positive electrode to the negative electrode during overcharge, and it enables bubbles or pockets of high-concentration oxygen gas to accumulate. However, a sufficiently high rate of overcharge can lead to popping events simply as a result of high rates of oxygen gas streaming from the positive to the negative electrodes.

The damage from popping occurs whenever these events take place, and it is typically associated with fracturing of the nickel electrode sinter structure, fragmentation of the Zircar separator, melting and fragmentation of the negative electrode, and melting of the gas screens (particularly at their edges). This kind of physical damage is cumulative during the life of the nickel-hydrogen cell. When popping damage accumulates sufficiently, short circuits can form as a result of conductive particles dislodged from the electrodes, or as a result of separator damage. Such short circuits are not typically seen initially as high-rate shorts, but they begin as unusually high cell-leakage current and eventually progress to more extensive shorts as further damage accumulates.

12.3.2 Poor Charge Efficiency
The charge efficiency of nickel-hydrogen cells can be strongly influenced by the cell environment and how the cell is operated. The most significant environmental factor that can affect charge efficiency is temperature. The charge efficiency of nickel-hydrogen cells drops exponentially as the temperature rises, becoming a major factor influencing performance above about 10°C. At temperatures above 20°C, cell capacity is markedly reduced, and at temperatures of 30°C or more, a cell may deliver less than half of its normal capacity. Because reduced charge efficiency can cause increased heat evolution, feedback exists between cell thermal control and charge efficiency that can cause thermal runaway.

Less obvious factors that influence charge efficiency are related to the prior history of a cycling nickel-hydrogen cell. For example, cells that are fully discharged for the purposes of reconditioning a battery will exhibit decreased charge efficiency during their initial recharge. Similarly, cells that are repeatedly cycled to a maximum state of charge that is less than 100% will gradually develop extremely poor charge efficiency for recharging that portion of the active material that is not normally discharged during each cycle. This deactivation of material that is not repeatedly recharged is a major cause of much of the history-dependent variability in nickel-hydrogen cell performance, and it can be a key factor in triggering cell-to-cell capacity divergence during the operation of nickel-hydrogen batteries.

12.3.3 Thermal Gradients
As discussed previously, the performance of nickel-hydrogen cells is quite sensitive to temperature. However, during the operation of nickel-hydrogen batteries, the thermal gradients to which a cell is exposed are often more important than the absolute temperature at which the cell is operated. The area where thermal gradients

12.3 Issues Related to the Cell Operating Environment

can have a pronounced effect on performance is the cell overcharge region. The overcharge rate is quite sensitive to temperature, varying by up to an order of magnitude for a 10–15°C shift in temperature.

If a cell is operated with the electrodes at the top of the stack 10°C warmer than those at the bottom of the stack, the electrodes at the top of the cell can go into overcharge long before the remainder of the cell is fully charged. This is particularly true of cells where both positive and negative leads emerge from the cell at the top ("rabbit ear" terminals), because the electrodes at the top of the cell are also at a higher potential as a result of differential lead resistance from the top to the bottom of the cell. These conditions can make cells with significant thermal gradients much more prone to performance problems and accelerated degradation stemming from overcharge, nickel electrode swelling, active-material extrusion, and popping damage.

Thermal gradients can also degrade cell performance as a result of condensation of water in the cell, a process that can be driven by thermal gradients. When a cell contains 31% potassium hydroxide electrolyte, a gradient of 9°C between the temperature of the electrolyte in the warmest part of the cell and the temperature of the coldest surface in the cell is sufficient to initiate water condensation on the cold surfaces. This process can lead to extensive dry-out of the warmer regions in a cell throughout long periods of operation if the cell design does not include wicking mechanisms sufficient to move the condensed liquid back to the warmer regions in the cell. While wall wicks can serve this role within a single electrode stack, they will not redistribute liquid electrolyte between two stacks (as in a dual-stack design), or across gaps in the wall wick at weld rings. In addition, wall wicks decrease their wicking rate capability as temperature drops, completely losing their wicking ability if the water or electrolyte freezes.

Cells containing 26% potassium hydroxide electrolyte are even more prone to water condensation issues than are the higher-concentration cells. With 26% potassium hydroxide, only a 6°C thermal gradient within a cell is required to initiate water condensation. It is clear that the successful use of nickel-hydrogen cells requires a thermal design that keeps thermal gradients below critical thresholds that are dictated by the wicking rates within the cell, the electrolyte concentration, and the cell operating temperature.

12.3.4 Electrolyte Freezing

Electrolyte freezing can become a significant issue in nickel-hydrogen cells during operation at low temperatures, in spite of the fact that the –40°C freezing point for 26% potassium hydroxide (–62°C for 31% potassium hydroxide) appears at first glance to be low enough that this should not be an area of concern. The first issue with electrolyte freezing has to do with the large gradients in electrolyte concentration that can arise during high-rate cell operation. For example, in the previous section the condensation of water was described in response to thermal gradients. Condensed water can have a freezing point just below 0°C if it does not contain large amounts of dissolved potassium hydroxide. If such condensate freezes it is totally immobilized, and can accumulate within the cell pressure vessel as a large block of ice that cannot become active again until it melts and is allowed to redistribute through the cell. This immobilization of water can contribute to cell failure as the result of stack dry-out.

Another situation in which electrolyte can freeze is during low-temperature (less than about –10°C) cell operation when the cold cell is required to discharge at a high rate. During high-rate discharge, a significant amount of water is generated by the electrochemical reactions in the hydrogen electrode. This water production reduces the electrolyte concentration in the negative electrode precipitously, which can result in freezing of the water in the hydrogen electrode. Freezing of the water produced during discharge can increase the resistance and drop the cell discharge voltage 0.1 to 0.2 V or more. However, this is typically a transient drop in discharge voltage, from which the cell recovers as the thermal dissipation in the added electrolyte resistance heats the negative electrodes and melts the frozen water.

Another process that can lead to electrolyte freezing and the failure of a nickel-hydrogen cell can occur during long periods of low-temperature cell operation when little or no discharge of the cell is performed. When a nickel-hydrogen cell is operated at temperatures lower than –5 to –10°C, its charge efficiency is very high, and the nickel electrode can be charged to a very high state of charge by the gradual conversion of a significant fraction of the active material to γ-NiOOH. Because the γ-NiOOH structure incorporates 1/3 of a potassium hydroxide molecule per nickel site, the formation of large amounts of this phase can significantly reduce the electrolyte concentration to a level where freezing is possible. Periodic discharge, which releases potassium hydroxide that is tied up in the γ-NiOOH phase, can actually help prevent electrolyte freezing in this situation. Clearly, higher concentrations of potassium hydroxide and greater quantities of electrolyte in a cell can make this situation less likely.

The mechanism described here for electrolyte freezing in operational cells is not generally regarded as one that causes catastrophic failures in real nickel-hydrogen cells, because once the cell is warmed the electrolyte can melt and normal cell performance is expected to be regained. However, as described below, it is possible for such electrolyte freezing to propagate into a catastrophic cell failure by causing short-circuiting in the cell.

In a real nickel-hydrogen cell, the thermal environment as well as the electrolyte concentration environment can vary significantly from one end of the electrode stack to the other. This variation can occur as a response to the thermal environment, overcharge within the cell, or previous cycling or operational history. For these reasons, as the temperature of a real cell is decreased it is likely that the electrolyte will initially freeze only at one end of the cell. If the cell is being cooled from its top end by heat pipes, the electrolyte at the top end of the stack will initially freeze. If the cell is being cooled from its bottom end as a result of being mounted to a cold base plate, the electrolyte at the bottom end of the stack will initially freeze.

As an example of the dynamics that would be expected once electrolyte freezing begins, consider a cell that is in trickle charge and is being cooled from its top end by a heat pipe assembly. If the cell temperature becomes too cold, then, because the stack units at the top of the cell are the coldest, their electrolyte will freeze. This will effectively stop all electrochemistry in the electrodes where the electrolyte has frozen, thus stopping all further changes in their nickel electrode phase composition and electrolyte concentration. The trickle charge current will thus be forced to be completely carried by the remaining unfrozen stack units, thus increasing the trickle charge current density on the remaining active electrodes. The higher charge rate will result in the formation of more γ-NiOOH, and a further drop in the electrolyte concentration in the remaining active stack units. The resulting reduction in electrolyte concentration will cause additional stack units to experience electrolyte freezing, which will stop all changes in their composition and force the current to be further concentrated toward the bottom of the cell.

This process will continue until eventually only the bottommost stack units remain unfrozen. They will remain unfrozen because all the trickle charge current has been concentrated into them, and the resulting heat generation is sufficient to keep them from freezing. These bottommost stack units will be exposed to extremely high overcharge rates as a result of all the cell current being concentrated into them. This will significantly increase the risk of damage and cell short-circuiting as a result of popping events. In addition, the extremely high levels of γ-NiOOH formation that will occur in the bottommost nickel electrodes will cause tremendous expansion of the electrodes and the electrode stack. When this stack expansion is combined with that resulting from electrolyte freezing, it is possible that compressive damage to the stack components (negative electrodes and core) may occur. The cascade freezing process described here can thus lead to catastrophic cell failure in a period of time that is quite short compared to the normal lifetime of a nickel-hydrogen cell.

12.4 Battery-Specific Issues

A number of issues that can arise during the operation of nickel-hydrogen batteries may not generally be associated with problems when the cells in the battery are tested individually, but are associated with the operation of a number of cells as a series string in the battery environment. Here some of the most significant such issues are described.

12.4.1 Cell Capacity Imbalance

When nickel-hydrogen cells are operated as a series string, capacity imbalance between the cells can develop because of cell-to-cell variability in the cell components, cell thermal interfaces, and the thermal environment of each cell. These cell-to-cell variations lead to differences in cell self-discharge rates and charge efficiencies. For these reasons, each cell can eventually settle into operation at a different state of charge than the other cells in the battery. As long as the cell having the lowest state of charge is not too low compared to the battery discharge requirements, this cell-to-cell divergence may not create a problem with battery operation. Overcharge of the battery will tend to reduce cell-to-cell capacity divergence (as long as it does not drive the cell temperature up significantly relative to the other cells), while undercharge typically magnifies the capacity divergence. However, overcharge is also a major factor contributing to battery cell wear-out. Thus, the correct overcharge level is the least amount that keeps the lowest-capacity cells in the battery acceptably charged.

For high cycle-life applications, such as low Earth orbit cycling profiles, the minimum cell state of charge in a battery may often be as low as 80–90% of full charge. Because the required depth of discharge is generally less than 40% in these applications, this range is quite acceptable. For lower cycle-life applications, such as geosynchronous cycling profiles, the greater available recharge time typically allows all the cells in a battery to be maintained at a state of charge well above 90–95%.

12.4.2 Thermal Control

Tests that are performed on individual nickel-hydrogen cells often do not evaluate the thermal control issues that may exist in batteries consisting of large numbers of cells that can generate significant waste heat. As discussed earlier, each cell within the battery must be thermally maintained such that its average operating temperature is neither too cold nor too hot (i.e., between about $-10°C$ and about $10°C$). If the thermal control system is not effective for all the cells in this regard, the performance of the entire battery can become controlled by the single hottest or coldest cell in the battery.

Thermal control of batteries can be either active or passive. Passive thermal control systems simply use passive conductive thermal balance between the cell heat dissipation and the radiator cooling to keep cells cold. This type of system typically uses heaters on the cells and battery to keep the average cell temperature in the battery above a minimum set point. While this type of thermal control system is relatively simple, in higher-power battery systems it can allow significant variations in cell temperatures, as well as thermal gradients both between and within cells. Active thermal control can provide improved regulation of cell temperature variations as well as thermal gradients between and within cells, particularly in higher-power systems. Active thermal control systems for batteries typically involve some features—such as heat pipes (variable or constant conductance), louvers, and heaters—that provide additional complexity and are not necessarily required in passive systems.

With any of these types of thermal control systems, if thermal performance becomes degraded or if the batteries produce more heat than expected, significant degradation of battery performance can occur. The typical degradation issues associated with thermal control system problems are from high or low temperature extremes either in individual cells or across a portion of the battery, or from thermal gradients across cells within the battery.

The proper functioning of the battery thermal control system requires that a number of thermal interfaces maintain good thermal conductivity. Such interfaces include that between the cell pressure vessel and its thermal sleeve, the sleeve-to-mounting-plate interface, and the path from the cell mounting plate to the thermal radiator. If any of these interfaces loses its integrity as a result of vibration, shock, thermal stresses, or poor workmanship, the thermal control system will not effectively control the battery temperature. In addition, the radiator must be kept exposed to a cold environment, the radiator surfaces must remain free of contaminants, and the battery heaters must function correctly if the radiator is to effectively cool the batteries. If any of these requirements are not maintained, battery thermal control will suffer as well.

12.4.3 Cell-Case Isolation

The pressure vessel in a nickel-hydrogen cell is electrically isolated from the positive and negative cell terminals. However, because electrolyte is always in contact with the inner wall of the cell pressure vessel as well as its positive and negative electrodes, any loss of external electrical isolation between the cell case and the electrode terminals will create a potential current path involving parasitic corrosion reactions at the interior surface of the cell pressure vessel. If the loss of isolation involves external moisture that is in contact with the pressure vessel, then external corrosion reactions can also occur on the surface of the pressure vessel.

Because Inconel metal is passivated in air and in alkaline electrolyte, such reactions typically occur at an extremely low rate for an individual cell. However, when such reactions can be driven by the higher potential created in a battery by a number of cells in series, the parasitic corrosion reactions can become very significant. During several years of operation, sufficient corrosion of the Inconel pressure vessel has been seen in batteries that have lost cell-case isolation with the result that the pressure vessel has developed pinhole leaks. Hydrogen leaks will rapidly cause all cell capacity to be lost, and over time can result in high cell impedance caused by electrolyte dry-out. Typically these kinds of issues have been experienced during ground testing of battery packs, and they result from water condensation compromising the cell-case isolation. However, any situation that results in loss of electrical case isolation would be expected to similarly compromise battery life.

12.5 Cycling and Life-Related Issues

A number of performance issues in nickel-hydrogen cells manifest themselves as the result of degradation processes that gradually occur to change cell performance as it is cycled throughout its operational lifetime. These processes must be understood to follow the accumulation of wear-related effects that will eventually limit the lifetime of a nickel-hydrogen battery cell. Here, these processes are discussed in the context of nickel-hydrogen cell performance and lifetime.

12.5.1 Nickel Corrosion

The relatively high-surface-area metallic nickel sinter that makes up the porous current collector in sintered nickel electrodes is maintained at a highly oxidizing potential throughout the entire operating life of the nickel electrode in a nickel-hydrogen cell. Because this potential is considerably greater than that where nickel metal oxidizes, the only thing that makes this sintered structure stable is the compact oxide layer that forms on its surface in alkaline electrolyte to passivate it against continued nickel corrosion. However, no passivation of a metal surface is perfect, and the nickel does continue to slowly oxidize during the many years of nickel-hydrogen cell operation. Cycling of the nickel electrode accelerates this corrosion process somewhat, because the volume changes in the active material and the evolution of oxygen gas exert forces on the passivation layer, causing it to repeatedly fracture. Such fractures are healed by the formation of a new compact passivation layer, but at the cost of additional metal oxidation.

Nickel corrosion is probably the one process that will almost certainly limit the ultimate cycle life and calendar life of nickel-hydrogen cells. The amount of corrosion that can be tolerated before electrode failure occurs is highly dependent on the required depth of discharge and discharge rate, as well as operating temperature, sinter structure and density, and sinter thickness. A typical amount of corrosion seen in 80% porous sintered plaque, at the point where cell failure occurs at 60% depth of discharge in a low Earth orbit cycling profile, is loss of about 30% of the total metal present in the sinter, based on cell destructive physical analysis results. For less-stressful cycling conditions, corrosion levels up to 40–50% may be tolerated before the nickel electrode degradation associated with corrosion results in cell failure.

Nickel corrosion is a key degradation process in the nickel-hydrogen cell because it is a factor contributing to many of the other degradation signatures seen in aged cells. Principal among these are increases in hydrogen pressure, active-material extrusion, and nickel electrode swelling, as well as the range of secondary problems that these physical changes can produce (separator dry-out, core fracturing, etc.). While the other degradation signatures are not solely the result of corrosion of the nickel sinter, as will be discussed in later sections, nickel sinter corrosion is a key contributing process that can be slowed but not halted in nickel-hydrogen cells.

Nickel sinter corrosion can be slowed somewhat at the beginning of cell life by manufacturing the sinter with a hard passivation layer prior to manufacturing the nickel electrodes. However, after several years of cell operation the beneficial effect of this added passivation disappears, and the corrosion rate becomes the same as if no added "hard" passivation were employed.

Nickel corrosion does tend to be less rapid at lower operating temperatures; however, the corrosion rate also displays significant variability between different lots of cells. The source or sources of this variability have not been identified; however, the corrosion rate is likely to be influenced by trace materials present in the sinter (such as alloyed carbon) as well as by foreign materials in the electrolyte (such as chloride, nitrate, borate, silicate, sulfate, etc.).

Operational variables associated with cell cycling have also been seen to influence the nickel corrosion rate. Corrosion seems to be significantly accelerated by greater amounts and rates of overcharge. The detailed mechanisms for how overcharge accelerates nickel corrosion have not been clearly identified, but they are likely linked to the higher voltages (and thus more-oxidizing environment) associated with more overcharge and with higher-rate overcharge. In addition, the hydrostatic and gas pressures associated with oxygen evolution are likely to contribute to repeated fracturing of the passivation layers that protect the nickel sinter from corrosion. The rate of corrosion also increases as the cell depth of discharge increases, again probably as the result of passivation layer fracturing in response to the greater active-material volume changes associated with higher depth-of-discharge operation. While lower-temperature cell operation can decrease the corrosion rate, if operation at low temperatures entails the repeated formation of more of the higher volume γ-NiOOH phase in the active material, it can significantly increase the corrosion rate as a result of greater active-material volume changes.

12.5.2 Nickel Electrode Swelling

The sintered nickel electrodes in nickel-hydrogen cells undergo a steady expansion of their thickness as they repeatedly charge and discharge. This swelling is relatively rapid if the electrodes are cycled in an uncompressed configuration. However, in a nickel-hydrogen cell, the electrode stack exerts a compressive force that counters the tendency of the nickel electrodes to swell with repeated use. Thus, as nickel electrodes swell throughout their cycle life, they tend to slowly expand the electrode stack and compress the other stack components (separators, gas screens, and hydrogen electrodes). If a stack from an extensively cycled nickel-hydrogen cell is disassembled, significant force can suddenly be released as the compression that has built up from swelling of the nickel electrodes is relieved. After the stack compression is released, the thickness of the nickel electrode structure can easily expand up to 50% or more in response to its accumulated internal stresses. For well-cycled nickel electrodes, the expansion seen after stack disassembly can sometimes result in irregular swelling, blistering, warping of the electrodes, and delamination of the sinter from its embedded nickel grid.

As mentioned earlier, nickel corrosion is a significant factor that contributes to nickel electrode swelling. Nickel corrosion results in the formation of pure nickel hydroxide as a corrosion product, which can be oxidized to NiOOH that is free of the cobalt additive normally present in active material. These corrosion products occupy a significantly greater volume within the sinter than the metal from which they formed, and thus will push against the overall sinter structure and cause it to gradually swell. The pure nickel oxyhydroxide formed as a corrosion product is significantly less compressible than the normal cobalt-doped active material in the nickel electrodes, which adds to the tendency of such corrosion products to cause swelling of the nickel electrode structure.

The construction of sintered nickel electrodes for nickel-hydrogen cells involves design features intended to accommodate the accumulation of corrosion products during many years of operation without causing catastrophic swelling. The loading level typically used in such electrodes is 1.6 to 1.7 g per cubic centimeter of void volume, which is significantly lower than loading levels typically used in nickel electrodes intended for less cycle-life–intensive applications. The lower loading level leaves significant void volume within the pores of the sinter to accommodate corrosion products formed throughout many years of operation without resulting in large amounts of swelling.

In applications that involve a very large number of charge/discharge cycles and a very large throughput capacity, an additional mechanism causing swelling can become more significant than corrosion. The large volume changes in the cycling active material will repeatedly flex the sinter structure and the active material that fills it as it expands and contracts during charge and discharge. This breathing motion of the sinter structure exerts a net force that will slowly deform the structure to cause it to expand into regions of lower density. The nickel electrode structure thus expands to push into the separator, causing the structure to gradually swell over time. This type of swelling is reduced by higher levels of cobalt additive in the nickel electrodes, which makes the active-material crystallites more compressible, thus transferring less expansion force to the sinter structure. Additionally, design or operation of nickel-hydrogen cells in ways that reduce the cycling of γ-NiOOH, which is a higher-volume phase compared to β-NiOOH, can reduce nickel electrode swelling rates considerably. Operation at low temperatures with greater amounts of overcharge also tends to promote the formation and cycling of γ-NiOOH. Design changes that can influence the rate of formation of γ-NiOOH include changes in potassium hydroxide electrolyte concentration, as well as the use of additives such as cobalt.

12.5.3 Cell Dry-Out
Nickel electrode swelling is one of the principal causes of increase in cell resistance arising from dry-out of the separator. The capillary forces between porous cell components tend to draw electrolyte from the separator into the nickel electrode pores. Therefore, if the nickel electrode swells significantly, it can draw a large fraction of the electrolyte out of the separator. Particularly for cell designs

with a single layer of separator, which provides less electrolyte volume margin, cell dry-out can cause a resistive depression of the cell discharge voltage plateau. This tends to be an issue primarily for cells that contain significantly less than 3.4–3.5 g of electrolyte per A h of cell capacity. If the fraction of the separator void volume that contains electrolyte drops much below 25–30%, the cell voltage performance during high-rate discharge can drop precipitously to cause premature cell failure.

This type of failure arises when the current passing through the cell reaches the limit that can be supported by ionic diffusion and migration in the relatively dry separator, and is referred to as the "diffusion limited" current. In this situation, cell failure occurs because the concentration of ions needed for the electrochemical reactions at the positive or negative electrode falls essentially to zero in response to the high current flow, and it cannot be diffusively replenished fast enough to support the discharge rate.

Cell dry-out tends to be a failure mode that is most likely when compressible separators, such as nylon, polypropylene, or asbestos, are used in the cell. Such separators can have much of their electrolyte squeezed from them by nickel electrode expansion, and it is not unusual for cells containing this type of separator to exhibit the depressed discharge voltage plateau late in life that is typically associated with electrolyte dry-out. With noncompressible separators, such as Zircar, the cell designs having lower electrolyte concentrations and lower total amounts of electrolyte tend to be most susceptible to dry-out–induced degradation of cell performance after many years of cell operation.

12.5.4 Active-Material Extrusion

The volume expansion and contraction of the active material in the nickel electrode during charge and discharge that causes electrode swelling also results in the gradual extrusion of particles of active material from the pores of the sinter onto the surface of the nickel electrode and into the separator pores, at locations where they contact the nickel electrode surface. This extrusion process occurs because each time the active material goes though a breathing cycle of expansion and contraction, the crystallites experience a slight net movement from regions of higher density toward regions of lower density.

Because the surface of the nickel electrode and the separator that it contacts is a much lower-density region than the interior of the nickel electrode, particles of active material will gradually extrude and accumulate as a layer on the electrode surface. The surface layer of particles can slowly move into the pores of the separator under the influence of the convective forces resulting from gas and electrolyte transport through the separator.

The extrusion process has several effects on cell performance. First, the extruded material is not used effectively for energy storage, largely because it contains no electronically conductive matrix to enable charge transport processes.

This means that extruded material largely ceases to contribute to cell capacity. Second, if these extruded active-material particles fill the separator pores with a sufficiently high density, they can produce a relatively high-resistance conductive path that bridges the positive and negative electrodes.* This path produces current leakage during cell operation, which forces an increased amount of overcharge to maintain cell capacity. Increased overcharge produces increased stress and accelerates the extrusion, ultimately resulting in a short circuit having a leakage rate high enough to make the cell fail in the battery configuration where the other cells cannot tolerate the excessive overcharge.

It is clear from test data on different cell designs that these failure modes resulting from active-material extrusion—capacity loss and cell short-circuiting—are indeed sensitive to cell design details. For example, cells with two layers of Zircar separator are much more robust relative to the short-circuit failure mode than cells with a single layer of Zircar separator are. This difference simply stems from the greater distance required for the conductive particle bridge in designs with two layers of separator. Additionally, 26% potassium hydroxide electrolyte produces extrusion rates that are significantly lower than for 31% potassium hydroxide electrolyte, primarily as a consequence of the larger volume changes experienced by the active material when cycled in 31% potassium hydroxide (because of more facile γ-NiOOH formation). Thus, cell design details can significantly alter the likelihood of cell failure brought on either by capacity loss or by short-circuiting.

12.5.5 Stack Compression Issues

As nickel electrodes swell during the operational lifetime of a nickel-hydrogen cell, it is the function of the cell core assembly to hold the electrode stack together without allowing either too little or too much compression of the components stacked onto the core. While some of the nickel electrode swelling, particularly early in life, is accommodated by the Belleville washers used in the stack assembly, most of the swelling must be accommodated by the combination of increased compressive force (push-back) generated by the core, and compression of the gas screens and hydrogen electrodes.

As nickel electrode swelling increases, it can potentially approach levels where the force on the plastic core assembly is great enough to either fracture the core itself or strip the end nuts that are threaded onto the end of the core to maintain the stack compression. If either of these events occurs, the electrode stack will lose its compression and the cell will either develop a short circuit or suffer a large capacity loss, depending on the detailed physical movements of the electrodes and leads as a result of the core failure.

---

*Active-material particles extruded from the nickel electrode typically have some electronic conductivity, because they consist primarily of the oxidized NiOOH phases.

The probability of this particular failure mode is minimized in nickel-hydrogen cells by designing the core assembly with enough strength to withstand the most extreme nickel electrode expansion at end of life. Because the amount of expansion possible at end of life is difficult to precisely predict, cores should generally be designed with at least a factor-of-two strength margin. With this design approach, the core can tolerate situations of anomalous force without breaking, such as during the formation of large amounts of the high-volume γ-NiOOH phase late in life if a cell is exposed to overcharge, particularly at low temperatures.

12.5.6 Pressure Growth

Pressure growth in nickel-hydrogen cells is the normal consequence of gradual corrosion of the nickel sinter in the nickel electrodes, as has been previously discussed. Hydrogen gas is a product of this corrosion process. This type of corrosion-induced pressure increase appears as a nearly linear rise in pressure during years of cell cycling and operation. Pressure growth rates (and the corrosion rate) can vary significantly from cell lot to cell lot. The typical pressure change during the lifetime of a nickel-hydrogen cell may involve an early-life walkdown in pressure as the cell state of charge stabilizes, followed by a long, relatively linear pressure increase throughout most of the cell life, sometimes interrupted by a more rapid drop-off in pressure at the end of life. Figure 12.3 shows this pressure change profile for the lifetime of a typical nickel-hydrogen cell.

The end-of-life drop-off in pressure is typically caused by capacity loss through the formation of short circuits caused by conductive active-material particles extruded into the separator from the nickel electrodes. Typically, it is

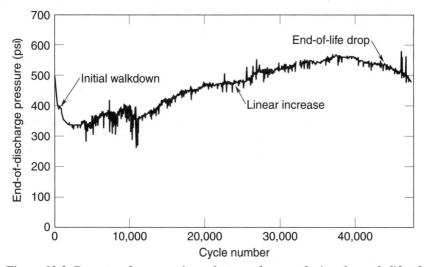

Figure 12.3. Pressure change regimes that may be seen during the cycle life of a typical nickel-hydrogen cell.

not possible to realize any real long-term recovery of cell performance once such short-circuit-induced capacity losses begin to occur.

The pressure trend shown in Fig. 12.3 can be significantly influenced by drift in strain gauge pressure measurements, as has been previously discussed. The possibility of such drift always introduces uncertainty in attempts to quantitatively interpret pressure changes over time. The best methods to calibrate actual pressure changes are to examine the lowest pressure reached by a cell after a complete low-rate discharge, as well as the highest pressure achievable at the point where the cell begins to generate significant overcharge heat during recharge.

12.5.7 Charge Efficiency Issues

As nickel electrodes degrade both chemically and structurally throughout their cycle life, the degradation acts to significantly affect their charge efficiency (the ability to readily store energy during recharge and bring the nickel electrode to a high state of charge), as well as their discharge efficiency (the ability to effectively discharge the stored capacity). These changes in efficiency result in an effective loss of usable capacity as a cell ages. Charge efficiency degrades because the energy storage process during recharge loses much of its ability to compete with oxygen evolution (overcharge). Cell recharge voltages tend to increase to make oxygen evolution much more likely at lower states of charge. Discharge efficiency degrades because of the formation of cobalt-depleted corrosion layers, degraded active-material conductivity, extrusion of active material, and loss of conductive sinter. Poor or degraded discharge efficiency is often accompanied by the formation of an extensive second plateau at the end of discharge.** These efficiency issues, which stem from the full range of chemical and physical degradation processes that occur in a cell during its life, contribute primarily to failure modes related to capacity loss in nickel-hydrogen cells.

12.5.8 Separator Degradation

The separator that is most commonly used in modern nickel-hydrogen cell designs is made of a zirconium ceramic fabric referred to as Zircar. This material is relatively inert in the alkaline electrolyte environment of the nickel-hydrogen cell and thus is not prone to chemical degradation. However, a Zircar separator is subject to physical damage from popping events (explosive oxygen/hydrogen recombination events) during overcharge. It can also be damaged by expansion of the nickel electrodes, which can result in compression of the separator. In

---

**A discharge plateau about 0.3 V below the normal discharge plateau of the nickel electrode occurs when the active material immediately in contact with the metallic current collector becomes depleted during discharge. The depleted layer acts as a p-type semiconductor, and it requires about a 0.3 V potential across it to develop good electronic conductivity. This 0.3 V drop enables electronic flow through the depleted layer to continue discharging material some distance from the metallic current collector, and thus a lower plateau is seen until the bulk active material becomes depleted.

addition, the separator can become filled with conductive particulates extruded over life from the surface of the nickel electrodes. These physical separator degradation modes can ultimately limit the lifetime of nickel-hydrogen cells as the result of short circuits that can develop through the separator. This type of failure mode is more likely to be found in cell designs using a single layer of Zircar fabric separator, as opposed to those using dual layers of separator.

Other types of separator material have been used in nickel-hydrogen cells throughout the years: nonwoven nylon, asbestos, and composites. However, in general these alternate separator materials have not performed as well as Zircar material. Polymer separators are compressible and thus prone to dry-out problems. Polymer separator materials can also melt and fuse when exposed to thermal events associated with rapid recombination of hydrogen and oxygen. If the separator fuses throughout any appreciable fraction of its area, it can significantly increase in resistance.

Asbestos separators were used for many years in early nickel-hydrogen cell designs. The asbestos, while compressible, does not dry out, because its fibers are so small. The wetted asbestos separator acts as an ionically conductive gasket that forces all oxygen from the nickel electrode out to the edges of the stack, where recombination can occur. This makes asbestos-separated cells unable to handle substantial overcharge without significantly deforming and compressing the separator, or in more extreme situations, popping holes through the asbestos.

Asbestos, a fibrous mineral consisting primarily of magnesium silicates, is also chemically attacked by the alkaline electrolyte, resulting in the breakdown of the fiber structure and the accumulation of silicate ions in the cell electrolyte. High levels of silicate in the electrolyte can result in high cell resistance. For these reasons, and because of concerns about the carcinogenic properties of asbestos, asbestos-separated cells were no longer produced after about 1998.

12.6 References

[12.1] A. H. Zimmerman, "The Role of Anionic Species in Energy Redistribution Processes in Nickel Electrodes in Nickel Hydrogen Cells," in *Hydrogen Storage Materials, Batteries, and Electrochemistry, Proc. Vol. 91-4, The Electrochem. Soc. Inc.* (Pennington, NJ).

[12.2] A. H. Zimmerman and M. V. Quinzio, "Model for Predicting the Effects of Long-Term Storage and Cycling on the Life of NiH_2 Cells," *Proc. of the 2003 NASA Battery Workshop* (Huntsville, AL, 20 November 2003).

13 Diagnostic Methods and Destructive Physical Analysis for Cells

While nickel-hydrogen battery cells are capable of operating reliably for many years, a number of problems that can occur either when the cells are built, during storage, or during operation can compromise their performance or life. Diagnosis of the root causes for cell performance degradation depends on carrying out the electrical, physical, and chemical tests that are appropriate for discerning the underlying cause or causes of the observed cell behavior. This is not always a straightforward procedure, because the common performance problems of low capacity, low voltage, high impedance, short-circuiting, and poor charge efficiency can be caused by almost any of a large number of chemical and physical processes, as well as by charge management issues with the cells.

This chapter summarizes the diagnostic procedures that have been developed to differentiate various underlying problems. Any analysis of problems in battery cells should start with the nondestructive tests that can provide useful signatures related to the problem. The first section of the chapter deals with such tests, which must be completed before beginning any destructive physical analysis of a cell. This section is followed by a discussion of commonly utilized techniques for cell destructive physical analysis studies. After cell destructive physical analysis, a large number of analyses are available to address particular degradation or other issues associated with particular cell components, including the nickel electrodes, hydrogen electrodes, electrolyte, separators, and other cell hardware; these are discussed in detail. The last section of the chapter discusses an expert system that has been developed[13.1] to help guide the diagnostic process, through the use of several examples describing typical diagnostic sequences.

13.1 Cell-Level Diagnostic Tests

Cell-level diagnostic tests are typically performed on the individual cells within a single pressure vessel before the integrity of the pressure vessel is compromised in any way. In all cases they are fully nondestructive, and in several cases they may be performed on all the pressure vessel units that are series-connected in a battery without disturbing the battery integrity.

13.1.1 X-Ray Measurements
X-ray measurements allow a wide variety of fabrication, degradation, or internal structure-related problems in nickel-hydrogen cells to be detected. Two methods are commonly used for x-ray measurements on cells. The first uses standard x-ray instruments that record several fixed exposures on film, typically taken with the cell oriented in different positions to allow inspection of the required internal

components. The second method involves the use of a high-resolution x-ray instrument that can record real-time views of the internal cell components as the cell is moved or rotated in the x-ray beam. These real-time x rays are typically recorded on a videocassette that may be inspected and analyzed as needed. Specialized high-resolution x-ray imaging instruments are also available for recording 3-D tomographic (CT scan) images. A CT scan provides a full three-dimensional image of the nickel-hydrogen cell—an image made up of a number of 2-D slices, one that allows the features of components throughout the cell structure to be examined as needed.

When carrying out x-ray inspections or measurements on nickel-hydrogen cells, it is important to limit total x-ray exposure time if one plans to electrically operate the cell after x-ray exposure. When the hydrophobic Teflon layer on the hydrogen electrodes is exposed to x rays, the Teflon is gradually degraded such that it loses its hydrophobic properties. With sufficiently high x-ray exposure levels, the hydrophobic side of the negative electrodes begins to become wettable, blocking the ready movement of hydrogen gas to the platinum catalyst material, thus making it impossible to discharge the cell at a high rate. X-ray damage of the negative electrodes is typically not an issue if total x-ray exposure time is limited to less than about 45 min to 1 h.

X-ray inspections are widely used to verify that cells are properly constructed and free of flaws after they are manufactured. Such inspections are a key part of the quality control program needed to ensure the high reliability required for satellite applications of nickel-hydrogen batteries. X-ray inspections may be used to detect flaws in pressure vessel welds, cell terminal assembly welds, or fill tube welds. Weld flaws typically either occur as a result of bubbles, contaminants, or particulates, or they result from inadequate weld material flow. These types of flaws are not detectable by external inspection, but they can be found by high-resolution x-ray examination. Because a leak of hydrogen gas from the pressure vessel will result in cell failure, it is necessary to locate any significant flaw or void in the cell pressure vessel welds to ensure reliability. One must either reject the cell or rework the weld so that it meets appropriate acceptance criteria.

Inspection of cell fill tube welds is particularly important for ensuring cell reliability, because traces of potassium hydroxide contamination are frequently found in the fill tube from the electrolyte activation process. This potassium hydroxide contamination will melt and vaporize at the temperatures reached during welding, leaving bubbles and voids within the weld area. If these voids are sufficiently large, or are interconnected by crevices, they can occasionally lead to cell leaks. Such voids are readily seen using high-resolution x-ray inspection, such as real-time or CT scan methods. Figure 13.1 depicts an example that illustrates the detail that can be seen with such inspection methods. It shows a CT scan of a fill tube containing large weld voids and interconnected crevices within the weld that resulted from potassium hydroxide contamination.

13.1 Cell-Level Diagnostic Tests   417

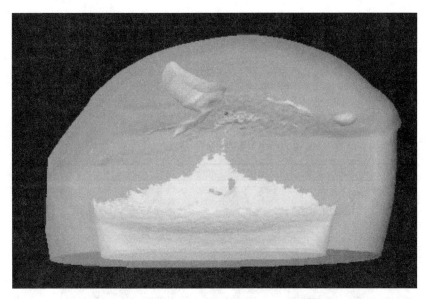

Figure 13.1. X-ray CT scan of the fill tube weld in a nickel-hydrogen cell. The fill tube is approximately 0.187 in. across. The large, light region at the bottom is the fill tube crimp, which is welded at its upper end. The bubbles and interconnected crevices in the weld region are weld voids from potassium hydroxide contamination; they can cause leaks.

The construction of high-capacity nickel-hydrogen cells often requires the precise stacking and alignment of more than 100 stack units on a supporting core that is affixed within the pressure vessel. X-ray inspection of these component stacks after insertion into the pressure vessel can detect any stack components that are misaligned or have had their edges bent during construction or insertion into the pressure vessel. Such defects can result in cell short circuits either early in life or later, when the cells are exposed to vibration. In addition, x-ray inspection of the component stacks in completed nickel-hydrogen cells can verify that the correct number of separator layers and electrodes have been built into the cell. High-resolution x-ray inspection of the leads where they are welded to the plates and where they are bent into the lead bundle is a key inspection technique that, as shown in Fig. 13.2, is also very important to ensure high reliability.

The lead welds can loosen or break as a consequence of either vibration-induced stress or excessive pulling of the leads during construction. If the leads are pulled up or pushed down in the core during fabrication, they can make close contact with adjacent electrodes, a condition that could lead to an incipient short circuit at some future time in the life of the cell. Clearly this condition should be avoided through careful inspection.

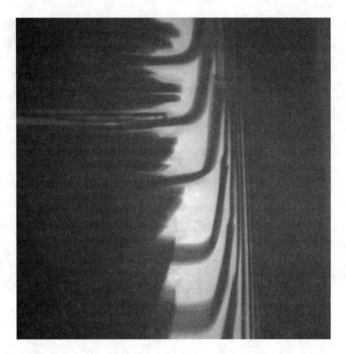

Figure 13.2. X-ray image of lead attachment to the negative plates in a cell, and the routing of the leads.

Conductive particles can flake from the edges of the nickel or the hydrogen electrodes, from either the stress of inserting the electrode stack into the pressure vessel or popping damage* at the edges of the electrodes. Either condition can be detected by x-ray inspection of the gap between the electrode stack and the internal wall of the pressure vessel. Inspection can be performed immediately after the cells are activated, after they have completed acceptance testing, or at any later time as a response to performance problems. It is standard practice to reject any cells that show any x-ray evidence of high-density (metallic) particles that could cause an internal short circuit.

The oxygen produced during cell overcharge can in some situations reach concentrations where it can undergo local ignition in regions within the electrode stack in addition to the popping sometimes seen at the edges of the plate stack.

---

* "Popping" in nickel-hydrogen cells is an audible sound produced by ignition of a bubble or jet of oxygen gas that may impinge upon the platinum catalyst during overcharge, when the cell contains a high pressure of hydrogen gas. It often occurs at the edges of the stack as a result of excessive overcharge, or in a very wet electrode stack that can facilitate the formation of oxygen bubbles during overcharge.

Ignition in a local area can result from voids in the nickel electrodes that channel oxygen into localized jets that impinge on platinum catalyst particles. Local ignition can also result from variations in nickel electrode oxygen evolution rates through the stack, either from thermal gradients or mismatched electrodes—variations that can concentrate most oxygen evolution on one or two nickel electrodes within a cell.

Damage from localized oxygen recombination tends to be accentuated by high-rate overcharge, particularly at low temperatures where the cell can enter overcharge quite abruptly. The damage from localized ignition at the surface of a nickel electrode, which disrupts the separator and hydrogen electrode in that area, can typically be detected by high-resolution x-ray inspection of the gas screen space on the backside of the hydrogen electrodes. Such damage, which can lead to internal cell short circuits, typically appears as high-density debris or particles extending into the gas screen region. Cell lots containing cells that show evidence of unusually high losses in a charge-retention test should be screened using high-resolution x-ray inspection for this kind of damage.

The compression on the electrode stack in a nickel-hydrogen cell is normally set to a prescribed preload level during fabrication, which is maintained by a

Figure 13.3. Belleville washer for maintaining stack compression in nickel-hydrogen cells.

Belleville washer that is placed at one end of the core that supports the stack, as shown in Fig. 13.3. The Belleville washer is typically partially compressed by the preload. When the cell is activated or vibrated, some reduction in the Belleville compression may occur as the components shift or compress. As the nickel electrodes expand during life, the Belleville compression will increase until it is eventually fully flattened. X-ray inspection of the Belleville washer can determine whether it has either relaxed or flattened from its initial compression. Significant Belleville washer relaxation is evidence of a stack that probably has insufficient compression and would be subject to plate movement that could break leads during vibration. Unexpected early-life compression or flattening of the Belleville washer may indicate defective nickel electrodes, excessive electrolyte concentration, or anomalous electrode damage for other reasons. It is normal to observe fully flattened Belleville washers after a cell has performed 25% or more of its expected cycle life. The Belleville washer shown in Fig. 13.3 is in the unflattened state expected at the beginning of cell life.

After a nickel-hydrogen cell has been cycled for many years and the Belleville washer has become fully flattened, the force exerted on the stack begins to increase. This force is absorbed by increasing compression of the stack components, as indicated in Fig. 13.4, as well as by elongation of the plastic core structure. High-resolution x-ray measurements may be used to determine the amount of stack elongation that has resulted from the stack swelling throughout cell life.

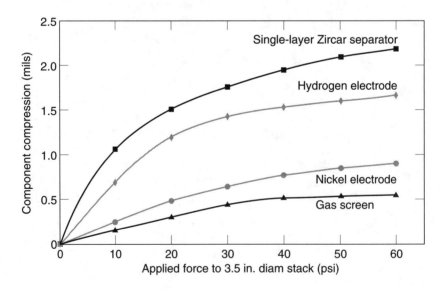

Figure 13.4. Compression of stack components in a 3.5 in. diam nickel-hydrogen cell as a function of applied force.

These measurements can provide an estimation of the degradation experienced by the nickel electrodes, which are primarily responsible for stack swelling.

A nickel-hydrogen cell may experience unusual increase in internal resistance if the stack is not sufficiently compressed to maintain the nickel electrodes in complete contact with the separator, as indicated in Fig. 13.5. In the back-to-back stacking arrangement, nickel electrodes are commonly constructed with their internal supporting nickel grid on the side that faces away from the separator. If the electrodes are allowed to stand open-circuited after a full recharge, gradual phase changes in the active material can make the nickel electrodes cup and partially lose contact with the separator.

Such deformations, which have been seen during 72 h charge retention tests, can result in up to a 35% increase in cell impedance and depressed discharge voltage. This issue, which is most likely to be seen in cell designs having compressible separators (e.g., those made of asbestos or nylon), can be detected by high-resolution x-ray imaging of the plate stack in the charged state during open-circuit stand. Typically the elevated cell impedance disappears as the cell is discharged as shown in Fig. 13.5, because the differential expansion and stress from active-material volume changes disappears as the active material is discharged.

Electrolyte pooling in nickel-hydrogen cells is highly undesirable. It can lead to popping damage in the stack or at the edges of the plates, and it can block access of hydrogen gas to the platinum catalyst in the gas screens. High-resolution

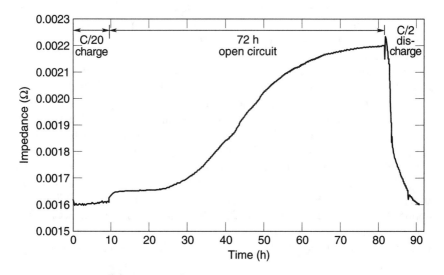

Figure 13.5. Impedance increase seen during the 72 h charged stand of a 76 A h nickel-hydrogen cell with asbestos separator, and subsequent impedance decrease upon discharge.

x-rays can directly detect pools of electrolyte in the dome of the pressure vessel or in the gas screens. If a pool of electrolyte is detected in the pressure vessel dome by x-ray inspection, it is likely that the cell will be prone to popping damage in any gravitational configuration where the pool is forced to the edge of the cell stack.

The appropriate x-ray inspection of nickel-hydrogen cells, either for quality control or diagnostic purposes, is potentially one of the most useful methods for screening a cell for the wide range of problems discussed in this section. High-resolution x rays are a key measurement tool that should be used to inspect any cell prior to disassembly and analysis.

13.1.2 Standard Capacity Tests

Nickel-hydrogen cell capacity performance is typically assayed using a group of relatively standardized test procedures that provide an indication of the performance of a cell when it is new that can be compared to the performance data of similar cells that have been previously built. Later in the storage or cycle life of a cell, these same standardized tests can be carried out to indicate how the performance of a cell has changed or degraded during its life. Similarly, if a cell fails, standardized capacity tests can indicate the level of degradation and the signatures that are associated with that particular failure.

Standard capacity tests are conducted at a constant temperature throughout the temperature range of $-10°C$ to $+30°C$. Typically, cell temperature (or average temperature, if a number of cells are being tested) is maintained as close as possible to the desired temperature set point during these capacity tests. The results from standard capacity tests are often used to match cells of similar performance into batteries.

Standard capacity tests typically involve an initial full discharge of each cell individually to a fully discharged state, either at a constant low current (e.g., $C/100$ to 0.1 V) or by resistive letdown to a low voltage such as 0.1 V. Each cell is then recharged to the fully charged state. In most modern ManTech cell designs, this recharge typically involves a $C/10$ charge for 16 h, which is sufficient to significantly overcharge the cell and bring it to its full capacity. In some cases, reduced recharge is applied (e.g., 12 h at $C/10$) to minimize the stresses that overcharge applies to a cell. In some cell designs that do not tolerate medium-rate overcharge well, such as those using asbestos separator, recharge regimes that apply a $C/10$ recharge for 9 h then $C/20$ recharge for 14 h have been successfully used. Each of these less stressful variants on the standard recharge procedure will result in somewhat lowered capacity but can provide a consistent measure of cell capacity.

Following recharge, cells are typically discharged at a $C/2$ rate to obtain cell capacity to 1.0 V, or any other desired voltage level. In some instances, open-circuit periods of up to 1 h have been inserted between the end of recharge and the start of discharge to allow for thermal stabilization and recombination of oxygen generated during overcharge.

The standard capacity of nickel-hydrogen cells is generally quite sensitive to temperature, and therefore it is customarily measured throughout a range of temperatures. The typical temperatures at which standard capacity measurements are obtained are 20, 10, 0, and −10°C. In some situations, standard capacity measurements are performed at 30°C, although at temperatures this high the capacity is quite low, typically only 50–60% of nominal. Because of the sensitivity of capacity to test temperature, any significant variations in temperature between individual cells during standard capacity tests should be taken into account when matching cell capacities or otherwise interpreting subtle capacity variations.

13.1.3 Charge-Retention Tests

Charge-retention tests are designed to detect any low-rate short circuits that may be present in a cell—ones that can result from conductive particles lodged in the separators or at the edges of the plate stack. Because nickel-hydrogen cells have a relatively high self-discharge rate, the observation of a stable voltage for the nearly fully discharged cell (like the observation that has been made in the past for nickel-cadmium cells) does not provide a useful charge-retention test. Two types of charge-retention test that are effective have been developed for nickel-hydrogen cells.

The first method uses a standard capacity test at 10°C (although 0 and 20°C have also been used), but it inserts a 72 h open-circuit stand period between the charge and discharge portions of the procedure. Charge retention is defined as the percentage of the standard capacity retained after this open-circuit stand period. Cells without any internal short circuits or other losses should retain 82–86% of their standard capacity in this test. Tests at higher temperatures will result in lower charge retention as a result of increased self-discharge at elevated temperatures.

The most sensitive method for detecting internal short circuits in a group of cells is not to look at the percent capacity retained, but to examine the distribution of cell voltages at the end of the open-circuit period, as shown in Fig. 13.6. A group of cells with no internal short circuits will exhibit a distribution of voltages at the end of the open-circuit stand period that is extremely tight, typically having a range of ±3 mV or less.

Any cell that does have internal short circuits will exhibit a voltage several standard deviations lower than those of the remaining cells, assuming that only a small number of the cells have internal shorts. If the majority of the cells have even low-rate internal short circuits, a quite wide distribution of cell voltages will be seen at the end of the open-circuit period. When interpreted in this way, the standard charge-retention test provides an extremely sensitive indicator for even very low-rate internal cell shorts.

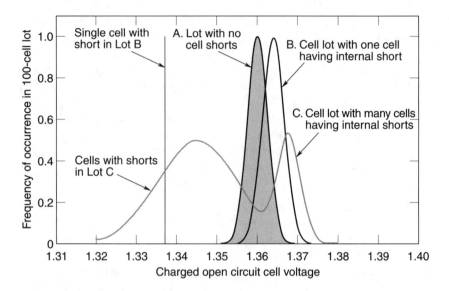

Figure 13.6. Distributions of cell voltages at the end of a 72 h open-circuit period in a charge-retention test for 100-cell groups or lots. Lot A is typical of a group of cells having no low-level internal cell short circuits. Cell lot B has a single cell with an internal short that makes its voltage fall well below the distribution of the other cells. Lot C has a large number of cells with low-rate internal short circuits, which display a broad distribution at voltages that are lower than normal.

The second method for detecting internal cell short circuits involves measurement of the current/voltage behavior of the cell at low voltages. One starts with a fully discharged cell at zero volts, then gradually increases the cell voltage to 1.0 V while the steady-state cell current is measured. If a cell has an internal short circuit, significantly more current will be required to hold it at voltage. Examples of these leakage current signatures are shown in Fig. 13.7. This test is extremely sensitive for nickel-precharged cells, for which only microamps of current are typically required to maintain voltage, and which are thus very sensitive to even very low parasitic currents.

13.1.4 Precharge Tests

Evaluation of the type of precharge in nickel-hydrogen cells is crucial for choosing the appropriate storage regime for a group of cells or a battery.[13.2] Cells with remaining active nickel capacity after the hydrogen gas has been completely discharged (termed nickel-precharged cells) can be stored in a fully discharged state for many years. However, cells with insufficient active nickel capacity to discharge all the hydrogen gas in the cell (hydrogen-precharged cells) must be

13.1 Cell-Level Diagnostic Tests

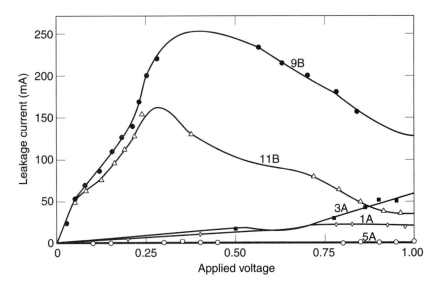

Figure 13.7. Leakage current as a function of voltage for nickel-precharged cells (group A) and hydrogen-precharged cells (group B). Cells 9B, 1A, and 3A have low-rate internal short circuits, as shown by the excess current observed at 1 V.

actively maintained at voltages greater than about 1.2 V to prevent degradation. Fortunately, simple tests are available to detect the type of precharge that is in a cell.

The most definitive test to determine the type of precharge in a nickel-hydrogen cell involves the application of a low reversal current to the fully discharged cell. Before this is done, however, the cell must be properly prepared for the test by ensuring that all possible hydrogen has been discharged. This is best done by carrying out a standard capacity cycle at either 10 or 20°C, followed by resistive letdown with a 1 Ω resistor until the cell voltage is below 0.005 V. It is critical that this standard capacity test be performed with no open-circuit stand period between the recharge and the discharge portions of the cycle, because some active nickel electrode capacity can become isolated during charged open-circuit stand periods.

After full discharge, a C/40 reversal current is applied to the cell for 5 min. Cells containing nickel precharge will drop well below –0.1 V, while cells with hydrogen precharge will remain just below zero volts, as shown in Fig. 13.8. In this test, cells with only small amounts of active nickel precharge may actually drop to voltages as low as –1.6 V during the reversal. If this type of behavior is seen, it is likely that these cells may soon transition to the hydrogen-precharged condition if they are subjected to further charge/discharge cycles.

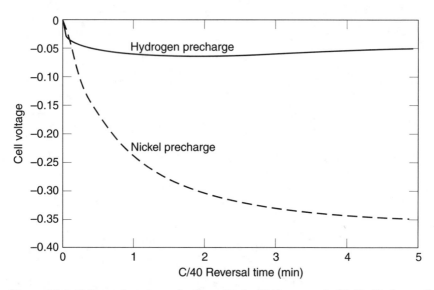

Figure 13.8. Voltage signatures during a 5 min C/40 reversal of fully discharged nickel-hydrogen cells with hydrogen precharge and with nickel precharge.

It is possible to actually measure the amount of active nickel precharge in a cell by continuing the C/40 reversal until the cell voltage falls to below −1.2 V. The amount of precharge is equal to the ampere-hours of capacity discharged during the reversal. This reversal procedure converts the active nickel precharge into oxygen gas within the pressure vessel. While the oxygen gas is reasonably stable in the cell, it can be converted back into active nickel precharge by recharging the cell at a C/40 rate until the cell voltage rises to 0.7 V.

The type of precharge can also be recognized based on the current/voltage behavior of cells at positive voltages. When a fully discharged cell with nickel precharge is allowed to stand open-circuited, it will typically settle at a voltage of 0.2 to 0.3 V, meaning that the partially charged nickel electrode is 0.2 to 0.3 V greater than the oxygen potential on the platinum catalyst electrode. However, when a fully discharged cell with hydrogen precharge is allowed to stand, it can recover to more than 0.9 V, depending on the hydrogen pressure, then gradually fall to near zero volts. Care must be taken to not mistake a cell that is gradually dropping towards zero volts, but happens to be in the 0.2 to 0.3 V range, for a nickel-precharged cell.

The steady-state current/voltage behavior such as that shown in Fig. 13.7 can also indicate the precharge. In general, hydrogen-precharged cells need a much higher current to hold the voltage between 0 and 1.0 V because of the self-discharge of hydrogen gas on the nickel electrode. Monotonically increasing current is needed to maintain an increasing cell voltage until eventually the catalytic activity of the nickel electrode passivates at about 0.2 V in the hydrogen-precharged cell, or until all oxygen gas is reduced in a nickel-precharged cell at 0.4 to 0.5 V. At levels greater than each of these voltage ranges, the current needed to hold the cell voltage will drop significantly, thus providing an indication of the precharge in the cell.

13.1.5 AC Impedance and Resistance

The ac impedance of a nickel-hydrogen cell[13.3] is dominated at low frequencies by the capacitive and diffusive response of the nickel electrode, which has a Warburg impedance** response[13.4] as a function of frequency. At higher frequencies (several hundred Hz), the hydrogen electrode also exhibits a capacitive response. At frequencies above 1 kHz, nickel-hydrogen cells typically exhibit an inductive response that arises from the internal geometry of the leads and electrode stack. The electrolyte resistance is best determined from the purely resistive crossover between capacitive and inductive behavior that typically occurs near 500 Hz.

Clearly, it is possible to extract significant information from the ac impedance of a nickel-hydrogen cell. However, this method is not widely used to diagnose cell problems for the simple reason that most nickel-hydrogen cell problems originate with the nickel electrodes and are the most difficult to characterize by ac impedance as a result of the very low frequency response and the dynamic nature of the nickel electrode impedance.

The resistance of a nickel-hydrogen battery cell generally refers to impedance at frequencies equal to or lower than that where the cell has no reactive impedance component, which is typically at approximately 500 Hz. At frequencies lower than this, the cell displays a capacitive reactance that results from the polarization of the electrodes as a result of current flow. The impedance at approximately 500 Hz thus provides an indication of the pure resistance of the electrolyte and the internal cell leads. Lower frequency measurements give higher resistance, because they include the polarization resistance of the electrodes as well.

The term "dc resistance" typically refers to measurements based on monitoring the cell voltage change in response to a change in current—for example, during cell discharge. Clearly, the resistance measured in this way depends on the time after the current change at which the cell voltage response is measured. Measurement after the first several milliseconds essentially provides the pure resistance of the electrolyte and the internal leads. Measurement after 100 milliseconds or 1 sec will give higher resistance, corresponding to roughly 10 Hz or 1 Hz ac measurement frequencies respectively. The value of dc resistance is best measured based on a current change while the cell is either charging or discharging, while ac

---

** The Warburg impedance is a constant-phase circuit element having a 45 deg phase angle, and which describes diffusion in an infinite medium.

impedance can be measured under any conditions for which the cell state does not change significantly during the course of the measurements.

13.1.6 Dynamic Calorimetry

Dynamic calorimetry is a technique that has been developed[13.5] to measure the heat generated by a device that is dynamically changing with time. This is the case with nickel-hydrogen battery cells during the high-rate charge or discharge operation that they experience during normal operation. A wide range of thermal effects may be probed using dynamic calorimetry, including electrochemical heat, pressure-volume work, oxygen evolution and recombination thermal effects, and resistive heating in the leads.

Dynamic calorimetry works by monitoring the thermal field produced over the surface of a cell using small precision thermal sensors. Each thermal sensor protects an area of several square centimeters from contact with a precisely regulated thermal bath, in which the cell is tested. The areas that are covered by a sensor will warm if the cell is producing heat, or will cool if the cell is absorbing heat. The amount of heat and the precise location of the heat within the cell can be determined from the time dependence of the surface thermal profile with the use of a finite-element thermal model that includes all the internal cell components. Figure 13.9 shows a typical thermal response for a nickel-hydrogen cell measured by dynamic calorimetry.

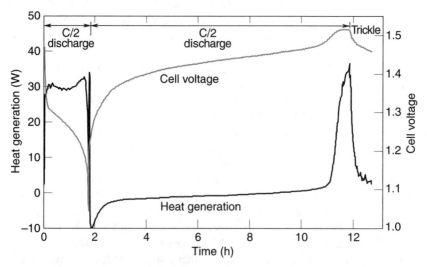

Figure 13.9. Heat generation measured from a nickel-hydrogen cell using dynamic calorimetry at a temperature of 10°C. The heat was measured during an electrical cycle consisting of discharge at a C/2 rate, recharge for 10 h at a C/10 rate, and trickle charge for 1 h at a C/100 rate.

The quality of the thermal measurements produced with dynamic calorimetry is directly proportional to the precision of the constant temperature bath in which the cell is tested. State-of-the-art thermal baths can provide ±0.0001°C temperature control, which can give highly precise calorimetry data. The accuracy of the calorimetry results depends on careful calibration, as well as a high-fidelity thermal model of the cell, the attached sensors, and the thermal environment.

The thermal coupling of the cell to the bath is typically adjusted such that the maximum cell heat generation produces no more than a 1°C temperature rise. Thus, high heat-dissipation rates can be measured by placing the cell directly in the thermal bath, while lower heat-generation rates may require some thermal isolation between the cell and the thermal bath to achieve about a 1°C temperature rise. This method can be used to measure the heating from self-discharge at levels down to microwatts, as well as heating rates up to 40–50 W during high-power cell operation.

Dynamic calorimetry can be used for a number of measurements other than evaluating simple heat generation as a function of time. Measurement of the thermal transients that occur when heating is abruptly started or interrupted can be used to evaluate the thermal conductivity of key thermal components within the cell. For example, this method can determine the thermal conductivity of the gap between the electrode stack and the internal wall of the pressure vessel. This gap is partially bridged by the edges of the separators, which are wetted with electrolyte and which make partial contact with the wall wick on the inner pressure vessel surface. The thermal conductivity of this gap cannot be precisely known and must be measured using a method such as dynamic calorimetry. In a dynamic calorimetry model, the wall gap thermal conductivity is varied until the value is found that correctly gives the thermal transient observed on the surface of the cell when heat generation from cell discharge is abruptly stopped.

The cell thermal model that is used to compute heat-generation rates within the cell based on the dynamic calorimetry data naturally provides the thermal gradients throughout all the internal cell components as a function of time while the cell is operating. The accuracy of the computed internal thermal profile is dictated by the fidelity of the thermal model. A highly accurate cell thermal model can enable the internal cell thermal gradients to be determined very accurately indeed. These results can be used to validate that internal thermal gradients are not great enough to cause condensation of water, and that the top of the electrode stack does not run too hot relative to the bottom of the stack.

The heat capacity of a battery cell may also be determined using dynamic calorimetry. For heat capacity measurements, the cell is fitted with a heater and several thermal sensors and placed in an adiabatic enclosure. A known heat-generation rate is applied to the cell using the heater, and the temperature rise of the cell is monitored. The finite-element model of the cell, heater, sensor, and enclosure is then run with the known heater power, and the correct heat capacity of the cell is determined by matching the observed temperature rise. This method automatically takes into account the heat leaks through the heater power wires and thermal sensor wires, as well as the thermal leaks that are always present for any cell enclosure.

13.1.7 Acoustic Monitoring Test

Nickel-hydrogen battery cells are capable of recombining the oxygen gas produced at the nickel electrodes during overcharge in two different ways. The first, which is desired, is to electrochemically recombine the oxygen with hydrogen at the platinum catalyst surfaces of the negative electrodes. The second way occurs if the oxygen concentration in the gas that contacts the platinum catalyst exceeds the level required to initiate and propagate gas-phase combustion. This oxygen recombination process will initiate a chain reaction and rapidly burn the oxygen-rich gas region. The rapid burn typically causes an audible popping sound resulting from the shock wave that emanates from the burning region. The shock wave can do significant damage to the electrodes and separator, potentially resulting in short circuits within the cell.

A common mechanism for this popping process involves the accumulation of a pool of electrolyte at the edge of the plate stack if the cell is held in a horizontal position. Bubbles of essentially pure oxygen can form in this pool at the edges of the nickel electrodes and grow until a bubble touches the platinum electrode, whereupon it immediately burns. High-rate overcharge can also produce streams of oxygen gas that can initiate popping within the electrode stack. Acoustic monitoring tests can determine the susceptibility of cells to popping damage by directly detecting the sounds produced from the shock waves as oxygen gas burns in the hydrogen atmosphere of the cell.

Acoustic monitoring of a nickel-hydrogen cell is done by fitting the cell pressure vessel with an ultrasonic sensor, typically tuned to sounds at about 50 kHz using a lock-in amplifier. Working at high frequencies eliminates much of the acoustic interference present in the environment at lower frequencies. However, the cell still must be enclosed in an anechoic box or chamber to minimize external interference.

The output of the sensor can be fed into a phase-sensitive detection system that will produce an electrical signal proportional to the amplitude envelope of the high-frequency noise. This envelope, which can persist for hundreds of milliseconds, can be readily detected and recorded with a threshold-sensitive data-acquisition system. The amplitude envelope of the sound recorded from a typical popping event in a nickel-hydrogen cell during overcharge is indicated in Fig. 13.10.

An acoustic monitoring test is typically designed such that a cell is put through a typical worst-case operating cycle with peak overcharge duration and current, and with the worst-case thermal gradient on the cell in the expected orientation relative to gravity. Under these conditions, the acoustic monitoring test can verify that the cell design is not subject to popping damage that could reduce its reliability.

Acoustic monitoring tests can also involve a high stress cycle that involves overcharge at rates higher than normal. A stress cycle such as this can evaluate

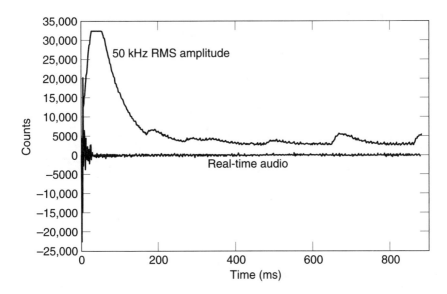

Figure 13.10. Acoustic signals accompanying a popping event recorded in a nickel-hydrogen cell several minutes after ending overcharge.

the amount of margin between the expected overcharge conditions and the much higher overcharge rates to which the cell is tested. If popping is heard in any cell under conditions that are close to those expected during worst-case operation, either the cell design should be changed to make it more tolerant of overcharge or the charge control system should be modified to reduce the overcharge stress.

13.1.8 Electrochemical Voltage Spectroscopy

Electrochemical voltage spectroscopy is a method originally developed to evaluate electrochemically active state density as a function of voltage for lithium insertion reactions in polymeric electrodes such as polyaniline or polypyrrole.[13.6] Electrochemical voltage spectroscopy can also be applied to a battery cell to provide the density of electrochemically active sites as a function of the voltage difference between the positive and negative electrodes.

The electrochemical voltage spectroscopy technique involves scanning the potential of the cell at a rate slow enough that the cell is always near its equilibrium potential. An excellent signal-to-noise ratio is maintained by integrating the charge into or out of the cell as a function of voltage. Typical scan rates are 1–2 μV/sec.

The electrochemical voltage spectroscopy measurement provides a direct spectrum of the density of active charge sites over the full range of cell voltages that are accessible. Figure 13.11 shows a typical electrochemical voltage spectroscopy spectrum for a nickel-precharged nickel-hydrogen cell in the voltage

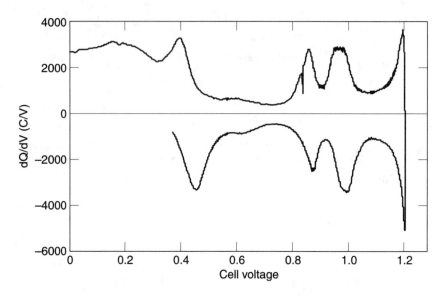

Figure 13.11. Electrochemical voltage spectroscopy measurements for a 76 A h nickel-hydrogen cell with nickel precharge. The cell voltage was initially near 0.4 V and was scanned up to 1.2 V at 2 µV/sec, then back to 0.0 V. The peak near 0.4 V corresponds to the oxygen reaction and the peak near 1.2 V to the hydrogen reaction on the hydrogen electrode.

range between 0.0 and 1.2 V, where peaks from oxygen, platinum oxide, and nickel reduction processes may be seen on the platinum catalyst electrode. In a nickel-precharged cell, the nickel electrode remains at a relatively constant voltage throughout the scan, while the platinum catalyst voltage is scanned from the highly oxidizing potential of the nickel electrode when the cell voltage is low, down to near the hydrogen evolution potential when the cell voltage is at 1.2 V.

For nickel-hydrogen cells, the electrochemical voltage spectroscopy method can assay the phases present in the nickel electrode, detect platinum complexes that can accumulate during storage, detect electroactive contaminants, or measure the oxygen or hydrogen evolution I/V behavior of the cell. Each of these electrochemically active materials produces a peak in the electrochemical voltage spectroscopy spectrum at its characteristic redox potential. Materials that have fully reversible electrochemistry give a positive peak when the voltage is scanned downwards and a negative peak at essentially the same potential when the voltage is scanned upwards.

In a nickel-hydrogen cell, the position of an electrochemical voltage spectroscopy peak for a given electrochemically active compound depends on the potential of both the positive and the negative cell electrodes. If the cell contains hydrogen gas, the voltage scale for the electrochemical voltage spectroscopy spectra corresponds to the nickel electrode relative to the hydrogen potential. If the cell has a nickel precharge and is discharged to the point where it has no hydrogen gas, the electrochemical voltage spectroscopy voltage scale suddenly changes to correspond to the voltage of a shifting negative electrode relative to that of the partially charged nickel electrode. Typically the potential of the open-circuited platinum electrode will correspond to the oxygen potential at whatever low levels of oxygen gas are present in the cell after the hydrogen gas has become exhausted.

13.1.9 Pressure Vessel Inspection Methods

The pressure vessel of a nickel-hydrogen cell must be highly reliable, because any leak of hydrogen gas will result in cell failure, as well as potentially being a safety hazard. It is normal procedure to inspect the Inconel pressure vessel for any indication of a flaw that could propagate into a leak during the life of the cell.

The most commonly used inspection method is dye penetrant inspection. This method involves swabbing the pressure vessel surface that is to be inspected with a dye solution that is then wiped from the surface. Any surface flaw will entrap some of the dye, which can then be seen by inspection under ultraviolet light. Any flaws seen by this method can be microscopically inspected and analyzed to determine the nature of the flaw and whether it could compromise the reliability of the pressure vessel.

Other methods for detecting pressure vessel flaws, such as eddy current and ultrasonic probes, are capable of detecting flaws in the pressure vessel or its welds that are not visible on the outer surface. Such flaws can include occluded bubbles or particles in the heat-affected zones of the welds, as well as the cracks or pits that can occur on the surfaces of the pressure vessel. The walls of nickel-hydrogen cell pressure vessels can vary in thickness from 15 to 45 mils, depending on the vessel design and the amount of localized thinning that may occur during the hydroforming process by which the pressure vessels are made. The detection of even small flaws is most critical for the thinnest pressure vessels to ensure high reliability.

No flaw detection method is 100% certain of finding all the flaws above the critical size that could lead to pressure vessel failure. Inspection methods and inspectors are typically certified to a prescribed level; for example, an inspector may be certified to find 99.9% of the flaws greater than 10 mils in length. Clearly the probability of a flaw escaping detection will also depend on the total number of flaws present in the pressure vessels that make up the cells in a battery. If the inspector described here finds only one flaw during inspection of 100 pressure vessels, there is only 0.1% chance that a critical flaw may have escaped detection. However, if the same inspector finds 25 flaws during inspection of the 100 pressure vessels, there is a 2.5% probability that a critical flaw was missed. To guarantee high battery reliability, a threshold for the number of flaws detected must be set that will differentiate between a good lot and a bad lot of pressure vessels based on the statistics of the inspection process. In the example given here, high reliability may require that the entire lot of pressure vessels be rejected if more than 5 critical flaws (flaws capable of causing pressure vessel failure) are found during inspection.

13.2 Destructive Physical Analysis Techniques

Destructive physical analysis of a nickel-hydrogen cell involves opening the pressure vessel, inspecting the electrode stack and other internal parts, disassembling the electrode stack, and weighing and inspecting all the electrodes and separators in the stack. However, it is recommended that an analysis of the gas within the fully discharged cell be performed before the destructive physical analysis is done. This requires puncturing the wall of the pressure vessel to draw out a sample of the gas that it contains for analysis and to measure the total pressure of this gas within the pressure vessel.

A gas analysis can help pinpoint the type and amount of active precharge in a cell and identify inert gas or organic contaminant levels. The gas analysis procedure, which is described in this section, does not affect any other components within the cell for a planned destructive physical analysis, and typically should be part of a standard destructive physical analysis.

Before a cell is disassembled, it is also possible to carry out a test that evaluates the margin of extra electrolyte contained in the cell above the minimum level where cell failure would occur as a result of separator dry-out. This test requires that the cell pressure vessel be punctured, allowing water vapor to be drawn from the cell to force eventual dry-out. If this test is performed, the distribution of electrolyte in the components will be modified, and later analysis of electrolyte concentration in the cell must be corrected for the water withdrawn. This electrolyte dry-out test, which is also described in this section, is generally not part of a routine destructive physical analysis procedure.

Destructive physical analysis of a nickel-hydrogen cell first involves removal of the electrode stack from the pressure vessel. This is done by milling or cutting the pressure vessel away from the weld-ring and by cutting off the terminals if they have a Ziegler compression seal. Cells with Teflon terminal seals do not require cutting of the terminals. At this point, an accurate weight is obtained for the cell and terminals to enable correction to be made for any evaporation of water from the electrolyte during destructive physical analysis. After removal of the electrode stack from the pressure vessel, the inner wall wick on the pressure vessel is inspected for damage or any particulates, and the stack is inspected for any damage to the edges of the separators or electrodes. The stack length should be measured in four locations around the perimeter of the end plates, and the compression of the Belleville washer should be documented. Photographic documentation of the condition of all these components is recommended.

The compression of the plate stack is released by removal of the nut at the end of the core. If this nut can be removed by hand, the stack probably has insufficient compression to properly hold all components together. However, it is more typical that the polysulfone nut threaded onto the end of the core must be broken to loosen the core compression. The core nut can be readily broken loose by dripping several drops of acetone onto it, making sure that all personnel are clear of the stack. The acetone will crack the core nut and allow it to be easily removed. If there is high stack compression, as in a cell that has been extensively cycled, the core nut may fly upwards as it breaks and the stack compression is released. The electrodes and separators may then be removed from the core and the electrode leads cut at the weld tabs.

All the components are inspected, weighed, and sealed into bags for further analysis at a later time. Typically, the separators are left on the surfaces of the nickel electrode pairs, and the gas screens are left between the pairs of hydrogen electrodes until these components are rinsed free of electrolyte. At the end of the destructive physical analysis, the weights of all the individual electrodes and separators are combined with the weight of all the other pieces of cell hardware to determine whether any net loss of water occurred during the destructive physical analysis process by evaporation. Typically, no evaporative correction to the electrolyte concentration is needed if the destructive physical analysis is performed in less than 1 h in approximately 40–50% humidity.

Measurement of the electrolyte distribution through the cell stack typically involves separating both the positive electrodes (with separators) and the negative electrodes (with gas screens) into five groups covering the length of the stack. Cells with dual stacks would thus have ten groups of positives and ten groups of negatives, five for each stack. The electrolyte is extracted and analyzed separately for each of these groups of electrodes to obtain electrolyte concentration and wetness profiles through the stack.

The electrodes should be inspected and photographed as required either during or after stack disassembly. The condition and wetness of the positives should be noted, along with any blisters, damage, or discoloration. The amount of black active material extruded from the positives into the separators should be documented. Any damage to the negatives, including holes or edge damage, should be documented, along with any burned or melted spots at the edges of the gas screens. The amount of electrolyte trapped in the gas screens, as well as any discolored or translucent regions on the Teflon side of the negatives, should be noted.

13.2.1 Gas Analysis

A quantitative gas analysis for each nickel-hydrogen cell that is subjected to destructive physical analysis provides the most effective way to quantify the precharge in a cell, check for inert gas contamination, and detect organic contamination. This analysis is done by initially discharging the cell fully—for example, by letting it down through a 1 Ω resistor for 16 h or to less than 5 mV. The cell is then placed in an evacuated chamber that is fitted with a precision pressure transducer; this chamber has an accurately calibrated volume with the cell in place. The chamber volume calibration involves measuring the pressure change when a known pressure is expanded from the chamber into an auxiliary calibrated volume that has been evacuated. The cell is then punctured while sealed in the evacuated chamber, and the rise in chamber pressure is measured. A sample of the gas in the cell is then withdrawn for analysis into an evacuated stainless steel gas-sampling bottle. The total of the cell and chamber volume is then determined by adding inert gas, and expanding a known pressure from the chamber that contains the punctured cell into the auxiliary calibrated volume, which yields the internal cell volume by difference. The total internal cell pressure may be determined from the volumes of the cell and chamber, and the measured pressures.

The gas sample obtained from the cell is analyzed via mass spectroscopy using a residual gas analyzer. This analysis provides the relative fractions of the various gases in the cell, from which one can calculate the gas partial pressures in the cell. Typically, the cell will contain either significant hydrogen or significant oxygen gas, but not both. This is because the platinum catalyst will allow these gases to react and consume each other until one is completely exhausted. If hydrogen gas is found, the cell has hydrogen precharge and the hydrogen partial pressure gives the amount of hydrogen precharge. If oxygen gas is found, the cell is nickel-precharged and the amount of nickel precharge must be determined by discharge of the nickel electrodes from the cell.

Other gases that may be found in a cell include inert gases such as water vapor, nitrogen, argon, helium, or fluorocarbons, as well as carbon dioxide and methane. The most common source of inert gases is atmospheric contamination during cell activation, which can give up to about 12 psi of nitrogen, and an argon level that is about 1% of the nitrogen. Higher levels of argon can result from the addition of argon during or prior to cell activation, coupled with incomplete evacuation of the argon before the cell is activated. Helium traces often result from pressure vessel leak testing, and fluorocarbons from solvents. If significant levels of these inert gases are present in a cell, its high-rate capacity will be decreased markedly by accumulation of the inert gas in the gas screens during discharge, thus preventing high-rate access of hydrogen gas to the platinum catalyst in the negative electrode. Nitrogen levels of 12 psi can result in a 5–10% loss of high-rate capacity. More than several psi of inert gases in a cell signify a problem with the cell activation procedure.

Gases such as carbon dioxide and methane result from organic contamination in the cell components. Organic materials, if not inert, may be oxidized on the nickel electrode to produce carbon dioxide, which will maintain equilibrium with carbonates in the electrolyte. Generally no more than trace levels of carbon dioxide are found, both because it reacts readily with the potassium hydroxide electrolyte and because the reducing hydrogen environment typically present in the cell will hydrogenate most organics to ultimately produce methane gas. Methane levels more than 0.5 to 1.0 psi in a cell can indicate problems with excessive organic contamination, which can result in increased cell impedance at low temperature and accelerated loss of nickel precharge during storage.

13.2.2 Electrolyte Dry-Out Test
The electrolyte dry-out test[13.7] evaluates the electrolyte margin in a nickel-hydrogen cell in excess of the minimum amount needed to maintain adequate conductivity through the separator during cell operation. To perform this test, one attaches a valve to the cell, preferably at the fill tube, to allow water to be removed from the cell. The weight of the cell with the attached valve is obtained before starting the test. The performance of the cell is measured by carrying out two cycles at 10°C: The first one is a stabilization cycle and the second one, a characterization cycle.

The stabilization cycle involves recharge for 16 h at a C/10 rate followed by discharge at a C/2 rate to 1.0 V, discharge to 0.5 V at a C/10 rate, and resistive letdown (1 Ω) to 0.1 V. The characterization cycle involves recharge for 16 h at C/10, then discharge at C/2 with a 5 min C-rate discharge pulse every 15 min during the C/2 discharge. If the voltage drops below 0.5 V during the C-rate pulse, the current should be set back to C/2. If the voltage drops below 0.5 V during the C/2-rate discharge, the characterization discharge is terminated. Following the high-rate discharge, the cell is discharged to 0.5 V at a C/10 rate, then resistively let down to 0.1 V.

After the two-cycle sequence, the cell is weighed and then connected to a vacuum line with a liquid nitrogen trap. A vacuum is drawn on the cell until several percent of its electrolyte weight has been withdrawn as water vapor. The cell is then weighed to accurately determine how much water was removed. The two-cycle characterization sequence described here is then repeated (the impedance of the cell during the high-rate discharge in this test is determined by the change in voltage when the current is changed from C-rate back to the C/2 rate).

The sequence of water removal and performance characterization is repeated until the cell resistance shows a dramatic increase during the C-rate discharge. When this occurs, the current through the cell at the C-rate has become limited by the diffusion of electrolyte ions through the separator, indicating that the separator has dried to the point where it is unable to support high-rate discharge. The amount of water removed gives the volume of electrolyte in the cell above the minimum needed to support high-rate discharge. Robust cell designs are expected to have significant electrolyte margin to accommodate nickel electrode corrosion and swelling throughout long-term operation while maintaining reliable performance.

13.3 Nickel Electrode Diagnostic Tests

13.3.1 Thickness Measurements

A change in the thickness of the nickel electrodes that are removed from a cell, as compared to their thickness when they were new, provides an indication of electrode degradation. The nickel electrodes increase in thickness for several reasons. First, the volume changes of the active materials in the electrodes as they are cycled induces strain on the sinter structure and net growth in the thickness of this structure during accumulated cycles. Charging to a higher state of charge (γ-NiOOH) causes greater volume change, and thus more electrode expansion. Second, high-rate overcharge can accelerate expansion from pressure within the sinter caused by high-rate oxygen evolution. Finally, gradual corrosion of the nickel metal in the sinter structure of the nickel electrode will form the higher-volume nickel oxyhydroxide corrosion products, thus tending to inflate the sinter structure with added material. In addition to expansion of the sinter structure, each of these processes can produce added swelling by the extrusion of active material from the pores of the sinter to form surface deposits.

The thickness of nickel electrodes is normally measured to a precision of 0.1 mils. Measurements are made approximately halfway between the inner and outer diameters of the electrode at four locations equally spaced around the electrode circumference. The electrode thickness is taken as the average of the four measurements.

The thickness of each electrode from a cell is typically measured in this way, and the thickness is plotted over the full length of the electrode stack. Thickness may be measured either before or after rinsing the electrolyte from the nickel electrodes. Any systematic variations in electrode thickness from one end of the stack to the other can indicate significant thermal gradients during cell operation, or concentration gradients over the length of the stack. For recirculating stack cell designs, it is normal to see a thickness variation develop over the length of the stack as a result of the electrolyte concentration gradients induced by the water recirculation process within the cell.

13.3.2 Full Electrode Resistance

Nickel electrodes contain a plaque substrate consisting of a nickel metal grid that supports the nickel sinter. This structure has high conductivity when new. However, electrode expansion and corrosion can fracture the sintered contacts between sinter particles and their contact with the grid structure. These changes can significantly increase electrode resistance.

The resistance of a nickel electrode is measured after the electrode is rinsed and dried. Two opposing edges of the electrode are clamped to copper contacts, and nickel-plated probes are screwed into the sinter structure just inside each copper contact. A current of 1 A is passed through the electrode via the copper clamps, and the voltage drop across the electrode is measured between the probes. To obtain the electrode resistance, divide the voltage across the electrode by the current.

These measurements are best interpreted by comparing the resistance of electrodes removed from a cell to the resistance of new electrodes. If no changes in resistance are seen, there is little reason to suspect significant structural degradation in the sintered plaque.

13.3.3 Precharge (Active and Inactive)

When nickel electrodes are removed from a cell, they may be partially charged. If the fully discharged cell contains no hydrogen gas, this residual charge is termed nickel precharge. Part of the nickel precharge exists as electrochemically dischargeable material, which is termed "active precharge." Every nickel electrode also contains some inactive material that cannot be electrochemically discharged, which is termed "inactive precharge." The amounts of each of these types of precharge can be measured for samples of nickel electrodes using electrochemical and chemical analysis procedures.

The active precharge is measured by placing one-half of a nickel electrode in a flooded test cell that contains 31.0% potassium hydroxide electrolyte, an Hg/HgO reference electrode, and a nickel sheet counterelectrode. The active precharge is then discharged from the nickel electrode at a current density of 2 mA/cm^2 until the electrode voltage falls to –0.5 V *vs.* the reference electrode. Discharge is then continued after a 30 sec open-circuit period at a current density of 0.2 mA/cm^2 until the electrode voltage again falls to –0.5 V. The total capacity from both these discharges provides the active precharge.

After discharge of the active precharge, the inactive precharge can be determined by measuring the remaining capability of the active material to oxidize ferrous ions to ferric ions. Any nickel or cobalt species in the active material that have an oxidation state greater than 2.0 will be detected by this procedure. This procedure is carried out on a sample of pure active material isolated from the one-half electrode after discharge of the active precharge. The active material is isolated by grinding the electrode into a fine powder, then stirring the powder in a water solution with a large magnetic stirring bar. The stirring bar will hold all the magnetic sinter and grid particles, which can be lifted out of the solution with the stirring bar, leaving behind a suspension of pure active material. The powder attached to the stirring bar is dislodged, ground a second time, and then magnetically separated again.

The combined active-material suspension from both magnetic separations is filtered through a Millipore filter, dried, and weighed. A one-half electrode sample typically provides about 1.5 g of purified active material for further analysis. It is important that no nickel metal particles are present in the active material, because the metal can be partially oxidized during the following analysis to give an incorrect oxidation state.

To determine the oxidation state of the active material, a sample is reacted with a ferrous ammonium sulfate solution.[13.8] The ferrous solution should be freshly mixed, and standardized by conducting a blank titration with standard 0.1 N potassium permanganate to a pink end point. The potassium permanganate oxidizes ferrous to ferric species, as does the residual charge in the active material. An identical amount of ferrous solution should be mixed with about 0.5 g of the purified active material, stirred for 1 h to allow reaction, and then back-titrated with standard potassium permanganate. The difference between the blank titration and the sample titration is a result of the oxidation of ferrous ions by the active material, and it can be used to determine the oxidation state of the active material.

The interpretation of the active precharge measurements requires that the amount of active precharge set for the cell when it was activated be known. The "as manufactured" precharge level can generally be obtained from the cell manufacturer. Analysis of a new cell can verify the initial level of both active and inactive precharge. After a cell has been stored for a significant period of time or cycled extensively, a lower level of active nickel precharge is typically expected. Eventually, the active nickel precharge will disappear completely, to be replaced by hydrogen precharge. The measurements described here can quantify the loss of active precharge and determine whether the active precharge has been converted to inactive precharge, or whether it has been consumed by oxidation of other materials within the cell.

13.3.4 X-Ray Diffraction

X-ray diffraction analysis of active-material powder that has been isolated from discharged nickel electrodes can be used to identify modified forms of active material that may be electrochemically inert and thus contribute to decreased cell capacity. The x-ray diffraction pattern of the discharged material should normally exhibit only nickel hydroxide (β-Ni(OH)$_2$). In situations where the electrodes are well cycled and contain significant levels of residual capacity that cannot be easily discharged, it is possible that low-level x-ray diffraction peaks may also be seen for some γ-NiOOH phase. If nickel sinter particles are not completely isolated from the active material, x-ray diffraction patterns from nickel metal will also be detected. This is probably the most sensitive method for detecting traces of nickel metal in active-material powder.

Under some conditions, electrochemically inactive phases of active material may form in the nickel electrode and result in capacity loss. For example, cobalt oxide spinel compounds (Co_2O_3 or Co_3O_4) can form if the discharged nickel electrode is exposed to hydrogen gas at reducing potentials.[13.9] In some cases, these cobalt compounds can be detected from their x-ray diffraction patterns. However, in other situations the spinel structures are present in such finely divided form that they do not produce well-defined x-ray diffraction patterns.

13.3 Nickel Electrode Diagnostic Tests

Another electrochemically inactive compound that has been observed in electrodes that have anomalously low capacity is nickel oxide hydroxide (Ni_2O_3H),[13.10] which exhibits the well-defined x-ray diffraction pattern shown in Fig. 13.12.

Nickel oxide hydroxide, which was first isolated from extensively cycled nickel electrodes from nickel-hydrogen cells,[13.11] has also been synthesized using a hydrothermal process.[13.12,13.13] Normal low-temperature operating conditions for nickel-hydrogen cells do not appear to result in formation of nickel oxide hydroxide. Nickel oxide hydroxide appears to form in cells that are operated at high overcharge voltages, generally with extensive overcharge at temperatures of 15–20°C or higher.

Nickel oxide hydroxide has also been seen to form in extensively cycled nickel electrodes after removal from a cell as a result of extended storage in a desiccated environment. It is likely that this compound can form from normal active material by a dehydration process. It is likely that traces of platinum-containing oxide materials play a catalytic role in this dehydration reaction.

13.3.5 Thermogravimetric Analysis

Thermogravimetric analysis is typically performed on a sample of active-material powder that has been isolated from discharged nickel electrodes. This analysis provides the characteristic decomposition temperature for the active materials based on the temperature where the sample loses weight. Normal discharged nickel

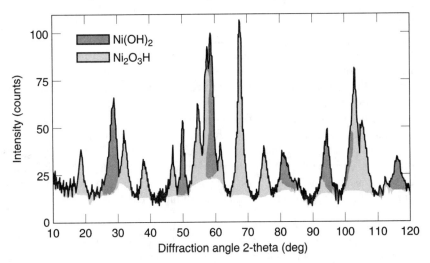

Figure 13.12. X-ray diffraction pattern of nickel electrode active material that was 30% low in capacity, showing a mixture of the normal nickel hydroxide and nickel oxide hydroxide. Reprinted courtesy of NASA.

hydroxide material decomposes at relatively high temperatures, and traces of residual charged phases can give weight loss at lower temperatures along with the loss of chemisorbed water. Thermogravimetric analysis provides a good method for quantifying the amount of nickel oxide hydroxide in an electrode.[13.10] This nickel oxide hydroxide compound has a well-defined decomposition temperature that is greater than that of residual charged material, but well below the temperature where nickel hydroxide dissociates.

13.3.6 Particulate Contamination Analysis

The analysis of nickel electrodes for particulate contaminants may be done using either purified active material or a sample of electrode that includes sinter and grid. Heating in a 90–95°C solution of 15% nitric acid for 1 h dissolves the electrode material and leaves behind a residue containing insoluble particulate contaminants. The undissolved residue may be weighed to determine the total weight of particulate contaminants, which will include highly stable cobalt- or nickel-containing spinel-type oxides that may have formed from normal active materials. The types of particulate contaminants are best analyzed in a scanning electron microscope. The particles are collected on a Millipore filter, coated with gold to eliminate surface charging, then located in the scanning electron microscope image based on their morphology. Their composition is analyzed by EDAX (Energy Dispersive Analysis by X-ray).

The most commonly observed contaminating particles are fragments of Zircar or asbestos separator, depending on the separator type used in the cell design. These fragments are normally present at a low level, but can be found in large quantities if significant popping damage or other separator damage has occurred. Other common contaminants include platinum oxides (in nickel-precharged cells only), cobalt or nickel-cobalt spinels (in hydrogen-precharged cells), iron oxide particles, silicate particles (dirt), platinum catalyst particles, and aluminum or titanium oxide particles. Occasionally, other contaminating particles that can originate from the electrodes or from the cell-manufacturing facility may be seen, such as cadmium hydroxide particles or lanthanum-containing oxide particles.

Extensive analysis of these electrode residues can detect a wide range of particulate contaminants that essentially fingerprint the environment in which the electrode was built and in which the cell was assembled. Figure 13.13 shows several types of particulate contaminants found in nickel electrodes.

13.3.7 Active-Material Extrusion

As a nickel-hydrogen cell is repeatedly cycled throughout its lifetime, the nickel electrodes undergo repetitive expansion and contraction of the active material as it is charged and discharged. The repetitive stress on the structure of the nickel electrode causes it to swell; however, the stress also tends to cause slow movement of the active material in the pores of the sintered electrode from regions of high density into regions of lower density.

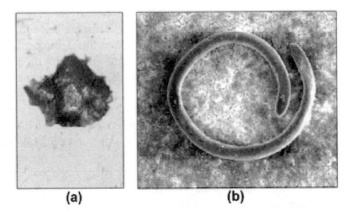

Figure 13.13. Foreign particles seen in nickel electrodes. Particle (a) is a greenish inorganic crystallite about 20 µm across that was rich in titanium, chlorine, oxygen, and silicon, with smaller amounts of magnesium and aluminum. The organic fiber in (b) was a "yellowish" color, was about 40 µm across, and showed significant sulfur content as well as carbon and oxygen.

This results in gradual extrusion of active material from the pores to build up a deposit of extruded active material on the surface of the electrode and in the separator. The black extruded material is an electrically conductive powder consisting primarily of charged active-material particles. Thus, if the extruded material penetrates and fills all the layers of separator material sufficiently, it can cause a current-carrying path between the electrodes.

This can place a parasitic load on the cell late in life, which can eventually lead to cell failure by loss of capacity relative to other cells in a battery. The amount of active material that has extruded into the separator thus provides an indication of how close a cell is to its end of life from a short-circuiting mechanism. The distribution of extruded material is also critical, because a few areas of separator that have very heavy localized deposits of extruded material can be sufficient to initiate a short circuit in those local regions.

The amount of extruded material per unit of separator area can be analyzed by dissolving the active material from the portion of separator that is of interest. If the extruded material is relatively uniform, an entire layer of separator should be analyzed. Localized deposits can be analyzed in smaller regions cut from the separator. In cells with two layers of separator, each layer should be analyzed separately, because short-circuiting is not likely until the second layer also contains significant extruded material. The active material in the separator sample can be dissolved by heating the separator sample in 15% nitric acid at 90°C for 1 h. The dissolved material is filtered from the insoluble Zircar particles and other insoluble residues, and then assayed for nickel and cobalt by either spectroscopic or gravimetric analysis procedures.

13.3.8 Flooded Utilization

The flooded utilization of a nickel electrode is determined by cutting one or more samples of standard size, typically 1 cm^2, from the electrode. The capacity of each sample is measured in a test cell that is flooded with 31% potassium hydroxide electrolyte using a series of standardized charge and discharge cycles. Utilization is defined as the measured capacity as a percent of the theoretical capacity. Theoretical capacity is determined by chemical analysis of the sample for total nickel and cobalt in the active material after the flooded cycling, and it is defined as one equivalent of charge per mole of nickel in the active material.

Utilization is typically about 90–100% for electrodes when they are exposed to minimal overcharge, and can be as high as 110–120% for cycles involving significant overcharge. The utilization is generally a function of the thickness (or swelling) of the electrode, density of the sinter, amount of sinter corrosion, active-material loading level, cobalt additive concentration, and the uniformity of the sinter and the active material in the sinter.

Flooded utilization analysis requires that a sample, for example 1 cm^2 in size, be cut from a rinsed and dried nickel electrode that is to be analyzed. The precise area and thickness of the sample are measured, and it is weighed in its initial air-dried condition. The electrode sample is then sandwiched between two sheets of nickel expanded-metal ("ex-met") grid that are spot-welded around the edge of the sample. The ex-met grid maintains electrical contact with the electrode sample through a nickel lead that is spot-welded to the top of the grid.

The electrode sample is then placed in a plastic test cell that contains approximately 250 cm^3 of 31% (by weight) potassium hydroxide electrolyte and a nickel sheet counterelectrode. An Hg/HgO reference electrode[***] is situated to one side of the sample electrode such that it makes contact with the electrolyte in the test cell at the edge of the sample. Such positioning minimizes the ohmic voltage drop between the reference electrode and the nickel test electrode, and it keeps the body of the reference electrode from blocking the direct current path between the test electrode and the counterelectrode.

[***] An Hg/HgO reference electrode for use in flooded cells can be made by sealing one end of a 1/4 in. polyethylene tube and filling the bottom 1/2 in. of the tube with an HgO amalgam. A platinum wire with the bottom 1/2 in. twisted into a spiral to maintain better contact with the amalgam is pushed into the amalgam, with the other end of the platinum wire extending from the open end of the tube. A fine hole is punched through the side of the tube just above the level of the amalgam, and a wick made of nonwoven nylon fabric is pushed into the hole to provide an electrolyte path. The tube is filled with potassium hydroxide of the desired concentration, leaving at least 1 in. of empty space at the top of the tube. The top of the tube is sealed around the platinum wire by heating it until it softens, then compressing the softened polyethylene around the wire. The reference electrode should be stored in a potassium hydroxide solution at all times. It is used by positioning the wick close to one edge of the working electrode.

13.3 Nickel Electrode Diagnostic Tests

The test cell is connected to a constant current source with the test electrode connected to the positive source and the counterelectrode to the negative source terminal. The electrode voltage is monitored relative to the reference electrode. All tests are normally done at room temperature, typically 23±1°C.

The electrical cycle used for flooded utilization tests initially involves a standard discharge cycle to determine the amount of active precharge in the nickel electrode sample. The standard discharge sequence involves discharge at 10 mA/cm^2 to a voltage of –0.5 (vs. Hg/HgO reference), a 30 sec open-circuit voltage recovery period, discharge at 2 mA/cm^2 to a voltage of –0.5, another 30 sec open-circuit period, then a final discharge at 0.2 mA/cm^2 to a voltage of –0.5. A typical flooded utilization test involves four cycles as follows:

1. Charge at 2 mA/cm^2 for 16 h, and then do a standard discharge. This cycle is intended to stabilize electrode charge efficiency and performance.

2. Charge at 2 mA/cm^2 for 16 h, and then do a standard discharge. This cycle is intended to measure electrode performance under conditions of limited overcharge.

3. Charge at 2 mA/cm^2 for 32 h, and then do a standard discharge. This cycle is intended to measure electrode performance under conditions of extensive overcharge.

4. Charge at 2 mA/cm^2 for 16 h, and then do a standard discharge. This cycle is intended to repeat the cycle 2 measurements to evaluate performance repeatability.

Following the four electrical cycles defined here, the electrode sample is rinsed in flowing deionized water for several hours to remove all potassium hydroxide electrolyte, then allowed to air-dry for 16 h while still sandwiched in the ex-met grid holder. When dry, the sample is removed from the ex-met holder and weighed. The sample is then dried in a weighing bottle at 90°C for 1 h in vacuum, and then allowed to cool with the weighing bottle closed before obtaining the weight of the dried electrode sample.

Chemical analysis of the sample first involves manual separation of the sinter from the embedded nickel grid, then weighing the grid. The sinter is mixed with a small amount of deionized water and ground into finely powdered slurry. The powder is quantitatively rinsed with 10% acetic acid and then deionized water into about 90 cm^3 of 10% acetic acid, to which is added about 0.75 g of hydrazine sulfate. The hydrazine sulfate will reduce any residual charged active material to nickel hydroxide, which is soluble in acetic acid.

The mixture is stirred in a covered, nitrogen-filled beaker at room temperature for 1 h with a magnetic stirring bar, to which all the nickel metal particles adhere. After 1 h, the active material should be dissolved, and the solution can be quantitatively filtered and rinsed into a 250 cm^3 volumetric flask, using the magnetic stirring bar to retain all metal particles. Any insoluble nonmagnetic residue is collected on the filter, dried, and weighed. Such residues should be a small percentage of the total electrode weight.

The amount of nickel in the sinter particles must also be determined analytically rather than simply by weight, because the fine metal particles are always covered with oxide layers. The nickel metal residue is dissolved in 100 cm^3 of 10% nitric acid at 65°C for 1 h and diluted to 250 cm3 in a second volumetric flask.

Each of the solutions collected in volumetric flasks is best analyzed for both nickel and cobalt using titrimetric or gravimetric methods that provide approximately 0.01% accuracy, which is significantly better than the 0.1 to 1% accuracy that can be achieved with inductively coupled plasma or other spectrometric analysis methods. Analysis for nickel may be performed by titrating for nickel with EDTA,[13,14] or nickel may be gravimetrically determined as the dimethylglyoxime chelate.[13,15] Cobalt can be gravimetrically determined as the 1-nitroso-2-napthol chelate[13,15] after removal of all the nickel by chelation with dimethylglyoxime.

Other elements are generally only present as trace materials, and can be spectroscopically analyzed either in the purified active-material powder or in the solutions containing the dissolved active material or sinter. Platinum is the only other element normally seen at levels greater than 500 ppm, particularly when analyzing electrodes removed from cells having nickel precharge that have been stored for a year or more.

The amount of nickel found in the active material can be used to determine the theoretical capacity of the nickel electrode. Utilization is determined for cycles 2–4 in the electrical sequence described here. Cycle 2 provides the utilization in the high charge efficiency region before the onset of oxygen evolution. Cycle 3 provides the upper limit attainable for the utilization when the electrode is extensively overcharged. Cycle 4 should reproduce the utilization seen in cycle 2, although frequently the utilization is slightly lower for cycle 4 because of minor extrusion of small amounts of active material from the freshly cut edges of the sample during the extended overcharge of cycle 3.

The results of the flooded utilization analysis provides a number of key parameters that are important in the performance of the nickel electrode, including hydration level, nickel hydroxide loading, cobalt loading, sinter density, thickness, insoluble residues, and grid weight. For example, the dependence of utilization on the active-material loading level, cobalt concentration, and sinter density may be determined for a group of electrodes to learn whether variations in any one of these parameters, or other parameters, are primarily responsible for variability in utilization and capacity. Figure 13.14 shows a typical correlation of utilization with the chemical and physical properties of a group of nickel electrodes.

Comparison of the weight of nickel sinter and nickel hydroxide per unit of electrode area provides the most direct measure of accumulated sinter corrosion during long-term cell operation, particularly when compared to an unused electrode having no electrochemically induced sinter corrosion. Sinter corrosion, and the

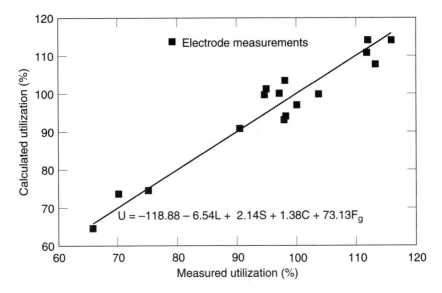

Figure 13.14. Comparison of measured nickel electrode utilization with linear model prediction. L is loading level in g/cm³ void volume, S is % sinter porosity, C is the % cobalt additive in the active material, and F_g is the fraction of the active material charged to the γ-NiOOH by the recharge protocol used in the utilization test.

concomitant buildup of additional nickel oxyhydroxides, constitutes a key long-term degradation mode that influences the performance of all nickel-hydrogen cells. Insoluble residues that are found in the nickel electrode, if significant, are generally associated with a loss of capacity, because they indicate the formation of phases that either are inactive or cannot be readily discharged to nickel hydroxide.

13.3.9 Inorganic Contaminant Analysis

A number of inorganic contaminants or additives other than cobalt can affect the performance of nickel electrodes. Most metallic contaminants can be determined by spectroscopic analysis of a dissolved sample of a nickel electrode. Dissolution can be done by heating in nitric acid, or a mixture of nitric and hydrochloric acids if necessary, followed by neutralization with sodium hydroxide solution. Metallic contaminants of most interest include iron, zinc, cadmium, lithium, aluminum, silicon, platinum, zirconium, yttrium, hafnium, copper, calcium, magnesium, and boron. Their effects on nickel electrode performance are as follows:

- Iron contamination typically comes from rust and steel particles from manufacturing equipment and plumbing. Iron contamination greater than 100 ppm can catalyze the evolution of oxygen gas in nickel electrodes, thus decreasing their charge efficiency.

- Zinc, which is typically present as the hydroxide, has been reported to decrease the resistance of the nickel active material, thus improving performance. However, zinc hydroxide is soluble in alkaline electrolyte, and in some situations has been reported to block sinter porosity. At levels less than 500 ppm, zinc is not expected to degrade electrode performance.
- Cadmium can significantly increase the oxygen evolution overpotential, thus significantly improving the charge efficiency of nickel electrodes. However, cadmium, platinum, and copper can form metallic deposits or dendrites at low cell potential, which can result in short circuits that may make it difficult to charge a cell after discharge to low voltage.
- Lithium can improve capacity by stabilizing the formation of γ-NiOOH. However, the greater volume change associated with the facile formation of this phase generally results in reduced cycle life.
- Aluminum and silicon are usually associated with mineral contaminants or dirt that has accumulated in the cell. Asbestos separator, however, provides a source of a range of silicate minerals and can lead to significant silicate levels. Silicates can result in elevated impedance, particularly at temperatures less than 10°C.
- Zirconium, yttrium, and hafnium all originate from Zircar separator fragments in the nickel electrode, and are insoluble and electrically nonconductive.
- The presence of significant calcium, magnesium, or boron (borate) typically indicates exposure to tap water and can be associated with reduced electrode capacity and elevated impedance.

Other contaminants of interest include chloride, sulfate, and nitrate anionic species. At levels of approximately 100 ppm, sulfate has been reported to cause the gradual leaching of cobalt additives from the active material in the nickel electrode, and a reduction in cycle life.[13,16] Chloride and nitrate can increase the corrosion rate of the sinter structure when present at levels greater than about 50 ppm. Sulfate may be measured gravimetrically by precipitation as barium sulfate. Chloride can also be measured gravimetrically by precipitation as silver chloride.

13.3.10 Organic Contaminant Analysis

Organic contaminants in nickel electrodes can range from polymers and surfactants to oils and other hydrocarbons. These contaminants, if present at sufficiently high level, can accelerate the loss of nickel precharge as the nickel electrode oxidizes the contaminants. They can also lead to high electrode impedance at low temperatures. Organic contaminants in nickel electrodes are evaluated by two separate extraction procedures, the first of which assays polar species, and the second of which measures saturated organic species such as oils.

Nickel electrodes are often fabricated with a polymer binder coated on the edges of the electrode to reduce the likelihood of particles breaking from the edges during manufacturing, handling, or operation in the cell. If edging polymer has been applied to an electrode, it must be removed before analysis for organic constituents. Edging material can be removed by cutting approximately 1/8 in. from the inner and outer edges before carrying out further analysis. After removing the edges (if necessary), one should measure the area of the remaining nickel electrode sample to allow determination of the amount of organic contaminants per unit electrode area.

The initial extraction for organic contaminants is performed by soaking the nickel electrode in spectrographic-grade acetone for several hours, then filtering the acetone solvent through a glass fiber filter (or any filter that will not introduce detectable organics) to remove any particulates that may have broken from the electrode. The solvent is then evaporated to dryness in a weighed beaker, and the total weight of dissolved organics is obtained by measuring the beaker's increase in weight. Best results are obtained by subtracting a blank correction from this weight gain, which is the weight gain measured for an identical beaker of solvent that was not exposed to a nickel electrode.

The identity of the organic contaminants (assuming measurable levels are found) is determined by Fourier transform infrared spectroscopy of the organic residue, which can be dissolved in several cubic centimeters of acetone and then transferred to a Fourier transform infrared sample plate. Typically the acetone extraction detects polymeric and surfactant contaminants. Acetone extraction material levels of 0.2–0.5 mg/in^2 or less are typical of nickel electrodes that have functioned well in nickel-hydrogen cells.

After extraction with acetone, the nickel electrode sample is similarly extracted with hexane, which will dissolve oils and other saturated hydrocarbon contaminants. Fourier transform infrared analysis of the hexane extract typically shows pump oils and tar-based contaminants. Hexane extraction material levels of 0.2 mg/in^2 or less are typical of nickel electrodes that have functioned well in nickel-hydrogen cells.

13.3.11 Electrochemical Voltage Spectroscopy

Electrochemical voltage spectroscopy was originally developed[13.6] to analyze the distribution of electrochemically active sites for storing lithium in conductive polymer materials. However, it can be used to map the voltage spectrum of electrochemically active states in any electrode or battery cell,[13.17] as was described earlier in this chapter for the nickel-hydrogen cell. For nickel electrodes the electrochemical voltage spectroscopy method can assay the amounts of β-NiOOH and γ-NiOOH phases present in a nickel electrode after any given recharge profile. The electrochemical voltage spectroscopy scan also indicates shifts in recharge potentials that can influence charge efficiency, shifts in the oxygen evolution behavior of the nickel electrode, and the presence of other active phases in the nickel electrode such as platinate complexes that can form under some conditions.

A standard electrochemical voltage spectroscopy scan that provides an excellent diagnostic method for verifying normal electrochemical behavior for nickel electrodes uses an electrochemical voltage spectroscopy voltage scan at 2 μV/sec over the operating range of the nickel electrode. The standard scan is typically carried out in a test cell flooded with 31% potassium hydroxide electrolyte and fitted with an Hg/HgO reference electrode and a nickel sheet counterelectrode. The voltage between the nickel test electrode and the reference electrode is controlled during the electrochemical voltage spectroscopy scan, and the current into the nickel electrode is integrated over time to give the electrochemical voltage spectroscopy response. The voltage is stepped upwards or downwards in 1 mV steps to produce the electrochemical voltage spectroscopy scan. A recommended diagnostic electrochemical voltage spectroscopy scan involves scanning from the initial potential of the electrode down to 0.1 V (*vs.* Hg/HgO), then scanning up to 0.5 V and then back to 0.2 V. The sweep to 0.5 then back to 0.2 V is then repeated after the first scan has conditioned the electrode to provide a standardized response that is relatively independent of its prior history, as shown in Fig. 13.15.

During the positive-going voltage scan, the nickel electrode is recharged. The voltage peaks in the electrochemical voltage spectroscopy spectrum indicate recharge of the phases that can be formed in the nickel electrode. The β-NiOOH phase typically forms at about 0.45 V and the γ-NiOOH phase forms along with oxygen evolution near 0.5 V. The peaks seen in the negative-going scan indicate

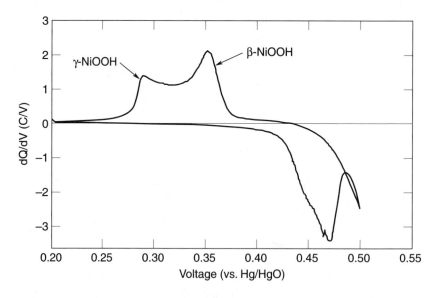

Figure 13.15. Stabilized electrochemical voltage spectroscopy response for a nickel electrode showing the amounts of β-NiOOH and γ-NiOOH.

the amounts of the β-NiOOH and γ-NiOOH phases present in the electrode. The β-NiOOH phase discharges at about 0.35 V, and the γ-NiOOH phase discharges at about 0.28 V. A well-performing nickel electrode should be largely converted into the γ-NiOOH phase by the standard scan described here, displaying only a very small β-NiOOH peak at 0.35 V. Electrodes that have degraded capacity, that are well cycled, or that have significant corrosion of the nickel sinter often have recharge voltage peaks appearing at elevated potentials, and discharge little γ-NiOOH at 0.28 V. Electrochemical voltage spectroscopy scans throughout various voltage ranges, or at different temperatures or scan rates, can be used to diagnose any of a number of processes that can degrade the performance of nickel electrodes.

13.3.12 Oxygen Overpotential Test

The overpotential for evolution of oxygen gas from the nickel electrode is a key factor in the charge efficiency of the nickel-hydrogen cell. A higher oxygen evolution potential results in better charge efficiency, because the oxygen reaction is always competing with the charge storage reactions in the nickel electrode. The reversible potential for oxygen evolution is about 0.3 V *vs.* the Hg/HgO reference; however, the kinetics for this reaction are so poor that appreciable oxygen generation is typically not seen until the voltage level is greater than 0.45 V. Because the overpotential is so high for oxygen evolution, the relationship between current and voltage follows the classical exponential (Tafel) dependence.

Oxygen evolution is readily measured for a sample cut from a nickel electrode (typically 1 cm^2 in size) or for a full electrode that is charged in a test cell flooded with 31% potassium hydroxide electrolyte. The test cell is equipped with an Hg/HgO reference electrode that is positioned close to one edge of the nickel electrode, and a nickel sheet counterelectrode. The nickel electrode is charged 16 h at 2 mA/cm^2 and discharged to 0.0 V (*vs.* the Hg/HgO reference) to stabilize its performance. The electrode is then charged 24 h at 2 mA/cm^2 to bring it to a high state of charge where oxygen is beginning to be produced at significant rates. The charge current is then set at the following levels for time periods long enough to achieve the steady-state oxygen potential, which is taken as the voltage at the end of each charge step.

1. 0.2 mA/cm^2 for 8 h.
2. 0.5 mA/cm^2 for 4 h.
3. 1.0 mA/cm^2 for 2 h.
4. 2.0 mA/cm^2 for 1 h.
5. 5 mA/cm^2 for 0.5 h.
6. 10 mA/cm^2 for 0.25 h.

The extended stabilization times in the sequence defined above, particularly at the lower rates, allow the large surface pseudocapacitance associated with the oxygen evolution reaction to reach steady-state conditions at each current.

The rate of oxygen evolution is quantified by plotting the logarithm of the current against the electrode voltage, which typically provides a linear relationship having a slope that is referred to as the "Tafel slope" in mV/decade of current. The oxygen evolution reaction can exhibit Tafel slopes ranging from 30 to more than 50 mV/decade.

A second parameter that defines the oxygen evolution behavior is the current at some reference potential, such as the reversible potential, 0.301 V $vs.$ Hg/HgO (referred to as the "exchange current"), or at some other reference potential such as 0.45 V. In some cases, the slope of the oxygen evolution current is seen to shift from near 30 mV/decade at low currents, to 50–60 mV/decade at higher currents. The shift in slope is attributed to a change in the rate-limiting mechanism.[13.18,13.19]

13.3.13 Stress Tests

Stress tests on nickel electrodes have been historically used to rapidly demonstrate the capability of electrode samples to operate successfully through a regimen of high-rate cycles. Good stress cycle performance indicates the desired level of electrode quality, while poor performance indicates unacceptable quality. The following discussion includes a number of different types of stress-test cycles that have been historically used.

The typical stress-test cycles performed by most nickel electrode manufacturers to demonstrate the quality of their electrodes involves applying from 200 to 1000 cycles involving 10C charge and discharge rates. A typical recharge involves applying the 10C rate to return 120% of the rated capacity, and then discharge at 10C to depletion, which may be detected by the voltage at the counterelectrode. This test is carried out using a full electrode that is placed in a container with 31% potassium hydroxide electrolyte and a nickel counterelectrode.

The high charge and discharge rates used in this test generate significant heating, making it necessary to run this test with cooling of the test cells, typically by running cooling coils through a bath in which the test cells are placed. It is also necessary to regularly add water to the test cells to replace water lost to the electrolysis reactions taking place during cycling. An electrode is considered to have successfully passed this test if it delivered the minimum required capacity during the last discharge and did not develop excessive swelling (i.e., no more than 3 mils) or blistering during the cycling.

A second type of stress test has been used not only to verify good electrode performance, but also to demonstrate that electrodes are adequately formed and have good charge efficiency. In the first stress test described here, a full electrode is operated in a flooded test cell containing 31% potassium hydroxide, which is held at about 25°C in a cooling bath. However, this test cycles the electrode to failure using an 80% depth-of-discharge cycle and requires a minimum number of cycles before electrode failure. The charge and discharge rate in this test is

5C, and recharge involves 105% return of the discharged capacity. The initial recharge returns 120% of the rated electrode capacity at a C/10 rate.

With the relatively low charge return ratio used in this test, nickel electrodes must have relatively good charge efficiency to pass. Those that are not fully formed, or electrodes that are nonuniformly loaded, can fail after less than 100 cycles in this test as a result of poor charge efficiency.

Poorly formed electrodes will eventually become fully formed by this stress test; thus an electrode may either fail or come close to failure early in life, only to recover and give good life with continued cycling. This type of behavior indicates a need for more formation in the electrode production procedures. After a large number of cycles in this test, nickel electrodes will fail as a result of wear-out damage from the cycling stresses. It is not uncommon for a good electrode to operate for more than 1000 cycles in this test, as indicated in Fig. 13.16.

A third type of stress test is performed on small samples of nickel electrodes cut from the electrodes that are to be evaluated. These electrode samples, which are typically 1 cm^2 in size, are cycled in test cells containing about 250 cm^3 of 31% potassium hydroxide and an Hg/HgO reference electrode. The electrode is charged at 2 mA/cm^2 for 24 h and then discharged at 10 mA/cm^2 to determine its initial capacity. The electrode is then cycled, with 100% of the initial measured capacity returned each cycle at the 10 mA/cm^2 rate, and discharged to 0.0 V at the 10 mA/cm^2 rate.

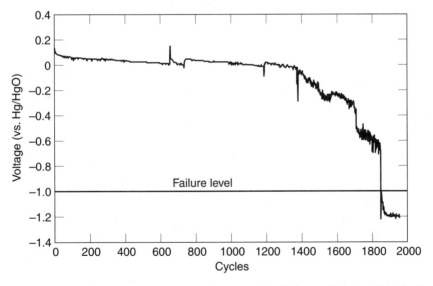

Figure 13.16. Stress test of a well-performing nickel electrode using 80% depth of discharge and 105% charge return each cycle. Charge and discharge rates are 5C.

The electrode is continually cycled in this manner until its discharge capacity drops to less than 50% of its initial level. Because the stresses, which are largely a result of the overcharge, increase significantly as the high-rate electrode discharge capacity decreases, for a good electrode this test gives a stable capacity with cycling for 1000 or more cycles followed by a relatively rapid drop-off in capacity at end of life.

13.3.14 Scanning Porosimetry (Pore Size Distributions)
The distribution of pore sizes in nickel electrodes can be determined by classical methods such as mercury intrusion porosimetry. Mercury porosimetry can provide an accurate distribution of the pore size in the active material that fills the pores in the sinter that makes up the conductive framework of the nickel electrode. However, the distribution of pores in the sinter, which controls how the sinter is filled with active material and the uniformity and utilization of the material, in many cases cannot be accurately measured by mercury porosimetry.

The pore structure of nickel sinter often involves a surface skin that has significantly smaller pore sizes than the interior of the sinter, which in some cases can include large voids or cracks. Mercury intrusion measurements must force mercury through the small pores of the skin to fill the larger pores in the interior. The pore volume associated with the larger interior pores thus appears to be associated with the more restrictive pore sizes at the surface.

The inaccuracies that can result when mercury intrusion porosimetry is applied to nonuniform sinter structures can be eliminated through the use of a technique called scanning porosimetry.[13.20] This technique uses piezoelectric scanning technology developed for scanning tunneling microscopy; however, the resolution of the scanning stages is reduced to the approximately 0.1 μ resolution associated with the smallest pore features in nickel sinter. X, Y, and Z piezoelectric translation stages scan a metal tip over the surface of a potted and cross-sectioned nickel sinter or electrode sample and map out the distribution of conductive features in the sinter.

Scanning porosimetry can operate in two distinct modes. Mode 1, or the imaging mode, scans the tip over the surface of the sinter in the region to be scanned and provides an image of the surface conductivity, which closely mirrors the sinter particles. The pores correspond to the regions between the conductive particles.

Mode 2, which is termed a line-scan mode, allows pore size distributions to be quantified. The probe is scanned from one edge of the cross-sectioned electrode to the other, detecting each sinter particle and its size as the probe traverses the cross-sectional thickness. One such scan will detect many hundreds of pores between these conductive particles.

In the line scan mode, about 40–50 scans are typically carried out over a portion of the cross-sectioned electrode approximately 1 in. long. This process detects a total of 20,000 to 30,000 pores during one scan. These pores may be

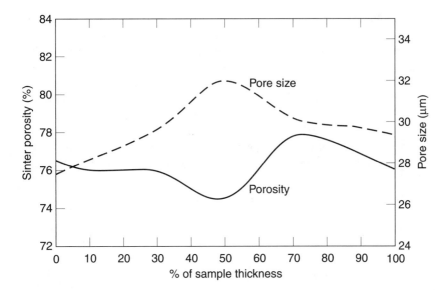

Figure 13.17. Example of porosity and pore size measurements through the thickness of sintered plaque for a nickel electrode, as measured by scanning porosimetry. This sintered plaque was made by a slurry process.

sorted into bins that encompass specific fractions of the electrode cross-section, and within each bin the distribution of pore sizes may be statistically determined. This analysis provides the pore size distributions in different locations through the thickness of the electrode, as well as the average pore size, pore volume, sinter porosity, and pore uniformity in each bin. Figure 13.17 shows typical distributions of pore size and porosity through the thickness of a nickel electrode.

Scanning porosimetry provides a method to evaluate the uniformity of the sintered pore structure in nickel electrodes. Measurements of this kind typically show that more-uniform pore structures, particularly those without large internal voids, have significantly better utilization. Large internal voids that are loaded with active material when the nickel electrodes are fabricated will have low utilization because the charge stored within these large regions is not in close enough proximity to conductive surfaces to be efficiently discharged. If large internal voids are not loaded, loading levels elsewhere in the structure will be too high and cycle life will be compromised by the large localized stresses where the excessive loading exists. All these types of internal structures that act to control electrode performance can be examined and diagnosed with the help of scanning porosimetry.

13.3.15 Cobalt Gradient Analysis

Nickel electrodes in nickel-hydrogen cells contain active material that consists of nickel hydroxide doped with cobalt hydroxide. The cobalt hydroxide is typically present at a level that is either 5 or 10% of the total metal hydroxide in the active material, with the cobalt concentration depending on the process by which the electrode is made. When the nickel electrode is properly made, the cobalt additive is dispersed uniformly through the nickel hydroxide crystal structure, resulting in reduced charge and discharge voltages and improved capacity. As a result of corrosion of the underlying nickel metal sinter, or as a result of other processes that can cause the cobalt to move relative to the nickel hydroxide in which it is normally dispersed, cobalt gradients can develop in the active-material deposits.

Typical cobalt gradients exhibit a region of depleted cobalt near the surface of the metal sinter particles, along with enhanced cobalt levels in the active material situated far from any metal current-collecting surfaces.[13.9] A depleted cobalt region near the current-collector surfaces can cause significant loss of capacity by reducing charge efficiency and by increasing residual capacity that cannot be discharged at high rate and usable voltage.

A second type of cobalt gradient can be built into a nickel electrode when it is initially loaded with active material. This type of gradient results when the active material in the center of the sintered electrode thickness contains a significantly different cobalt concentration than does the active material at or near the surfaces of the electrode. This situation is caused by the differential solubility of nickel and cobalt hydroxide complexes that can form as the active materials are deposited in the sinter.[13.21] It does not necessarily reduce capacity, but it can result in differential stresses through the electrode structure that can result in sinter fragmentation, blistering, plaque warping, and early failure. Analysis for cobalt gradients is thus an important part of diagnosing potential causes of capacity loss or failure in nickel electrodes.

Analysis of nickel electrodes for gradients in cobalt concentration requires that the nickel electrode sample be potted and cross-sectioned, then coated with a conductive material such as carbon or gold to prevent surface charging in a scanning electron microscope. Nickel electrodes must be fully discharged to yield a well-potted and smoothly polished cross section. Nickel electrodes that contain significant charge will oxidize most potting materials and give off oxygen gas, resulting in black ooze that distorts a polished electrode cross section. To ensure that the potting material fills even the finest active-material porosity, the liquid potting material should be added to a nickel electrode sample that is in vacuum. When the sample is covered with liquid potting material, the vacuum is released to force potting material into all the pores before the potting material is allowed to solidify and cure.

Analysis of the variability of cobalt through the thickness of the nickel electrode is done in the scanning electron microscope using EDAX. This analysis is typically performed in several locations, in each location traversing a line of analysis spots from one side of the electrode to the other side. Analysis at one spot consists of focusing the EDAX analysis area on a spot of standard size (i.e., 2 μ square) and locating this spot in a region of active material that is well separated from any sinter particles. This eliminates any contribution of nickel metal to the measured level of nickel. EDAX measurements for nickel and cobalt can then provide the percentage of cobalt in that particular region.

By plotting the cobalt percentage as a function of distance along the scan from one edge of the electrode to the other, one obtains a cobalt concentration profile, as indicated in the example in Fig. 13.18. Concentration profiles from different regions of the same electrode sample may be averaged to yield an average cobalt profile through the sample, as well as statistics on the variability of the profile within the sample. Cobalt gradients through the electrode thickness can also be measured directly by chemical analysis as described in the following section as part of a loading-level gradient analysis.

Scanning electron microscopy also enables detection of gradients in the concentration of cobalt additive as a function of distance from the individual sinter particles in the nickel electrode. There are several ways to perform this analysis. One method is to focus a small EDAX analysis area on the active material, and to measure the cobalt level in the active material as a function of distance at 1 µ intervals from a sinter particle surface out to active material 10 µ or more distant from sinter surfaces. This analysis can provide a typical cobalt profile in one localized spot.

To detect cobalt gradients on a more global scale, EDAX elemental mapping is performed on a given region of sintered electrode cross section at a magnification

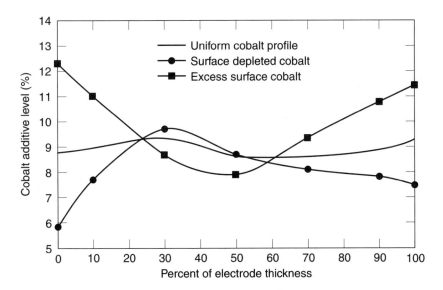

Figure 13.18. Variations in cobalt additive level seen through the thickness of nickel electrodes, measured by scanning electron microscopy/EDAX analysis of cross-sectioned samples.

that provides good resolution of the distances between individual sinter particles. The nickel distribution may then be displayed as a blue image and the cobalt distribution as a red image, for example. An overlay of these two images should show uniform cobalt color in the active material relative to the nickel color, and bright blue nickel intensity on the sinter particles. Image processing allows the bright nickel sinter regions to be masked off, and the resulting difference between the red intensity and the blue intensity will display any significant cobalt gradients around all the sinter particles.

It must be kept in mind that with this method it is difficult to detect cobalt gradients within about 1 μ of the sinter surfaces because of spreading of the EDAX analysis beam. Scanning electron microscopy instruments with high resolution and reduced EDAX beam-spreading enhance the sensitivity of this analysis method to thin layers of depleted cobalt.

13.3.16 Loading-Level Gradient Analysis

The uniformity of loading through the thickness of a sintered nickel electrode can be a key variable influencing capacity and utilization of the electrode. Ideally, a loading level of 1.6 to 1.8 g/cm^3 of void volume is desired throughout the entire thickness of the electrode. In real electrodes, however, there are often very significant variations in loading level, as illustrated in Fig. 13.19. There are several methods that can be used to measure the loading-level gradients through the thickness of an electrode.

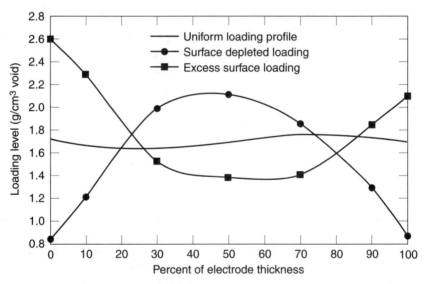

Figure 13.19. Variations in loading level seen through the thickness of nickel electrodes, as measured using a scanning electron microscopy/EDAX analysis of cross-sectioned electrode samples.

The easiest way to detect loading gradients involves obtaining the total EDAX intensity of nickel and cobalt in active-material pockets (well away from the sinter particles) through the thickness of a cross-sectioned electrode in the scanning electron microscope. If the nickel and cobalt intensities are scaled to the average proportions of these elements in the electrode (as measured by a bulk chemical analysis), then the total intensity will be proportional to the loading level. The total loading profile through the electrode thickness can be normalized to give the correct average loading, and the resulting variation in loading is obtained. A similar procedure can be followed using an ion microprobe analysis method as an alternative to EDAX.

Another method that relies purely on chemical analysis methods can also be used to evaluate variability not only in loading level through the thickness of an electrode sample, but also in sinter porosity and cobalt concentration. This analysis method uses a nickel electrode sample that has been vacuum-potted in polystyrene (or any other plastic that is soluble in acetone). The electrode is shaved into approximately 10 layers of known thickness using a microtome, each layer being about 10% of the electrode thickness. Each microtomed layer is collected, the plastic binder dissolved away with acetone, and the amounts of nickel metal, nickel hydroxide, and cobalt hydroxide determined by chemical analysis of the residue. From this analysis, the variation in loading level, cobalt level, and sinter porosity is directly obtained in each of the 10 layers through the thickness of the electrode (averaged over the sample area). While this method provides the most quantitative results, it is also the most labor-intensive method for measuring gradients through nickel electrodes.

13.4 Hydrogen Electrode Diagnostic Tests

13.4.1 Platinum Catalyst Adhesion Analysis
The platinum black catalyst that is coated on the side of the negative electrode that faces the separator is sintered with a Teflon binder to the underlying nickel screen and Gore-Tex backing. Several conditions can result in poor adhesion of the platinum black to the nickel screen. Poor adhesion can cause problems with cell performance as a result of particles of platinum black becoming dislodged and causing short circuits or poor conductivity to the nickel screen. Poor adhesion can be caused either by inadequate sintering or by the presence of a thin Teflon coating between the nickel screen and the platinum black, as indicated in Fig. 13.20.

A relatively simple test to verify adequate platinum black adhesion has been found to be highly effective for screening suspected negative electrodes. This test involves scraping the flat edge of a stainless steel spatula against the platinum black surface of the negative electrode. Adhesion can be classified as good, moderate, or poor from this test based on how readily the platinum black is removed from the underlying screen. Good adhesion means that it is difficult to remove

460 Diagnostic Methods and DPA for Cells

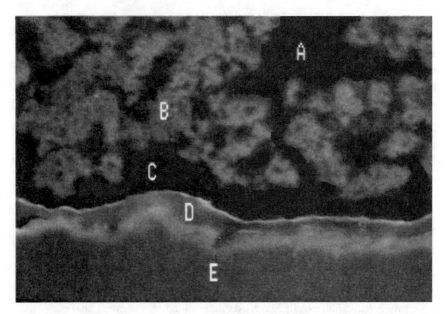

Figure 13.20. Teflon layer (D) on the nickel grid (E) in the cross section of a hydrogen electrode. The platinum catalyst layer consists of catalyst particles (B) and pores (A). The catalyst layer is not well bonded to the grid surface (note void region C at the interface) because of the Teflon film.

the platinum black layer; moderate adhesion means that the platinum black can be peeled back with moderate scraping force; and poor adhesion means that little force is required to peel it off. This qualitative classification may be calibrated by applying the test to electrodes that are known to be good. If poor adhesion is detected by this test, it may be verified by shaking a negative electrode over a piece of white paper, whereupon particles of platinum black will be dislodged. Good adhesion leaves no platinum black particles on the white paper.

13.4.2 Hydrogen Overpotential Test

The negative electrode used in nickel-hydrogen cells contains sufficient platinum black catalyst that at beginning of life it has a relatively low overpotential. At a C/2 rate, the overpotential for either hydrogen generation or recombination is typically about 20 mV. Significant changes in the active surface area of platinum black must occur before the overpotential changes significantly. The electrochemical activity of the platinum black for the hydrogen reaction provides the basis for a test to verify the proper operation of the negative electrode by measuring the overpotential as a function of current density. The electrochemical activity (overpotential) of the negative electrode has been seen to increase as a result of oxidation and replating of the platinum, which reduces surface area, or the presence

13.4 Hydrogen Electrode Diagnostic Tests

of contaminants that can poison the catalytic activity for the hydrogen reaction. Catalysts such as mercury, cadmium, or iron at levels of 50 ppm or higher can produce a noticeable poisoning effect and increase the hydrogen overpotential.

A test for screening the negative electrode hydrogen overpotential involves placing a test electrode, or a portion of a test electrode (e.g., one-half of the electrode), in a sealed test cell that is pressurized with 50 psi of hydrogen gas. The test electrode sample is placed in a fixture that holds it in a horizontal position with the platinum black side down in a plastic reservoir that is filled with 31% potassium hydroxide electrolyte. The electrolyte level is adjusted until the electrolyte contacts the bottom side of the electrode, but does not cover the upper, hydrophobic Gore-Tex–covered surface. A test fixture at the bottom of the plastic reservoir holds a counterelectrode consisting of a discharged nickel electrode that has approximately twice the surface area of the hydrogen test electrode. This nickel counterelectrode is maintained in a state-of-charge range where it will not evolve significant amounts of either oxygen or hydrogen gas that could block access of electrolyte to the platinum black surface of the test electrode. A reference electrode consisting of a small piece of negative electrode with the black side down and in contact with the electrolyte is situated close to one edge of the test electrode.

The test cell, after being pressurized with 50 psi of hydrogen, is sealed and the following sequence of constant current steps is applied to the test electrode. Positive, or anodic, current will recombine gas, while negative, or cathodic, current will evolve hydrogen gas. The voltage of the test electrode relative to the reference electrode at the end of each current step is taken to be the steady-state hydrogen voltage.

1. Apply an anodic current of +0.1 mA/cm^2 for 4000 sec.
2. Apply an anodic current of +0.2 mA/cm^2 for 3000 sec.
3. Apply an anodic current of +0.5 mA/cm^2 for 1600 sec.
4. Apply an anodic current of +1 mA/cm^2 for 800 sec.
5. Apply an anodic current of +2 mA/cm^2 for 500 sec.
6. Apply an anodic current of +5 mA/cm^2 for 400 sec.
7. Apply an anodic current of +10 mA/cm^2 for 200 sec.
8. Apply an anodic current of +20 mA/cm^2 for 100 sec.
9. Apply an anodic current of +50 mA/cm^2 for 100 sec.
10. Apply a cathodic current of −50 mA/cm^2 for 100 sec.
11. Apply a cathodic current of −20 mA/cm^2 for 100 sec.
12. Apply a cathodic current of −10 mA/cm^2 for 100 sec.
13. Apply a cathodic current of −5 mA/cm^2 for 200 sec.
14. Apply a cathodic current of −2 mA/cm^2 for 250 sec.
15. Apply a cathodic current of −1 mA/cm^2 for 400 sec.
16. Apply a cathodic current of −0.5 mA/cm^2 for 500 sec.

17. Apply a cathodic current of -0.2 mA/cm^2 for 800 sec.
18. Apply a cathodic current of -0.1 mA/cm^2 for 1000 sec.

Following this sequence, the nickel counterelectrode should be fully discharged at 1 mA/cm^2 relative to the platinum test electrode before another test electrode sample is evaluated.

The anodic and cathodic polarization curves are plotted as the logarithm of the current as a function of the measured potential, after correction of the potential for the uncompensated resistance between the test electrode and the reference electrode. The anodic and cathodic overpotentials at a standard rate corresponding to C/2 for the cell, which is typically 15–20 mA/cm^2, may then be read off of the plot.

Comparison of new hydrogen electrode samples known to be good with other test electrodes enables normal overpotentials to be confirmed, or increased overpotentials to be detected. If increased overpotentials are detected, they may be correlated with independent measurements of surface area (by BET analysis) or contaminant levels (by chemical analysis) to isolate the root cause of the degraded performance. Because there is significant excess platinum surface area in the negative electrodes, even significant changes in surface area or catalytic activity may result in less than a 10 mV increase in overpotential at the C/2 rate, a change that may not be readily detected in an operating nickel-hydrogen cell.

13.4.3 Inorganic and Organic Contaminant Analysis

Inorganic contaminants that remain in rinsed hydrogen electrodes may be detected by dissolution of a sample of the hydrogen electrode in aqua regia. This will dissolve all platinum and other materials, while leaving behind Teflon and any carbon additives, which may be weighed. The dissolved hydrogen electrode sample is neutralized with high-purity sodium hydroxide and evaporated to dryness. The residue, which contains significant amounts of platinum compounds, may be spectrographically analyzed for metallic contaminant levels as well as the total amount of platinum.

Organic contaminants in hydrogen electrodes are typically surfactants such as Triton-X-100, or oil-based hydrocarbons. The presence of surfactants at a sufficiently high level can result in the loss of hydrophobicity of the Teflon-coated side of the hydrogen electrode. If the normally hydrophobic surface of the hydrogen electrode becomes wetted with electrolyte, hydrogen gas will be unable to react with the platinum catalyst at high rates, and the cell discharge voltage will be low. Excessive levels of oily contaminants can result in poor wetting of the platinum black–coated side of the hydrogen electrode, which causes elevated cell impedance. Organic contaminants in hydrogen electrodes are evaluated by two separate extraction procedures, the first of which assays polar species such as surfactants, and the second of which measures saturated organic species such as oils.

The first extraction for organic contaminants involves soaking the hydrogen electrode in spectrographic grade acetone for one hour, then filtering the acetone solvent through a glass fiber filter (or any filter that will not introduce significant organics) to remove any particulates that may have become dislodged from the electrode. The solvent is then evaporated to dryness in a weighed beaker, and the total weight of dissolved organics is obtained by the beaker's weight increase. Best results are obtained by subtracting a blank correction from this weight gain; the blank is the weight gain measured for an identical beaker of solvent that was not exposed to a hydrogen electrode.

The identity of the organic contaminants (assuming measurable levels are found) are established by Fourier transform infrared spectroscopy of the organic residue, which can be dissolved in several cubic centimeters of acetone and then transferred to an Fourier transform infrared spectroscopy sample plate. Typically, the acetone extraction detects some residual surfactant contaminants in hydrogen electrodes.

After extraction with acetone, the hydrogen electrode is similarly extracted with hexane, which will dissolve oils and other saturated hydrocarbon contaminants. Fourier transform infrared spectroscopy analysis of the hexane extract will indicate the chemical composition of the extracted contaminants.

13.4.4 Scanning Electron Microscopy Cross-Sectional Analysis

Scanning electron microscopy imaging of potted and cross-sectioned hydrogen electrode samples can be used to analyze the internal structure of hydrogen electrodes. Electrode samples about 1/2 by 1 in. in size are typically used for this analysis. To ensure that the potting material fills all the pores in the hydrogen electrode structure, add the liquid potting material to the electrode sample while it is in vacuum. When the sample is covered with liquid potting material, the vacuum is released to force potting material into all the pores before the potting material is allowed to solidify and cure. After potting is completed, the sample is cross-sectioned and polished, then coated with a conductive material such as carbon or gold to prevent surface charging in the scanning electron microscope.

The hydrogen electrode consists of a nickel metal grid that is coated on one side with a mixture of Teflon particles and platinum black catalyst particles. On the other side is a hydrophobic layer of porous Teflon. This structure is sintered at a temperature sufficiently high to bond the particles to each other and to the screen on the one side to form a porous platinum black layer, and to also bond the Teflon sheet on the other side to the nickel screen. Inadequate sintering or improper preparation of the screen surface can result in poor adhesion of these materials to the nickel screen. Adhesion problems of this kind, or nonuniform platinum loading, may be detected by examination of cross-sectioned hydrogen electrodes. Scanning electron microscopy examination of hydrogen electrode cross sections can also evaluate whether the platinum catalyst loading is uniformly composed of the desired high-surface-area particles of platinum black. Figure 13.21 shows examples of poor adhesion and nonuniform platinum particles in cross sections of hydrogen electrodes.

464 Diagnostic Methods and DPA for Cells

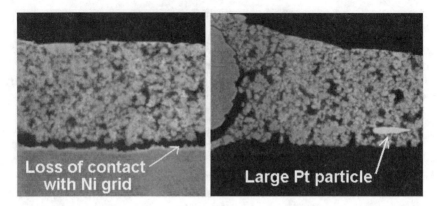

Figure 13.21. Cross sections of hydrogen electrodes from nickel-hydrogen cells. The image on the left shows separation and poor adhesion of the porous platinum catalyst layer to the nickel grid. On the right, a large, nonporous platinum particle can be seen in the porous platinum catalyst layer.

13.4.5 Pore Size Distribution Measurements

The distribution of pore sizes in hydrogen electrodes is typically bimodal. The pores in the hydrophobic Teflon sheet and those between the catalyst particles are on the order of 10 μ or more in size, while the pores within the individual platinum black particles are less than 1 μ in size. The distribution of pore sizes can be accurately measured by mercury intrusion porosimetry, and the size distribution of platinum black particles can be measured using scanning porosimetry.[13.20]

Mercury intrusion porosimetry only measures pore sizes averaged over a sample of several square inches and thus is not sensitive to localized nonuniformities in the pore structure. Scanning porosimetry, on the other hand, can directly provide localized measurements of the conductive platinum black particles and can provide particle size distributions in any region of the electrode structure. These two methods thus provide complementary information regarding the distribution of pores and the distribution of catalyst particles in hydrogen electrodes.

13.5 Electrolyte and Separator Analysis

13.5.1 Electrolyte Distribution and Quantity Analysis

Before the electrolyte or other cell components can be analyzed, the electrolyte must be quantitatively rinsed from the electrodes and the separators. Typically, the quantities of electrolyte rinsed from the separators and positive electrodes in each of five regions for each stack in a cell are combined for analysis, and those rinsed from the gas screen and negative electrodes in the same regions are also combined for a separate analysis. Thus, in a dual-stack cell there are typically 10 samples of electrolyte from the positives and separators, and 10 samples of electrolyte from the negatives and gas screens. In a single-stack cell, only half this number of separate electrolyte samples are obtained. If analysis of the electrolyte requires an accurate carbonate assay, all cell disassembly and rinsing operations must be performed in a glove box containing a carbon dioxide–free atmosphere.

Electrolyte is rinsed from each group of components by weighing the wet components, then placing the components in a closed, nitrogen-filled plastic container that allows the components to lay flat, and adding sufficient deionized water to cover the components. After waiting 1 h minimum, one decants the solution from the components and adds fresh deionized water and nitrogen. This rinsing procedure is continued until the pH of the rinse solution is less than 8.0. All rinsings from each group are combined, and particulate residues that may have been dislodged during the rinsing process are filtered from the solution. The components are spread out on paper towels and allowed to dry in air at 40–60% relative humidity for at least 16 h before they are weighed. The dry weight is the sum of the component weight and the weight of particulate residue, and the electrolyte weight is the loss in weight as a result of rinsing.

The distribution of electrolyte quantity through each stack in a cell is generally analyzed by determining the electrolyte weight per repeating stack unit. In a back-to-back stack this would involve dividing the total electrolyte weight in a group of positives and separators by the number of repeating positive pair/separator units in the group. The amount of electrolyte per stack unit is then plotted as a function of relative position in the stack. An ideal electrolyte quantity distribution has the same amount of electrolyte in each portion of the stack. However, it is not unusual to see somewhat greater amounts of electrolyte in regions of a cell that have been in the "down" position relative to gravity for long periods of time. If electrolyte quantity gradients become sufficiently large, significant popping damage may be seen around the outer edges of the stack in its wetter regions.

13.5.2 Concentration Analysis

The hydroxide concentration of the electrolyte is determined by titration of an aliquot from each rinse solution with standard hydrochloric acid. The level of hydroxide is determined at an end-point pH of 8.0, and the level of carbonate is given by a pH 4.0 end point. If the electrolyte and components were not protected from air during cell disassembly and rinsing, the caustic concentration of the electrolyte is taken as the sum of the hydroxide and the carbonate. If care was taken to prevent exposure to atmospheric carbon dioxide, the measured hydroxide and carbonate levels should accurately reflect those present in the cell before disassembly. Typically, carbonate levels in the electrolyte of nickel-hydrogen cells should be less than 0.5% by weight.

Most nickel-hydrogen cells are activated with either 26% or 31% (by weight) potassium hydroxide electrolyte. Ideally, the concentration that the cell was activated with should be the same as that determined by analysis, when averaged over the entire cell. However, a number of internal cell processes can change the electrolyte concentration. Concentration may be increased by corrosion of nickel in the positive electrode, or by any burned plastic materials in the cell (results of popping). Electrolyte concentration may be reduced by the formation of γ-NiOOH, which contains one potassium atom for every three nickel atoms in the crystal lattice. Electrolyte concentration may also be significantly reduced by oxidation of platinum from the negative electrodes during the storage of nickel-precharged cells, and the incorporation of the resulting platinate species into the nickel electrode active material.[13.22]

13.5.3 Inorganic and Organic Contaminant Analysis

Many inorganic contaminants can be rinsed from the nickel electrodes, the hydrogen electrodes, or the separator with the electrolyte. These contaminants will be in the electrolyte rinse solution, and they may be assayed by neutralizing an aliquot of the rinse solution with nitric acid. The neutralized electrolyte, which is primarily potassium nitrate, may be evaporated to dryness, ground into a finely divided mixture, and spectrographically analyzed for metallic contaminants. Chloride and sulfate contaminants may be assayed gravimetrically by precipitation from the dissolved potassium nitrate as silver chloride or barium sulfate, respectively.

The inorganic constituents of the separator material may be determined by direct spectrographic analysis of the Zircar or asbestos materials, depending on the type of separator used in the cell. Most cells built after about 1995 use Zircar separator, which is a yttria-stabilized zirconium oxide ceramic fabric. Analysis of this material will indicate its precise composition, which can be somewhat variable. It is not uncommon to find that a significant amount of the approximately 15% yttrium used to stabilize the zirconium oxide has been replaced by hafnium. No undesirable reactivity or other negative aspects of hafnium, as opposed to yttrium, have been reported for nickel-hydrogen cells.

Analysis of the inorganic constituents of the asbestos separator material used in older nickel-hydrogen cells is more difficult. While spectrographic analysis of asbestos samples indicates the average inorganic composition, it does not indicate its stability in the electrolyte of the nickel-hydrogen cell. Asbestos is composed of a mixture of fibrous magnesium silicate minerals, with more than 200 individual magnesium silicates having been detected in natural asbestos. While some of these silicate phases are actually soluble in water, typically these have been removed from separator materials used in nickel-hydrogen cells by the washing and reconstitution processes involved in preparing sheets of asbestos fiber mats. The remaining silicate phases are all reactive with potassium hydroxide electrolyte, with the silicate fibers slowly converting to hydroxide powders. The key analyses thus involve determining the amount of silicate in the electrolyte by spectrographic analysis as discussed here, and evaluating the reactivity of the asbestos to the electrolyte. If the asbestos has not been used in a cell, it is essential to verify that it does not react rapidly with alkaline electrolyte to result in the accumulation of soluble silicate in the electrolyte. Levels of dissolved silicate greater than 0.5% in the electrolyte (by weight) have been found to result in increased cell resistance, particularly at low temperatures.

The reactivity of asbestos separator materials may be determined by cutting samples approximately 1 in.^{2} in size from random locations that cover the entire lot of material to be used. The mineral composition of asbestos can vary significantly over large areas of material, primarily because small differences in fiber size or density for the different silicate phases influence the deposition of the fibers onto the mat as it is produced. The samples may be analyzed for stability by exposure to 31% potassium hydroxide electrolyte at room temperature for a 1-week period, and the amount of silicate replaced by hydroxide determined from the measured weight loss. Because each silicate phase has a different reactivity and activation energy for reaction with potassium hydroxide, any attempts to accelerate the measurement at higher temperatures are not recommended because an unknown number of silicate phases present in unknown amounts are reacting, each with a different rate constant and activation energy.

It is also difficult to extrapolate initial reaction rates to longer times, because the reactivity of asbestos will always decrease with continued exposure to potassium hydroxide. This is so because the more reactive silicates are more rapidly attacked. The composition of these more reactive silicates can be determined by analysis of the dissolved materials for metal content. Generally, a wide variety of metals other than magnesium are present at various levels within the silicate minerals found in asbestos. The ideal asbestos separator for nickel-hydrogen cells should be prerinsed with potassium hydroxide to remove all the most reactive silicate minerals, leaving a material that does not react with potassium hydroxide at a significant rate at room temperature.

Separators can be analyzed for organic contaminants after the potassium hydroxide has been rinsed by sequential extraction; first with acetone for polar organic species, then with hexane for oily organics. Electrolyte that has been rinsed from nickel-hydrogen cell components can be analyzed for organic contaminants by liquid extraction with methylene chloride. One performs this analysis by shaking an electrolyte aliquot with methylene chloride repeatedly in a glass container, then allowing the phases to separate. The methylene chloride layer, which will dissolve the organics, is separated from the aqueous phase. The methylene chloride is evaporated, leaving behind a residue of extracted organic contaminants that can be weighed. It is recommended that a blank analysis be performed using distilled water rather than an electrolyte sample, and that the weight of organics in the blank, which should be very low, be subtracted from that found for the sample. The organic contaminants in the residue remaining after evaporation of the methylene chloride can be identified using Fourier transform infrared analysis.

13.5.4 Separator Pore Size Distribution

The distribution of pore sizes in separator materials may be measured by mercury intrusion porosimetry. The two types of separator most commonly used in nickel-hydrogen cells have very different distributions of pore sizes. The asbestos separator material used in the COMSAT cell design produced until about 1995 is composed of fine magnesium silicate fibers that are extremely hydrophilic and have very fine pores. Asbestos separator material thus becomes fully saturated with electrolyte, forming a gas-impermeable layer between the nickel and hydrogen electrodes that channels all undissolved oxygen out to the edges of the electrode stack during overcharge. This channeling forces all gas-phase oxygen recombination with hydrogen to occur at the edges of the stack, a situation that can result in popping damage or separator damage if a cell is overcharged at a high rate. As asbestos ages in cells, some of the silicate fibers become converted to compressed magnesium hydroxide powder deposits, which tend to be even more wettable than the asbestos starting material.

Zircar separator material for nickel-hydrogen cells typically consists of bundles of relatively fine zirconia fibers that are either woven or knit into a fabric, as shown in Fig. 13.22. The fiber-bundle structure results in a bimodal distribution of pore sizes.

Figure 13.22. Structure of woven Zircar fabric, showing bundles of ceramic zirconia fibers.

The finer pores within the fiber bundles tend to wet and retain electrolyte, while the larger spaces between the bundles tend to allow the relatively free movement of oxygen gas or electrolyte from the surface of the nickel electrode to the surface of the hydrogen electrode. Oxygen recombination can thus occur directly without allowing the buildup of high oxygen concentrations at the surface of the platinum electrode, assuming that the larger voids in the Zircar material do not become flooded with electrolyte.

Other types of separator materials have been used with limited success in nickel-hydrogen cells, such as nonwoven nylon, Zirfon,**** polysulfone, polybenzimidizole, and various composite separators[13.24] that combine layers of differing materials. Many of these separators attempt to engineer the sizes of the pores such that the gas-carrying pores remain free of electrolyte, while the ionically conducting pores remain wetted at all times. While some of these materials have given good performance in cells, the performance of cells containing alternate separators has at best only approached that of cells containing Zircar separator.

13.6 Expert Systems for Nickel-Hydrogen Cell Diagnostics

Diagnosis of the root causes of performance problems in battery cells is often difficult because each of the component characteristics, physical processes, and chemical processes in most battery cells has a strong interaction with many other variables in the cell. In nickel-hydrogen battery cells this is particularly true, because of the coexistence of solid, liquid, and gaseous active phases, each of which can provide coupling between the components and processes that occur within the cell. In nickel-hydrogen cells, as is the case for most battery cells, diagnostic procedures in the event of performance problems typically involve the combination of some electrical tests (hopefully perceptive), cell disassembly and inspection, and physical/chemical analysis of the cell components.

All laboratories that perform such destructive physical analysis procedures have a somewhat different set of procedures and analyses that are followed. Some analyses may be either more or less appropriate for any given cell diagnostic procedure, depending on the performance signatures of that cell. However, normally any given laboratory utilizes a standard destructive physical analysis regimen, and it is assumed that a broad range of analysis procedures will yield the key information needed to diagnose the root cause for the cell problems.

A more efficient approach to the destructive physical analysis of nickel-hydrogen battery cells has been developed. This approach defines a destructive physical analysis in terms of applying the optimum diagnostic procedures from a large toolbox of procedures, and performing only those analyses that are most

---

**** Zirfon is a polysulfone-bonded zirconium oxide material. For data on its use in nickel-hydrogen cells, see Jamin et al.[13.23]

appropriate for the observed behavior of the battery cell. Thus, with this approach, the details of each specific destructive physical analysis procedure are generally different from all other destructive physical analysis procedures.

The most appropriate procedures to be used for a given destructive physical analysis are those that have the greatest likelihood of yielding the root cause of the cell performance problems or degradation modes. Determination of the most appropriate analysis associated with each root cause (and its performance signatures) is based on a combination of performance modeling, experience, and the known symptoms exhibited by the cell. This approach to the perceptive destructive physical analysis of nickel-hydrogen battery cells has been codified into an expert system to guide the diagnostic analysis of nickel-hydrogen cells.[13.1]

The Expert System for Battery Cell Analysis (ESBCA) provides a software package that allows a user to interactively design a destructive physical analysis procedure to best fit the observed behavior of a battery cell, or to determine the most likely root cause for any collection of observed battery cell symptoms. This software system is designed to operate on a PC using a Windows 95 (or higher-level) operating system. The ESBCA system, which continues to be upgraded as knowledge of the nickel-hydrogen battery improves, considers more than 36 different root causes for nickel-hydrogen battery problems or degradation, and it has a toolbox of more than 60 different analysis procedures at its disposal to differentiate between these root causes.

The ESBCA system saves each destructive physical analysis in an analysis file, which contains the cell design characteristics, destructive physical analysis analysis results, and electrical or chemical analysis symptoms. As additional analyses are performed during the course of a destructive physical analysis procedure, the results are added to this analysis file, enabling the expert system to zero in on the root cause for cell performance problems.

This expert system includes three analysis tools that allow this system to effectively guide the destructive physical analysis of a nickel-hydrogen cell. The first is a list of the observed performance symptoms and analysis results that serves as the basis for evaluating the most likely root causes for the cell performance problems. The second analysis tool translates the symptom list into a list of the most likely root causes, which are displayed in descending order of likelihood that the cause fully accounts for the observed symptoms. Typically, one uses this tool to obtain the most likely explanations for cell behavior, after which one uses the third analysis tool, which indicates the preferred analysis procedure that is best suited to either confirm or reject any possible root cause.

The normal course of a cell destructive physical analysis that is guided by this expert system involves three phases. During the initial phase, in which only very general symptoms are known (e.g., capacity is low by 30%), numerous root causes are likely with nearly equal probability. The second phase of a destructive physical analysis involves collecting data from a sequence of analysis procedures

13.6 Expert Systems for Nickel-Hydrogen Cell Diagnostics

to eliminate (or confirm) the most likely root causes. During this intermediate phase the user strives to eliminate all but one particular root cause scenario. The final phase of the destructive physical analysis is reached when all root causes but one have been eliminated to a satisfactory probability. The use of this expert system has been described in detail with a number of examples of actual destructive physical analysis case studies,[13.1] demonstrating how the user can quickly determine, with a high degree of certainty, a root cause for specific cell problems with a minimal amount of analysis effort. What follows are descriptions of two such case studies.

13.6.1 Expert System Analysis: Failed Cell from 60% Depth-of-Discharge Life-Test

This example of an expert system analysis is based on an 81 A h ManTech cell with a composite separator consisting of one layer of Zircar and one of asbestos. The cell failed after about 10,000 cycles at 60% depth of discharge in a ground life-test. Tests after failure indicated a cell capacity that was about 45% low, and good charge retention capacity. Providing this relatively basic information to the expert system, along with information about the swelling of the positives and the positive corrosion, yielded the evaluation in Fig. 13.23. The expert system indicated that to a very high probability, the cell had problems caused by separator dry-out.

When the fact that the discharge voltage plateau had dropped more than 100 mV at end of life was added to the list of symptoms in the expert system, the probability of cell dry-out increased to more than 97%. At present this analysis provides the most reliable evaluation available for the root cause for the failure of this cell. While the additional possible root causes listed in Fig. 13.23 could be eliminated with further analysis, such effort probably is not warranted in this situation simply to increase confidence from 97% to 100%. This example shows how a large amount of destructive physical analysis effort on cell components can be eliminated when the expert system is correctly applied, with little loss in the confidence that the correct root cause has been established.

13.6.2 Expert System Analysis: Failed Cell from 75% Depth-of-Discharge Life-Test

A 48 A h cell failed after about 1000 cycles in a 75% depth-of-discharge life-test at 20°C. The cell exhibited about a 30% capacity loss, good charge-retention behavior, only 8% swelling of the positive plates, and 5.5% corrosion of the nickel sinter in the positive plates (from pressure growth in the cell). This cell was quite puzzling because none of the traditional failure modes for nickel-hydrogen cells seemed to account for its behavior. It should be pointed out that all cells in this test pack had appeared to degrade in the same way, and thus this was no special behavior associated with an outlying cell.

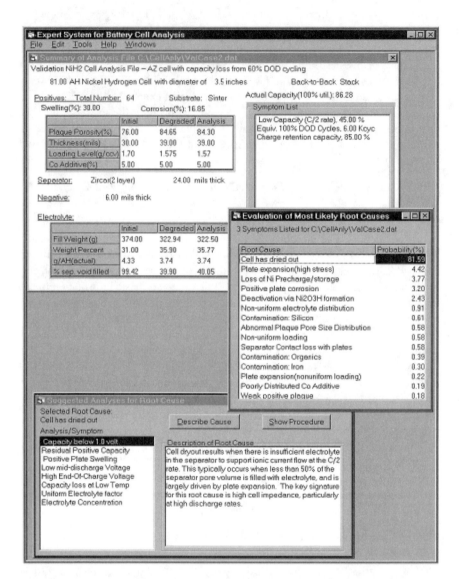

Figure 13.23. Output of expert system indicating nickel-hydrogen cell failure as a result of separator dry-out. Reprinted courtesy of NASA.

When this information was supplied to the expert system, the evaluation in Fig. 13.24 was obtained. Plate expansion resulting from high stress was listed as the most likely root cause for cell failure, with hydrogen precharge if the cells were stored as another possible problem. Because the plate expansion was not

13.6 Expert Systems for Nickel-Hydrogen Cell Diagnostics

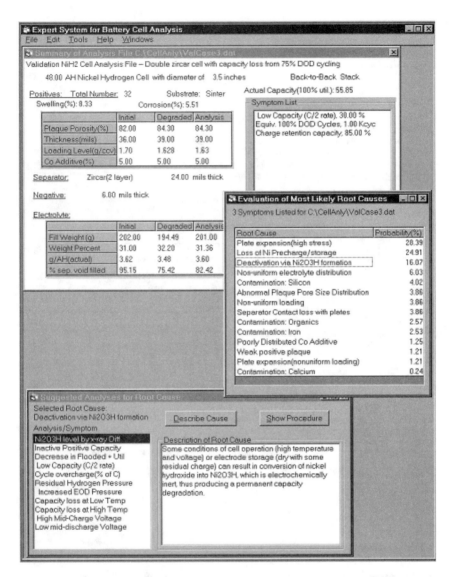

Figure 13.24. Interim output of expert system, indicating nickel-hydrogen cell failure possibilities, with the three most likely possibilities being nickel plate swelling, storage without nickel precharge, and active-material deactivation. Reprinted courtesy of NASA.

very great, and the cells had not been stored but had failed during cycling, both these root causes were not given a high probability. More interesting, the third root

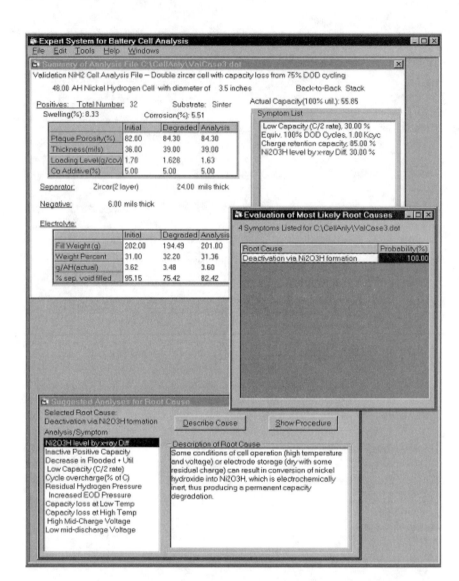

Figure 13.25. Final output of expert system, concluding that nickel-hydrogen cell failure resulted from active-material deactivation through nickel oxide hydroxide formation. Reprinted courtesy of NASA.

cause listed was a process that could deactivate nickel electrode active material so that it could no longer be charged or discharged. At the time of this destructive physical analysis, no such process had been identified, but deactivation has long been recognized as a possible scenario for capacity loss.

After extensive evaluation of electrodes from this cell, it was discovered that active material had in fact been converted from the normal nickel hydroxide to a new material, Ni_2O_3H, which was electrochemically inert. A technique was developed that used x-ray diffraction to assay the amount of this compound in active material taken from nickel electrodes, as is described in the interim analysis of Fig. 13.24. This assay indicated that about 30% of the active material had undergone this conversion to inactive material. When this analysis result was added to the list of symptoms for this cell, the final expert system evaluation in Fig. 13.25 was obtained, which indicated a 100% probability that Ni_2O_3H formation was the root cause of cell failure. This provided a quite definitive root cause for the failure of this cell, in spite of the fact that the process whereby active material is deactivated was not clearly understood.

As shown by the preceding two examples, the ESBCA can very efficiently guide the diagnosis of performance problems or degradation in nickel-hydrogen battery cells. In all cases where this software has been used, it has rapidly guided the user to a conclusive root cause for the observed cell symptoms. The interactive combination of the ESBCA system with the destructive physical analysis results in a major reduction in the destructive physical analysis and test effort. In several instances, over an order of magnitude reduction in effort was realized relative to the actual destructive physical analysis testing that was ultimately required to confidently identify a root cause. This expert system offers an approach that maximizes the probability of rapidly arriving at the cause of nickel-hydrogen cell performance problems and provides an interactive procedural framework for nickel-hydrogen cell destructive physical analysis.

13.7 References

[13.1] A. H. Zimmerman and L. H. Thaller, "Expert System for Nickel Hydrogen Battery Cell Diagnostics," *Proc. 1998 NASA Battery Workshop*, NASA/CP-1999-209144, pp. 297–316.

[13.2] L. H. Thaller and A. H. Zimmerman, *Nickel-Hydrogen Life Cycle Testing* (The Aerospace Press, El Segundo, CA, and the American Institute of Aeronautics and Astronautics, Reston, VA, 2003), pp. 58–63.

[13.3] A. H. Zimmerman, S. W. Donley, J. Matsumoto, T. Poston, and H. F. Bittner, *Proc. 33rd International Power Sources Symp.* (1988), p. 503.

[13.4] J. E. B. Randles, *Discuss. Faraday Soc.* **1**, 11 (1947).

[13.5] M. V. Quinzio and A. H. Zimmerman, "Dynamic Calorimetry for Thermal Characterization of Battery Cells," *Proc. of the 17th Annual Battery Conference on Applications and Advances*, IEEE 02TH8576, ISBN 0-7803-7132-1 (2002), pp. 281–286.

[13.6] J. H. Kauffman, T. C. Chung, and A. J. Heeger, "Fundamental Electrochemical Studies of Polyacetylene," *J. Electrochem. Soc.* **131**, 2847–2856 (1984).

[13.7] L. H. Thaller, M. V. Quinzio, and G. A. To, "Volume Tolerance Characteristics of a Nickel Hydrogen Cell," *Proc. of the 14th Annual Battery Conference on Applications and Advances* (Long Beach, CA, 12–15 January 1999) pp. 329–334.

13.8 G. Halpert and V. Kunigahalli, *Procedures for Analysis of Nickel Cadmium Cell Materials*, NASA Rep. X-279, Rev. A (1980).

13.9 A. H. Zimmerman and R. Seaver, "Cobalt Segregation in Nickel Electrodes during Nickel Hydrogen Cell Storage," *J. Electrochem. Soc.* **137**, 2662 (1990).

13.10 A. H. Zimmerman, M. Quinzio, G. To, P. Adams, and L. Thaller, "Nickel Electrode Failure by Chemical De-activation of Active Material," *Proc. 1998 NASA Battery Workshop*, NASA/CP-1999-209144, pp. 317–328.

13.11 R. McEwan, "Crystallographic Studies on Nickel Hydroxide and the Higher Nickel Oxides," *J. Phys. Chem.* **75**, 1782 (1971).

13.12 C. Greaves, A. Malsbury, and M. Thomas, "Structure of the Mixed Conductor Ni2O3H by Powder Neutron Diffraction," *Solid State Ionics* **18/19**, 763 (1986).

13.13 A. Malsbury and C. Greaves, "The Crystallographic and Magnetic Structure of Ni2O3H," *J. Solid State Chem.* **71**, 418–425 (1987).

13.14 R. F. Hoskins and M. A. Malati, *Experimental Inorganic/Physical Chemistry* (Horwood Publishing, Chichester, UK, 1999), p. 214.

13.15 *Standard Methods of Chemical Analysis*, 6th edition, vol. 1, N. H. Furman, ed. (D. Van Nostrand Co., Princeton, NJ, 1962).

13.16 A. H. Zimmerman, "The Role of Anionic Species in Energy Redistribution Processes in Nickel Electrodes in Nickel Hydrogen Cells," in *Hydrogen Storage Materials, Batteries, and Electrochemistry, Proceedings Vol. 91-4*, The Electrochemical Society, Inc. (Pennington, NJ).

13.17 L. Thaller and A. Zimmerman, "Electrochemical Voltage Spectroscopy for Analysis of Nickel Electrodes," *Proc. of the 15th Annual Battery Conference on Applications and Advances* (January 2000), pp. 165–172.

13.18 B. E. Conway and T. C. Liu, "Experimental Evaluation of Adsorption Behavior of Intermediates in Anodic Oxygen Evolution at Oxidized Nickel Surfaces," *J. Chem. Soc., Faraday Trans.* **1**, 83, 1063–1079 (1987).

13.19 A. H. Zimmerman and P. K. Effa, in *Extended Abstracts of the Fall 1984 Meeting of the Electrochemical Society, Vol. 84-2*, The Electrochemical Society (Pennington, NJ), p. 136.

13.20 A. H. Zimmerman, G. A. To, and M. V. Quinzio, "Scanning Porosimetry for Characterization of Porous Electrode Structures," *Proc. of the 17th Annual Battery Conference on Applications and Advances*, IEEE 02TH8576, ISBN 0-7803-7132-1 (2002), pp. 293–298.

13.21 C. H. Ho, M. Murthy, and J. W. Van Zee, "Studies of the Co-deposition of Cobalt Hydroxide and Nickel Hydroxide," *Proc. 1996 NASA Battery Workshop*, NASA Conf. Pub. 3347, p. 289 (1997).

13.22 A. H. Zimmerman, "Effects of Platinum on Nickel Electrodes in Nickel Hydrogen Cells," *J. Power Sources* **36**, 253 (1991).

13.23 T. Jamin, N. Tassin, J. Bouet, A. Delahaye, P. Vermeiren, and M. Schautz, "Current Data on Zirfon®-2 Separator Material and Foamy Positive Evaluation under Cycling in NiH_2 Boiler Plate," *Proc. of the 5th European Space Power Conference* (Tarragona, Spain, 21–25 Sept. 1998), p. 689.

[13.24] W. W. Lee, R. A. Sutula, C. R. Crowe, and W. A. Ferrando, "Electrochemical Behavior of Thin Nickel Electrode with Composite Substrate," *Proceedings of the Symposium on the Nickel Electrode*, R. G. Gunther and S. Gross, eds., The Electrochemical Society, Pennington, NJ 82-4 (1982), p. 243.

Index

A

activation
  in cell dynamics 178, 182, 183
  and degradation and failure modes 385, 387, 394, 416, 436
  in electrical and thermal performance 66
  in evolution of cell designs 17, 23
  and hydrogen electrode 152
  in modeling 251, 255, 259
  and nickel electrode 85, 102, 138, 142–144
  and separators and electrolyte 167
  in test experience 355
  in thermal management and reliability 317, 319
additives 35, 120–137, 175
  cadmium 35, 119, 120–122, 447–448, 461
  cobalt
    in advanced nickel electrode concepts 145–146
    in cell dynamics 178
    and degradation and failure modes 386, 393, 397–398, 408, 412
    in diagnostics and DPA 440, 442–444, 446, 448, 455–458, 459
    in electrical and thermal performance 60
    in evolution of cell designs 18, 24–25, 35
    in first-principles models 223–225, 240–243, 245–246, 248
    in nickel electrode chemistry 87, 96, 97–99, 104, 111, 115, 119
    in nickel electrode construction 79, 86
    role of, in nickel electrode 120, 123–130
    and separators and electrolyte 176
    in storage models 251–252, 256–258, 260, 261, 262
    in test experience 344
    in wear-out models 274
  iron 447
  lithium 35, 120, 122–123, 175, 447–448, 449
  platinum 143
    in cell dynamics 177, 179–180, 181, 184, 185, 191
    in cell-level diagnostics 416, 419, 421, 426, 430, 432–433
    in charge management 304
    and degradation and failure modes 385–387, 390–391, 395–397, 399
    in DPA techniques 436
    in electrical and thermal performance 64
    in electrolyte and separator analysis 466, 469
    in evolution of cell designs 33, 35
    and hydrogen electrode 149–153, 155–157, 159, 459–464
    in modeling 210, 211, 213–214, 215–216, 251–252, 258
    and nickel electrode 85, 87, 108, 120, 122, 129–132, 441, 442, 446, 447–448
    in test experience 372–373, 379
Air Force Flight Experiment 5, 17

B

Belleville washers 20, 22, 33, 410, 420, 434
Butler-Volmer equation 39, 41, 43, 153, 207
bypass devices 317

C

calorimetry 59–61, 109, 119, 193, 243–244, 310–311, 428–430
capacity
  accompanying oxygen evolution 103–104
  in advanced nickel electrode concepts 144, 146
  in cell models
    first-principles 205, 228, 230, 237–238, 241–243, 248–250
    storage 251–252, 253–254, 258
    wear-out 263–264, 265–272, 278–282
  in charge management 289, 293–294, 300, 303–304
  and degradation and failure modes 386–387, 391, 392, 394, 395, 398, 409, 410
  in diagnostics and DPA
    cell-level diagnostics 415, 422–423, 424–426
    DPA techniques 436
    expert systems 471
    nickel electrode diagnostics 439, 444, 446–447, 448, 452–454, 458
  in electrical and thermal performance 66–67
  and electrolyte concentration dependence 164
  increase of, by gradient nickel electrode 145
  isolated, discharge of 114
  loss of
    in cell dynamics 183, 196
    in charge management 301, 305, 306
    and degradation and failure modes 387, 396, 398, 399, 400, 405, 410, 411–412

480 Index

capacity, loss of (*continued*)
    in diagnostics and DPA 440, 456, 471, 474
    in first-principles models 225, 234, 244, 248
    and nickel electrode 80, 84, 86, 90, 111, 132
    in storage models 252, 256–262
    in test experience 332–333, 336–338, 356, 357, 360, 379
    in wear-out models 267–268, 271, 275, 279–282
measurement of 125, 131, 260, 262, 297, 316, 343, 423
of nickel electrode 142
    in nickel electrode chemistry 92–93, 108, 110–111, 115
    in nickel electrode construction 84–85
role of, in nickel electrode 124–127, 129, 133
and separators and electrolyte 163–164, 174–175
temperature dependence of 43, 45, 138, 178, 313, 400
in test experience
    accelerated vs. real-time 366
    electrolyte concentration 357–358
    pressure behavior 348–351, 353
    recharge behavior 340, 344–345
    special design features 375, 378
    specialized life tests 380
in thermal management and reliability 313–314, 319, 324, 326–327, 328
capillary forces 17, 101, 117, 159, 166, 168, 202, 209–210, 211, 213, 227, 408
catalysts 3, 170
    and cell dynamics 177–182, 184–186, 189–191, 202
    in charge management 304
    and degradation and failure modes 385–387, 390–391, 395–396, 399
    in diagnostics and DPA
        cell-level diagnostics 416, 419, 421, 426–427, 430, 432
        DPA techniques 436
        hydrogen electrode diagnostics 459, 460–462, 463, 464
        nickel electrode diagnostics 442, 447
    in electrical and thermal performance 64–65
    in evolution of cell designs 12, 27, 33, 35
    and hydrogen electrode 149–153, 155–161
    in modeling 209–210, 211, 213–214, 215–216, 265
    and nickel electrode 85, 108, 122, 129, 138
    in test experience 379
cell designs 4–5
    AF/HAC 13, 18–21, 22–24, 320, 328

back-to-back 23–25, 32, 34, 44, 150, 185, 188, 236, 248, 265, 274, 365–368, 373, 378, 421
bipolar stacks 32–33
Boeing 361
    in cell dynamics 184–185, 186, 188, 190, 191, 198, 202
    in charge management 293
common pressure vessel (CPV) 26, 325–326, 331, 375
    with compressible separators 421
COMSAT 13, 15–18, 22, 23, 25, 37, 45–47, 166, 319, 328, 331, 468
    and degradation and failure modes 408, 410, 412–413
dependent pressure vessel (DPV) 31–32
    in diagnostics and DPA 430, 437, 442, 470
dual-anode 32, 44, 150, 185, 236, 275, 331, 377–378
dual-layer 413
dual-stack 25, 265, 331, 366, 375, 401, 435, 464
EaglePicher 361
    in electrical and thermal performance 65
Gates Energy Products 361
    guidelines for 324–327
    with high-bubble pressure separators 188
Hughes 361
ManTech 13, 21–23, 46, 282, 319, 323, 328, 331, 352, 370, 377–378, 422, 471
    in modeling 234–237, 238, 240, 246, 247, 250, 263–264, 274–275
    and nickel electrode 71
recirculating 18–19, 150, 365–367, 401, 438
    and separators and electrolyte 163–164, 169–170, 174–175
separators in 361
single pressure vessel (SPV) 22, 26, 30–31, 326, 331, 376
    specialized features 375–379
    and substrate corrosion rate 115
    in test experience 334, 363
    in thermal management and reliability 312–313, 318, 323
    wear-out rates of 321
Yardney 361
centrifuging 24
charge management 5, 289–307
    adaptive or self-optimizing 289, 297–300
    adiabatic charging 305–306
    capacity recovery 306–307
    cell capacity balance 303
    cell capacity imbalance 305
    cell-level 290, 415
    electrolyte redistribution 306
    pressure-based 289, 294–297

Index   481

ratchet charging 289, 300–302
recharge ratio 289, 290, 292, 293–294, 295, 297–300, 344–345
reconditioning 289, 292, 296–297, 302–304
safe recharge ratios 293
software-based 294
during storage 304–305
thermal environment 289, 293, 295, 300–301, 306
trickle charge 291–292, 293, 295, 305
  in cell dynamics 180, 184, 200–201
  and degradation and failure modes 396, 403
  in modeling 232, 253, 277–278, 280–284
  and nickel electrode 132
  in test experience 340, 346, 347, 371
V/T charge control 289–295, 298, 300, 342
charge transfer 97, 99, 110, 122, 163, 206, 211, 214–215, 245
  distribution of, through microscopic layers 218
  reactions 90–91, 102, 206, 218
  sites 89–92, 135–136, 214, 218, 219
cobalt additive
  in advanced nickel electrode concepts 145–146
  in cell dynamics 178
  chemistry of 79, 86, 87, 96, 99, 104–105, 120, 123–129, 134, 138, 145, 223–225, 256–258
  concentration of 60, 145, 223, 245–246, 397, 459
  and degradation and failure modes 386, 393, 397–398, 408, 412
  in diagnostics and DPA 440, 442, 443, 444, 446, 448, 455–458, 459
  distribution of 224, 240–242, 248, 256
  effect of
    on conductivity 224
    on electrochemical processes 242
    on voltage level 126, 397
  in evolution of cell designs 18, 24–25
  gradient to improve performance 111, 145, 225, 240–243
  in modeling 223–225, 251–252, 256–258, 260, 261, 262, 274
  and modeling effect of gradients 225, 240–243
  in nickel electrode chemistry 87, 96, 97–99, 104, 115, 119
  in nickel electrode construction 79, 86
  reduction in charge efficiency and 225
  reduction of 251, 257, 258, 259
  role of, in nickel electrode 120, 123–129, 130
  and separators and electrolyte 176

  in superlattices 125
  in test experience 344
common pressure vessel cell tests 375
compressibility of hydrogen 56
computer-aided cell design 234–240
  key design features in 235
  life-testing of 238
  and optimization of cycle life 235
  and performance 234
condensation 197–198, 237, 300, 314–315, 325, 334, 339, 343, 401–402, 405, 429
conductivity 71, 75, 95, 197, 211, 235, 437, 438, 459
  of active material 124, 142
  electrical 81, 258
  electronic 89–91, 95–97, 97–99, 110–111, 127–129, 138, 210, 218, 222–224, 228, 337, 399, 409–410
  enhancers to 145
  gradients of, in pores 221
  ionic 90, 95–97, 127, 163–164, 169, 170, 172, 202, 210, 219, 246, 325, 357, 413, 469
  of sinter 245, 248, 253
  thermal 315, 405
contaminants
  atmospheric 436
  in cell dynamics 182, 183
  chloride and nitrate 87, 120, 133–134, 448, 466–467
  copper and iron 87, 120, 132, 158, 386
  and degradation and failure modes 385–387, 389–390, 392, 405
  in diagnostics and DPA 416, 432, 434, 461–462
  gaseous 170, 386–387, 435–436
  in hydrogen electrode 152–153, 158–159, 176
  inorganic 447–448
  iron 461
  in nickel electrode 75–76, 120–137, 142
  organic 133, 135–137, 153, 170, 385, 435–437, 448–449
  particulate 170, 442
  silicate 87, 120, 133, 135–137, 169, 176, 386, 413, 442, 448
  sulfate 87, 133–134, 386, 448, 466
  in thermal management and reliability 316, 318–319
contractors and testing agencies 370
  The Aerospace Corporation 32, 208, 331
  Ford Aerospace/Loral 331
  Lockheed-Martin Marietta 331
  Martin Marietta 6
  NASA 6, 11, 28–30, 265, 331, 361, 362
  TRW/Northrop-Grumman 331

482 Index

contractors and testing agencies (*continued*)
    Wright-Patterson Air Force Laboratories 18
core 22, 27, 33, 237, 387, 388, 403, 410–411,
    417, 420, 435
cost 146
    cell 13, 264
    space system 8, 11
Crane test packs
    3001C 345
    3003L 377
    3004L 377
    3005L 375
    3006L 375–376
    3214E 58, 384
    3314E 275, 365–366
    3600X 265, 268, 271
    3601X 265, 268
    3602X 265, 268
    3603X 248, 251, 265, 268, 272, 340
    3604G 358
    3604X 251, 265, 268, 272, 341, 352, 370
    3605X 265, 268, 272, 350
    3731E 355
    3761E 354, 355
    3765E 354, 355
    3831E 265, 268
    3835E 265, 268
    5000H 368
    5002E 248, 251, 274–275, 352, 365–366, 375
    5002G 366
    5002H 367
    5004L 337
    5402G 366
    5402H 367
    5631W 355
    5635W 355
    5661W 355
    5665W 355
    5735E 355
current gradients 153, 196
cycle life 5, 10, 323
    in cell dynamics 196
    in cell models
        first-principles 227, 235, 237, 240, 247,
            249–250
        storage 253
        wear-out 263–284
    and degradation and failure modes 392, 395,
        404, 406, 407, 412
    in diagnostics and DPA 422, 448, 455
    in electrical and thermal performance 37
    electrolyte concentration, dependence on
        356–358
    in evolution of cell designs 30, 35
    and hydrogen electrode 149, 176
    and nickel electrode 81, 84, 92, 121–122,
        123, 126, 144–145
    prediction of 205
    requirements for many years of 115
    temperature, dependence on 355
    in test experience
        discharge voltage signature 336
        electrolyte concentration 356, 362–368,
            370
        pressure behavior 352–353
        recharge behavior 340–341, 345
        specialized life tests 380
        storage effects 371–375
        temperature 353–356

D

degradation modes 5, 7, 10, 383–413, 399
    active-material dehydration 360
    active-material extrusion 99, 201–202, 210,
        247–249, 321, 357, 398, 401, 409–412,
        438, 442–443, 446
    capacity loss 332–333
    cell capacity imbalance 289, 292, 403–404
    in cell dynamics 196–197, 201, 202
    corrosion 384, 397–399, 405–408, 411, 412,
        437, 444, 446, 451, 466, 471
        in charge management 296, 304
        in first-principles models 210, 224, 235,
            240–241, 247–249
        in nickel electrode 79, 81–83, 111,
            114–116, 128, 133–134, 142–143,
            144
        in storage models 251–254, 259–262
        in test experience 338, 344, 348, 373
        in thermal management and reliability
            315, 321
        in wear-out models 267
    in diagnostics and DPA 415, 422, 425, 438,
        439, 470, 475
    dry-out 401, 402, 405, 408–409, 413
        in cell dyamics 197
        in charge management 306
        in diagnostics and DPA 434, 437, 471
        in evolution of cell designs 33
        and hydrogen electrode 164, 175
        in modeling 236
        in test experience 332, 333–334,
            338–339, 343
        in thermal management and reliability
            315
    electrode swelling 391, 392, 398, 401,
        407–408, 409, 410
        in cell dynamics 201
        in diagnostics and DPA 421, 437, 438,
            442–443, 444, 452
        in modeling 245, 247–248

Index   483

nickel electrode 80, 86, 144, 146
   in test experience 332, 338, 357
   in thermal management and reliability 321
in evolution of cell designs 20–21, 26
high-current pulsing 380
and hydrogen electrode 163
mineral separators 176
in modeling 240–241, 255, 271, 284–285
phase separation 125
popping 401
porosity changes 210
storage-related 114, 142, 151, 205, 251–252, 259
in test experience 337, 348, 354–355, 375, 377
in thermal management and reliability 325
destructive physical analysis (DPA) 5, 250, 251, 253, 260, 262, 356, 372–373, 380, 406, 415–475
   active precharge 434
   chemical precharge analysis 436
   cobalt gradient analysis 455–458
   compression measurement 419–420, 434–435
   cross-sectional analysis 463
   documentation of 434
   dry-out test 434, 437
   electrochemical voltage spectroscopy 431–433, 449–451
   electrode resistance 438–439
   electrolyte concentration analysis 420, 434–435, 438, 466
   electrolyte distribution 464–465
   expert systems 415, 469–475
   extrusion analysis 442–443
   flooded utilization 444–447
   gas analysis 434, 435–437
   hydrogen overpotential analysis 460–462
   inorganic contaminant analysis 447–448, 462–463, 466–467
   loading analysis 458–459
   loading-level gradient analysis 457–459
   organic contaminant analysis 448–449, 462–463, 466–467
   oxygen overpotential test 451
   particulate contamination analysis 442
   physical inspection procedures 433–434, 469
   platinum catalyst adhesion analysis 459–460
   pore size distributions 454–455, 464, 468–469
   scanning porosimetry 454–455, 464
   stress tests 452–454
   thermogravimetric analysis 441–442
   thickness measurement 438
   x-ray diffraction 440–441
diagnostic procedures 10, 415–434, 469–471
   ac impedance 427–428
   acoustic monitoring 430–431
   active precharge and 439–440, 445
   calorimetry 428–430
   charge-retention statistics and 423
   electrochemical voltage spectroscopy 431–433, 449–451
   expert systems 469–475
   leakage current measurements in 424
   precharge tests 424
diffusion-limited current 164, 166, 334, 409
dipole polarization layer 49–50
DPA. *See* destructive physical analysis
dynamics of cells 4, 177–203
   charge efficiency 170, 177–180, 196
   in advanced nickel electrode concepts 145, 146
   in charge management 292, 300, 303, 305
   and degradation and failure modes 386, 396, 397, 400, 402, 404, 412
   in diagnostics and DPA 415, 445–446, 448, 449, 451, 452–453, 456
   in electrical and thermal performance 54–55
   in nickel electrode-hydrogen gas interactions 138
   in modeling 223–225, 234–235, 249, 253
   in nickel electrode chemistry 102–103, 105
   in nickel electrode construction 83–85
   role of, in nickel electrode 121, 122, 123, 126, 132, 133
   in test experience 332, 344–345, 348–349, 357
   in thermal management and reliability 309–311, 313
   gas management 151–153, 173, 180–185
   hydrogen parasitic reactions 178–179
   platinum reactions 157, 179–180
   self-discharge 9, 177–178, 180, 192, 201
   in charge management 289–291, 292, 293–294, 297, 301, 305–306
   in degradation and failure modes 396, 404
   in diagnostics and DPA 423, 426, 429
   in electrical and thermal performance 52–53
   in first-principles models 207, 210, 225–226, 227, 231–234
   and hydrogen electrode 149
   in nickel electrode chemistry 102–104, 105–106, 110, 113, 114
   in nickel electrode-hydrogen gas interactions 137, 142

484 Index

dynamics of cells, self-discharge (*continued*)
    role of, in nickel electrode 121, 122, 132
    in storage models 255–256
    in test experience 344, 346, 348, 349, 351, 371
    in thermal management and reliability 327
    in wear-out models 267, 278, 282–284
  thermal dynamics 88, 91, 93, 101, 114, 117–120, 157–158, 177, 192–201, 206, 252, 335, 373, 400, 404–405

E

EDAX. *See* Energy Dispersive Analysis by X-Ray
edge-popping 190
electrical performance 4–5
electrical phenomena
  charge efficiency 177–180, 196
    in advanced nickel electrode concepts 145, 146
    in charge management 292, 300, 303, 305
    in degradation and failure modes 386, 396, 397, 402, 404, 412
    in diagnostics and DPA 415, 445–446, 448, 449, 451, 452–453, 456
    in electrical and thermal performance 54–55
    in modeling 223–225, 234–235, 249, 253
    in nickel electrode chemistry 102–103, 105
    in nickel electrode construction 83–85
    in nickel electrode-hydrogen gas interactions 138
    role of, in nickel electrode 121, 122, 123, 126, 132, 133
    and separators and electrolyte 170
    in test experience 332, 344–345, 348–349, 357
    in thermal management and reliability 309–311, 313
  charge memory effect 208, 244–245
  heat generation 38, 65
  hysteresis 93–95
  I/V curves 38–41
  performance 85–86, 469–470, 475
  power response 44–45
  reconditioning 41, 58
  resistance 97–99, 234, 236–238, 333–334, 425, 427–428, 438–439, 448, 462
  second plateau discharge 53, 107–111, 207, 223, 304, 332–333, 336–338, 386, 399
  self-discharge 9, 52–53, 149, 177–178, 180, 192, 201
    in charge management 289–291, 292, 293–294, 297, 301, 305–306
    in degradation and failure modes 396, 404
    in diagnostics and DPA 423, 426, 429
    in first-principles models 207, 210, 225–226, 227, 231–234
    in nickel electrode chemistry 102–104, 105–106, 110, 113, 114
    in nickel electrode-hydrogen gas interactions 137, 142
    role of, in nickel electrode 121, 122, 132
    in storage models 255–256
    in test experience 344, 346, 348, 349, 351, 371
    in thermal management and reliability 327
    in wear-out models 267, 271–272, 278, 282–284
  voltage behavior 99, 102–103, 109–111
    in accelerated vs. real-time testing 361, 364, 367, 369
    in cell-level diagnostics 415, 423–424, 425–427, 431–433
    in discharge voltage signature testing 332–340
    in expert systems for cell diagnostics 471
    in first-principles models 215, 226, 228–230, 234–235, 237–239, 241–243, 248–249, 251
    in hydrogen electrode diagnostics 462
    in nickel electrode diagnostics 439, 444–445, 448, 451, 452
    in pressure behavior testing 350–351
    in specialized design features testing 375–377
    in storage effects testing 375
    in storage models 255
    in wear-out models 265, 267
electric vehicles 11, 12–13, 22
electrochemical voltage spectroscopy (EVS) 449–451
electrolyte 35, 163–176
  activity coefficient of 170
  and cell dynamics 177–182, 186–187, 189–190, 193, 197–202
  in cell models
    first-principles 206, 210, 213, 216, 218, 219–220, 227, 237, 247–248
    storage 251, 252, 258
    wear-out 274
  in charge management 300, 306
  concentration of 164–165, 170–173, 176
    in cell dynamics 181, 183–184, 200–202
    and degradation and failure modes 394, 401, 402–403, 408, 409

Index   485

in diagnostics and DPA 420, 434–435, 465–466
and hydrogen electrode 152, 157, 158, 161
in modeling 211, 212, 221, 235–236, 241, 245–246, 247
and nickel electrode 92, 115, 123
in test experience 331, 356–358, 366, 371, 373
in thermal management and reliability 314, 321, 323, 326
conductivity of 175, 176, 247, 334
in degradation and failure modes
   cycling issues 406–410, 412–413
   manufacturing problems 389–390
   operating environment issues 399, 401, 405
   storage issues 395, 396, 397
density of 170, 211
in diagnostics and DPA
   cell-level diagnostics 415, 416, 420–422, 429, 430
   DPA techniques 434–435, 436–437
   electrolyte and separator analysis 464–465, 465–466, 466–467, 468–469
   hydrogen electrode diagnostics 461, 462
   nickel electrode diagnostics 438, 439, 444–445, 448, 451, 452
distribution of 159–161, 174, 307, 435
draining of 152
in electrical and thermal performance 46–50
in evolution of cell design 16–18, 19–20, 23–26, 30, 32, 33, 35
and foreign particles 76
freezing of 173, 236, 245–247, 307, 313–314, 327, 338–340, 357–358, 380, 401–403
gradients 19, 25, 48, 161, 401
and hydrogen electrode 149, 153, 157, 158–159
internal resistance of 43
ionic conductivity of 170, 172
management of 151–153, 201–202
in nickel electrode chemistry 90, 92–93, 94–95, 100–101, 103, 104, 116–117
in nickel electrode construction 71, 86
in nickel electrode-hydrogen gas interactions 138
in nickel electrode storage 142–143
quantity of 174, 246, 247, 325, 385, 394, 409, 465
resistance of 427
role of, in nickel electrode 120, 122–123, 130–132, 133, 134, 135–137
in test experience 333–334, 352, 365–366, 368, 373, 376, 378–379
in thermal management and reliability 318–319
transport of 159–161
vapor pressure of 173
viscosity of 152, 166, 170–171, 173, 211, 246–247
end plates 16, 22, 31, 434
energy density 8, 30, 31, 34
Energy Dispersive Analysis by X-ray (EDAX) 442, 456–458, 459
enthalpy of reaction 59, 109, 117, 155, 157, 309–310
entropy of reaction 42, 157, 311
environment, thermal 28
EVS. See electrochemical voltage spectroscopy

F

failure modes 5, 383–413
   capacity loss 272, 278, 412
   cell-case isolation loss 405
   cell voltage drop 339
   compressive damage to stack components 403
   conductive particles 387
      shedding of 391, 400
   conductive particulates 413
   in diagnostics and DPA 475
   dry-out 408–409
   electrolyte freezing after long-term overcharge 245
   freezing 201
   gas leaks 433
   loss of capacity 443
   manufacturing-related 383–394
   in modeling 245–247, 267–272, 285
   and separators and electrolyte 163–164, 166, 168, 175
   short circuit 216, 275, 340, 365
   stack core fracture 387, 410
   in test experience 348, 350, 356, 360, 378–379
   in thermal management and reliability 325
   transfer of water 197
   vibration and shock 388–389, 405
fill tubes 22, 318–319, 389–390, 416, 437
finite elements 206, 235, 247
   cell as assembly of 191, 209, 210, 241
   gas space 209, 216
   hydrogen electrode 209
   interfaces 209, 211–213, 219, 223, 227, 230, 240
   nickel electrode 209, 221, 232–233
   separator 209
   solid electronic conduction 209–210
   subelements 219, 221–222, 223
   thermal model 428
   thermal transport in 209, 213–214

freezing of liquids 199–201
fuel cells 3, 11, 149, 157

G

gamma phase
    formation of 246
    modeling formation 230–231
    volume changes 221
gas screens
    in cell dynamics 181–182, 183, 186, 190
    and degradation and failure modes 387, 400, 407, 410
    in diagnostics and DPA 419, 422, 435, 436, 464–465
    in evolution of cell designs 18–20
    and hydrogen electrodes 150, 152–153
    in modeling 209, 216, 237
    and separators and electrolyte 164
geosynchronous orbit 3, 5, 7–8, 11, 17–18, 23–24, 27, 65, 253, 276–284, 291, 304, 331, 340, 353, 394, 404
    and profile tests 370–371
gravitational forces 74

H

heat generation
    in cell dynamics 193, 195, 196, 201
    in charge management 303–304
    and degradation and failure modes 403, 404, 412
    in diagnostics and DPA 452
    in electrical and thermal performance 38, 59–63, 65
    in evolution of cell designs 28
    in modeling 207, 210, 211, 243–244
    and nickel electrode 138
    in test experience 335, 337, 338–339, 379
    in thermal management and reliability 309–312
hydroformation 16, 20, 21, 318–319, 389–390, 433
hydrogen electrodes 4, 149–161
    and cell dynamics 179–191
    in charge management 297, 303
    construction of 150–151, 158
    and degradation and failure modes 385–386, 387, 396, 402, 407, 410
    design of 149–150
    in diagnostics and DPA 415, 416, 418–419, 427, 435, 462–463, 466, 468–469
    in electrical and thermal performance 49–50, 53, 60, 63–64
    in evolution of cell designs 18–20, 32, 33, 35

    gas management of 151–153, 180–185
    in modeling 209–210, 210–211, 213, 214, 215–216, 232, 235–236
    and nickel electrodes 87, 118–119, 122, 129–130, 132, 140
    overpotential at 153–154, 156–157, 158, 215–216, 460–462
    oxygen recombination at 151, 153–156, 158, 168, 211, 225, 244, 311–312
    pore size in 159–161, 182, 202, 247
    porosity of 206, 213
    reactions in 149, 151–152, 153–157, 157–158
    and separators and electrolyte 163, 166, 170
    sintering in construction of 150–151, 390–391
    in test experience 338, 372–373, 378–379
    in thermal management and reliability 321
    thermodynamics of 157–158
hydrogen gas
    in cell dynamics 177, 178, 185, 189–190, 193, 195, 201
    in cell models
        first-principles 207, 211, 214, 215–216, 224, 226, 232, 237, 240–241
        storage 255, 256, 262
    in charge management 303
    and degradation and failure modes 389, 397, 398, 399, 411
    in diagnostics and DPA 426, 432–433, 436, 439, 440, 459–464
    in electrical and thermal performance 62, 63
    and hydrogen electrode 149–150, 157, 159–161
    and nickel electrode 87, 108, 110–111, 114, 129–130, 137–142
    in test experience 338–339, 376, 379
    in thermal management and reliability 309–312, 318–319, 320, 328
hydrogen gas flow 180–185, 202, 416, 421
    ice formation and 183–185
    inert gases and 153, 183
    liquid electrolyte and 152, 181–182

I

impedance
    ac 49, 427–428
    in cell dynamics 184
    and cell resistance 48–50
    and degradation and failure modes 405
    in diagnostics and DPA 415, 421, 437, 448, 462
    in electrical and thermal performance 39, 46, 48–52, 58, 66
    in evolution of cell designs 27, 30, 32, 33, 34
    in modeling 214, 215–216, 217, 218, 228,

Index   487

235–237, 246, 275
and nickel electrode 96, 136
and separators and electrolyte 169
in test experience 332, 333–334, 338–339, 342–343, 378
in thermal management and reliability 313, 315, 317
transient response 49–50
impregnation 71, 83, 85–86, 133, 142, 393
   active-material 76
   alcoholic 18, 78–79
   aqueous 15, 17, 24–25, 78–79
   corrosion during 79, 81, 82, 133
   electrochemical 18
   of nickel electrodes 77–83
   uniform 76, 392–393
Inconel 16, 20, 21, 318–319, 389–390, 405, 433
   pressure vessels 327–328
inductance 49–50
interface matrix 211–212
isolation switches 317

L

latent heat 192–194
life-test experience 324, 331–381
   capacity loss in 278, 379, 471
   capacity walk-down in 349
   cell diameter and 366–368
   charge-control effects in 351–353
   CPV tests in 375–376
   deconditioning in 334–336
   depth of discharge in 331, 356, 384, 471
      in accelerated vs. real-time testing 359–360, 360–361, 362–364, 365, 370
      in charge-control methods testing 351–353, 356
      in discharge voltage signature testing 334–335
      in recharge behavior testing 342
      in specialized design features testing 375, 378–379
      in specialized life tests 380
      in storage effects testing 375
   discharge voltage degradation in 332–340
   electrolyte concentration in 356–358
   electrolyte dry-out in 333–334
   geosynchronous tests in 276–284, 370–371
   high-temperature tests in 360, 379–380
   impedance changes in 332, 333–334, 338–339, 342–343, 378
   life variability in 369–370
   overcharge tolerance in 363
   pressure shifts in 347–351
   recharge effects on life, seen in 340–346
   recharge rate in 282, 331, 340–341

   recharge ratio effects in 331
      in accelerated vs. real-time testing 360–361, 363, 369
      in charge-control methods testing 352–353
      in pressure behavior testing 349, 350
      in recharge behavior testing 344–345
      in specialized design features testing 375, 378
      in specialized life tests 380
      in temperature testing 354
   reconditioning in 334–336
   second plateau cycling in 338
   separator type in 362
   short-circuit failures in 279
   sinter type in 364–365
   SPV battery tests in 376–377
   stack design in 365–366, 378
   storage and cycle life in 371–375, 379
   temperature and 278, 331, 384
      in accelerated vs. real-time testing 359–360, 369, 370
      in electrolyte concentration testing 357–358
      in recharge behavior testing 342, 344–345
      in specialized design features testing 375–376, 378
      in specialized life tests 379–380
      in temperature testing 353–356
   thermal gradient tests and 380
   trickle charge in 338–339, 340, 346, 347, 371
life testing 26, 28, 67, 205, 238–239, 241, 250–251, 263, 264, 300, 321, 323, 384
   by computer simulation 205, 209, 247–251, 284
lithium-ion technology 6, 8–9, 144
loading level 151, 252, 321, 325, 392–393, 408, 444, 446, 455, 458–459
load leveling 11, 13, 22
low Earth orbit 3, 5, 7–8
   and profile tests 360–370
   in cell models
      first-principles 235–237, 239, 248
      storage 253
      wear-out 263, 265, 267, 274, 276–278, 284
   in charge management 291, 293, 295, 304
   and degradation and failure modes 384, 394, 404, 406
   in electrical and thermal performance 65
   in evolution of cell designs 11, 17, 18, 21, 32
   in test experience 331, 340, 342, 345, 353, 358, 371, 380

M

manufacturers 6, 10, 65, 71, 238, 326–327, 362, 363, 367, 373–374, 383, 393, 394, 440
  AT&T/Bell Labs 15, 17, 18
  Boeing 331
  EaglePicher 375–376
    Colorado Springs 18, 22, 331
    Joplin 17, 22, 25–27, 28–30, 31, 32, 331
  Gates Energy Products 28–30, 331
  Hughes 18, 331
  Intelsat 15, 17
  Johnson Controls 22, 30–31, 376
  Saft 28–30
  Tyco 15, 17
  Yardney 21–22, 25, 28, 331
mass balance 25
microporous substrates 146, 213
microsatellites 9
models 205–285
  active-material conductivity 207, 218, 222–223, 242, 245
  applications of, to performance 229–251
  cobalt additive 223–225, 240–243
  continuous mechanics 206–208
  corrosion 251, 252–254
  emergent behaviors from 243–247, 284
  empirical 4, 251, 284–285
  finite-difference method 206
  finite-element 206, 208–229, 232, 240–241, 243–244, 247, 429
  first-principles 4, 191, 205–251, 284–285, 331
  hydrogen electrode 215–216
  life 205, 209, 210, 247
  nickel electrode 99, 207–208, 217–223, 230, 256–258
  particle size 221–222
  performance 4, 205, 208–209, 226–228, 234–240, 241–242, 284–285, 470
  self-discharge 225–226, 255–256
  storage 251–263, 276, 372
  thermal 196, 309, 311, 429
  validation of 4, 228–229, 259
  wear 205, 263–284

N

nanosatellites 9
Nernst equation 41, 43, 64–65, 118
nickel-cadmium technology 3–4, 5–7, 9, 291
  in cell dynamics 192
  in diagnostics and DPA 423
  in evolution of cell designs 12, 15, 17, 22, 24–25, 34
  and nickel electrode 71, 78, 119–120, 122, 128, 137, 138, 145
nickel electrodes 3–4, 9, 71–146
  additives in 120–137, 223–225, 447–448
  alpha phases in 92–93, 100, 104, 105, 114
  beta phases in 88, 91–93, 97–98, 100–101, 102, 104–106, 112–114, 127, 146
  and cell dynamics
    charge efficiency 177–180
    hydrogen gas management 180, 181, 184
    oxygen gas management 185, 186, 188, 189–190, 191
    pressure fluctuations 202–203
    thermal management 193–194, 197, 200, 201
  and cell electrolyte concentration 152
  in cell models
    first-principles 208–212, 214, 217–218, 221–231, 235–237, 240–242, 244–248
    storage 251–258, 260–262
    wear-out 274
  in charge management 303, 304–305, 306
  conductivity in 222–223
  construction of 71–86
  corrosion in 79, 81–83, 114–116, 176, 271, 296, 332–333, 333–334, 395, 437, 438
  crystalline disorder in 104–106
  deactivation of 395, 397
  in degradation and failure modes
    cycling issues with 406, 407–408, 408–409, 410–411, 412–413
    manufacturing problems with 387, 391–392
    storage issues with 395–399
  depletion layer in 113–114, 145
  in diagnostics and DPA 438–459
    cell-level diagnostics 415, 418–421, 425–427, 430, 432–433
    DPA techniques 435, 436
    electrolyte and separator analysis 466, 468–469
    expert systems 474, 475
    nickel electrode diagnostic tests 438–442, 448–451, 455–457
  in electrical and thermal performance 47, 49–51, 52, 53, 60–61, 63–65
  electrochemical reactions in 87
  in evolution of cell designs 12, 18–20, 22, 23–25, 32, 33, 34, 35
  formation of 40, 71, 83–85, 393–394
  gamma phases in 91, 97–98, 100–101, 102–103, 105–106, 111–113, 115, 121, 125, 127, 130, 143, 146
  and hydrogen electrode 149, 150
  hydrogen gas interactions with 87, 232

kinetics of 255
lightweight quality of 144–145, 264
loading level of 80–82, 248, 392–393, 458–459
pasted 145
pore size in 77, 116–117, 146, 202, 211, 220–221, 247, 454–455
porosity of 71, 115, 116–117, 124, 193
   in cell dynamics 202
   in degradation and failure modes 385, 391–392, 394, 408
   in diagnostics and DPA 442
   in evolution of cell designs 15
   in modeling 206, 218–221, 227, 230, 240–242, 244–245, 247, 252
   in thermal management and reliability 312
proton diffusion behavior of 95–97, 227–228
reconditioning of 111–113
self-discharge rate within 231–234
and separators and electrolyte 163, 165, 168, 169–170
state-of-charge gradients in 230
storage and 75–76, 114, 120, 130–131, 134, 142–144
stress testing of 85, 391, 392, 452–454
in test experience
   accelerated vs. real-time 363–365, 368
   discharge voltage signature 332–333, 334, 338
   electrolyte concentration 357
   pressure behavior 348–349
   recharge behavior 343–344
   special design features 378
   specialized life tests 380
   storage effects 372–373
   temperature 354, 356
in thermal management and reliability 310–312, 321, 325, 327
thermal stability of 114, 177
thermodynamics of 117–120, 158
thickness of 394, 407, 438
and transport processes 78, 102, 109
uniformity of 77, 86
utilization testing of 85–86, 444
volume changes of 80, 83, 86, 91, 93, 99–101, 105–106, 123, 146
nickel-iron batteries 12
nickel-metal-hydride technology 8, 34, 71, 78, 128, 137, 138, 145
nickel oxyhydroxide 3
   in advanced nickel electrode concepts 146
   in charge management 306
   and degradation and failure modes 397–398, 402–403, 408, 410, 411
   in diagnostics and DPA 438, 440, 447, 448, 449–451, 466
   in electrical and thermal performance 40–41
   in modeling 207–208, 210, 217, 221, 230–231, 242
   in nickel electrode chemistry 88, 90–93, 96–98, 100–103, 106, 109, 111–115, 117–118
   in nickel electrode-hydrogen gas interactions 140
   in nickel electrode storage 143
   role of, in nickel electrode 120–121, 122, 124–125, 126, 127, 130, 133
   and separators and electrolyte 174–175
   in test experience 354, 356, 357–358, 373
nickel plaque 72–78, 85–86, 88, 114, 133, 248, 252–253, 391–393, 438
NTS-2 flight test 5, 17

O

Ostwald ripening 210
   and order/disorder transitions 105
overcharge 5, 7–9, 9–10
   in cell dynamics 180, 181, 186, 188, 189–190, 193, 196–197, 200–201
   in cell models
      first-principles 206–207, 231, 233, 235–237, 239, 244–247, 248
      storage 252, 255
      wear-out 271, 275, 277–278, 282
   in charge management 289–291, 293, 297, 300, 301, 303, 305, 306–307
   and degradation and failure modes 391, 393, 399–400, 401, 402, 404, 408, 410, 411, 412–413
   in diagnostics and DPA 418–419, 422–423, 430–431, 438, 441, 444–446, 454, 468
   in evolution of cell designs 26–27, 35
   heat generation from 118, 138, 191, 289, 291–292
   and hydrogen electrode 153–154
   and nickel electrode 83, 85–86, 91, 100, 111–112, 115, 117, 120–121, 122
   reactions to 101–104, 143, 339, 363
   and separators and electrolyte 166, 168, 174
   steady-state 191
   in test experience
      accelerated vs. real-time 359–360, 361, 362, 364–365, 369, 370
      charge-control methods 351–353
      discharge voltage signature 332
      pressure behavior 348, 349, 350
      recharge behavior 340–341, 341–344, 344–345
      special design features 378–379
      specialized life tests 380

490 Index

overcharge, in test experience (*continued*)
    temperature 354–355
  in thermal management and reliability
    310–311, 313, 322, 326, 328
  voltage behavior during 154
oxygen gas
  in cell dynamics 177, 180–181, 185–191,
    193–194, 199
  and degradation and failure modes 399–400,
    406
  in diagnostics and DPA 426–427, 430, 433,
    436, 447, 451, 456, 461
  edge-popping 188, 189
  in electrical and thermal performance 47, 57
  evolution of 102–103, 114, 133
  flow 150, 168–169, 186, 244
    in high-bubble pressure separator 168,
      188
    in low-bubble pressure separator 166,
      186–187
  gradients 197, 199
  and in-stack popping 190
  in modeling 211, 214, 232, 244, 261
  and nickel electrode 87, 108, 110, 117, 130
  popping caused by 61
  recombination of 32
    in cell dynamics 186, 187, 188–191, 192
    in charge management 297
    and degradation and failure modes 399,
      412–413
    in diagnostics and DPA 419, 423, 430,
      468–469
    in electrical and thermal performance
      61–62
    and hydrogen electrode 149, 153–156,
      158
    in modeling 207, 210, 215–216, 244, 246
    and nickel electrode 118, 137–138
    in test experience 379
    in thermal management and reliability
      310–311, 325, 328
  recombination wall wicks 191, 311
  in thermal management and reliability 310

P

performance 5–10
  electrical 4, 37–67
  thermal 4–5, 37–67
picosatellites 9
popping
  in cell dynamics 186, 188–191, 194, 197, 201
  in charge management 307
  and degradation and failure modes 399–400,
    403, 412–413
  in diagnostics and DPA 418, 421–422,
    430–431, 442, 465, 466, 468
  in evolution of cell designs 23, 25, 30
  in modeling 246–247
  and separators and electrolyte 166, 168,
    174–175
  in test experience 340
  in thermal management and reliability 313,
    328
pore size, in nickel electrodes 116
porosity 209–210, 211–213, 227, 456
  of hydrogen electrode 150, 159–161, 206,
    211, 216, 463, 464
  of nickel electrode 71, 115, 116–117, 124,
    144, 193
    in cell dynamics 186
    and degradation and failure modes 385,
      391–392, 394, 408
    in diagnostics and DPA 442, 454–455
    in electrical and thermal performance 57
    in modeling 206, 227, 230, 241–242,
      244–245, 247, 252
    in thermal management and reliability
      312
  of separators 164
  of sinter
    and degradation and failure modes
      391–392, 393, 409
    in diagnostics and DPA 448, 454–455,
      459
    in evolution of cell designs 34
    in modeling 206, 221, 228, 233, 245, 250
    and nickel electrode 71, 83–84, 116, 144
    in thermal management and reliability
      321, 325
  weld 389–390
precharge 9, 63–65, 439–440, 445–446
  evaluation of 424, 434, 435
  hydrogen 143, 241, 247, 252, 255, 256,
    261–263, 300, 303, 305, 396–398,
    424–427, 436, 440, 442, 472
  loss
    during cycling 425, 440
    during storage 398, 440
  nickel 111, 114, 120, 130–131, 135, 143
    in cell dynamics 179, 195
    in charge management 304–305
    and degradation and failure modes
      395–397, 398, 399
    in diagnostics and DPA 424–427,
      431–433, 436–437, 439–440, 442,
      446, 448
    in evolution of cell designs 33
    and hydrogen electrode 156, 159
    in modeling 224, 237, 252, 258–261
    in test experience 338, 372–373
  pressure behavior 347–351

and excess oxygen 348
growth during cycling 348–351, 411–412
pressure/volume work 56–57
walkdown 349, 411
pressure fluctuations 194–196, 202–203, 383
pressure monitoring
    drift observed in 58, 296, 348, 383–385, 412
    strain gauges 58–59, 290, 295–296, 309, 347, 383–385, 412
pressure vessels 3, 9
    in cell dynamics 183, 190–191, 192, 193, 195, 201, 202
    condensation in 237
    contaminants in 318–319
    and degradation and failure modes 388–389, 402, 405
    in diagnostics and DPA 415, 417, 422, 426, 429, 430, 434
    domes of 27, 56, 62–63, 181, 183, 197, 295, 311, 339, 347, 383, 389, 422
    in electrical and thermal performance 44, 60, 66–67
    in evolution of cell designs 16–17, 19–20, 25–27
    flaws in 318–319, 389, 416, 433
    and hydrogen electrode 149, 152
    inspection of 433–434
        using dye-penetrant methods 318, 389, 433
        using eddy current 433
        using helium leak test procedure 319, 389, 436
        statistics of 433
        ultrasonic cleaning during 319, 433
        using x-ray screening methods 318–319, 416
    leaks in 30, 67, 318–319, 389, 405, 436
    in modeling 207
    and nickel electrode 142
    reliability of 389–390
    in test experience 361, 366, 379
    in thermal management and reliability 310–312, 315, 318–319, 321, 323, 325–326
    volume changes in 57
    weldments in 318, 389, 416, 433
proton diffusion coefficient 96–97, 228

R

ray-tracing model 220
reliability 3, 5, 8–9, 316, 317, 320, 321, 324–327
    battery 30, 272–273, 289, 292
    cell 205, 272–273, 321–324
    cell redundancy 322
    changes over life 323
    and degradation and failure modes 387, 388

    in diagnostics and DPA 415, 416–417, 430, 433, 437
    in evolution of cell designs 11–13
    and hydrogen electrode 151, 157
    in modeling 264, 285
    pressure vessel 318, 389–390
    and separators and electrolyte 170
    in space system development 71
    of SPV, relative to IPV 326
    in test experience 353, 360, 369, 371, 378–379
    wear accumulation and 292
resistance matrix 214, 227
reversal 5, 7–9, 83
    as precharge test 425–426

S

safety 259, 327–328, 433
    cell rupture 318, 327–328
    hydrogen venting 328
    issues 9
    standards 7
    thermal runaway 114, 197, 201, 291–292, 301, 327–328, 400
satellites 3, 5, 6–7
    in cell dynamics 196, 202
    in charge management discussion 291, 293, 295, 300, 301, 304, 305, 307
    commercial 6
    Cosmos-1691 190
    Cosmos-1823 191
    in diagnostics and DPA 416
    Ekran-2 190
    in electrical and thermal performance 65
    in evolution of cell designs 11, 13, 22–24, 26–27, 30–31, 32
    Hubble Space Telescope 7
    Intelsat IV 17
    Intelsat V 17
    Iridium 6, 31, 376
    in modeling 235, 264, 277–278, 282, 285
    noncommercial 6
    small 9
    space stations 11, 28–30, 362
    Teledesic 31
    in test experience 331, 336, 351, 353, 360, 378, 379
    in thermal management and reliability discussion 327
scanning electron microscopy (SEM) 220, 393, 442, 456–458, 459, 463
Schottky barrier layer 110
self-discharge 9
    catalytic 140, 178
    in cell dynamics 177–178, 180, 192, 201

492 Index

self-discharge (*continued*)
   in charge management 289–291, 292, 293–294, 297, 301, 305–306
   and degradation and failure modes 386, 396, 404
   in diagnostics and DPA 423, 426, 429
   by direct reaction of hydrogen gas 207
   by direct reaction with hydrogen 138–140, 178
   in electrical and thermal performance 52–53
   gradients 231–234
   and hydrogen electrode 149
   in modeling 207, 210, 225–226, 227, 267, 278, 282–284
   in nickel electrode chemistry 102–104, 105–106, 110, 113, 114
   in nickel electrode-hydrogen gas interactions 137
   in nickel electrode storage 142
   oxygen recombination and 137–138
   role of, in nickel electrode 121, 122, 132
   in test experience 344, 346, 348, 349, 351, 371
   in thermal management and reliability 327
SEM. *See* scanning electron microscopy
separators 4, 163–176
   asbestos 166–170
      in cell dynamics 186, 188
      and degradation and failure modes 386, 409, 413
      in diagnostics and DPA 421, 422, 442, 448, 466–467, 468, 471
      in electrical and thermal performance 46–47
      in evolution of cell designs 17–18, 22, 25, 33
      and nickel electrodes 133, 136–137
      in test experience 343, 361
   bubble pressure 166, 168, 186–188, 190
   in cell dynamics 177, 182, 184, 186–187, 188, 191, 202
   composite 25, 28, 33, 188, 265, 361–362, 363, 413, 469, 471
   compressibility of 164, 166, 168–169
   compressible
      asbestos 334
      nylon 334
   and degradation and failure modes 387, 394, 407–408, 409, 411, 412–413
   in diagnostics and DPA
      cell-level diagnostics 415, 417, 419, 421, 423, 429, 430
      DPA techniques 434–435, 437
      expert systems 464–469
      hydrogen electrode diagnostics 459
      nickel electrode diagnostics 443

   in electrical and thermal performance 58
   electrolyte retention of 150, 164
   in evolution of cell designs 18–20, 23–27
   and hydrogen electrode 152
   in modeling 206, 209, 211, 214, 232–234, 235–236, 247–248
   and nickel electrodes 81, 99
   nylon fabric 24, 33, 409, 413, 421, 469
   in test experience 331, 333–334, 343, 357
   in thermal management and reliability 319, 321, 323, 325, 328
   Zircar 164–166, 166–170, 175
      in cell dynamics 186, 187, 188, 189
      and degradation and failure modes 400, 409, 410, 412–413
      in diagnostics and DPA 442, 443, 448, 466, 468–469, 471
      in electrical and thermal performance 37, 46
      in evolution of cell designs 22, 24–27, 28–30, 32, 33
      in modeling 235–236, 241, 248, 265, 274
      in test experience 334, 343, 352, 357, 361–363, 365, 366, 368, 370, 371, 373, 378
      in thermal management and reliability 323
   Zirfon 24, 33, 469
short circuits 9
   in cell dynamics 180, 186, 188, 190, 201–202
   and degradation and failure modes 387, 389, 400, 402, 410, 411, 413
   in diagnostics and DPA 415, 417–419, 423–424, 430, 443, 448, 459
   in electrical and thermal performance 58, 61–62
   in evolution of cell designs 23, 25, 32, 35
   and hydrogen electrode 158
   in modeling 216, 245–247, 254, 267–268, 271, 275, 279–281
   and nickel electrode 81, 85, 116, 133, 135, 141
   and poor formation of gamma phase 357
   from popping 246
   and separators and electrolyte 163, 168
   in test experience 332, 339–340, 350–351, 356, 357, 365, 380
   in thermal management and reliability 314, 317, 327–328
sinter
   in advanced nickel electrode concepts 145–146
   in cell dynamics 178
   in cell models
      first-principles 206, 210, 218, 220–221, 223, 224, 228, 230, 231, 241–242, 245, 247–250

storage 252–254, 255, 256, 258
in charge management 304
defects in 75, 77
in degradation and failure modes
    in battery-specific issues 400
    in cycling issues 406–409, 411, 412
    in manufacturing problems 391, 392, 393
    in storage issues 395, 397, 398, 399
in diagnostics and DPA 438–440, 442, 444–446, 448, 451, 454–455, 458–459, 471
dry 24, 34, 72–77, 363–364, 364–365, 366, 368, 391–392
in nickel electrode chemistry 96–97, 99–100, 108–111, 114–115
in nickel electrode construction 71, 77–83, 85–86
in nickel electrode-hydrogen gas interactions 140
in nickel electrode storage 143
passivation 76, 115, 140–142
pore size of 71, 116, 228, 230, 233, 454–455
porosity of 86
role of, in nickel electrode 124, 128, 133–134, 140
slurry 25, 34, 72–77, 78, 233, 363–364, 364–365, 391–392, 445
strength of 364
in test experience 333, 337–338, 343–344, 348, 354, 356, 373
in thermal management and reliability 321, 325
uniformity of 34, 391–392
software for cell design 324–327
state-of-charge gradients 230, 255, 404
Stokes-Einstein equation 170
storage 9
    active 9
    in cell dynamics 179–180
    and degradation and failure modes 394–399, 395, 398, 412
    in diagnostics and DPA 415, 422, 424, 437
    discharged 76, 106
    in modeling 240–241, 260–263, 271
    of nickel electrodes 120, 142–144
    platinum plating and 131–132, 395–397
    precharge changes and 114
    in test experience 371–375, 379
storage modeling 251–263
    capacity loss in 252–254, 256–262
    cobalt reduction in 257, 258
    nickel corrosion 251–254, 259–261, 395
    nickel electrode active-material reduction in 256–258

passivation layers 252–254
platinum corrosion 399
self-discharge in 255–256
storage environment timeline 252
validation 259, 373

T

Teflon
    in cell dynamics 181, 182
    and degradation and failure modes 385, 390–391
    in diagnostics and DPA 416, 435, 459, 462, 463, 464
    in evolution of cell designs 20, 22, 25
    and hydrogen electrode 149, 150, 159
    in modeling 213
    in thermal management and reliability 320–321
terminal boss 320, 321
terminal seals
    ceramic 321, 327
    compressive force and 319–320, 390
    with Teflon compression fittings 319, 320, 327–328, 390, 434
    temperature tolerance of 320
    Ziegler 319–320, 390, 434
terminals, rabbit-ear 27–28, 191, 331, 401
testing 4–5, 9–10, 13, 35, 45
    accelerated life 331, 359–371
    acceptance 5, 65–67, 387
    capacity verification 307, 422, 423
    charge-retention 423–424
    cycle life and 331, 336, 340–341, 351–358, 361–376, 379, 380
    leak 436
    performance degradation and 415
    qualification 5, 65–67
    real-time life 331, 359–371
    resistance 67
    standard LEO screening 361, 366, 368
    thermal 428
        cycling 67, 316, 320
        vacuum 316
thermal behavior of cells 59–63, 207, 244, 326, 404, 428
thermal control systems 18, 27, 63, 209, 265, 306, 309, 311, 312–313, 315–316, 321, 326, 404–405
thermal effects on voltage 310
thermal environments 5, 28, 196–198, 209, 289, 293, 295, 300–301, 306, 316, 402, 403, 429
thermal gradients 246–247
    in cell dynamics 196–201, 202
    in charge management 300, 306

494 Index

thermal gradients (*continued*)
   and degradation and failure modes 400–401, 404–405
   in diagnostics and DPA 419, 429, 430, 438
   in electrical and thermal performance 56–57
   in evolution of cell designs 18, 26
   in modeling 237
   and nickel electrode 81
   in test experience 334, 339, 343, 379–380
   in thermal management and reliability 312–315, 325–326
thermal management 5, 192–201, 303–304, 309–316, 366, 404
thermal performance 4–5
thermal pressure vessel work 60, 62–63
thermal requirements 312–315
thermal runaway 114, 197, 201, 291–292, 301, 327–328, 400
thermal sleeves 237, 315–316, 405
thermal stability of active material 114
thermal transients 429
thermoneutral potential 117–120, 158, 310
thermoneutral voltage 309–310
tortuosity 211–213
transport processes 338
   convection 164, 166, 180, 181, 183, 191, 197, 207, 209–210, 213–214, 227, 232, 244, 387, 399, 409
   ionic diffusion 179, 184, 206, 209, 409, 437
   oxygen gas 199, 206–207, 214, 216, 236, 376
   proton diffusion 90, 95–97, 99, 206, 209–210, 212, 219, 227–228, 230, 337

V

virtual life testing 250–251
   pack 3603X simulation 251
voltage gradients 337

W

wall coatings
   catalytic 27
   hydrophobic 17, 26
   wall wicks 315
      in cell dynamics 191, 198–199, 202
      in charge management 306
      and degradation and failure modes 389, 394, 401
      in diagnostics and DPA 429, 434
      in evolution of cell designs 17, 19, 25–27
      in modeling 209, 237, 265
      in test experience 331, 339, 366, 379
wear-out models 205, 251, 263–284, 324, 371
   calendar life in 265, 267, 272, 277, 353, 371
   capacity-loss failure in 267, 275, 278
   cycle-life prediction in 263, 264–265, 274–276, 278, 352–353
   depth of discharge in 263–265, 267–268, 271, 274–282, 352–353
   design variables' role in 263–276
   excess recharge voltage in 267, 271
   geosynchronous profiles in 276–284, 370–371
   multiparameter 205, 265–272, 274–276
   multivariable regression in 268, 353
   orbital data comparison in 278, 282
   overcharge amount in 265, 267, 271, 277, 353, 371
   overcharge rate in 265, 267, 277, 353
   short circuit failure in 267, 279
   single-parameter 205, 263–265
   statistics of 272–274
   temperature dependence of 264–265, 267–268, 271, 277–279, 282–284, 353
   trickle-charge wear and 277, 284
   validation of 274–276
weight
   active material 441
   battery 27, 300, 316, 378
   cell 27, 32, 235–236, 240, 264, 325, 365, 394, 437
   cell and terminals 434
   component 465
   contaminants 442
   electrode 144–145, 445
   electrolyte 170, 437, 465
   in electrolyte concentrations 202
   foreign particles 75
   grid 446
   nickel electrodes 392
   organic contaminants 449
   particulate residue 465
   potassium hydroxide electrolyte 357
   pressure vessel 45
   space system 8, 11, 71
welding 318–319, 416
   e-beam 16, 21
   laser 16
   TIG 16, 22
weld rings 388–389, 401, 434

X

x-ray tests 415–422, 440–441
   component deformation 420
   CT scans 416
   for electrolyte pooling 422
   high-resolution 389, 416, 421
   lead alignment 417
   real-time scans 416

stack compression 420
weld inspection by 318–319, 390, 416

Z

zero gravity 17